A
GEOGRAPHICAL BIBLIOGRAPHY
FOR
AMERICAN LIBRARIES

A

GEOGRAPHICAL BIBLIOGRAPHY

FOR

AMERICAN LIBRARIES

Edited by

Chauncy D. Harris
Editor in Chief

Salvatore J. Natoli
Richard W. Stephenson
Harold A. Winters
Wilbur Zelinsky

with assistance from
Susan Fifer Canby
Phillip J. Parent
Steven S. Stettes

A Joint Project
of the
Association of American Geographers
and the
National Geographic Society
1985

Library of Congress Catalog Card Number: 85-11284
International Standard Book Number: 0-89291-193-X

Library of Congress Cataloging-in-Publication Data
Main entry under title:
A Geographical Bibliography for American Libraries.
"A joint project of the Association of American
Geographers and the National Geographic Society."
Includes index.
1. Geography—Bibliography.   I. Harris, Chauncy D.
(Chauncy Dennison), 1914-   . II. Association of
American Geographers.   III. National Geographic Society (U. S.)
Z6001.G44   1985   (G116)   016.91   85-11284
ISBN 0-89291-193-X

Available from
THE ASSOCIATION OF AMERICAN GEOGRAPHERS
1710 Sixteenth Street, N.W.
Washington, D. C. 20009

INTRODUCTION

The purpose of this **Geographical Bibliography for American Libraries** is to assist libraries in the United States, Canada, and other countries to identify, select, and secure publications of value in geography that are appropriate to the purposes and resources of each collection. This Bibliography attempts to make a critical and well-informed selection of the most useful publications. The entries are arranged by major categories: general aids and sources; history, theory, and methodology of geography; the fields of physical geography; the fields of human geography; applied geography; regional geography; and publications suitable for school libraries, each with numerous subdivisions.

Because this work is compiled particularly for use by American libraries, works in English have been emphasized but publications in other languages have not been excluded. Since the predecessor volume, **A Geographical Bibliography for American College Libraries** (1970), still provides an excellent guide to works published before 1970, the current bibliography focuses on publications for the period 1970-1984. Earlier works of abiding value have been included selectively (as in Part II, Section 1, History of Geography). Among the sections of this bibliography for which there was no separate treatment in the 1970 bibliography are dissertations, geography and map librarianship, coastal areas, natural hazards, historical geography, historical cultural geography (Old World), New World prehistory, cultural ecology, cultural geography, social geography, behavioral geography, environmental perception, communication, international trade and location, technological hazards, development, applied geography, planning, geography in education, and publications suitable for school libraries.

The entries and annotations for each of the sections of this bibliography were prepared by contributors who are specialists in the topics covered, as indicated in the Table of Contents and at the head of each section or subsection. These individuals, however, cannot be held accountable for the exact form and content of their sections since in the editing process some deletions and additions were made in order to provide better balance, some entries were moved among the sections, and in cases in which a publication was suggested by several contributors the work is cited in the most appropriate place or places.

An attempt has been made also to stress items which can be acquired currently by American libraries and can be entered as separate identifiable publications in the card catalogue, i.e. books, monographs, bibliographies, serials, or atlases. With few exceptions we have excluded individual articles in journals, fugitive materials, pamphlets, brochures, ephemeral materials, newsletters, individual maps, or works with limited or controlled distribution (such as reports by some government agencies). But these limitations have been overridden if a work is judged by a contributor to have exceptional value, for example, as the best orientation to a field (as in Part V, Section 1, General Applied Geography).

Cooperation has provided the basis for this bibliography. First of all, co-operation between the Association of American Geographers and the National Geographic Society in agreeing to produce this volume, in appointing the editorial committee, and in providing administrative and technical support at all stages. Secondly, co-operation among 71 geographers and librarians in selecting entries, in writing annotations, and in editing. The contributors have drawn on about 1500 **years** of professional reading, teaching, research, consultation, and writing in the specialized areas for which they have taken responsibility. Thirdly, co-operation of individuals at all stages in their careers from the editor-in-chief, who has already retired after half a century in geography, bibliography, and university administration, to the two National Geographic Society geography interns, who are just beginning their professional careers after graduating from college. The contributors include many senior professors in major universities and, happily, also many younger assistant professors with close knowledge of emerging trends and potential future directions of their chosen fields. Fourthly, co-operation among geographers, librarians, and publication specialists in deciding what materials are significant, which ones are appropriate to recommend to libraries ranging from vast research collections to small school libraries, and how best to process this information into a publication of practicable size and cost, useful arrangement, informed and informing annotations, adequate bibliographic data for identification, selection, and ordering, helpful index, and attractive format. Finally, this bibliography has built upon a valuable legacy of two previous bibliographies sponsored by the Association of American Geographers. The first of these was the pioneer work **A Basic Geographical Library: A Selected and Annotated Book List for American Colleges,** compiled and edited by Martha Church, Robert E. Huke, and Wilbur Zelinsky (Washington, D.C.: Association of American Geographers, Commission on College Geography, Publication no. 2, 1966). This was superseded by **A Geographical Bibliography for American College Libraries,** compiled and edited by Gordon R. Lewthwaite, Edward T. Price, Jr., and Harold A. Winters (Washington, D.C.: Association of American Geographers, Commission on College Geography, Publication no. 9, 1970), long out of print but not replaced as a guide to geographical works published before 1970.

As aids for positive identification and easy ordering we have tried to provide full first names of authors (having wasted much time ourselves in card catalogues or bibliographic sources trying to identify an author listed merely as J. Adams or D. Smith, which leaves a vast array of cards or lines to be searched). Information is also provided on the International Standard Book Number, the Library of Congress card number, and, for books confirmed as in print, the price.

The impulse for this bibliography came from the widespread consensus that in the fifteen years since the publication of **A Geographical Bibliography for American College Libraries,** the discipline of geography had evolved rapidly, new points of view had found expression, modern research methodologies had been developed, fresh fields had emerged, and a large number of books, serials, atlases, and bibliographies had been published.

The initiative for launching this bibliography came from Salvatore J. Natoli, Educational Affairs Director of the Association of American Geographers. On March 15, 1984, while Acting Director of the Association of American Geographers, he met with Barry C. Bishop, Special Assistant to the President and Vice Chairman of the Committee for Research and Exploration of the National Geographic Society, and with Susan Fifer Canby, Director of the Library of the National Geographic Society. They developed a plan for the Association of American Geographers to appoint a committee to select contributors and to edit suggested entries for a revision of **A Geographical Bibliography for American College Libraries** and for the National Geographic Society to provide the services of one or more National Geographic Society geography interns to work full-time for several months in bibliographical confirmation, editorial checking, and composition of the bibliography under the general oversight of Barry C. Bishop, head of the National Geographic Society internship program, and under the direct supervision of Susan Fifer Canby.

On April 21, 1984, the council of the Association of American Geographers approved the plan and appointed the following committee: Chauncy D. Harris, chairman, Wilbur Zelinsky, Harold A. Winters, A. David Hill, and Richard W. Stephenson (**AAG Newsletter,** vol. 19, no. 7, August 1, 1984, p. 9). The **AAG Newsletter** (vol. 19, no. 6, June 1, 1984, p. 1, 8) carried a notice of the appointment of the committee and invited contributions.

With support from the National Geographic Society, the Committee met at the Society on July 20, 1984, discussed the scope of the volume, the form of entry, the organization of topics, editing procedures, and the individuals to be invited to prepare the entries and annotation for each section. It was decided that the contributors would be named and recognized, unlike the previous two works in which, in order to encourage frankness, anonymity was assured. It was thought that the user could evaluate the recommendations and annotations more discriminatingly if he or she knew which qualified scholar had prepared them. The chairman of the editorial committee invited the suggested contributors to participate. Most of them accepted promptly and submitted entries and annotations during the autumn of 1984, or in early 1985. Since it was anticipated that most of the entries would be new, for books published in the last fifteen years, it was decided at this meeting to treat the bibliography not as a new edition of the 1970 bibliography but as an entirely new work with a different title. The work was expanded to include publications suitable for school libraries, public libraries, specialized libraries, corporate libraries, individual geographers, scholars in other fields, specialists in related problems, and the general public, as well as for college and university libraries.

Salvatore J. Natoli agreed to act as liaison for the Committee with the Association of American Geographers and Barry C. Bishop, with the National Geographic Society. In addition Robert E. Dulli, also at the National Geographic Society, assisted in the administration of the program and the production of the volume. After that meeting, A. David Hill, an original member of the Committee, withdrew from further active participation in its work since he was to spend the academic year 1984-1985 in research out of the country.

The Committee met again in Washington, D.C., February 1 and 2, 1985, with the participation of the two National Geographic Society geography interns assigned to the library, Phillip J. Parent from San José State University, and Steven S. Stettes from Southwest Missouri State University, and with other personnel of the Society. The focus of the meeting was to go over all steps in the bibliographic checking and editing of the entries and their inputting into the ATEX text processing system for composition and indexing. At this meeting the Committee reviewed the entries submitted by the contributors and made some adjustments in the number of entries by sections to achieve better balance. At this time Chauncy D. Harris was asked to become editor-in-chief. The Committee held its final meeting in Detroit, Michigan, April 22, 1985.

The Committee proved to be a particularly congenial and effective group. Wilbur Zelinsky brought enormous enthusiasm and verve, wide knowledge of publications in human geography, and the experience of participation in the compilation of the original **A Basic Geographical Library: A Selected and Annotated Book List for American Colleges** in 1966. Harold A. Winters aided with his familiarity with the task of compiling such a bibliography from his work as one of the editors of **A Geographical Bibliography for American College Libraries** in 1970. He also helped with judicious advice and with broad professional understanding of the literature in physical geography. Richard W. Stephenson was sensitive to the demands of users based on his service as head of the Reference and Bibliography Section of the Geography and Map Division of the Library of Congress. He also made available his special knowledge of atlases, maps, and cartography. Salvatore J. Natoli emphasized educational needs. Susan Fifer Canby provided help on library requirements. Phillip J. Parent and Steven S. Stettes reflected student desires and also brought enthusiasm, freshness, energy, perseverance, and technical competence to the tasks of checking, editing, and composing the text.

Chauncy D. Harris

Editor-in-Chief

July 12, 1985

LIST OF CONTRIBUTORS

Ronald F. Abler, Department of Geography, Pennsylvania State University, University Park, Pennsylvania 16802. Currently Director, Geography and Regional Science Program, National Science Foundation, 1800 G Street NW, Washington, D.C. 20550.
IV. 17. Communication

John S. Adams, Department of Geography, University of Minnesota, Minneapolis, Minnesota 55455.
IV. 20. Urban Geography.

Charles S. Aiken, Department of Geography, University of Tennessee, Knoxville, Tennessee 37996.
VI. 3. J. The United States: The South

Roger G. Barry, Department of Geography, University of Colorado, Boulder, Colorado 80309.
VI. 17. Polar Areas.

Sanford H. Bederman, Department of Geography, Georgia State University, Atlanta, Georgia 30303.
VI. 13. Africa South of the Sahara.

Karl W. Butzer, Department of Geography, University of Texas at Austin, Austin, Texas 78712.
IV. 3. Historical Cultural Geography (Old World).

Alvar W. Carlson, Department of Geography, Bowling Green State University, Bowling Green, Ohio 43403.
VI. 10. Southeast Asia.

Kenneth E. Corey, Department of Geography and Institute for Urban Studies, University of Maryland, College Park, Maryland 20742.
V. 2. Planning.

John C. Dixon, Department of Geography, University of Arkansas, Fayetteville, Arkansas 72701.
III. 5. Soils.

Roger M. Downs, Department of Geography, Pennsylvania State University, University Park, Pennsylvania 16802.
IV. 8. Behavioral Geography.

Fillmore C. F. Earney, Department of Geography, Earth Sciences, and Conservation, Northern Michigan University, Marquette, Michigan 49855.
III. 6. Oceans and Lakes.

Susan Fifer Canby, Director of the Library, National Geographic Society, Washington, D.C. 20036.
Supervision of bibliographical checking by the National Geographic Society Geography Interns assigned to the Library.

J. Keith Fraser, Executive Secretary and General Manager, the Royal Canadian Geographical Society, Wilbrod Street, Ottawa, Ontario, Canada K1N 6M8.
VI. 4. Canada.

John W. Frazier, Department of Geography, State University of New York at Binghamton, Binghamton, New York 13901.
V. 1. General Applied Geography.

Owen J. Furuseth, Department of Geography and Earth Sciences, University of North Carolina at Charlotte, Charlotte, North Carolina 28223.
IV. 13. Agricultural Geography.

Contributors

Reginald G. Golledge, Department of Geography, University of California, Santa Barbara, California 93106.
IV. 8. Behavioral Geography.

William L. Graf, Department of Geography, Arizona State University, Tempe, Arizona 85287.
III. 2. Geomorphology.

John L. Harper, Department of Geography, Humboldt State University, Arcata, California 95521.
III. 8. Water Resources.

Chauncy D. Harris, Department of Geography, University of Chicago, 5828 University Avenue, Chicago, Illinois 60637.
I. 1. Geographic Bibliographies. I. 2. Geographical Serials. I. 6. Geographical Dictionaries. I. 7. General Statistical Sources. I. 8. Encyclopedias and Handbooks. I. 9. Biographical and Institutional Directories. I. 10. Dissertations. VI. 7. U.S.S.R. VI. 9. B-E. Japan. Review of geographical serials in other sections. Editorial and bibliographic checking. Indexing.

R. Cole Harris, Department of Geography, University of British Columbia, Vancouver, British Columbia, Canada V6T 1W5.
VI. 4. Canada.

George W. Hoffman, Department of Geography, University of Texas at Austin, Austin, Texas 78712. Currently Secretary, East European Program. The Woodrow Wilson Center, Smithsonian Institution Building, 1000 Jefferson Drive S.W., Washington, D.C. 20560.
VI. 6. Europe

Donald G. Janelle, Department of Geography, University of Western Ontario, London, Ontario, Canada N6A 5C2.
IV. 17. Communication.

John R. Jensen, Department of Geography, University of South Carolina, Columbia, South Carolina 29208.
II. 7. Remote Sensing, Air Photo Interpretation, and Photogrammetry.

Douglas L. Johnson, Graduate School of Geography, Clark University, Worcester, Massachusetts 01610.
VI. 16. Arid Lands.

Ronald J. Johnston, Department of Geography, University of Sheffield, Sheffield S10 2TN, United Kingdom.
II. 4. Philosophy, Theory, and Methodology.

Roger E. Kasperson, Graduate School of Geography, Clark University, Worcester, Massachusetts 01610.
IV. 23. Technological Hazards.

Andrew M. Kirby, Department of Geography, University of Colorado, Boulder, Colorado 80309.
IV. 11. Political Geography.

George Kish, Program in Geography, University of Michigan, Ann Arbor, Michigan 48109.
II. 1. History of Geography. II. 2. Geographical Exploration.

Gregory W. Knapp, Department of Geography, University of Texas at Austin, Austin, Texas 78712.
IV. 5. Cultural Ecology.

Jeffrey A. Lee, Department of Geography, Arizona State University, Tempe, Arizona 85287.
III. 2. Geomorphology.

George K. Lewis, Department of Geography, Boston University, 48 Cummington Street, Boston, Massachusetts 02215.
VI. 3. H. The United States: The Northeast.

Gordon R. Lewthwaite, Department of Geography, California State University, Northridge, California 91330.
VI. 14. Oceania

David F. Ley, Department of Geography, University of British Columbia, Vancouver, British Columbia, Canada V6T 1W5.
IV. 7. Social Geography

Alan M. MacEachren, Department of Geography, University of Colorado, Boulder, Colorado 80309.
II. 5. Cartography

George Macinko, Department of Geography and Land Studies, Central Washington University, Ellensburg, Washington 98926.
III. 11. Conservation and Environmental Management.

Ian R. Manners, Department of Geography, University of Texas at Austin, Austin, Texas 78712.
VI. 12. Southwest Asia and North Africa.

Geoffrey J. Martin, Department of Geography, Southern Connecticut State College, New Haven, Connecticut 06515.
II. 1. F. History of Geography: Modern: United States. II. 3. Biographies.

Tom L. Martinson, Department of Geography, Ball State University, Muncie, Indiana 47306.
VI. 5. Latin America.

James E. McConnell, Department of Geography, State University of New York at Buffalo, Buffalo, New York 14260.
IV. 18. International Trade and Location.

Marvin W. Mikesell, Department of Geography, University of Chicago, 5828 University Avenue, Chicago, Illinois 60637.
IV. 6. Cultural Geography.

E. Willard Miller, Department of Geography, Pennsylvania State University, University Park, Pennsylvania 16802.
III. 9. Minerals.

Lisle S. Mitchell, Department of Geography, University of South Carolina, Columbia, South Carolina 29208.
IV. 21. Recreational Geography.

Robert D. Mitchell, Department of Geography, University of Maryland, College Park, Maryland 20472.
IV. 2. Historical Geography. VI. 2. E. North America: Historical Geography. VI. 3. G. The United States: Historical Geography.

David Morton, Librarian, Natural Hazards Research and Applications Information Center, Institute of Behaviorial Science, University of Colorado, Boulder, Colorado 80309.
III. 12. Natural Hazards.

Peter O. Muller, Department of Geography, University of Miami, Coral Gables, Florida 33124.
IV. 12. General Economic Geography.

Salvatore J. Natoli, Association of American Geographers, 1710 Sixteenth Street, N.W., Washington, D.C. 20009.
VII. Publications suitable for School Libraries. Support in planning, preparation, and production of the volume.

Contributors

John E. Oliver, Department of Geography and Geology, Indiana State University, Terre Haute, Indiana 47809.
III. 3. Climatology.

Clifton W. Pannell, Department of Geography, University of Georgia, Athens, Georgia 30602.
VI. 8. Asia: General. VI. 9. East Asia.

Phillip J. Parent, Department of Geography, San José State University, San José, California. Geography Intern, National Geographic Society, Washington, D.C. 20036, January 28 - August 2, 1985. Currently Publications Account Manager, Association of American Geographers, 1710 Sixteenth Street, N.W., Washington, DC 20009.
Bibliographic checking, inputting of entries into ATEX system, indexing, composition, and layout.

Albert J. Parker, Department of Geography, University of Georgia, Athens, Georgia 30602.
III. 4. Biogeography.

Kathleen C. Parker, Department of Geography, University of Georgia, Athens, Georgia 30602.
III. 4. Biogeography.

Martin J. Pasqualetti, Department of Geography, Arizona State University, Tempe, Arizona 85287.
III. 10. Energy

Philip W. Porter, Department of Geography, University of Minnesota, Minneapolis, Minnesota 55455.
IV. 19. Development.

Larry W. Price, Department of Geography, Portland State University, Portland, Oregon 97207.
VI. 18. Mountain Geography.

Norbert P. Psuty, Center for Coastal and Environmental Studies, Rutgers University, New Brunswick, New Jersey 08903.
III. 7. Coastal Areas.

Gerald F. Pyle, Department of Geography and Earth Sciences, University of North Carolina at Charlotte, Charlotte, North Carolina 38223.
IV. 22. Medical Geography.

John Rees, Department of Geography, Syracuse University, Syracuse, New York 13210.
IV. 14. Industrial Geography.

Thomas F. Saarinen, Department of Geography and Regional Development, University of Arizona, Tucson, Arizona 85721.
IV. 9. Environmental Perception.

Joseph E. Schwartzberg, Department of Geography, University of Minnesota, Minneapolis, Minnesota 55455.
VI. 11. South Asia.

James W. Scott, Department of Geography and Regional Planning, Western Washington University, Bellingham, Washington 98225.
VI. 3. K. The United States: The West.

Paul D. Simkins, Department of Geography, Pennsylvania State University, University Park, Pennsylvania 16802.
IV. 10. Population Geography.

Richard W. Stephenson, Geography and Map Division, Library of Congress, Washington, D.C. 20540.
I. 3. Atlases. I. 4. Map Guides. I. 5. Gazetteers. I. 11. Geography and Map Librarianship. II. 6. History of Cartography. VI. 1. The Ancient and Medieval Worlds. VI. 3. B. The United States: Atlases. Review of atlases in other sections.

Steven S. Stettes, Geography Intern, National Geographic Society, January 28 - August 2, 1985. Currently Department of Geosciences, Southwest Missouri State University, 901 South National, Springfield, Missouri 65802.
Bibliographic checking, inputting of entries into ATEX system, indexing, composition, and layout.

Joseph P. Stoltman, Department of Geography, Western Michigan University, Kalamazoo, Michigan 49008.
V. 3. Geography in Education.

Grant Ian Thrall, Department of Geography, University of Florida, Gainsville, Florida 32611.
II. 8. Quantitative Methods.

B. L. Turner II, Graduate School of Geography, Clark University, Worcester, Massachusetts 01610.
IV. 4. New World Prehistory.

Connie Weil, Department of Geography, University of Minnesota, Minneapolis, Minnesota 55455.
VI. 15. The Tropics.

James O. Wheeler, Department of Geography, University of Georgia, Athens, Georgia 30602.
IV. 15. Transportation.

Jessie H. Wheeler, Jr., Department of Geography, University of Missouri-Columbia, Columbia, Missouri 65211.
VI. 3. I. The United States: The Middle West and Great Plains.

Harold A. Winters, Department of Geography, Michigan State University, East Lansing, Michigan 48824.
III. 1. General Physical Geography. V. 4. Military Geography. VI. 2. C. North America: Physical Geography. VI. 3. D. The United States: Physical Geography.

Wilbur Zelinsky, Department of Geography, Pennsylvania State University, University Park, Pennsylvania 16802.
IV. 1. General Human Geography. VI. 2. North America: B. General, and D. Human Geography. VI. 3. The United States: A. Bibliographies. B. Atlases. C. Statistics. E. Human Geography. and F. Economic Geography.

ACKNOWLEDGEMENTS

Particular gratitude is expressed for help from the following members of the staff of the National Geographic Society, Washington, D.C. 20036:

Barry C. Bishop, Vice Chairman, Committee for Research and Exploration; Assistant to the President; Director, National Geographic Society Student Intern Program, for support in all stages of the planning, preparation, and production of the volume.

Jolene M. Blozis, Index Editor, for advice on the index.

Richard A. Bredeck, Assistant Systems Manager, Typographic, for advice on the use of the computer in typesetting and format.

Margaret Cole, for coordination of printing between the National Geographic Society and the printer, R. R. Donnelley and Sons Co.

Robert E. Dulli, President's Office, for the preparation of entry forms, coordination of the layout design and printing of the volume, and for general administrative support.

Bernard G. Quarrick, Assistant Systems Manager, Typographic, for coordinating all typesetting production, and also for advice on formatting and general systems operation.

Lyle Rosbotham, Art Director, National Geographic Research, for help in design.

Marta Strada, Supervisor of Cataloging, and her staff in the Library, for advice on bibliographic checking.

And in the Association of American Geographers, 1710 Sixteenth St. N.W., Washington, D.C. 20009:

Robert T. Aangeenbrug, Executive Director, for administrative support.

FORM OF ENTRIES

The form of entry generally follows that of the Library of Congress (LC) or the National Union Catalogue (NUC).

The basic information provided in each entry is as follows:

(1) Author(s), editor(s), or compilers
(2) Title and subtitle
(3) Place of publication
(4) Publisher
(5) Date of Publication
(6) Number of pages
(7) International Standard Book Number (ISBN) for books or International Standard Serial Number (ISSN) for periodicals and other serials
(8) Price in United States dollars, if the book has been confirmed as in print in the United States. The price may be given in another currency if the work has been confirmed as in print in another country but not in the United States
(9) Library of Congress card number (LC)

In addition, where appropriate, the following information is given:

(1) Translator and original language, title, and date if a book has been translated.
(2) Series or conference title
(3) Edition, if other than first
(4) Date of first edition and number of pages. If the author or editor, title or publisher varies, this may also be noted
(5) Number of volumes, if more than one
(6) Availability of a reprint edition

Each entry has an annotation prepared by a specialist in the particular field.

Entries are numbered. These numbers are used both in cross-references and in the index.

A bullet after the entry number indicates that the publication is considered suitable for school libraries.

The index is in a single alphabetical sequence including authors, short titles, and major subjects

The form of listing and bibliographical details have been confirmed, when possible, in the following sources:

National Union Catalogue: Pre-1956 imprints. London: Mansell, 1968-1981. 754 v.
_____. Author list. 1958-1962. New York: Rowman and Littlefield, 1963. 50 v.
_____. Author list. 1963-1967. Ann Arbor, MI: J.W. Edwards, 1969. 59 v.
_____. Author list. 1968-1972. Ann Arbor, MI: J.W. Edwards, 1973. 104 v.
_____. 1973-1977. Totowa, NJ: Rowman and Littlefield, 1978. 135 v.
_____. 1978. Washington, DC: Library of Congress, 1979. 16 v.
_____. 1979. Washington, DC: Library of Congress, 1980. 16 v.
_____. 1980. Washington, DC: Library of Congress, 1981. 18 v.
_____. 1981. Washington, DC: Library of Congress, 1982. 15 v.
_____. 1982. Washington, DC: Library of Congress, 1983. 21 v.
_____. Books. 1983 Jan.-Dec. Name Index. Title Index. Register. Washington, DC: Library of Congress, Microfiche.
_____. Books. 1984 Jan.-Nov.. Name Index. Title Index. Register. Washington, DC: Library of Congress, Microfiche.

Form of Entries

The prices for books in print generally, but not invariably, have been taken from the following sources.

Books in print 1984-1985. New York; London: R.R. Bowker, 1984. v. 1-3. Authors. v. 4-6. Titles

_____. Supplement 1984-1985. 2 v.

British books in print on microfiche. April 1985. London: J. Whitaker.

Canadian books in print. Author and title index, 1985. Toronto; Buffalo, NY; London: University of Toronto Press, 1985. 1003 p.

International books in print 1984. English-language titles published outside the United States and the United Kingdom. Part 1. Author-title list. München; New York; London: K.G. Saur, 1983. 2 v. 1737 p.

Australian books in print 1984. Port Melbourne, Victoria, Australia: D.W. Thorpe, 1984. 636 p.

New Zealand books in print 1984-1985. Port Melbourne, Victoria, Australia: D.W. Thorpe; Auckland, New Zealand: Ray Richards, 1984. 190 p.

TABLE OF CONTENTS

PART I. GENERAL AIDS AND SOURCES

1. GEOGRAPHIC BIBLIOGRAPHIES

Chauncy D. Harris

A. BIBLIOGRAPHIES OF BIBLIOGRAPHIES AND REFERENCE MATERIALS

Brewer, James Gordon. The Literature of geography: a guide to its organization and use. 2nd ed. London: Clive, Bingley; Hamden, CT: Linnet, 1978. 264 p. (1st ed., 1973. 208 p.) ISBN 0-208-01683-X. $25. LC 78-16852.   1
Discussion of the principal guides to the literature of geography, to reference works, and to the major substantive works in each of the fields of geography. Covers the scope, structure, and use of the literature of geography; organization of geographical literature in libraries; bibliographies and reference works; periodicals; monographs; textbooks and collections; cartobibliography; sources of statistics; government publications; history of geography; techniques and methodology; physical geography; human geography; and regional geography.

Durrenberger, Robert W. Geographical research and writing. New York: Crowell, 1971. 246 p. ISBN 0-690-32301-8. LC 77-136033.   2
Part one covers research and writing. Part two covers aids to geographical research: general guides and bibliographies; special indexes, abstracts, and bibliographies; sources of statistical information; map sources; sources of photographs; and a list of periodicals.

Harris, Chauncy D. Bibliography of geography. Part I. Introduction to general aids. (Research paper no. 179). Chicago, IL: University of Chicago, Department of Geography, 1976. 276 p. ISBN 0-89065-086-1. $10. (pbk). LC 76-1910.   3
Primarily a bibliography of geographical bibliographies, reference works, and source materials. Covers general bibliographies of bibliographies, comprehensive current bibliographies of geography, comprehensive retrospective bibliographies of geography, specialized bibliographies of geography, books, serials, government documents, dissertations, photographs, maps and atlases, gazetteers, place-name dictionaries, encyclopedias, statistics, and methodology of geography. Extensive comparative analysis of current and retrospective bibliographies of geography.

Josuweit, Werner. Studienbibliographie Geographie. Bibliographien und Nachschlagewerke. (Wissenschaftliche Paperbacks. Studienbibliographien). Wiesbaden, West Germany: Franz Steiner Verlag, 1973. 122 p. LC 74-323769.   4
Balanced coverage of bibliographies and reference works in German, English, French, and other languages, but with some emphasis on works in German.

Martinson, Tom L. Introduction to library research in geography: an instruction manual and short bibliography. Metuchen, NJ: Scarecrow Press, 1972. 168 p. ISBN 0-8108-0495-6. LC 72-2012.   5
An introduction to bibliographies and reference works for research in geography with emphasis on works in the United States in English. Contains a brief introduction to bibliography.

Sheehy, Eugene P., with the assistance of Keckeisen, Rita G. and McIlvaine, Eileen. Guide to reference books. 9th ed. Chicago, IL: American Library Association, 1976. 1015 p. ISBN 0-8389-0205-7. $40. LC 76-11751.
    First supplement: 1980. 305 p. ISBN 0-8389-0294-4. $25. (pbk). LC 79-20541.
    Second supplement: 1982. 243 p. ISBN 0-8389-0361-4. $15. (pbk). LC 82-1719.   6•

Geographic Bibliographies

Standard American guide to reference materials. Section CL, Geography, p. 572-595, is particularly strong on gazetteers, geographical names, and atlases. Also useful for good coverage of related social sciences, sociology, anthropology, statistics, economics, political science, history and area studies, and of earth sciences. The tenth edition is in preparation. The first edition entitled Guide to the study and use of reference books *(1902. 104 p.) was by Alice Bertha Kroeger. Later editions were edited by Isadore Gilbert Mudge and Constance Mabel Winchell.*

Walford, Albert John, with the assistance of Harvey, Joan M. and Taylor, L.J.
Walford's guide to reference material. V.2. Social and historical sciences, philosophy, and religion. 4th ed. London: Library Association; Phoenix, AZ: Oryx Press, 1982. 812 p. (1st ed., 1959. 543 p.). ISBN 0-85365-564-2. $67.50. LC 80-489414.   7
The standard British library reference guide. Of particular geographic value are sections 908, Area studies, p. 365-433; 91, Geography, exploration, travel, p. 434-447; and 912, Atlases and maps, p. 448-469.

B. SELECTIVE GUIDES TO THE GEOGRAPHIC LITERATURE

Arnim, Helmuth. Bibliographie der geographischen Literatur in deutscher Sprache. (Bibliotheca bibliographica Aureliana, 21). Baden-Baden, West Germany: Librairie Heitz, 1970. Reprinted Baden-Baden: Koerner, 1980. 177 p. ISBN 3-87320-021-X.   8
A key aid in identifying and checking the vast German geographic literature.

Blotevogel, Hans H. and Heineberg, Heinz, eds. Bibliographie zum Geographiestudium. Paderborn, West Germany: Ferdinand Schöningh, 1976-1981. 4 v. LC 76-471276.
    Vol. 1: 1976. 240 p. ISBN 3-506-71117-2. DM 19.80.
    Vol. 2: 1976. 352 p. ISBN 3-506-71118-0. DM 23.80.
    Vol. 3: 1980. 304 p. ISBN 3-506-71119-9. DM 23.80.
    Vol. 4: 1981. 372 p. ISBN 3-506-71121-0. DM 26.80.   9
International and comprehensive in scope but of particular value in full coverage of the geographical literature in German. Volume one covers theory, teaching, methods of geography, physical geography, and geoecology. Volume two covers cultural geography, social geography, regional planning, developing countries, and statistical sources. Volume three covers regional geography of Europe, including the Soviet Union. Volume four covers the regional geography of Asia, Africa, the Americas, Australia, Oceania, the oceans, and the polar regions.

Dion, Louise. Introduction aux ouvrages de référence en géographie: choix d'ouvrages de la collection de la Bibliothèque de l'Université Laval. (Guides bibliographiques, 4). Québec: Bibliothèque de l'Université Laval, 1970. 108 p. LC 73-351304.   10
Particularly valuable for coverage of Canadian and French publications.

Goddard, Stephen, ed. A Guide to information sources in the geographical sciences. London: Croom Helm; Totowa, NJ: Barnes and Noble Imports, 1983. 273 p. ISBN 0-389-20403-X. $27. LC 83-6065.   11
Good coverage of topics included: systematic fields (geomorphology, historical geography, agricultural geography, and industrial geography); regions (Africa, South Asia, the United States, and the Soviet Union); and tools (maps, atlases, and gazetteers; aerial photographs and satellite information; statistical materials and the computer; and archival materials--governmental and otherwise).

Lewthwaite, Gordon R.; Price, Edward T., Jr.; and Winters, Harold A., eds. A Geographical bibliography for American college libraries. A revision of A Basic Geographical Library: A Selected and Annotated Book List for American Colleges. (Commission on College Geography, publication no. 9). Washington, DC: Association of American Geographers, 1970. 214 p. LC 77-126351.   12
Of particular value for publications in the period 1945-1969 and for American works. For works before 1945 see Wright and Platt (entry 15).

Nolzen, Heinz. Bibliographie allgemeine Geographie: Grundlagenliteratur der Geographie als Wissenschaft (UTB 608). Paderborn, West Germany: Ferdinand Schöningh, 1976. 185 p. ISBN 3-506-99182-5. DM 12.80. LC 77-451587.   13
Guide to geographic literature particularly in German, though not limited to works in that language. Arranged by systematic fields of geography.

Webb, William, ed. Sources of information in the social sciences: a guide to the literature. 3rd ed. (1st and 2nd eds. ed. by Carl M. White) Chicago, IL: American Library Association, in press. (1st ed., 1964; 2nd ed., 1973). ISBN 0-8389-0405-X. LC 84-20494.   14
Chapter three, Geography, by Chauncy D. Harris, covers both substantive works, mainly of systematic fields, and bibliographic guides to the literature. Other chapters cover social science literature, history, economics and business administration, sociology, anthropology, psychology, education, and political science.

Wright, John Kirtland, and Platt, Elizabeth T. Aids to geographical research: bibliographies, periodicals, atlases, gazetteers, and other reference books. 2nd ed. New York: Columbia University Press for the American Geographical Society. 1947. Reprinted: Westport, CT: Greenwood Press, 1971. 331 p. (1st ed., 1923. 243 p.). ISBN 0-8371-3384-X. LC 73-106702.   15
The best bibliography of geographic bibliographies and reference works for the period up to 1945. Now somewhat dated. International in coverage and comprehensive in scope.

C. CURRENT GEOGRAPHICAL BIBLIOGRAPHIES

Bibliographie géographique internationale. 1891- . Quarterly plus annual index volume. (Annual 1891-1975/76). Laboratoire d'Information et de Documentation en Géographie (Intergéo), Centre National de la Recherche Scientifique, 191, rue Saint-Jacques, 75005, Paris, France. ISSN 0067-6993.   16
The most international and the best balanced bibliography of geography, both regional and systematic, with annotations for each entry. Mainly in French but with supplementary English table of contents and headings. Some annotations in English. Indexes include list of periodicals analyzed, and separate subject, place, and author indexes. Since this bibliography has been published since 1891 it provides the most comprehensive international inventory of geographical publications since the rise of the modern field of geography.

Current geographical publications: additions to the research catalogue of the American Geographical Society collection of the University of Wisconsin-Milwaukee library. 1- (1938-). 10 nos. a year. Indexes. American Geographical Society Collection at the University of Wisconsin-Milwaukee, Box 399, Milwaukee, Wisconsin 53201. ISSN 0011-3514.   17
A major current geographical bibliography organized both by region and by systematic field. Particularly strong for publications in and on the United States and for works in English, but international and comprehensive in coverage. Volumes 1-41, no. 2 (1938-February 1978) published by the American Geographical Society of New York, in New York.

Geo abstracts. 1966- . Seven separate series, each with 6 nos. a year. Geo Abstracts, Ltd., Regency House, 34 Duke St., Norwich, NR3 3AP, England.   18
The most comprehensive geographical abstracting service in English. Seven series:
    A. **Landforms and the Quaternary,** 1966- .
    B. **Climatology and hydrology,** 1966- .
    C. **Economic geography,** 1966- .
    D. **Social and historical geography,** 1966- .
    E. **Sedimentology,** 1972- .
    F. **Regional and community planning,** 1972- .
    G. **Remote sensing, photogrammetry, and cartography,** 1974- .

Geographic Bibliographies

Comprehensive annual indexes in one volume per year 1966-1971, in two volumes per year 1972- (volume one covers series A,B,E, and G; volume two covers series C,D, and F). Separate five-year cumulative indexes for each series.

Referativnyi zhurnal: geografiia. Svodnyi Tom (Vsesoiuznyi Institut Nauchnoi i Tekhnicheskoi Informatsii). 1956- . Monthly. Annual subject and geographical index and annual author index as separate volumes. Orders: Mezhdunarodnaia Kniga, Moscow, G-200, U.S.S.R. ISSN 0034-2378.   19
   The key bibliographic source for current geographic literature in Russian or on the Soviet Union. Detailed coverage of works in Russian and other languages of the Soviet Union. Each section edited by a specialist. Extensive signed abstracts.

D. RETROSPECTIVE GEOGRAPHICAL BIBLIOGRAPHIES

American Geographical Society. Research catalogue of the American Geographical Society. Boston, MA: G.K. Hall, 1962. 15 v. 10,453 p. ISBN 0-8161-0628-2. $1450.   20
   Detailed bibliography for the years 1923-1961 arranged by regional and systematic fields. The most comprehensive world-wide cumulative regional geographical bibliography for the years covered. Reproduction of cards of the Research catalogue of the AGS. Use facilitated by close study of the classification scheme and the table of contents.

American Geographical Society. Research catalogue of the American Geographical Society. First supplement. Boston, MA: G.K. Hall, 1972-1974. 4 v.: Regional (1972), 2 v. ISBN 0-8161-0999-0. $260.; Topical (1974), 2 v. ISBN 0-8161-1083-2. $265.   21
   Ten-year supplement covering the years 1962-1971 with improved table of contents, indexes, and headings and more legible reproduction of cards. Topical coverage greatly expanded.

American Geographical Society. Research catalogue of the American Geographical Society. Second supplement. Boston, MA: G.K. Hall, 1978. 2 v. Regional catalogue, 614 p. Topical catalogue, 686 p. ISBN 0-8161-0081-0. $260.   22
   Five-year supplement covering the years 1972-1976.

E. INTERNATIONAL GEOGRAPHICAL CONGRESSES

Kish, George, ed. Bibliography of International Geographical Congresses, 1871-1976. Boston, MA: G.K. Hall, 1979. 540 p. ISBN 0-8161-8226-4. $36.50. LC 79-10511.   23
   Lists nearly 7000 papers presented at the 1st to 23rd International Geographical Congresses (1871-1976). Arranged by congress and then by section. Bibliographical information on the published proceedings. Useful classed subject and author indexes.

F. BOOK REVIEWS

Van Balen, John. Geography and earth science publications: an author, title, and subject guide to books reviewed, and an index to the reviews. Ann Arbor, MI: Pierian Press, 1978. ISBN 0-685-38823-9. $62.50. (set). LC 78-52361.
   Vol. 1: 1968-1972. 313 p. ISBN 0-87650-090-4. $35.
   Vol. 2: 1973-1975. 232 p. ISBN 0-87650-091-2. $35.   24
   Bibliographic listing of books in geography and related earth sciences reviewed in geographical periodicals. Also a record of the reviews published in 21 periodicals in volume one and in 38 periodicals in volume two.

2. GEOGRAPHICAL SERIALS

Chauncy D. Harris

A. BIBLIOGRAPHIES

Harris, Chauncy D. Annotated world list of selected current geographical serials. (Research paper no. 194). 4th ed. Chicago, IL: University of Chicago, Department of Geography, 1980. 165 p. (1st ed., 1960, 14 p.) ISBN 0-89065-101-9. $10. LC 80-17561. 25
    Annotated list of 443 carefully selected current geographical serials considered to be of greatest continuing scholarly interest, usefulness, and accessibility. Entries are by 72 countries arranged in alphabetical order. Annotations on the nature of contents, editors, publishers, inclusion of abstracts, and language utilized. Includes study of serials most cited in geographical bibliographies.

Harris, Chauncy D. and Fellmann, Jerome D. International list of geographical serials. (Research paper no. 193). 3rd ed. Chicago, IL: University of Chicago, Department of Geography, 1980. 457 p. (1st ed., 1950. 124 p.). ISBN 0-89065-100-0. $10. LC 80-16392. 26
    Comprehensive inventory of 3445 geographical serials from 107 countries in 55 languages, by countries in alphabetical order. Includes both current and closed serials. Limited to geographical serials proper. Notes dates and the frequency of publication, listings in union catalogues, name changes, cumulative indexes, abstracts, and address of publication (if current).

B. SERIALS

    The 24 geographical serials here listed are those considered to have the most general international interest and value for American libraries. See also the relevant entries in other sections of this bibliography. Since subscription prices are subject to frequent changes they are not here reported.

Annales de géographie. 1- (1891-). Bimonthly. Librairie Armand Colin, 103 Boulevard Saint-Michel, 75240 Paris Cedex 05, France. ISSN 0003-4010. 27
    A key international journal with articles, notes, book reviews, and a chronicle of wide interest in all fields of geography. In French with abstracts in French and English. Cumulative tables of contents: 1891-1901. Cumulative indexes: 1891-1901, 1902-1911, 1912-1921, 1922-1931, 1932-1951, 1952-1961, 1962-1971, 1972-1981.

Area. 1- (1969-). Quarterly. Institute of British Geographers, 1 Kensington Gore, London, SW7 2AR, England. ISSN 0004-0894. 28
    Short articles, reports, comments, notes, and notices on a wide range of topics of contemporary geographic interest.

Association of American Geographers. Annals. 1- (1911-). Quarterly. Association of American Geographers, 1710 Sixteenth Street, N.W., Washington, D.C. 20009. ISSN 0004-5608. 29
    The leading scholarly geographical periodical of the United States with long and short articles, book reviews, texts of presidential addresses, memorials, and commentary. Cumulative indexes: 1-25 (1911-1935), 26-55 (1936-1965), 56-65 (1966-1975).

Cahiers d'outre-mer: revue de géographie de Bordeaux. 1- (1948-). Quarterly. Institut de géographie, Université de Bordeaux III, 33405 Talence, France. ISSN 0045-3765. 30
    Geographical articles of wide interest on all parts of the world but especially on Latin America, Africa, and Asia. Notes. Reviews. In French with abstracts in French and English. Cumulative indexes: 1-10 (1948-1957), 11-20 (1958-1967), 21-30 (1968-1977).

Geographical Serials

Canadian geographer. 1- (1951-). Quarterly. Canadian Association of Geographers, Burnside Hall, McGill University, 805 Sherbrooke Street West, Montréal, Québec, H3A 2K6, Canada. ISSN 0008-3658.

Leading scholarly geographical periodical of Canada. Devoted particularly but not exclusively to Canadian geography. Articles. Notes. Cumulative index: 1951-1967.

31

Chicago. University. Department of Geography. Research papers. 1- (1948-). Irregular. Department of Geography, University of Chicago, 5828 University Avenue, Chicago, Illinois, 60637. Each number has a separate ISBN and LC card number.

Research monographs and collected works, particularly in the fields of cultural, economic, historical, physical, and urban geography. Bibliographies.

32

Die Erde. Zeitschrift der Gesellschaft für Erdkunde zu Berlin. 1- (1839-). Quarterly. Gesellschaft für Erdkunde zu Berlin, Arno-Holz-Strasse 14, D-1000 West Berlin 41, Germany. ISSN 0013-9998.

Major international scholarly journal of long standing and high current value. Articles. Extensive reviews. Proceedings of the society. Each issue contains an interpreted aerial photograph. In German, with supplementary English in table of contents, and English abstracts. Cumulative indexes: 1853-1863, 1863-1901.

33

Erdkunde: Archiv für wissenschaftliche Geographie. 1- (1947-). Quarterly. Ferd. Dümmlers Verlag, Kaiserstrasse 31/37, Postfach 1480, D-5300 Bonn, Federal Republic of Germany. ISSN 0014-0015.

A leading international scientific periodical with a wide range of interests. Articles. Extensive reviews. In German, sometimes in English. Supplementary titles in table of contents in English and abstracts in English. Cumulative index: 1-17 (1947-1963).

34

Geoforum: the international multi-disciplinary journal for the rapid publication of research results and critical review articles in the physical, human, and regional geosciences. 1- (1970-). Quarterly. Pergamon Press, Headington Hill Hall, Oxford OX3 0BW, England, or Pergamon Press, Fairview Park, Elmsford, New York 10523. ISSN 0016-7185.

Articles on a wide range of topics in human, economic, and physical geography. Expensive.

35

Geographical analysis: an international journal of theoretical geography. 1- (1969-). Quarterly. Ohio State University Press, 2070 Neil Avenue, Columbus, Ohio, 43210. ISSN 0016-7363.

Articles on theory in geography, typically, but not necessarily, with mathematical analysis. Research notes and comments. Book reviews. Cumulative index 1-10 (1969-1978) in v. 10 (1978).

36

Geographical journal. 1- (1893-). 3 per annum. Royal Geographical Society, 1 Kensington Gore, London SW7 2AR, England. ISSN 0016-7398.

Original articles of broad geographic interest. Strong on geographic exploration. Extensive section of reviews. Cartographic survey. Notes. University and society news. Cumulative indexes: 1-20 (1893-1902), 21-40 (1903-1912), 41-60 (1913-1922), 61-80 (1923-1932), 81-100 (1933-1942), 101-120 (1943-1954), 121-130 (1955-1964), 131-140 (1965-1974).

37

Geographical magazine. 1- (1935-). Monthly. The geographical magazine, c/o Magnum Distribution Ltd., Watling Street, Milton Keyes MK2 2BW, England. ISSN 0016-741X.

The most successful geographical journal in English in combining sound scholarship with non-technical, popular, interesting, and well-illustrated short articles. Informative news, particularly of British geography and geographers. Book reviews. Cumulative indexes: 23-45 (1950-1972).

38 •

Geographical review. 1- (1916-). Quarterly. American Geographical Society, 156
Fifth Avenue, Room 600, New York, New York, 10018. ISSN 0016-7428.
*A key international journal with original articles, generally of broad interest. Book
reviews. Cumulative indexes: 1-15 (1916-1925), 16-25 (1926-1935), 26-35 (1936-1945),
36-45 (1946-1955), 46-55 (1956-1965).*

39•

Geographische Zeitschrift. 1-50 (1895-1944); 51- (1963-). Quarterly. Franz Steiner
Verlag, Friedrichsstrasse 24, Postfach 5529, D-6200 Wiesbaden, Federal Republic of
Germany. ISSN 0016-7479.
*Major scholarly periodical with both long and short articles and short notes. Book
reviews. In German, with supplementary English titles in table of contents, and ab-
stracts in English. Cumulative index: 1-50 (1895-1944).*

40

Geography. 1- (1901-). Quarterly. Geographical Association, 343 Fulwood Road,
Sheffield S10 3BP, England, or G. Philip and Sons, 12-14 Long Acre, London WC2E
9LP, England. ISSN 0016-7487.
*Substantial geographical articles. School geography. Association affairs. Book re-
views. Cumulative index: 1-54 (1901-1969).*

41•

**GeoJournal: international journal for physical, biological, and human geosci-
ences and their applications in environmental planning and ecology.** 1- (1977-).
Bimonthly. Akademische Verlagsgesellschaft, Postfach 1107, D-6200 Wiesbaden, Federal Republic of Germany. ISSN 0343-2521.
*International authorship and topics. Each issue is devoted in whole or in part to a
special theme with articles by leading specialists from around the world. Book reviews. Expensive but of high quality. In English. Abstracts.*

42

IGU bulletin (International Geographical Union). 1- (1950-). Semi-annual. Distrib-
uted in each country by the IGU National Committee for that country. In the United
States distributed by the Association of American Geographers, 1710 Sixteenth Street,
N.W., Washington, D.C. 20009. ISSN 0018-9804.
*Reports of activities of the commissions of the International Geographical Union.
Announcements of international congresses or regional conferences. News from
member countries. For further information write Prof. Leszek Kosinski, Secretary-
General, International Geographical Union, Department of Geography, University
of Alberta, Edmonton, Alberta, Canada T6G 2H4.*

43

Institute of British Geographers. Transactions. 1-66 (1935-1975); New Series 1-
(1976-). Quarterly. Institute of British Geographers, 1 Kensington Gore, London SW7
2AR, England. ISSN 0020-2754.
*The leading scholarly geographical periodical of Britain. Articles on a wide range
of topics. Presidential addresses. Obituaries. Cumulative index: 1-42 (1935-1967) in
42 (1967).*

44

National geographic. 1- (1888-). Monthly. National Geographic Society, Seventeenth
and M Streets, N.W., Washington, D.C. 20036. ISSN 0027-9358.
*Popular non-technical journal of very wide circulation with narrative and general-
interest articles, numerous colored and black-and-white photographs of superb quality, and large folded atlas-type general-reference and thematic maps. Captions to the
illustrations are exceptionally well done. Cumulative indexes: 1888-1946, 1947-1983.*

45•

National geographic research: a scientific journal. 1- (1985-). Quarterly. National
Geographic Society, 1145 17th St., N.W., Washington, D.C. 20036. ISSN 8755-724X.
*Articles of peer-reviewed scientific research authored primarily by recipients of
grants awarded by the National Geographic Society's Committee for Research and
Exploration. Covers a broad scope of geography including a variety of scientific disciplines ranging from anthropology to zoology. Color-illustrated with maps and
photographs.*

46

Norois: Revue géographique de l'Ouest et des pays de l'Atlantique Nord. 1- (1954-). Quarterly. Norois, 8 rue René-Descartes, F-86022 Poitiers, France. ISSN 0029-182X. 47
Articles particularly on the west of France, the Low Countries, British Isles, North Atlantic, and North America. Book reviews. Chronicles. Bibliographies. In French with abstracts and table of contents in both French and English. Cumulative indexes: 1-10 (1954-1963), 11-20 (1964-1973).

Petermanns geographische Mitteilungen (Geographische Gesellschaft der Deutschen Demokratischen Republik). 1- (1855-). Quarterly. VEB Hermann Haack, Geographisch-Kartographische Anstalt, 58 Gotha, German Democratic Republic. ISSN 0031-6229. 48
Long established scholarly journal. Long and short articles. News. Reviews. Statistics. Section on the Soviet Union. In German. Supplementary tables of contents and abstracts in English and Russian. Cumulative indexes: 1855-1864, 1865-1874, 1875-1884, 1885-1894, 1895-1904, 1905-1934.

Professional geographer. n.s. 1- (1949-). Quarterly. Association of American Geographers, 1710 Sixteenth Street, N.W., Washington, D.C. 20009. ISSN 0033-0124. 49
Short articles, views and opinions, professional notes, and numerous short book reviews. Cumulative index 1946-1966 in volume 19 no. 2 (March 1967).

Soviet geography 1- (1960-). Monthly. V.H. Winston and Sons, 7961 Eastern Avenue, Silver Spring, Maryland, 20910. ISSN 0038-5417. 50
Particularly valuable in providing English translations of articles in many fields from principal geographical journals of the Soviet Union. Articles on the Soviet Union by Western specialists. Informative news notes.

See also listings under serials in the index and especially the following entries: 192; 249-252; 255; 436-438; 461; 463-464; 466; 505; 543; 870-873; 898; 1131; 1167-1170; 1335-1337; 1468-1469; 1540-1542; 1813; 1823-1824; 1909-1910; 1915-1917; 2095-2111; 2138-2143; 2156-2157; 2187-2192; 2206-2210; 2276; 2288-2294; 2314-2315; 2369-2370; 2433-2438; 2488-2490; 2580-2584; 2666-2668; 2670; 2736-2737; 2800; 2844.

3. ATLASES

Richard W. Stephenson

See also the listings under atlases in the index.

A. KEY WORLD ATLASES

Atlante internazionale: del Touring Club Italiano. 8th ed. Milano, Italy: Touring Club Italiano, 1978. v. 1, 300 p. (volume of maps and 173 plates); v. 2, 1032 p. (gazetteer). Lira 170,000. LC 79-383311. 51
Comprehensive atlas of the world, consisting for the most part of physical-political maps. A separate volume in a smaller format lists more than 250,000 place names.

National Geographic atlas of the world. 5th ed. Washington, DC: National Geographic Society, 1981. 383 p. ISBN 0-87044-347-X. $44.95. LC 81-675324. 52•
Comprehensive atlas of the world consisting for the most part of general reference political maps in which relief representation has been subordinated to place names. Includes brief profiles to each country containing economic and political data and a 143-page index to place names.

The New international atlas. Chicago, IL: Rand McNally and Co., 1984. 568 p. (Update of 1980 ed.; 1st ed., 1969. 556 p.) ISBN 0-528-83150-X. $100. LC 80-51969. 53•

Detailed atlas of the world. The result of a cooperative effort involving scholars and cartographic houses in the United States, Latin America, Europe, and Japan. Introductory material, titles, and legends are written in English, German, Spanish, French, and Portuguese. An index provides the geographical coordinates for more than 160,000 place names. Population figures updated to January, 1984.

The Times atlas of the world. Comprehensive ed., 6th ed. London: Times Books in collaboration with John Bartholomew and Son Limited, 1980 (Reprint 1981), 227 p. ISBN 0-7230-0235-5. £45. LC 84-675148.   54•

One of the finest English language general atlases of the world. Features 123 colored plates and a detailed index indicating page and grid references as well as geographic coordinates for more than 200,000 place names. The attractively-drawn double-page regional maps feature relief by layer tints.

B. OTHER IMPORTANT WORLD ATLASES

Die Erde. Adolf Hanle, ed., Meyers Grosskarten-edition. Manneheim, West Germany: Bibliographisches Institut, 1978. 2 v. ISBN 3-411-01742-2. LC 79-368288.   55

Published on the 150th anniversary of the Bibliographisches Institut, the atlas consists of two volumes: an atlas of 87 maps and an accompanying place name gazetteer with more than 175,000 entries.

Fiziko-geograficheskii atlas mira. (Akademiia Nauk SSSR) Moscow: Akademii nauk SSSR i Glavnoe upravlenie geodezii i kartografii, 1964. 298 p. LC Map 65-116 rev.   56

Based on the contribution of nearly 200 Soviet cartographers and physical geographers, this is one of the most comprehensive and detailed physical world atlases published to date. The atlas emphasizes the Soviet Union and was published in Russian only, although an English translation of titles and legends appears in Soviet Geography: Review and Translation, v. 6, no. 5-6, May-June 1965, 403 p.

Goode's world atlas. Edward B. Espensahde, Jr., ed., 16th edition. Chicago, IL: Rand McNally and Co., 1983, 368 p. (1st ed., Goode's School Atlas, 1923. 96, 41 p.) ISBN 0-528-83125-9. $19.95. ISBN 0-528-63007-5. $15.95. (pbk). LC 81-51412.   57•

Sixteenth edition of the well-known atlas by Dr. J. Paul Goode, first published sixty years ago. Includes general reference maps showing relief by shading and layer tints, thematic maps, and a pronouncing index to place names. A relatively inexpensive handy and useful atlas.

The Great geographical atlas. London: Mitchell Beazley; Chicago, IL: Rand McNally, 1982. xvi, 256, 144 p. ISBN 0-85533-386-3. $75. LC 83-118754.   58•

The result of a cooperative effort involving three major cartographic publishing houses. Divided into four parts entitled "Encyclopedia Section," "International Map Section," "United Kingdon and Ireland Map Section," and "Geographical Information and International Map Index."

Hammond medallion world atlas. Maplewood, NJ: Hammond, Inc., 1984, 655 p. ISBN 0-8437-1250-3. $60. LC 82-81116.   59•

Principal maps in this atlas emphasize political subdivisions and populated places. Smaller maps depict relief and economic information. Separate place name indexes accompany each state or country map. Names are also included in a complete alphabetical index of more than 100,000 entries. Concludes with an "Atlas of the Bible Lands," "World History Atlas," and a "United States History Atlas."

Hawkes, Jacquetta, ed. Atlas of ancient archaeology. New York: McGraw-Hill; London: Heinemann, 1974. 272 p. ISBN 0-07-027293-X. $37.95. LC 73-22453.   60

More than 170 archaeological sites world-wide are profiled with brief descriptions. Arranged by area, the text is accompanied by 350 two-color maps, plans, and line drawings.

Map Guides

Philip's universal atlas. B.M. Willet, ed., 4th edition. London: George Philip, 1983. 400 p. ISBN 0-540-05430-5. £14.95. LC 84-675239.　　　　　　　　　　　　61 •
  A good general atlas containing useful thematic maps, general reference maps featuring relief by layer tints and shading, economic statistics for each country, and a 140-page index to place names.

Post, Jeremiah B. An Atlas of fantasy. Rev. ed. New York: Ballantine Books, 1979. 210 p. ISBN 0-345-27399-0. $8.95. (pbk). LC 79-63506.　　　　　　　　　　　62 •
  A marvelously entertaining atlas reproducing maps of imaginary places, "maps from the lore of various cults which have distinctive geographical beliefs, maps from the world of advertising..., disproportionate maps, the 'might have been' maps showing the war aims of the losing side, and the completely decorative map." This work pulls together in one volume many hard-to-find works of cartographic fantasy.

The Times atlas of world history. Geoffrey Barraclough, ed., Rev. ed. London: Times Books; Maplewood, NJ: Hammond, 1984. 360 p. ISBN 0-8437-1129-9. $75. LC 84-675088.　　　　　　　　　　　　63 •
  A superbly designed atlas using modern cartographic techniques to depict world history. Rather than emphasizing European history, the aim of this atlas "has been to present a view of history which is world-wide in conception and presentation and which does justice, without prejudice or favour, to the achievements of all peoples in all ages and in all quarters of the globe." Includes a 12-page chronology, a glossary, geographical index, and explanatory text.

The Times concise atlas of the world. London: Times Books, 1982. 276 p. ISBN 0-7230-0238-X. £18.50. LC 84-675146.　　　　　　　　　　　64 •
  Reduced version of the famous Times Atlas of the World. With the exception of those covering conurbations, the maps were compiled by the Edinburgh publishing house of John Bartholomew and Son Ltd.

The World atlas. A.N. Baranov, et al., eds., 2nd edition. Moscow: Glavnoe Upravlenie Geodezii i Kartografii, 1967. V. 1. Atlas, 250 plates; v. 2. Index-Gazetteers, 1010 p. LC 68-379.　　　　　　　　　　　65
  General world atlas, with some emphasis on the Soviet Union, containing well-made oversize physical-political maps in English and index maps. A large number of the maps are double-page plates and some of these have a third foldout page. Clarity and cartographic detail are of high order. (English edition of Atlas Mira, v.1, 2nd ed., v. 1, 1967. 250 p. and v. 2 Ukazatel' geograficheskikh nazvanii, 2nd ed., 1968. 533 p. 1st ed. 1954).

4. MAP GUIDES

Richard W. Stephenson

A. COMPREHENSIVE GUIDES

American Geographical Society. Map department. Index to maps in books and periodicals. Boston, MA: G.K. Hall, 1968. 10 v. ISBN 0-81610753-X. $990. LC 68-5087.
  First Supplement: 1971. ISBN 0-8161-0806-4. $110.
  Second Supplement: 1976. ISBN 0-8161-0995-8. $110.　　　　　　　　66
  Indispensable bibliography of the maps published in books and periodicals. Entries are arranged according to subjects and geographical-political divisions in one alphabet. Within each geographical division the arrangement is chronological.

Bibliographie cartographique internationale. Paris: A. Colin [etc.], 1936-1975. Annual. Closed 1975. LC 48-18629.　　　　　　　　　　67
  Important annual list, now closed, of maps, atlases, charts, and globes arranged by continent and country with subject and author index.

British Library. Map Library. Catalogue of printed maps, charts, and plans: ten year supplement, 1965-1974. London: British Museum Publications, 1978. 1380 columns. ISBN 0-7141-0366-7. £45. LC 79-307082.   68
   The first supplement to the British Museum Catalogue of Printed Maps, Charts, and Plans, *next entry. Contains entries for items cataloged between 1965 and 1974.*

British Museum. Department of printed books. Map room. Catalogue of printed maps, charts, and plans. Photolithographic edition complete to 1964. London: British Museum, 1967. 15 v., about 7500 p. ISBN 0-7141-0324-1. £300. LC 68-91645.   69
   A complete inventory of printed maps, atlases, globes, and related materials in the Map Room, and other important cartographic material in other collections of the British Museum at the end of 1964.

California. University. Bancroft Library. Index to printed maps. Boston, MA: G.K. Hall, 1964. 521 p. ISBN 0-8161-0704-1. $79. LC 65-4745.
   First supplement: 1975. ISBN 0-8161-1172-3. $110.   70
   Describes the extensive collection of printed maps of California and the American West acquired by Hubert H. Bancroft.

National Map Collection. Catalogue of the National Map Collection, Public Archives of Canada, Ottawa, Ontario. Boston, MA: G.K. Hall, 1976. 16 v. ISBN 0-8161-1215-0. $1550. LC 76-379551.   71
   Divided into three sections (area, author, and subject), this catalog describes some of the maps included in the National Map Collection, Public Archives of Canada.

National Union Catalog. Cartographic materials. Washington, DC: Library of Congress, 1983- . Quarterly. 6 parts (register and 5 indexes: name, title, series, subject, and geographic classification code). ISSN 0734-7634 48X. sn 82-6945.   72
   "Since the beginning of 1983, the cataloging of cartographic materials, both LC and contributed records, have been published as part of the microfiche National Union Catalog. In addition, the entire retrospective LC MARC MAP database, from 1969 to date, has been included. This segment, National Union Catalog. Cartographic Materials *provides, for the first time, a separate, continuing record of the cataloging of maps and atlases and the holdings of these cartographic resources by the libraries of the United States and Canada."*

New York (city). Public Library. Map Division. Dictionary catalog of the Map Division. Boston, MA: G.K. Hall, 1971. 10 v. ISBN 0-8161-0783-1. $990. LC 78-173371.   73
   Some 165,000 cards describing maps and other cartographic publications are reproduced in this publication. Included are 280,000 sheet maps and 6000 atlases in the New York Public Library's Map Division; manuscript maps in the custody of the Manuscript Division; early printed maps in the Rare Book Division; and the Phelps Stokes American Historical Views in the Prints Division.

U.S. Library of Congress. Geography and Map Division. A List of geographical atlases in the Library of Congress, with bibliographical notes. Washington, DC: Library of Congress, 1909-1974. 8 v. ISBN 0-8444-0117-X. LC 9-35009.   74
   Includes 10,254 atlases received by the Library to 1920, 2326 world atlases received between 1920 and 1955, 2647 atlases of Europe, Asia, Africa, Oceania, the polar regions, and the oceans received between 1920 and 1960, and 8181 atlases of the Western Hemisphere added to the collections between 1920 and 1969. Volumes one to four compiled by Philip L. Phillips; volumes five to eight compiled by Clara E. Le Gear. The most extensive bibliography of its kind.

Winch, Kenneth L. International maps and atlases in print. 2nd ed. London; New York: Bowker, 1976. 866 p. (1st ed., 1974, 864 p.) ISBN 0-85935-036-3. LC 77-357993.   75
   Important although somewhat dated list of single-sheet maps, large-scale topographic maps, and atlases available for sale.

Map Guides

Cummings, William P. The Southeast in early maps, with an annotated check list of printed and manuscript regional and local maps of southeastern North America during the colonial period. Chapel Hill, NC: University of North Carolina Press, 1962. 284 p. LC 62-5390.　76
> *An exhaustive study of the mapping of the American Southeast before the Revolutionary War. Includes 67 illustrations of maps.*

Karrow, Robert W., Jr., gen. ed. Checklist of printed maps of the Middle West to 1900. Boston, MA: G.K. Hall, (Originally from the Newberry Library, Chicago, Illinois). 1981-1983. 14 v. in 12. 4663 p. ISBN 0-8161-0344-5 $885. set. (Also available on microfilm). LC 81-131746.
> Contents:
> Vol. 1. **North central states region,** compiled by Patricia A. Moore.
> Vol. 2. **Ohio,** compiled by Stephen Gutgesell, James F. Monteith, and Arlene J. Peterson.
> Vol. 3. **Indiana,** compiled by Thomas Rumer.
> Vol. 4. **Illinois,** compiled by David A. Cobb.
> Vol. 5. **Michigan,** compiled by LeRoy Barnett.
> Vol. 6. **Wisconsin,** compiled by Michael J. Fox and Elizabeth Singer Maule.
> Vol. 7. **Minnesota,** compiled by Nancy Erickson.
> Vol. 8. **Iowa,** compiled by Diana J. Fox.
> Vol. 9. **Missouri,** compiled by Randolph K. Tibbits.
> Vol. 10. **North Dakota,** and 11. **South Dakota,** compiled by Eileen H. Dopson (1 v).
> Vol. 12. **Nebraska,** compiled by Helen Brooks; and 13. **Kansas,** compiled by Ann Hagedorn (1 v).
> Vol. 14. Subject, author, and title index.　77
> *Indispensable checklist to the printed maps of the American Middle West to 1900.*

Marshall, Douglas W., ed. Research catalog of maps of America to 1860 in the William L. Clements Library, University of Michigan, Ann Arbor, Michigan. Boston, MA: G.K. Hall, 1972. 4 v. ISBN 0-8161-1003-4. ISBN 0-8161-1227-4. $270. (Four reels microfilm). LC 73-157170.　78
> *Describes more than 10,000 maps and atlases of America in the William L. Clements Library.*

Sellers, John R. and Van Ee, Patricia Molen. Maps and charts of North America and the West Indies, 1750-1789: a guide to collections in the Library of Congress. Washington, DC: Library of Congress, 1981. 495 p. ISBN 0-8444-0335-0. LC 80-607054.　79
> *Important cartobibliography of more than 2000 maps, including almost 600 original manuscript drawings of North America from the years 1750 to 1789.*

Shirley, Rodney W. The Mapping of the world: early printed world maps, 1472-1700. (Holland Press Cartographica volume 9). London: Holland Press; West Orange, NJ: Saifer, 1983. 669 p. ISBN 0-87556-671-5. $100. LC 84-182964.　80
> *Definitive, well-illustrated cartobibliography of the more than 600 maps of the world published to 1700.*

U.S. National Archives. Guide to cartographic records in the National Archives. (National Archives publication no. 71-16). Washington, DC: National Archives, 1971. 444 p. LC 76-611061.　81
> *Describes the cartographic records consisting of more than 1,600,000 maps and 2,250,000 aerial photographs maintained in the Cartographic Branch of the National Archives. The guide is arranged by branch of government subdivided by record group.*

Wheat, Carl I. Mapping the Transmississippi West. San Francisco, CA: Institute of
Historical Cartography, 1957-1963. 5 v. in 6. LC 57-59400.   82
   Comprehensive cartobibliography of the American West, with numerous illustra-
tions. Volume one, "The Spanish Entrada to the Louisiana Purchase, 1540-1804;"
volume two, "From Lewis and Clark to Frémont, 1804-1845;" volume three, "From
the Mexican War to the Boundary Surveys, 1846-1854;" volume four, "From the Pa-
cific Railroad Surveys to the Onset of the Civil War, 1855-1860;" and volume five (two
parts), "From the Civil War to the Geological Survey."

C. SERIALS

Bibliographic guide to maps and atlases. 1- (1979-). Annual. G.K. Hall and Co., 70
Lincoln Street, Boston, Massachusetts, 02111.
   1979, 633 p. ISBN 0-8161-6874-1 $75.
   1980, 622 p. ISBN 0-8161-6890-3 $120.
   1981, 482 p. ISBN 0-8161-6960-4 $120.
   1982, 430 p. ISBN 0-8161-6976-4 $160.
   1983, 559 p. ISBN 0-8161-6997-7 $160.
   1984, 455 p. ISBN 0-8161-7010-X $175.   83
   A selection of titles mostly from the Library of Congress, Geography and Map Divi-
sion machine-readable map cataloging data base. Also includes selections from the
New York Public Library and serves as the latter's supplement to the Dictionary Catalog of the Map Division (entry 73).

The Map Collector. 1- (D 1977-). Quarterly. Map Collector Publications Ltd., Church
Square, 48 High St., Tring, Hertfordshire HP23 5BH, England. ISSN 0140-427X.   84
   Important journal containing articles about maps, map collecting, and map librar-
ies. Each issue includes the "Collectors' Barometer" which lists the important items
sold in auction and the prices paid, and the "Collectors' Marketplace" which con-
tains advertisements for rare or out of print maps available for sale.

D. DIRECTORIES OF MAP COLLECTIONS

Carrington, David K. and Stephenson, Richard W., compilers. Map collections in
the United States and Canada: a directory. 4th ed. New York: Special Libraries Association, 1985. 178 p. ISBN 0-87111-306-6. $35. LC 84-27571.   85
   Important guide to 804 map collections in the United States and Canada.

Ristow, Walter W., ed. World directory of map collections. Compiled by the Geography and Map Libraries Sub-Section. (IFLA publications no. 8). München, West Germany: Verlag Dokumentation; Hamden, CT: Shoe String Press, 1976. 326 p. ISBN 3-7940-
4428-2. $26. LC 76-381150.   86
   Compiled by the Geography and Map Libraries Sub-Section of the International
Federation of Library Associations, this directory describes 285 map collections in 45
countries. To be replaced by a much-expanded edition containing more than 700 en-
tries by **Wolter, John A.** and **Grim, Ronald E., eds.**, World directory of map collections. (IFLA publication). 2nd ed., München; New York; London: K.G. Sauer,
1985. ISBN 3- 598-20374-8. $36.50.

5. GAZETTEERS

Richard W. Stephenson

A. BIBLIOGRAPHIES

Meynen, Emil. Gazetteers and glossaries of geographical names of the member-countries of the United Nations and the agencies in relationship with the United Nations: bibliography 1946-1976. Wiesbaden, West Germany: Franz Steiner Verlag, 1984. 518 p. ISBN 3-515-04036-6. DM 98. LC 84-196392. 87
  Comprehensive bibliography of gazetteers published between 1946 and 1976 arranged alphabetically by country.

Sealock, Richard B.; Sealock, Margaret M.; and Powell, Margaret S. Bibliography of place-name literature: United States and Canada. 3rd ed. Chicago, IL: American Library Association, 1982. 435 p. ISBN 0-8389-0360-6. $30. LC 81-22878. 88
  Comprehensive bibliography of place-name literature including gazetteers of the United States and Canada.

B. GAZETTEERS AND INDEXES OF GEOGRAPHICAL NAMES

(See also Part I, Section 3, Atlases, entries 51-56. Many atlases have indexes.)

Geograficheskii entsiklopedicheskii slovar': geograficheskie nazvaniia (Geographical encyclopedic dictionary: geographical names). Main editor: A.F. Trëshnikov. Moskva: Sovetskaia Entsiklopediia, 1983. 527 p. LC 83-215968. 89
  World-wide gazetteer but of special value in providing the most up-to-date information on places in the Soviet Union in alphabetical order.

Manguel, Alberto, and Guadalupi, Gianni. The Dictionary of imaginary places. New York: Macmillan, 1980. 438 p. ISBN 0-02-546400-0. LC 80-11128. 90
  A sumptuous, lavishly illustrated gazetteer covering more than a thousand imaginary places created in the worlds of fiction and the stage by a host of distinguished authors. The text in this delightful volume treats each of these realms soberly and eruditely.

Seltzer, Leon E., ed. The Columbia Lippincott gazetteer of the world. With 1961 supplement. New York: Columbia University Press, 1962. 2148, 22 p. (First published in 1952). ISBN 0-231-01559-3. $200. LC 62-4711. 91•
  Despite its age, this is still the best and most detailed one volume gazetteer of the world in the English language.

The Times index-gazetteer of the world. London: Times Publishing Co., 1965. 964 p. LC 66-70286. 92
  The most comprehensive one-volume, world-wide locating index. Name, latitude, and longitude of 345,000 towns, villages, rivers, mountains, and other geographical features, about 198,000 of which are on the plates of The Times Atlas of the World.

U.S. Board on Geographic Names. Gazetteers (various titles). Washington, DC: Defense Mapping Agency (DMA Office of Distribution Services), 1955- . 165 v. to date. Prices vary. LC Card no. varies depending on volume. 93
  Invaluable alphabetical listing by country of names for populated places, cultural entities, and physical features as compiled from detailed topographic map series for each country.

U.S. Geological Survey. The National gazetteer of the United States of America. (Geological Survey Professional Paper 1200). Washington, DC: Government Printing Office, 1982- .
  New Jersey, 1982, 220 p. $7.50. LC 82-600282.
  Delaware, 1983, 101 p. $7.50. LC 83-600395. 94

The "National Gazetteers" are derived from the Geographic Names Data Base, which is part of a computerized Geographic Names Information System developed by the U.S. Geological Survey. In each gazetteer, geographic names are listed as one-line entries in alphabetical order followed by seven categories of information including the type of feature, geographic coordinates, elevation, and the 7.5 minute quadrangle on which the name appears. Gazetteers of other states are in various stages of preparation and publication.

Webster's new geographical dictionary. Springfield, MA: Merriam, 1984. 1376 p. (First published in 1972; Webster's Geographical Dictionary, 1949. 1293 p.) ISBN 0-87779-446-4. $17.50. LC 79-27006.   95 •

An important, English-language gazetteer of the world containing descriptions of 47,000 places, with 15,000 cross references and 218 maps. Population figures from the most recent censuses are included. The 1984 printing contains population data from censuses of 1980 or 1981. Generally the most widely useful and up-to-date world gazetteer.

6. GEOGRAPHICAL DICTIONARIES

Chauncy D. Harris

A. BIBLIOGRAPHY

International Geographical Union. Commission on International Geographical Terminology. Bibliography of mono- and multilingual dictionaries and glossaries of technical terms used in geography as well as in related natural and social sciences. Compiled by Emil Meynen. Wiesbaden, West Germany: Franz Steiner Verlag, 1974. 246 p. ISBN 3-515-01846-8. DM 62. LC 75-316465.   96

Lists 3211 dictionaries by fields of geography and related disciplines including 49 monolingual dictionaries of terms for geography as a whole and 38 bilingual or multilingual dictionaries for geography as a whole. Most entries are for specialized dictionaries in related fields.

B. IN ENGLISH

Johnston, Ronald J., et al., eds. The Dictionary of human geography. Oxford, U.K.: Basil Blackwell; New York: Free Press, a division of Macmillan, 1981. 411 p. ISBN 0-02-903550-3. $40. LC 81-15084.   97 •

The best reference work on terms in human geography in English, balancing the emphasis on terms in physical geography in Stamp and in Monkhouse.

Monkhouse, Francis J. A Dictionary of geography. 2nd ed. London: Edward Arnold, 1970. 378 p. (1st ed., 1965. 344 p.) ISBN 0-7131-5495-0. ISBN 0-7131-5659-7. (pbk). LC 70-503724.   98 •

Useful dictionary with about 4000 terms, particularly in physical geography. Clear definitions.

Schmieder, Allen A.; Griffin, Paul F.; Chatham, Ronald L.; and Natoli, Salvatore J. A Dictionary of basic geography. Boston, MA: Allyn and Bacon, 1970. 299 p. LC 74-94338.   99 •

Especially useful for the beginning student in geography. About 1000 terms in cultural, political, economic, and physical geography.

Stamp, Laurence Dudley, and Clark, Audrey N., eds. A Glossary of geographical terms. Based on a list prepared by a committee of the British Association for the Advancement of Science. 3rd ed. London; New York: Longman, 1979. 571 p. (1st ed., 1961) ISBN 0-582-35258-4. $37.50. LC 79-40238.   100

Geographical Dictionaries

Definitions of terms in modern geography with notes on the origin and development. Includes definitions from original authorities and from other dictionaries. Supplemented by Clark, Audrey N. Longman dictionary of geography: human and physical. *London: Longman, 1985. 724 p. ISBN 0-582-35261-4. £25. LC 85-4290. More than 10,500 entries.*

Whittow, John B., ed. The Penguin dictionary of physical geography. London: A. Lane; New York: Penguin, 1984. 591 p. ISBN 0-14-051094-X. $8.95. (pbk). LC 83-242697.
Brief definitions and descriptions of a large number of terms. Comprehensive and up-to-date.

101 •

International geographical terminology. Internationale geographische Terminologie. Deutsche Ausgabe. Herausgegeben im Auftrag des Zentralverbandes der deutschen Geographie von Emil Meynen (International Geographical Union. Commission on International Geographical Terminology). Wiesbaden: Franz Steiner, 1985. 1479 p. DM 288.
About 2400 specialized key terms in geography in alphabetical order with definitions in German and equivalent terms in English, French, Italian, Spanish, Russian, and Japanese. About 11,000 subterms. Terms also grouped by fields for English, German, and French. The product of international collaboration.

102

Quencez, G. Vocabularium geographicum. Luxembourg: Service des publications des communautés européennes, 1967; Brussels, Belgium: Presses Académiques Européennes, 1968. 298 p. LC 78-449102.
Best general comparative lists of geographical terms in major West European languages: English, French, German, Italian, Spanish, and Dutch. Includes 3100 basic terms arranged in 617 closely related concepts and grouped in 42 fields, with an alphabetical index for each language of all terms. Compiled with the aid of many international specialists.

103

Shchukin, Ivan Semenovich. Chetyrëkh iazychnyi entsiklopedicheskii slovar' terminov po fizicheskoi geografii (Four-language encyclopedic dictionary of terms in physical geography). Edited by A.I. Spiridonov. Moscow: Sovetskaia Entsiklopediia, 1980. 703 p. LC 80-508866.
Includes 5700 terms in physical geography in Russian with definitions in Russian and equivalents in English, German, and French. Separate alphabetical indexes of terms in English, German, and in French facilitate the use of the dictionary.

104

Cabanne, Claude, ed. Lexique de géographie humaine et économique. Paris: Dalloz, 1984. 431 p. ISBN 2-247-00534-9. FFr 82
Handy paperback dictionary of French terms in human and economic geography with clear definitions and a useful list of French abbreviations.

105

George, Pierre, ed. Dictionnaire de la géographie. 2nd ed. Paris, France: Presses Universitaires de France; New York: French and European Publications, 1974. 451 p. (1st ed., 1970). ISBN 0-686-57193-2. $47.50. LC 74-160566.
Comprehensive coverage of geographical terms and concepts in French by 20 specialists, ranging from physical to human geography. Modern clear definitions.

106

Herder Lexikon: Geographie. By Margit Klein and Johannes Klein. 8th ed. Freiburg, West Germany: Herder, 1981. 237 p. (1st ed., 1972. 237 p.) ISBN 3-451-16451-5. DM 22. LC 72-336954.
About 2400 main word entries for German terms in geography, with 400 diagrams and tables.

107

Kalesnik, Stanislav Vikent'evich, ed. Entsiklopedicheskii slovar' geograficheskikh terminov. (Encyclopedic dictionary of geographical terms). Moscow: Sovetskaia Entsiklopediia, 1968. 435 p. NUC 74-133114. 108

Best all-around dictionary of geographical terms in Russian. About 4200 terms defined by specialists in each field. Especially strong for physical geography.

7. GENERAL STATISTICAL SOURCES

Chauncy D. Harris

A. BIBLIOGRAPHIES

Goyer, Doreen S. International population census bibliography: revision and update, 1945-1977. New York: Academic Press, 1980. 576 p. ISBN 0-12-294380-5. $39.50. LC 79-25890. 109

Supplements the International Population Census Bibliography *of the University of Texas (next entry). Extends coverage of national and territorial censuses through 1977.*

Texas. University. Population Research Center. International population census bibliography. Austin, TX: University of Texas, Bureau of Business Research, 1965-1968. 7 v. LC 66-63578. 110

World-wide coverage of population censuses. V. 1. Latin America and the Caribbean; v. 2. Africa; v. 3. Oceania; v. 4. North America; v. 5. Asia; v. 6. Europe; and v. 7. supplement, 1968.

U.S. Library of Congress. Census Library Project. Statistical yearbooks: an annotated bibliography of the general statistical yearbooks of the major political subdivisions of the world. Prepared by Phyllis G. Carter. Washington, DC: Library of Congress, Card Division, 1953. 123 p. LC 53-60036. 111

Listing of the statistical yearbooks of 200 countries or areas with full bibliographical details. Dated but still valuable.

Wasserman, Paul, and O'Brien, Jacqueline. Statistics sources: a subject guide to data on industrial, business, social, educational, financial, and other topics for the United States and internationally. 8th ed. Detroit, MI: Gale Research, 1983. 2 v. 1875 p. ISBN 0-8103-0391-4. $195. LC 83-241783. 112

About 26,000 references arranged under 12,000 subject headings with finding guides to sources of information.

B. STATISTICS

United Nations. Statistical Office. Statistical yearbook. 1948- . Annual. United Nations Publications, Room A-3315, New York, New York 10017. LC 50-2746. ISSN 0082-8459. 113

The key international source for statistics in summary form and by countries and major world areas for many characteristics of the population and of the economy.

United Nations. Statistical Office. Demographic yearbook. 1948- . Annual. United Nations Publications, Room A-3315, New York, New York 10017. LC 50-641. ISSN 0082-8041. 114

Source for many statistics of characteristics of the population on an international basis by countries.

8. ENCYCLOPEDIAS AND HANDBOOKS

Chauncy D. Harris

A. GEOGRAPHICAL ENCYCLOPEDIA

Tietze, Wolf. Westermann Lexikon der Geographie. Braunschweig, West Germany: G. Westermann Verlag, 1968-1972. 5 v. LC 74-372550.   115
 The most widely useful geographic encyclopedia with authoritative articles on major regions and countries of the world and on the fields of geography by 125 leading German geographers. Extensive bibliographies.

B. GEOGRAPHICAL HANDBOOKS

Geographical digest. 1- (1963-). Annual. George Philip and Son Ltd., 12-14 Long Acre, London WC2E 9LP, England. LC 63-57664.   116
 Concise data on recent changes in the world on topics such as political and administrative divisions, names, population, agricultural and mineral production, manufacturing and mining, trade, engineering projects, communications, conservation, exploration, and natural catastrophes.

Sachs, M.Y., ed. Worldmark encyclopedia of the nations. 6th ed. New York: John Wiley and Sons, 1984. (1st ed., 1960). 5 v. ISBN 0-471-88622-X. $199.95. LC 83-26013.   117•
 Well-organized, brief, comparable data on 172 countries, arranged in a uniform outline of 50 topics. V. 1. United Nations; v. 2. Africa; v. 3. Americas; v. 4. Asia and Oceania; v. 5. Europe.

Statesman's year-book: statistical and historical annual of the states of the world. 1- (1864-). Annual. Macmillan Press Ltd., Little Essex St., London WC2R 3LF, England, or St. Martin's Press, 175 Fifth Avenue, New York, NY 10010. ISSN 0081-4601. LC 4-3776.   118•
 Concise up-to-date information on each country of the world, arranged by countries in alphabetical order.

U.S. Department of State. Office of the Geographer. International boundary study. 1- (1961-). Irregular. U.S. Department of State, c/o Bureau of Intelligence and Research, 2201 C St., N.W., Washington, D.C. 20520. ISSN 0502-0034.   119
 Authoritative studies of the offical status of international boundaries. Each number is devoted to a single country. Includes many maps.

9. BIOGRAPHICAL AND INSTITUTIONAL DIRECTORIES

Chauncy D. Harris

Association of American Geographers: directory 1982. Washington, DC: Association of American Geographers, 1982. 186 p. ISBN 0-89291-164-6.   120
 Name, mailing address, birth date, birth place, education, employment, specialty, and geographic area of interest of 5402 members. Earlier editions 1949, 1952, 1956, 1961, 1967, 1970, 1974, and 1978.

Association of American Geographers. Guide to departments of geography in the United States and Canada. 1968/1969- . Annual. Association of American Geographers, 1710 Sixteenth Street, N.W., Washington, DC 20009. ISSN 0072-8497. LC 68-59269.   121•
 Annual listing of departments of geography in the United States and Canada. Programs and facilities. List of staff at each institution with specialties. Alphabetical index of names of staff members (valuable as it is revised annually). 1968/69-1983/84 listed only graduate departments; 1984/85- also lists undergraduate departments.

Canadian Association of Geographers: directory. 1978- . Annual. Canadian Association of Geographers, Burnside Hall, McGill University, Montréal, PQ, H3A 2K6, Canada. ISSN 0707-3844.   122
Departments of geography with lists of faculty, research, and publications; government documents, and members, with addresses.

Meynen, Emil. Orbis geographicus 1980/1984. World directory of geography. Wiesbaden, West Germany: Franz Steiner Verlag, 1982. 962 p. ISBN 3-515-02910-9. ISSN 0030-4395. DM 88.   123
Part one lists geographical institutions, such as international associations, geographical societies, departments of geography in universities, and government agencies. Part two lists geographers by countries with biographical information. Worldwide in coverage. The only international directory of geographical institutions and geographers. Earlier editions, 1960; 1964/66; and 1968/1972 and 1968/1974 (2 v.) also have reference value.

10. DISSERTATIONS

Chauncy D. Harris

Browning, Clyde E. A Bibliography of dissertations in geography: 1901 to 1969. American and Canadian universities. (Studies in geography no. 1). Chapel Hill, NC: University of North Carolina, Department of Geography, 1970. 96 p. LC 73-632461.   124
Includes 1582 doctoral dissertations arranged by 23 subject categories. A regional classification records the entry numbers for large regions.

Browning, Clyde E. A Bibliography of dissertations in geography: 1969 to 1982. American and Canadian universities. (Studies in geography no. 18). Chapel Hill, NC: University of North Carolina, Department of Geography, 1983. 145 p. LC 82-620045.   125
Includes 2270 doctoral dissertations arranged by subject categories. Author index.

Comprehensive dissertation index, 1861-1972. Ann Arbor, MI: Xerox University Microfilms, 1973. 37 v. ISBN 0-8357-0080-1. LC 73-89046. (For prices write to: University Microfilms International, 300 N. Zeeb Road, Ann Arbor, MI 48106).   126
Volume 16, p. 1-166 covers Geography and Geology. A computer-generated index based on keywords in dissertation titles. Author index in v. 33-37. Annual supplements 1973- .

Stuart, Merrill M. A Bibliography of master's theses in geography: American and Canadian universities. Tualatin, OR: Geographic and Area Studies Publications, 1973. 275 p. ISBN 0-88393-001. ISBN 0-88393-002. (pbk). LC 73-16417.   127
Lists 5054 master's theses arranged by 27 subject fields, then alphabetically by author within each field. Regional classification index.

11. GEOGRAPHY AND MAP LIBRARIANSHIP

Richard W. Stephenson

Anglo-American Cataloguing Committee for Cartographic Materials. Cartographic materials: a manual of interpretation for AACR2. Chicago, IL: American Library Association, 1982. 258 p. ISBN 0-8389-0363-0. $40. LC 82-11519.   128
Detailed manual for the cataloging of cartographic materials in libraries. The manual is designed for use in conjunction with the second edition of Anglo-American Cataloguing Rules (Chicago: American Library Association, 1978). The manual is also of value to persons involved in cartobibliography.

Map Librarianship

**Association of Canadian Map Libraries. Association des cartothèques cana-
diennes.** Bulletin. 1- (1968-). Quarterly. Association of Canadian Map Libraries, National Map Collection, Public Archives of Canada, 395 Wellington St., Ottawa, Ontario
K1A 0N3, Canada. ISSN 0318-2851. LC 75-643248.   129
  *Valuable professional journal containing full length articles about maps and map
  librarianship, recent acquisitions, new publications, reviews, news and communications, reports, and association news.*

Drazniowsky, Roman, comp. Map librarianship: readings. Metuchen, NJ: Scarecrow
Press, 1975. 548 p. ISBN 0-8108-0739-4. $26. LC 74-19244.   130
  *Dated but still useful collection of 48 significant articles in the field of map
  librarianship.*

Farrell, Barbara, and Desbarats, Aileen. Guide for a small map collection. Ottawa,
Ontario: Association of Canadian Map Libraries, 1981. 88 p. ISBN 0-9690682-2-0. LC
83-104233.   131
  *Excellent practical guide to the development, operation, and care of a small map
  collection.*

Larsgaard, Mary. Map librarianship: an introduction. Littleton, CO: Libraries Unlim-
ited, 1978. 330 p. ISBN 0-87287-182-7. LC 77-28821.   132
  *Important survey of the techniques of map librarianship followed in the United
  States today.*

Nichols, Harold. Map librarianship. 2nd ed. London: Clive Bingley; Hamden, CT:
Shoe String Press, 1982. 272 p. (1st ed., 1976. 298 p.) ISBN 0-85157-327-4. $29.50. LC
82-126289.   133
  *Good overview of map librarianship, but oriented toward problems and needs of
  the map custodian working in a British library.*

Ristow, Walter W. The Emergence of maps in libraries. Hamden, CT: Linnet Books;
London: Mansell Publishing, 1980. 358 p. ISBN 0-208-01841-7. $27.50. LC 80-12924. 134
  *Brings together in one volume some 38 articles, memorials, and introductions writ-
  ten by Walter W. Ristow, retired chief of the Geography and Map Division, Library of
  Congress.*

Special Libraries Association. Geography and Map Division. Bulletin.
1- (N 1947-). Quarterly. N 1947-Je 1950 under earlier name: Geography and Map Group.
Mrs. Kathleen I. Hickey, Business Manager, Geography and Map Division, SLA, 9927
Edward Avenue, Bethesda, Maryland 20014. ISSN 0036-1607. LC 64-2334.   135
  *Important journal which serves as a "medium of exchange of information, news,
  and research in the field of geographic and cartographic bibliography, literature, and
  libraries." Each issue includes feature articles, association news, news notes, bibliographical citations to new maps, atlases, books and government publications, and reviews of selected books and atlases. Cumulative indexes: 1-70 (1947-1967), 71-102
  (1968-1975).*

Western Association of Map Libraries. Information bulletin. 1- (S 1969-). 3 per annum. Western Association of Map Libraries, University Library, University of California, Santa Cruz, California 95064. ISSN 0049-7282. LC 72-625238.   136
  *A significant journal containing feature articles pertaining to geography, maps,
  and map librarianship, association news, other news notes, citations to new mapping
  of Western North America, and atlas and book reviews.*

PART II. HISTORY, PHILOSOPHY, AND METHODOLOGY

1. HISTORY OF GEOGRAPHY*

See also History of Geographical Exploration

George Kish

*In this section the editorial committee reduced the number of entries recommended by the contributor. See also the following sections: Geographical Exploration; Biographies; and History of Cartography.

A. GENERAL WORKS

Claval, Paul. La pensée géographique, introduction à son histoire. (Publications de la Sorbonne. Series N.-S. Recherches, 2). Paris, France: Société d'Édition d'Enseignement Supérieur, 1972. 116 p. FFr 75. LC 73-352810.
Succinct re-evaluation of the development of geographic thought.
137

Dickinson, Robert E. and Howarth, Osbert, J.R. The Making of geography. Oxford, U.K.: Clarendon Press, 1933. 264 p. Reprinted: Westport, CT: Greenwood Press, 1976. ISBN 0-8371-8669-2. $20.75. LC 75-38379.
A chronological history of geography emphasizing the period preceding the 20th century. Still a useful introduction to the subject.
138 •

James, Preston E. and Martin, Geoffrey J. All possible worlds: a History of geographical ideas. 2nd ed. New York: John Wiley and Sons, 1981. 508 p. (1st ed. by Preston E. James. Indianapolis, IN: Odyssey Press, 1972. 622 p.) ISBN 0-471-06121-2. $31.95. LC 80-25021.
Excellent survey. Good bibliography. Good 'capsule' biographies of geographers. Chapters 13 to 18 deal with the United States.
139 •

Kish, George. A Source book in geography. Cambridge, MA: Harvard University Press, 1978. 453 p. ISBN 0-674-82270-6. $30. LC 77-25972
Annotated anthology of geographical writings, from ancient times to Humboldt and Ritter.
140 •

Peschel, Oskar R. and Ruge, Sophus. Peschel's Geschichte der Erdkunde bis auf Alexander von Humboldt und Carl Ritter. (Geschichte der Wissenschaften in Deutschland. Neuere Zeit. v. 4). 2nd ed. München, Germany: R. Oldenburg, 1877-1878. 2 v. 832 p. LC 2-9699 rev.
A well-balanced history of geography and exploration. Still an indispensable work.
141

B. ANCIENT

Berger, Ernst Hugo. Geschichte der wissenschaftlichen Erdkunde der Griechen. 2nd ed. Leipzig, Germany: Veit and Co., 1903. Reprinted: Berlin: De Gruyter, 1966. 662 p. ISBN 3-11-001357-6. DM 160. LC 4-21357 or LC 67-92919.
Survey of seven centuries of Greek geography. Includes a bibliography.
142

Stahl, William H. Roman science: origins, development, and influence to the later Middle Ages. Madison, WI: University of Wisconsin Press, 1962. 308 p. Reprinted: Westport, CT: Greenwood Press, 1978. 308 p. ISBN 0-313-20473-X. $30. LC 78-5597.
The basic reference work on Roman Science, including geography. Extensive bibliography and index. Covers both Roman and early Medieval times.
143 •

History of Geography

C. MEDIEVAL TO EARLY MODERN PERIOD

Beazley, Charles R. The Dawn of modern geography: a history of exploration and geographical science. London: H. Froude, 1897-1906. Reprinted: Gloucester, MA: Peter Smith, 1964. 3 v. ISBN 0-8446-1063-1. $60. LC 4-14818.   144 •
   Continues to be the most important work on medieval geography. Fully indexed. Volume one, from the conversion of the Roman Empire to A.D. 900; volume two, A.D. 900-1260; volume three, A.D. 1260-1420.

Broc, Numa. La Géographie de la Renaissance (1420-1620). (Ministère des Universités-Comité des Travaux Scientifiques, Mémoires de la Section de Géographie 9). Paris, France: Bibliothèque Nationale, 1980. 258 p. ISBN 2-7177-1543-6. FFr 110. LC 80-145005.   145
   Review of the changing geography and cartography of the period of the Great Discoveries.

Broc, Numa. La Géographie des philosophes. Géographes et voyageurs français au XVIIIe siècle. Paris, France: Editions Ophrys, 1975. 595 p. ISBN 2-7080-0413-1. FFr 156. LC 75-522806.   146
   Study of geography and discoveries of the 18th century Age of Enlightenment, with special emphasis on French practitioners. Good bibliography.

de Dainville, François. La Géographie des humanistes. Paris, France: Beauchesne et ses fils. 1940. 562 p. LC A42-4055.   147
   Study of the geography of the 16th century and the role of Jesuits in French education.

Taylor, Eva G.R. Tudor geography: 1485-1583. London: Methuen, 1930. 290 p. Reprinted: New York: Octagon Books, 1968. ISBN 0-374-97847-6. $23.50. LC 68-21957.   148
   Together with its companion volume on Late Tudor and early Stuart geography, (next entry), a fundamental work on English geography in the age of overseas expansion. Fully documented, essential bibliography.

Taylor, Eva G.R. Late Tudor and early Stuart geography, 1583-1650; a sequel to Tudor geography, 1485-1583. London: Methuen, 1934. 322 p. Reprinted: New York: Octagon Books, 1968. ISBN 0-374-97809-3. $23.50. LC 68-21958.   149
   Supplements and expands the look at English geography in the age of overseas expansion and travel.

D. NONWESTERN

Miquel, André. La Géographie humaine de monde musulman jusqu'au milieu du XIe siècle. Paris; The Hague, The Netherlands; Hawthorne, New York: Mouton, 1973-1975. 3 v.
   Vol. 1. **Géographie et géographie humaine dans la littérature arabe des origines à 1050.** (Civilisations et sociétés, no. 7). 1973. 420 p. ISBN 90-2797-264-8. $43. (pbk.)
   Vol. 2. **Géographie Arabe et representation du monde - la terre et l'étranger.** (Civilisations et sociétés, no. 37). 1975. 707 p. ISBN 90-2797-955-3. $96.25. (pbk).
   Vol. 3. **La Milien naturel.** 1981. 544 p. (Civilisations et sociétés, no. 68). ISBN 2-7193-0468-9. FFr 290.   150
   Basic work on Moslem general and regional geography.

Needham, Joseph, and Wang, Ling. Science and civilization in China. Vol. 3: Mathematics and the sciences of the heavens and the earth. Cambridge, U.K.; New York: Cambridge University Press, 1959. 877 p. ISBN 0-521-05801-5. $150. LC 54-4723.   151
   Part of the magisterial survey of the history of Chinese science. Geography and cartography are discussed on pages 497-591. Full bibliography.

E. MODERN

Beck, Hanno. Geographie: Europäische Entwicklung in Texten und Erläuterungen. Freiburg-München, West Germany: Karl Alber, 1973. 510 p. ISBN 3-495-47262-2. LC 73-319660.
Anthology of geographical writings, heavily emphasizing German authors. Excellent bibliography. Capsule biographies.

152

Brown, Eric H. ed. Geography: yesterday and tomorrow. Oxford, U.K.; New York: Oxford University Press, 1980. 302 p. ISBN 0-19-874096-4. $32.50. LC 80-40193.
A series of essays commissioned as part of the 150th anniversary celebrations of the Royal Geographical Society. The longest essay, by T.W. Freeman, covers the Society's history; the other thirteen deal with various aspects of human and physical geography.

153

Capel Saez, Horacio. Filosofía y ciencía en la geografía contemporánea. Barcelona, Spain: Barcanova, 1981. 464 p. ISBN 0-84-7533-009-5. Pesetas 1400.
History, epistemology, and sociology of geography. Part one deals with modern geography, part two with the 'organization' of geography in the 19th century, and part three with the evolution of geographical ideas. A work of major importance.

154

Corna Pellegrini, Giacomo, and Brusa, Carlo, eds. La Ricerca geografica in Italia, 1960-1980. Varese, Italy: Ask Edizioni, 1981. 1007 p. Lira 15,000.
A set of papers presented at a 1980 conference on the status of Italian geography; structure, organization, branches and specialities, and methodology and epistemology in Italian geography.

155

Corna Pellegrini, Giacomo, and Brusa, Carlo, eds. Italian geography, 1960-1980: general and physical geography. Varese, Italy: Ask Edizioni, 1982. [1984] 312 p.
Abridged version in English of La Ricerca Geografica Italiana *(preceding entry).*

156

Dalmasso, Étienne, ed. La Recherche géographique française. Paris, France: Comité National Français de Géographie, 1984. 265 p.
Reviews of recent trends in French geography.

157

Dickinson, Robert E. The Makers of modern geography. London: Routledge and Kegan Paul; New York: Praeger, 1969. 305 p. ISBN 0-7100-6775-5. £15. LC 76-435505.
This work, introduced by a discussion of Humboldt and Ritter's work, deals with Germany up to the 1960's.

158 •

Dunbar, Gary S. The History of modern geography: an annotated bibliography of selected works. (Bibliographies of the history of science and technology, vol. 9; Garland Reference Library of the Humanities, vol. 445). New York: Garland, 1985. 386 p. ISBN 0-8240-9066-7. $53.50. LC 83-48277.
Includes 1717 annotated entries on the history of modern geography from the rise in Western Europe in the mid-eighteenth century to its current world-wide distribution. Emphasis is on the academic discipline. Divided into three parts: general and topical; geography in various countries; and biographical works. Good coverage of Western languages.

159

Freeman, Thomas Walter. A History of modern British geography. London; New York: Longman, 1980. 258 p. ISBN 0-582-30030-4. $35. LC 79-41477.
A brief chronology of British geography since the 1880's, plus brief biographies of a number of leading practitioners.

160 •

Freeman, Thomas Walter. A Hundred years of geography. Chicago, IL: Aldine, 1963. 334 p. Reprinted: London: Duckworth, 1971. 335 p. ISBN 0-7156-0598-4. £5.95. LC 72-181348.

161 •

Survey of the development of the various branches of geography from the mid-19th to mid-20th century, emphasizing British geography. Short bibliographies are included.

Gerasimov, Innokenti Petrovich, ed. Soviet geography: accomplishments and tasks. (Occasional publication no. 1). Translated from Russian by Lawrence Ecker; English edition edited by Chauncy D. Harris. New York: American Geographical Society, 1962. 409 p. NUC 63-335.   162

English translation of Sovetskaya Geografiya: Itogi i Zadachi (Moscow: Geografgiz, 1960. 634 p.) with additional explanatory materials and aids. Basic survey of the principal fields of geography in the Soviet Union as of 1960.

Hall, D.H. History of the earth sciences during the scientific and industrial revolutions, with special emphasis on the physical geosciences. Amsterdam, The Netherlands; New York: Elsevier Scientific Pub. Co., 1976. 297 p. LC 77-357384.   163

Monograph dealing with the interaction between societal development and the earth sciences.

Johnston, Ronald J. and Claval, Paul, eds. Geography since the second world war: an international survey. London: Croom Helm; Totowa, NJ: Barnes and Noble Imports, 1984. 290 p. ISBN 0-389-20481-1. $27.50. LC 84-2896.   164

A collection of essays documenting the main trends in geography in eleven countries or groups of countries. The main themes are the diversity of practice but a gradual coming together.

Schultz, Hans Dietrich. Die Deutschsprachige Geographie von 1800 bis 1970: ein Beitrag zur Geschichte ihrer Methodologie. (Abhandlungen des Geographischen Instituts, Anthropogeographie: Band. 29). Berlin: Selbstverlag des Geographischen Instituts der Freien Universtät, 1980. 478 p. LC 80-104302.   165

A survey of German geography during the 19th and much of the 20th century. Important bibliography.

Steel, Robert W. The Institute of British Geographers: the first fifty years. London: Institute of British Geographers, 1984. 170 p.   166

A specially commissioned anniversary history of the Institute of British Geographers, the principal learned society for geographers in the United Kingdom. Covers the history and activities of the Institute, with appendices of texts and data.

F. MODERN: UNITED STATES   GEOFFREY J. MARTIN

Bladen, Wilford A. and Karan, Pradyumna P. The Evolution of geographic thought in America: a Kentucky root. Dubuque, IA: Kendall/Hunt, 1983. 149 p. ISBN 0-8403-3045-6. $13.95. (pbk). LC 83-81304.   167

Studies by five authors in the history of American geography with special reference to Kentucky. Photographic illustrations are numerous and unusual.

Blouet, Brian W. and Stitcher, Teresa, eds. The Origins of academic geography in the United States. Hamden, CT: Archon Books, Shoe String Press, 1981. 342 p. ISBN 0-208-01881-6. $37.50. LC 81-8091.   168

A collection of twenty essays by different authors grouped around the sectional titles: 'preacademic origins,' 'the profession,' 'the scholars,' 'the schools,' and 'the ideas.' The book contains a wealth of ideas and detail concerning the history of geography in North America.

Dickinson, Robert E., ed. Regional concept: the Anglo-American leaders. London; Henley, U.K.; Boston, MA: Routledge and Kegan Paul, 1976. 408 p. ISBN 0-7100-8272-X. £12.00. LC 76-369286.   169

An assessment of the work of many of the founding fathers of British and American geography of the last 150 years. Emphasis is placed on the development of the regional concept.

James, Preston E. and Jones, Clarence F., eds. American geography: inventory and prospect. Syracuse, NY: Syracuse University Press, 1954. 590 p. ISBN 0-8156-2013-6. $15. LC 54-9225.   170

An inventory of twenty-six subfields in American and Canadian geography published on the 50th anniversary of the Association of American Geographers. Each of the chapters has now become a significant repository of information for the history of North American geography in the first half of this century.

James, Preston E. and Martin, Geoffrey J. The Association of American Geographers: the first seventy-five years, 1904-1979. Washington, DC: Association of American Geographers, 1978. 279 p. ISBN 0-89291-134-4. $12. LC 78-74887.   171

A history of the birth and development of the Association over a 75 year period. Necessarily the study includes something of the history of North American geography. The whole is drawn essentially from original sources.

Warntz, William. Geography now and then: some notes on the history of academic geography in the United States. (American Geographical Society research series no. 25). New York: American Geographical Society, 1964. 162 p. LC 64-15416.   172

Advances the notion that the first cycle of academic geography in schools and colleges took place in the seventeenth and eighteenth centuries. This first cycle geography is investigated.

Wright, John K. Geography in the making: the American Geographical Society, 1851-1951. New York: American Geographical Society, 1952. 437 p. LC 52-11527.   173

A history of the American Geographical Society written on the occasion of its centennial; the study also embraces a meaningful part of the history of American geography.

2. GEOGRAPHICAL EXPLORATION*

George Kish

*In this section the editorial committee reduced the number of entries recommended by the contributor. See also the preceding section History of Geography.

A. BIBLIOGRAPHIES, ENCYCLOPEDIAS, AND COLLECTIONS

Cox, Edward G. A Reference guide to the literature of travel, including voyages, geographical descriptions, adventures, shipwrecks, and expeditions. (Publications in language and literature v. 9, 10, 12). Seattle, WA: University of Washington Press, 1935-1949. 3 v. Reprinted: New York: Greenwood Press, 1969. 401; 501; 732 p. ISBN 0-8371-2506-5. $95. LC 70-90492.   174

Bibliography of books in English, printed prior to 1800. Volume one deals with the Old World, volume two with the New World, and volume three with Great Britain.

Delpar, Helen, ed. The Discoverers: an encyclopedia of explorers and exploration. New York; London: McGraw-Hill, 1980. 471 p. ISBN 0-07-016264-6. $47.95. LC 79-9259.   175 •

Alphabetically organized, by types of exploration, major regions, major time periods, and explorers. Brief biographies, maps, and explorers' portraits.

Geographical Exploration

Encyclopedia of discovery and exploration. New York: Doubleday, 1973. 5 v.
   Volume 1. 488 p. ISBN 0-38504-352-X. LC 72-93387.
   Volume 2. 488 p. ISBN 0-38504-321-X. LC 72-93384.
   Volume 3. 488 p. ISBN 0-38504-335-X. LC 72-93388.
   Volume 4. 488 p. ISBN 0-38504-350-3. LC 72-93385.
   Volume 5. 488 p. ISBN 0-38504-286-8. LC 72-93389.   176•
   Volume one deals with the Glorious Age of Exploration, *volume two with the* Conquest of North America, *and volume three with* Pacific Voyages. *Volume four looks at* Exploring Africa and Asia, *and volume five with the* Last Frontiers. *All five volumes comprise a popular, well-written, well-illustrated reference work.*

Hakluyt Society. Works. Series 1. 1-100 (1847-1899). Series 2. 1- (1899-). Cambridge, U.K.; New York: Cambridge University Press.   177
   More than 250 volumes in two series dealing with "Original Narratives of Important Voyages, Travels, Expeditions, and Geographical Records." Each volume includes an introduction and copious notes. The volumes consist of English originals (or translations) of the narratives of important voyages.

Stefansson, Vilhjalmur, ed. Great adventures and explorations from the earliest times to the present, as told by the explorers themselves. Rev. ed. New York: Dial Press, 1952. 788 p. (1st ed., 1947. 788 p.) LC 52-13350.   178•
   Anthology of exploration from ancient times to the early years of the 20th century. Includes editorial comments.

B. ATLAS

Newby, Eric, ed. The [Rand McNally] World atlas of exploration. South Melbourne, Victoria, Australia: Macmillan; London: Mitchell Beazley; New York; Chicago, IL: Rand McNally, 1975. 288 p. ISBN 0-528-83015-5. LC 74-27897. LC 76-351385.   179•
   The best single-volume reference for the general reader. Chronological and area oriented. Excellent illustrations and maps.

C. GENERAL WORK

Baker, J.N.L. (John). A History of geographical discovery and exploration. Rev. ed. London: Harrap, 1937. 552 p. Reprinted: New York: Cooper Square, 1967. 552 p. (1st ed., 1931. 543 p.) ISBN 0-8154-0014-4. $30. LC 66-30785.   180•
   Standard work covering subjects from Greek to modern times.

D. SPECIAL TOPICS

Berg, Lev Semenovich. Ocherki po istorii russkikh geograficheskikh otkrytii. (Essays on the history of Russian geographical discoveries.) 2nd ed. Moskva: Izdatel'stvo Akademii Nauk SSSR, 1949. 465 p. (1st ed., 1946. 358 p. LC 50-17023).   181
   An overall summary of Russian geographical exploration and research. Also available in German translation: Geschichte der Russischen Geographischen Entdeckungen *(Leipzig: Bibliographisches Institut, 1954. 283 p.).*

Cary, Max. and Warmington, Eric H. The Ancient explorers. (Penguin books no. A420) Rev. ed. Harmondsworth, U.K.; Baltimore, MD: Penguin Books, 1963. 319 p. (1st ed., London: Methuen, 1929. 270 p.) NUC 64-67022.   182•
   A readable and informative survey of exploration in classical times. Includes many good maps.

Gvozdetskii, Nikolai Andreevich. Soviet geographical explorations and discoveries. Moscow: Progress Publishers; Woodstock, NY: Beekman, 1974. 342 p. ISBN 0-8464-0871-6. $17.95. LC 75-316828. (Sovetskie geograficheskie issledovaniia i otkrytiia. Moskva: "Mysl'," 1967. 390 p.). 183
Useful summary of Russian geographical exploration.

Hakluyt, Richard. The Principall navigations, voyages, traffiques, and discoveries of the English nation. London: George Bishop and Ralph Newberie, deputies to Christopher Barker, 1589. 2 v. 825 p. Facsimile reprint: New York: Kelley, 1970. 12 v. ISBN 0-678-00488-9. $275. (set). LC 70-75411. 184
The greatest book on English voyages. Includes Hakluyt's correspondence, of importance in his compiling this book.

Jones, Gwyn. The Norse Atlantic saga: being the Norse voyages of discovery and settlement to Iceland, Greenland, and America. London; New York: Oxford University Press, 1964. 246 p. LC 64-1311. 185 •
Perhaps the best interpretation of Norse voyages in the North Atlantic.

Kirwan, Laurence P. The White road: a summary of polar exploration. London: Hollis and Carter, 1959; New York: W.W. Norton, 1960. 374 p. LC 60-5847. 186 •
The best written, complete survey of polar exploration. Deals with the period from the Vikings to the voyage of the USS "Nautilus."

Morison, Samuel Eliot. The Great explorers: the European discovery of America. New York: Oxford University Press, 1978. 752 p. ISBN 0-19-502314-5. $25. LC 77-21831. 187 •
An abridged version of the two works by the leading American authority on the period: the northern voyages, A.D. 500-1600, and the southern voyages, 1492-1616. Eminently readable.

Parry, John H. The Age of reconnaissance: discovery, exploration, and settlement, 1450-1650. (The World Histories of Civilization series). Berkeley, CA: University of California Press, 1981. 365 p. ISBN 0-520-04234-4. $26.50. ISBN 0-520-04235-2. $8.95. (pbk). LC 81-51175. 188 •
The best general survey of all aspects of the Great Age of Exploration.

Parry, John H. The Discovery of the sea. Berkeley, CA: University of California Press, 1981. 279 p. ISBN 0-520-04236-0. $26.50. ISBN 0-520-04237-9. $8.95. (pbk). LC 81-51174. 189 •
The history of navigation, instruments, and achievements.

Sauer, Carl O. Northern mists. Berkeley, CA: University of California Press, 1968. 204 p. Reprinted: Berkeley, CA: Turtle Island Foundation, 1973. 204 p. ISBN 0-913666-00-9. $3.50. (pbk). LC 68-16757. 190 •
Excellent survey of recent research on pre-Columbian contacts between Europe and America.

Taylor, Eva G.R. The Haven-finding art: a history of navigation from Odysseus to Captain Cook. New augmented edition. London: Hollis and Carter for the Institute of Navigation; New York: American Elsevier, 1971. 310 p. LC 76-151853. 191 •
The best general survey of the history of navigation. By a leading British historian of the history of geography and discovery. A brief appendix by Joseph Needham adds important information on early Chinese navigation.

3. BIOGRAPHIES

Geoffrey J. Martin

A. SERIAL

Geographers: biobibliographical studies. Edited by T.W. Freeman. 1- (1977-).
Annual. Mansell Publishing Limited, 6 All Saints Street, London N1 9RL, England, or
H.W. Wilson Co., 950 University Avenue, Bronx, New York 10452. ISSN 0308-6992.　192•
　*An annual publication containing about twenty-five essays dealing with eminent
geographers from many countries of the world. Emphasis is on the last century, but
figures from much earlier periods are published. Careful work containing valuable biography, bibliography and index.*

B. BOOKS

Beck, Hanno. Alexander von Humboldt. Wiesbaden, West Germany: Franz Steiner,
1959-1961. 2 v. LC 61-42817 rev.
　Vol. 1. 1959. 303 p. ISBN 3-515-00019-4. DM 54.
　Vol. 2. 1961. 440 p. ISBN 3-515-00020-8. DM 78.　193
　*Among the most important publications marking the centenary of Humboldt's
death, this bibliography by the leading scholar of Humboldt's life and work occupies
a permanent place. Volume one covers 1769-1804 and volume two, 1804-1859.*

Beck, Hanno. Carl Ritter, Genius der Geographie. Berlin: Dietrich Reiner, 1979. 132 p.
ISBN 3-496-00102-X. DM 17.50. LC 80-458531.　194
　A short life of Ritter, marking the bicentennial of his birth.

Buttimer, Anne. The Practice of geography. London; New York: Longman, 1983.
298 p. ISBN 0-582-30087-8. $29.95. LC 82-13091.　195
　*A series of autobiographical reflections by senior geographers, setting their work in
the context of the intellectual traditions that influenced them.*

De Terra, Helmut. Humboldt: The life and times of Alexander von Humboldt,
1769-1859. New York: Knopf, 1955. 386 p. Reprinted: New York: Octagon Books, 1979.
386 p. ISBN 0-374-92134-2. $30.50. LC 78-27653.　196•
　*Important study of one of the nineteenth century German geographical giants.
Humboldt's contribution, still being discussed, helped to form and structure current
geographical thought.*

Dunbar, Gary S. Élisée Reclus: historian of nature. Hamden, CT: Archon Books,
Shoe String Press, 1978. 193 p. ISBN 0-208-01746-1. $17.50. LC 78-17346.　197
　*The first booklength biography in English of the French geographer and anarchist,
Élisée Reclus (1830-1905). Written largely from original sources, the work emphasizes
Reclus' contribution to the emergence of professional geography in France and the
relevance of geography to society.*

Eyre, J. Douglas (John), ed. A Man for all regions: the contributions of Edward L.
Ullman to geography. (Studies in geography no. 11). Chapel Hill, NC: University of
North Carolina, Department of Geography, 1978. 158 p. LC 77-620057.　198
　*An appreciation of the varied contributions of Edward L. Ullman and their impact
upon recent geographical thought and research in the form of essays by eight special-
ists. The editor contributes a biographical sketch.*

Kish, George. North-east passage: Adolf Erik Nordenskiöld, his life and times. Stockholm, Sweden: Almqvist & Wiksell; Amsterdam, The Netherlands: Nico Israel (A. Ash-
er), 1973. 283 p. ISBN 9-0607-2720-7. Swedish krona 70. Dutch florins 45. LC 73-175149.　199•

A biographical study of one of the world's nineteenth century Arctic pioneers. Special attention is given to Nordenskiöld's scientific studies of a Northeast Passage.

Leighly, John, ed. Land and life: a selection from the writings of Carl Ortwin Sauer. (California reprint series). Berkeley, CA: University of California Press, 1963, Reprinted: 1974. 435 p. ISBN 0-520-02633-0. $38.50. LC 63-21069.   200

A generous selection from essays written over a span of half a century on a surprising variety of themes by one of the most creative leaders in modern human geography. Introductory essay by the editor.

Lenz, Karl, ed. Carl Ritter: Geltung und Deutung. Berlin: Dietrich Reimer, 1981. 233 p. ISBN 3-496-00183-6. DM 78. LC 81-173891.   201

Fifteen essays, on the work and world-wide influence of Ritter, read at the bicentenary celebration in Berlin, 1979.

Lowenthal, David. George Perkins Marsh: versatile Vermonter. New York: Columbia University Press, 1958. 442 p. LC 58-11679.   202

A careful study (from original sources) of a man who was a learned student of nature. Nineteenth century attitudes toward resource management and the genesis of 'conservation' are important themes.

Martin, Geoffrey J. Ellsworth Huntington: his life and thought. Hamden, CT: Archon Books, Shoe String Press, 1973. 315 p. ISBN 0-208-01347-4. $25. LC 73-5682.   203 •

An exhaustive investigation of the career and contribution of Ellsworth Huntington. Special emphasis is placed on Huntington's quest for the triadic causation (climate, heredity, and culture) of civilization. Written from original sources.

Martin, Geoffrey J. Mark Jefferson: geographer. Ypsilanti, MI: Eastern Michigan University Press, 1968. 370 p. LC 68-22804.   204 •

A life and thought study of a luminary American geographer in the first half of the twentieth century that contributes to a broader history of the development of geography. The whole is drawn largely from first sources.

Martin, Geoffrey J. The Life and thought of Isaiah Bowman. Hamden, CT: Archon Books, Shoe String Press, 1980. 272 p. ISBN 0-208-01844-1. $25. LC 80-14650.   205 •

A prosopographic inquiry concerning the ways in which the career and contribution of Isaiah Bowman contributed to the development of geographical (and other) science. A significant contribution to the history of geography in North America.

Meinig, Donald W., ed. On geography: selected writings of Preston E. James. (Geographical series no. 3). Syracuse, NY: Syracuse University Press, 1971. 407 p. ISBN 0-8156-0084-4. $14. LC 77-170097.   206

Twenty-two essays by Preston James selected from the period 1922-1971, are reprinted under the headings: "On History and Theory;" "On Research and Writing;" "On the World and its Religions;" and "On Teaching." Preface, introduction, and commentaries provide a good perspective.

Parker, William H. Mackinder: geography as an aid to statecraft. Oxford, U.K.: Clarendon Press; New York: Oxford University Press, 1982. 295 p. ISBN 0-19-823235-7. $34.95. LC 82-3623.   207

The work is biographical but places substantial emphasis on ideas and ideals of Mackinder. Significant as a biography and as a contribution to the history of thought.

Rowley, Virginia M. J. Russell Smith: geographer, educator, and conservationist. Philadelphia, PA: University of Pennsylvania Press, 1964. 247 p.   208 •

A biography of one of the leading U.S. geographers of the earlier part of the twentieth century. Special emphasis is placed on Smith's contribution to economic geography and conservation.

Taylor, T. Griffith (Thomas). Journeyman Taylor: the education of a scientist. London: Hale; New York: Transatlantic, 1959. 352 p. LC 59-755.

209 •

Detailed study of the life and times of a large intellect at work. Filled with fascinating insights concerning the ways in which geographical science may be accomplished.

Ullman, Edward L. Geography as spatial interaction. Edited by Boyce, Ronald R. Seattle, WA: University of Washington Press, 1980. 231 p. ISBN 0-295-95711-5. $22.50. LC 79-6759.

210

A rich and varied collection of fourteen essays by the late Edward L. Ullman whose theoretical contributions were seminal in the evolution of modern geographical thinking about regional development, urbanization, transportation, and complementary spatial interactions. Valuable brief foreword by Chauncy D. Harris.

Wanklyn, Harriet G. Friedrich Ratzel: a biographical memoir and bibliography. Cambridge, U.K.; New York: Cambridge University Press, 1961. 96 p. LC 61-65229.

211 •

A brief biographical study of one of the leading nineteenth century German geographers. Helpful bibliographies.

Wright, John K. Human nature in geography. Cambridge, MA: Harvard University Press, 1966. 361 p. LC 66-10128.

212

A collection of fourteen essays (1925-1965) by one of America's historians of geography. Includes several essays on the history of American geography.

Zube, Ervin H., ed. Landscapes: selected writings of J.B. Jackson. Amherst, MA: University of Massachusetts Press, 1970. 160 p. ISBN 0-87023-054-9. $13. ISBN 0-87023-072-7. $6.95. (pbk). LC 78-103475.

213 •

A sample of the writings of one of the most thoughtful and influential commentators on the American landscape.

4. PHILOSOPHY, THEORY, AND METHODOLOGY*

Ronald J. Johnston

See also Part II, Section 1, History of Geography.

*In this section the editorial committee reduced the number of entries recommended by the contributor.

Amedeo, Douglas M. and Golledge, Reginald G. An Introduction to scientific reasoning in geography. New York: John Wiley, 1975. 431 p. Reprinted: Melbourne, FL: Krieger, 1985. ISBN 0-89874-764-3. LC 84-9656.

214

A basic text on positivist methodology in human geography covering both spatial theory and its applications.

Anuchin, Vsevolod Aleksandrovich. Theoretical problems of geography. Ed. by Fuchs, Roland J. and Demko, George J. Columbus, OH: Ohio State University Press, 1977. 331 p. ISBN 0-8142-0221-7. $16. LC 77-8437. Teoreticheskie Problemy Geografii (Moskva: Geografgiz, 264 p.)

215

Sharply challenges predominant Soviet views that geography is a family of disciplines with a sharp separation of physical and economic geography, with the assertion that geography is a single unified discipline.

Bennett, Robert J. and Chorley, Richard J. Environmental systems: philosophy, analysis, and control. London: Methuen; Princeton, NJ: Princeton University Press, 1978. 624 p. ISBN 0-691-08217-0. $100. LC 79-100507.

216

A major synthesis of the systems approach in geography emphasizing physical and human systems and their interlinkage. Separate sections deal with 'hard' (those capable of rigorous, quantitative analysis) and 'soft' (not mathematically tractable) systems and with their integration.

Browning, Clyde E. Conversations with geographers: career pathways and research styles. (Studies in geography no. 16). Chapel Hill, NC: University of North Carolina, Department of Geography, 1982. 139 p. LC 81-69844.
217•
Transcripts of interviews with the leading geographers, exploring the development of their careers.

Buttimer, Anne. Society and milieu in the French geographic tradition. (Monograph series, v. 6). Washington, DC: Association of American Geographers, 1971. 226 p. ISBN 0-89291-085-2. $4.95. LC 72-158112.
218
A detailed examination of French geography, with particular emphasis on the work and influence of P. Vidal de la Blache and J. Brunhes.

Claval, Paul. Géographie humaine et économique contemporaine. Paris, France: Presses Universitaires de France, 1984. 442 p.
219
A substantial review of human geography during the twentieth century from a French perspective.

Gale, Stephen, and Olsson, Gunnar, eds. Philosophy in geography. (Theory and Decision Library: no. 20). Dordrecht, Holland; Boston, MA: D. Reidel, 1979. 469 p. ISBN 9-02-770948-3. $40. LC 78-21037.
220
A wide-ranging series of essays on philosophical and methodological issues.

Gatrell, Anthony C. Distance and space: a geographical perspective. Oxford, U.K.: Clarendon Press; New York: Oxford University Press, 1983. 195 p. ISBN 0-19-874128-6. $24.95. ISBN 0-19-874129-4. $12.95. (pbk). LC 83-7829.
221
A modern treatment of the spatial perspective in geography emphasizing its positive contribution and available techniques.

Gould, Peter R. The Geographer at work. London; Boston, MA: Routledge and Kegan Paul, 1985. 320 p. ISBN 0-7102-0459-0. $39.95. ISBN 0-7102-0494-9. $16.95. (pbk). LC 84-24961.
222•
A personal review of geography written for a non-academic audience and emphasizing the many positive contributions of the discipline.

Gould, Peter, and Olsson, Gunnar, eds. A Search for common ground. London: Pion; New York: Methuen, 1982. 277 p. ISBN 0-85086-093-8. $25. LC 82-183939.
223
A selection of eleven commissioned essays concerned with current philosophical debates within human geography.

Gregory, Derek. Ideology, science, and human geography. London: Hutchinson, 1978; New York: St. Martin's Press, 1979. 198 p. ISBN 0-312-40477-8. $20. LC 78-11746.
224
A penetrating and seminal critique of positivism in human geography and an argument for the development of a critical science perspective which emphasizes the role of human agency in the creation, recreation, and alteration of the regional mosaic.

Guelke, Leonard T. Historical understanding in geography: an idealist approach. (Studies in historical geography no. 3). Cambridge, U.K.; New York: Cambridge University Press, 1982. 109 p. ISBN 0-521-24678-4. $27.95. LC 82-4356.
225
A full statement of Guelke's views on the philosophy of idealism as the correct one for geography, illustrated by historical studies of South Africa.

Philosophy, Theory, and Methodology

Haggett, Peter; Cliff, Andrew D.; and Frey, Allan. Locational analysis in human geography. 2nd ed. London: Edward Arnold; New York: John Wiley and Sons, 1977. 605 p. (1st ed., Haggett, Peter. Locational Analysis in Human Geography. 1965. 339 p.) ISBN 0-470-99207-7. ISBN 0-470-99208-5. $24.95. (pbk). LC 79-125491.   226

    A second, much expanded edition of Haggett's pioneering text which presents geography, basically human geography, as locational analysis. The first part discusses spatial systems in terms of six elements: interaction; networks; nodes; hierarchies; surfaces; and diffusion. The second part focuses on methods for analyzing systems and their components.

Hartshorne, Richard. The Nature of geography: a critical survey of current thought in the light of the past. (Reprinted from the Annals of the Association of American Geographers, v. 29 (1939), pp. 171-658 with the addition of "Abstract" and "Corrections and supplementary notes."). Washington, DC: Association of American Geographers, 1939 and later reprintings. Reprinted: Westport, CT: Greenwood Press, 1977. 482 p. ISBN 0-8371-9328-1. $38.75. LC 76-48691.   227

    An overview dealing with numerous aspects of geography and how those aspects differ from geography of the past.

Hartshorne, Richard. Perspectives on the nature of geography. (AAG Monograph series no. 1). Washington, DC: Association of American Geographers, 1959. 201 p. ISBN 0-89291-080-1. $4.95. LC 59-7032.   228

    Takes the previous entry one step farther by looking at individual perspectives of geography.

Harvey, David. Explanation in geography. London; New York: Edward Arnold, 1969. 521 p. ISBN 0-7131-5693-7. $19.95. (pbk). LC 79-470643.   229

    The first major work presenting geography as a scientific discipline embracing the positivist philosophy. Most of the chapters are concerned with methodological issues involved in the conduct of geographical research.

Harvey, David. The Limits to capital. Oxford, U.K.: Basil Blackwell; Chicago, IL: University of Chicago Press, 1982. 478 p. ISBN 0-226-31952-0. ISBN 0-226-31953-9. $12.95. (pbk). LC 82-40322.   230

    A seminal work combining an exegesis of Marx's economic theories and their development to incorporate the geography of capitalism.

Harvey, Milton E. and Holly, Brian P., eds. Themes in geographic thought. New York: St. Martin's Press; London: Croom Helm, 1981. 224 p. ISBN 0-312-79530-0. $26. LC 80-28423.   231

    A collection of ten essays in geographic philosophy. Subjects treated include positivism, pragmatism, functionalism, phenomenology, existentialism, idealism, realism, environmental causation, and Marxism. The essays relate most particularly to geographic thought post World War II.

Holt-Jensen, Arild. Geography: its history and concepts. Translated from Norwegian by Brian Fullerton. London: Harper and Row; Totowa, NJ: Barnes and Noble, 1982. 171 p. ISBN 0-389-20262-2. $10.95. (pbk). LC 82-175933. (Geografiens innhold og metoder 2nd ed., Bergen: Universitetsforlaget, 1977. 132 p.).   232

    A brief, introductory treatment of the history of geography and of philosophical issues germane to the practice of geography.

Jackson, Peter, and Smith, Susan J. Exploring social geography. London; Boston, MA: George Allen and Unwin, 1984. 239 p. ISBN 0-04-301169-1. $30. ISBN 0-04-301170-5. $9.95. (pbk). LC 83-25732.   233

    An excellent introductory text which emphasizes philosophical and methodological issues.

Johnston, Ronald J. Geography and geographers: Anglo-American human geography since 1945. 2nd ed. London; New York: Edward Arnold, 1983. 264 p. (1st ed., New York: John Wiley and Sons, 1979. 232 p.) ISBN 0-7131-6387-9. $14.95. (pbk). LC 83-126365.   234

    A general survey of human geography as practiced in North America and the United Kingdom since 1945, focusing on philosophical and methodological debates. It covers the growth of systematic studies, spatial science, and behavioral, humanistic, and radical geography.

Johnston, Ronald J. Philosophy and human geography: an introduction to contemporary approaches. London; New York: Edward Arnold, 1983. 152 p. ISBN 0-7131-6385-2. $12.95. (pbk). LC 82-226016.   235

    A brief introduction, aimed at undergraduate students, of the three main philosophies, positivist, humanistic, and structuralist, with which human geographers have been concerned in recent years. There are discussions of the epistemologies, ontologies, and methodologies of each, their relevance to human geography, and their use by geographers.

Ley, David, and Samuels, Marwyn S., eds. Humanistic geography: prospects and problems. Chicago: Maaroufa; London: Croom Helm; New York: Methuen, 1978. 337 p. ISBN 0-416-60101-4. $18.95. LC 78-52408.   236

    A seminal collection of essays introducing what is widely known as the humanistic approach in geography.

Mikesell, Marvin W., ed. Geographers abroad: essays on the problems and prospects of research in foreign areas. (Research paper no. 152). Chicago, IL: University of Chicago, Department of Geography, 1973. 296 p. ISBN 0-89065-059-4. $10. (pbk). LC 73-87829.   237

    Ten chapters by specialists reviewing the study of the world (excluding Europe and North America) by geographers and arguing for a greater commitment to foreign-area research.

Mitchell, Bruce, and Draper, Dianne L. Relevance and ethics in geography. London; New York: Longman, 1982. 222 p. ISBN 0-582-30035-5. $28.50. LC 81-19386.   238

    An exploration of ethical issues in geographical work, especially in applied geography. Parallels are drawn with other disciplines in which these issues have already been tackled, but the range of geographical examples is narrow.

Olsson, Gunnar. Birds in egg/eggs in birds. 2nd ed. London: Pion; New York: Methuen, 1980. 300 p. (1st ed., Michgan Geographical Publication no. 15, Ann Arbor, MI: University of Michigan Press, 1975. 532 p.) ISBN 0-85086-077-6. $19.50. LC 80-514198.   239

    An idiosyncratic exploration of philosophical issues in geography with particular emphasis on logic and language.

Paterson, John L. David Harvey's geography. London: Croom Helm; Totowa, NJ: Barnes and Noble Imports, 1984. 220 p. ISBN 0-389-20441-2. $27.50. LC 83-22293.   240

    A rare review of the writings of a living geographer, unfortunately only up to 1981. The book seeks to explicate Harvey's work and set it in context, with special reference to his shift from positivist spatial science to Marxian analysis.

Peet, Richard, ed. Radical geography: alternative viewpoints on contemporary social issues. Chicago, IL; Maaroufa; New York; London: Methuen; 1977. 387 p. ISBN 0-416-71240-1. $8.95. LC 76-55222.   241

    A collection of reprinted essays, many of them from the journal Antipode, *introducing 'radical' (largely Marxian) perspectives in a range of geographical issues.*

Sack, Robert D. Conceptions of space in social thought: a geographic perspective. London: Macmillan; Minneapolis, MN: University of Minnesota Press, 1980. 231 p. ISBN 0-8166-1012-6. $27.50. ISBN 0-8166-1015-0. $9.95. (pbk). LC 80-16894.   242

Cartography

> *A comprehensive review of various conceptions of space in the social sciences and in the humanities. The stress is on the subjective as well as the objective meanings of space.*

Stoddart, David R., ed. Geography, ideology, and social concern. Oxford, U.K.: Basil Blackwell; Totowa, NJ: Barnes and Noble, 1981. 250 p. ISBN 0-389-20207-X. $27.50. LC 81-166142.
243
> *A collection of essays on various aspects of the history and philosophy of geography. The central theme of many of the essays is the contextual approach to disciplinary history.*

Tinkler, Keith J. A Short history of geomorphology. London: Croom Helm; Totowa, NJ: Barnes and Noble, 1985. 336 p. ISBN 0-389-20544-3. $25. LC 84-24364.
244
> *A comprehensive history of this important subfield of physical geography.*

5. CARTOGRAPHY*

Alan M. MacEachren

*In this section the editorial committee reduced the number of entries recommended by the contributor.

A. BIBLIOGRAPHIES

Bibliographia cartographica, Zogner, Lothar, ed. 1- (1974-). Annual. Supersedes Biblioteca Cartographica (1957-72). K.G. Saur Verlag KG, Poessenbacherstr. 12B, Postfach 711009, 8000 Munich 71, West Germany. ISSN 0340-0409.
245
> *A major effort by volunteer contributors in almost 40 countries. Indexes major journals with more than 200 listings normally included. Most of the explanatory material is printed in English, German, and French.*

Geo abstracts G: remote sensing, photogrammetry, and cartography. 1- (1974-). Bimonthly. 1974-1977 as Geo Abstracts G: Remote Sensing and Cartography. Supersedes in part, Geo Abstracts D: Social Geography and Cartography (1966-73). Geo Abstracts Ltd., Regency House, 34 Duke Street, Norwich NR3 3AP, England. ISSN 0305-1951.
246
> *International bibliography with abstracts on remote sensing, photogrammetry, cartography, maps and atlases, projections, and map design and compilation.*

International Cartographic Association bibliography, 1956-1972. Enschede, The Netherlands: International Institute for Aerial Survey and Earth Sciences, 1974. 129 p.
247
> *References (847) cover all ICA communications including articles, papers, reports, addresses, reviews, and guides. Arranged by conference, commission or working group, national reports, exhibitions, and other ICA publications.*

U.S. Library of Congress, Geography and Map Division. The Bibliography of cartography. Boston, MA: G.K. Hall, 1973. 5 v. First Supplement, 1980. 2 v. ISBN 0-8161-0259-7 (5 v.) Out of print but available on microfilm from publisher for $560. First Supplement ISBN 0-8161-1008-5 $260. LC 73-12977.
248
> *Indispensable bibliography consisting of more than 120,000 card references to literature about maps, mapmaking, and mapmakers.*

American cartographer. 1- (1974-). Semi-annual. American Congress on Surveying and Mapping, 210 Little Falls Street, Falls Church, VA, 22046. ISSN 0094-1689.　　249
Journal of the American Cartographic Association of the American Congress on Surveying and Mapping. One of the premier English-language journals. Contains articles on all aspects of cartography, communications, technical notes, reviews, recent literature, and cartographic news. Articles. Reviews.

The Cartographic journal. 1- (June 1964-). Semi-annual. British Cartographic Society, Department of Geography, King's College, Strand, London WC2R 2LS, England. ISSN 0008-7041.　　250
Journal of the British Cartographic Society. One of the premier English language journals. Includes articles, reviews, recent literature, maps and atlas descriptions, the Society record and reports, and other news items.

Cartographica. 17- (1980-). Quarterly. Supersedes and continues numbering of: The Canadian cartographer, 1-16 (1964-79). Also incorporates Cartographic monographs, 1-24 (1971-79). University of Toronto Press, Toronto, Ontario M5S 1A6, Canada. ISSN 0317-7173.　　251
A leading international journal with articles in English and French. Combines former journal and monograph series of the Canadian Cartographic Association and serves as that organization's official publication. Generally two monographs or special topic issues per year and two standard journal issues with papers, reviews, notes, and recent literature.

Cartography. 1- (1954-). Semi-annual (1954-55 annual). Australian Institute of Cartographers, Box 1292, Canberra City, A.C.T. 2601, Australia. ISSN 0069-0805.　　252
Contains original papers contributing to the theory and practice of cartography including papers presented at, or arising from, meetings of the Institute. Also contains a news bulletin and reviews.

Comité français de cartographie. 1- (1958-). Quarterly. Comité Français de Cartographie, 39 ter rue Gay-Lussac, 75005 Paris, France.　　253
Journal of the French Committee on Cartography with articles, activities of the Committee, reviews, bibliographies, and national information. In French with occasional English abstracts.

Geo-processing: geo-data, geo-systems, and digital mapping. 1- (1981-). Quarterly. Elsevier Scientific Publishing Co., Box 211, Amsterdam, The Netherlands. ISSN 0165-2273.　　254
Directed to topics in spatial information systems. Covers the collection of data, editing, conceptual structures, data structures, and methodology for display and analysis. Systems emphasized include cadastral, geocoding, land use, resources, thematic and topographic mapping, and digital terrain models. Includes papers, surveys, reviews, and discussions.

International yearbook of cartography / Internationales Jahrbuch für Kartographie/ Annuaire international de cartographie. 1- (1961-). Annual. Published in cooperation with the International Cartographic Association, Meijerinksweg 9, Lonnecker, The Netherlands, by Kirschbaum Verlag, Siegfriedstr 28, Postfach 210209, 5300 Bonn 2, West Germany. ISSN 0074-9842. LC 61-49049.　　255
A major international publication including both papers read at international conferences, and independent papers. Particular themes are often emphasized. Articles and abstracts in German, English, or French.

Revista cartográfica. 1- (1952-). Semi-annual. Instituto Panamericano de Geografia e Historia, Servicios Bibliográficos, Ex-Arzobispado 29, Col. Observatorio, Deleg. Miguel Hidalgo, 11860 Mexico D.F., Mexico. ISSN 0080-2085.　　256
Cartographic journal on the Pan American Institute of Geography and History, Commission on Geography. Articles are in Spanish, English, Portuguese, or French.

Cartography

Society of University Cartographers. Bulletin. 1- (1966-). Semi-annual. Society of University Cartographers, Department of Geography, Portsmouth Polytechnic, Portsmouth PO1 3HE, England. ISSN 0036-1984.
    Covers all aspects of cartography. Included are a combination of articles, reviews of maps, atlases, and books, reports on drawing equipment, and announcements about publications, meetings, and other developments in the field.

257

Surveying and mapping. 1- (Oct. 1941-). Quarterly. American Congress on Surveying and Mapping, 210 Little Falls Street, Falls Church, VA 22046. ISSN 0039-6273.
    Until 1974 when The American Cartographer was begun, this was the major U.S. publication for cartography. Currently it includes only an occasional article of direct interest to cartographers.

258

C. DICTIONARIES

Edson, Dean T. and Denegre, Jean, eds. Glossary of terms in computer-assisted cartography. 3rd ed. Falls Church, VA: American Congress on Surveying and Mapping, 1980. 157 p.
    Produced in French and English by the Commission on Cartographic Terminology of the International Cartographic Association.

259

Meynen, Emil, compiler. Multilingual dictionary of technical terms in cartography. Wiesbaden, West Germany: Franz Steiner; New York: French and European Publications, 1973. 573 p. ISBN 3-515-00127-1. $88. (pbk). LC 74-154516.
    Glossary in 14 languages produced by the Commission on Terminology of the International Cartographic Association.

260

D. MAP READING

Blair, Calvin L. and Gutsell, Bernard V. The American landscape: maps and air photo interpretation. New York: McGraw-Hill, 1974. 62 p. ISBN 0-07-005597-1. LC 74-5482.
    A traditional introduction to landscape interpretation from topographic maps and air photos using U.S. examples. Emphasis is on physical landscapes but attention is also given to cultural features.

261 •

Dury, George H. Map interpretation. 4th ed. London: Pitman and Sons, 1972. 216 p. (1st ed., 1952. 203 p.) ISBN 0-273-36151-1. LC 72-172711.
    Somewhat updated edition of a popular map interpretation text. Still offers a traditional approach with emphasis on topographic maps.

262

Muehrcke, Phillip C. Map use: reading, analysis, and interpretation. rev. ed. Madison, WI: JP Publications, 1980. 469 p. (1st ed., 1978. 469 p.) ISBN 0-960-29781-2. $17.95. (pbk). LC 78-70573.
    First comprehensive introduction to all aspects of map use. An insightful well-illustrated text that uses examples from both common and specialized sources. Covers both traditional topics of distance, direction, and topographic mapping, and equally important topics such as statistical analysis of map patterns, interpretation of abstract map symbolization, and differences between maps and reality.

263 •

Riffel, Paul A. Reading maps: an introduction to maps using color stereo photographs. Northbrook, IL: Hubbard Press, 1973. 72 p. ISBN 0-8331-1300-3. $6.95. (pbk). LC 79-138628.
    Considers how maps and air photos can be used together. Included are: an introduction to scale; coordinates and contours; a discussion of map symbols and what they mean; applications of maps and photos; and an examinination of urban growth patterns.

264 •

Speak, Peter, and Carter, A.H.C. (Anthony). Map reading and interpretation: new edition with metric examples. London: Longman, 1981. 80 p. (1st ed., 1970. 79 p.) ISBN 0-582-33076-9. £4.50. (pbk). LC 82-198239.

265 •

> *A traditional introduction to topographic map reading using metric Ordnance Survey examples. Interpretation of both physical and cultural features.*

Tyner, Judith. The World of maps and mapping: a creative learning aid. New York: McGraw-Hill, 1973. 48 p. LC 73-157167.

266 •

> *A useful introduction to map reading covering a variety of map types with the emphasis on non-topographic maps. Approximately half of the book is devoted to map illustrations. Student activities using maps follow the different map categories.*

E. THEMATIC MAPS

Cuff, David J. and Mattson, Mark T. Thematic maps: their design and production. New York: Methuen, 1982. 169 p. ISBN 0-416-33500-4. $14.95. LC 82-8216.

267

> *A text devoted entirely to thematic mapping. Presents an introduction to symbolization, design, production, and reproduction. Emphasis is on practical guidelines for map making rather than cartographic theory. Concepts presented are well-illustrated but some relevant concepts are omitted or treated in too brief a manner.*

Dent, Borden D. Principles of thematic map design. Reading, MA: Addison Wesley, 1985. 398 p. ISBN 0-201-11334-1. $29.95. LC 83-9256.

268

> *A stimulating, well-illustrated text. Narrowest in scope of available texts. Map construction and reproduction techniques are considered only briefly, while both practical and theoretical aspects of design and symbolization of small scale thematic maps are covered in detail.*

Muehrcke, Phillip C. Thematic cartography. (Commission on College Geography resource paper no. 19) Washington, DC: Association of American Geographers, 1972. 66 p. LC 72-77214.

269

> *An introduction to a variety of analytical, modeling, and symbolization concepts and techniques in thematic mapping. Most basic concepts presented remain relevant, but attempts to relate symbology and map communication to analytic procedures and model building result in fragmented presentation.*

F. COMPUTER-ASSISTED CARTOGRAPHY

Auto-Carto I proceedings. (International Conference on Automation in Cartography, "Auto-Carto I" December 9-12, 1974, Reston, Virginia). Falls Church, VA: American Congress on Surveying and Mapping, 1976. 318 p. ISSN 0270-5133. LC 76-2164.

270

> *Proceedings from the first in a series of conferences on computer-assisted cartography. Papers and discussions transcribed from tape recordings. Most sessions consisted of panels on specific topics.*

Auto-Carto II proceedings. Kavalivnas, John C., ed. (International Symposium on Computer-Assisted Cartography). Falls Church, VA: American Congress on Surveying and Mapping, Cartography Division, 1975. 614 p. ISSN 0270-5133. LC 79-117745.

271

> *Papers from the second of a series of Auto-Carto meetings. Covers a very broad range of topics with approximately equal emphasis on strictly computer-related problems and basic cartographic research.*

Auto-Carto III proceedings. (International Conference on Computer-Assisted Cartography). Falls Church, VA: American Congress on Surveying and Mapping, 1979. 519 p. ISSN 0270-5133. LC 79-64463.

272

> *Proceedings of Auto-Carto III held in San Francisco. Includes papers presented as well as transcripts of discussions preceding and following the presentations.*

Cartography

Auto-Carto IV proceedings: applications in health and environment. Aangeenbrug, Robert T., ed. (International Symposium on Cartography and Computing). Falls Church, VA: American Congress on Surveying and Mapping and American Society of Photogrammetry, 1979. 2 v. 621 p.; 479 p. ISSN 0270-5133. LC 79-64463.   273

Papers from the fourth in a series of major international conferences. Primary focus is on applications of computer mapping in health planning and environmental resources. Includes session summaries and papers ranging from theoretical research to applications.

Auto-Carto V proceedings: environmental assessment and resource management. Forman, Jack, ed. (International Symposium on Computer-Assisted Cartography). Falls Church, VA: American Congress on Surveying and Mapping and American Society of Photogrammetry, 1983. 738 p. ISBN 0-937294-44-6. $18. ISSN 0270-5133. LC 83-204609.   274

Papers from the fifth Auto-Carto meeting held jointly with the International Society for Photogrammetry and Remote Sensing Commission IV Symposium. Although the meeting theme was environmental assessment and resource management, papers cover the entire field of cartography and its applications.

Carter, James R. Computer mapping: progress in the 80's. (Resource publications in geography). Washington, DC: Association of American Geographers, 1984. 86 p. ISBN 0-89291-175-1. $5. (pbk). LC 84-70007.   275 •

An overview of the current status of computer-assisted cartography. Topics covered include: map categories from a digital perspective, concepts and equipment, the nature of cartographic data, turnkey systems, software, potential of microcomputers, and a review of relevant literature.

Davis, John C. and McCullaugh, Michael J., eds. Display and analysis of spatial data. London; New York: John Wiley and Sons, 1975. 378 p. ISBN 0-471-19915-X. LC 74-3449.   276

An important collection of papers written by participants at the NATO Advanced Study Institute on Display and Analysis of Spatial Data in Nottingham, England, in 1973. Papers are grouped under three headings: theoretical aspects of spatial analysis; automated cartography; and practical applications of computer cartography.

Harvard Library of computer graphics mapping collection. Moore, Patricia A., ed.ed. Philadelphia, PA: University of Pennsylvania Press, 1981. 19 v. ISBN 0-8122-1180-4. $99.50. Set. (pbk).   277

This series represents a significant contribution to theory and applications of computer-assisted cartography. Contributions to the series are by staff of the Harvard Laboratory for Computer Graphics and Spatial Analysis and many other scholars. Topics covered in the series include: management's use of maps; software and data bases; urban, regional, and state applications; natural resource mapping; educational applications; thematic map design; hardware; and cadastral mapping.

Marble, Duane, ed. Brassel, K. and Wasilenko, M., coordinators. Computer software for spatial data handling. Prepared by the International Geographical Union Commission on Data Sensing and Processing for the USGS. Reston, VA: U.S. Geological Survey, 1980. Three volume series including v. 3, Cartography and Graphics. 1042 p. LC 81-601526.   278

A comprehensive survey of computer software dealing with both cartographic and other spatial data processing procedures. Contains sections on several classes of mapping.

McEwen, Robert B.; Witmer, Richard E.; and Ramey, Benjamin S., eds. USGS digital cartographic data standards. (Geological survey circular 895). Reston, VA: U.S. Geological Survey, 7 v.

Vol. A. **Overview and USGS activities,** 1983. 20 p. LC 83-600139.
Vol. B. **Digital elevation models,** 1983. 40 p. LC 83-600191.
Vol. C. **Digital line graphs from 1:24,000 scale maps,** 1984. 79 p. LC 83-600190.
Vol. D. **Digital line graphs from 1:2,000,000 scale maps,** 1983. 40 p. LC 83-600189.
Vol. E. **Land use and land cover digital data,** 1983. 21 p. LC 83-600188.
Vol. F. **Geographic names information system,** 1983. 25 p. LC 83-600187.
Vol. G. **Digital line graph attribute coding standards,** 1984. 31 p. LC 83-600186.

279

A series of reports produced to provide details concerning standards used in compiling digital cartographic data. Available free from the USGS, the reports describe both the theoretical and practical basis of various standards and the specific data formats used for different data. Individual reports cover overall USGS standardization efforts, digital elevation models, digital line graphs at various scales, land use land cover data, and the geographic names data base.

Moellering, Harold, ed. Issues in digital cartographic data standards. (Report series on committee activities). Columbus, OH: National Committee for Digital Cartographic Data Standards of the American Congress of Surveying and Mapping. 1- (1982-). Irregular. 5 v.

280

Report of the National Committee for Digital Cartographic Data Standards. First two reports contain reprints of papers on proposed standards presented at special sessions of ACSM and Auto-Carto meetings. The third report reviews committee progress during its first year, the fourth reports on alternatives examined, and the fifth is a bibliography for digital cartographic data standards.

Monmonier, Mark S. Computer-assisted cartography: principles and prospects. Englewood Cliffs, NJ: Prentice-Hall, 1982. 214 p. ISBN 0-13-165308-3. $27.95. LC 81-14380.

281

The first text on computer-assisted cartography. Attempts to cover basic computer algorithms, characteristics of raster versus vector map display, cartometry, map projections, data structures, and computer-assisted map design. Some concepts are well presented and illustrated but the book's brevity leads to superficial coverage of others.

Peuquet, Donna J. and Boyle, A. Raymond. Raster scanning, processing, and plotting of cartographic documents. Williamsville, NY: SPAD Systems, INC., 1984. 122 p. ISBN 0-913913-01-4. $35.95. (pbk).

282

An in-depth analysis of current technology for raster scanning of cartographic documents and the implications for data processing and subsequent cartographic output.

Proceedings of the 1976 workshop on automated cartography and epidemiology. Hyattsville, MD: U.S. Department of Health, Education, and Welfare; Public Health Service; Office of Health Research, Statistics, and Technology; National Center for Health Statistics. (Now part of the U.S. Dept. of Health and Human Services). 1979. 80 p. LC 80-601081.

283

Proceedings of a workshop designed to coordinate development of automated cartographic techniques for epidemiological study, exchange ideas on current applications, and identify avenues for potential progress. Papers include both practical applications and theoretical work in analytical mapping.

Cartography

Taylor, D.R. Fraser (David), ed. The Computer in contemporary cartography. (Progress in contemporary cartography, v. 1). Chichester, U.K.; New York: John Wiley and Sons, 1980. 252 p. ISBN 0-471-27699-5. £41.25. LC 79-42727.   284

First in a series designed to review progress in cartographic theories, methods, and empirical research. Emphasis in this volume is on applications of computer mapping in government agencies. Useful in bringing together work from widely scattered sources.

Tufte, Edward R. The Visual display of quantitative information. Cheshire, CT: Graphics Press, 1983. 197 p. $34. LC 83-156861.   285 •

A superb guide to theory and practice in the design of statistical graphs, charts, maps, and tables. Handsomely illustrated.

Yoeli, Pinhas. Cartographic drawing with computers. (Computer Applications special issue, vol. 8, 1982. 137 p.). University of Nottingham, Department of Geography, Nottingham NG7 2RD, England. ISSN 0308-4221.   286

Provides a practical introduction to mapping with computers with emphasis on large scale mapping. Several FORTRAN subroutines are included and described for tasks such as map unit shading, line symbol drawing, etc. The book complements a set of mapping subroutines available on tape from the same source.

G. MAP PROJECTIONS

Maling, Derek H. Coordinate systems and map projections. London: George Philip and Son, Ltd., 1973. 255 p. LC 73-330722.   287

Projections are addressed as a series of coordinate transformations from geographical coordinates to rectangular or polar coordinates. A fairly comprehensive text requiring a moderate background in mathematics, it is intended to cover certificate requirements in Britain for surveying, cartography, and planning.

McDonnell, Porter W., Jr. Introduction to map projections. New York: Marcel Dekker, 1979. 174 p. ISBN 0-8247-6830-2. $19.75. LC 79-13560.   288

Text-reference work on map projections for students in geography, cartography, and surveying. Projections are presented grouped by property. Formulas and tables for construction are included.

Richards, Peter, and Adler, Ron K. Map projections for geodesists, cartographers, and geographers. Amsterdam, The Netherlands: North-Holland, 1972. 174 p. LC 79-182493.   289

An intermediate level discussion of projections. A knowledge of elementary spherical trignometry, differential calculus, and integral calculus is presupposed. Computational formulas are presented in an attempt to provide a basis from which computer programs can be developed.

Snyder, John P. Map projections used by the U.S. Geological Survey. (Geological Survey bulletin 1532). Washington, DC: Government Printing Office, 1982. 313 p. $8. LC 81-607569.   290 •

Technical manual describing all map projections used by the U.S.G.S. in both the past and present. Aspects of projections presented include history, features, usage, and formulas. A significant information source for those using U.S.G.S. maps as well as for projection use and construction in general.

H. GENERAL WORKS

Campbell, John. Introductory cartography. Englewood Cliffs, NJ: Prentice-Hall, 1984. 406 p. ISBN 0-13-501304-6. $33.95. LC 83-15990.   291 •

An introductory text with a strong emphasis on large-scale mapping. Contains excellent overviews of map projections principles and computer techniques. One fourth is devoted to data collection and display for large scale topographic type maps (surveying, photogrammetry, and terrain mapping). Only 40 pages on thematic mapping.

Coulson, Michael R.C., ed. The Introductory cartography course at Canadian Universities. Ottawa, Canada: Canadian Cartographic Association, 1980. 92 p.　　292
Report on panel discussion about introductory, university level, cartography courses. An informative description of twelve courses with associated comments from instructors representing a broad cross section of approaches.

Greenhood, David. Mapping. Rev. ed. Chicago, IL: University of Chicago Press, 1964. 289 p. (1st ed.," Down to Earth: Mapping for Everybody," New York: Holiday House, 1944. 262 p.) ISBN 0-226-30696-8. $20. ISBN 0-226-30697-6. $8.95. (pbk). LC 63-20905.　　293 •
An excellent easy to read introduction to maps and mapping for beginners and non-professionals. Although somewhat dated, most basic concepts presented remain relevant and are often more clearly presented here than in more recent texts.

Hodgkiss, A.G. (Alan Geofrey). Maps for books and theses. Newton Abbot, U.K.: David and Charles; New York: Pica Press, 1970. 267 p. LC 72-476928.　　294 •
A useful text/reference book to assist non-cartographers in creating thematic maps. Although ignoring cartographic theory and advocating symbolization methods of mixed quality, some sound practical advice is presented. Attention is given to tools, material and equipment, statistical maps, map design, specialized maps, and map reproduction.

Lawrence, George R.P. Cartographic methods. 2nd ed. London; New York: Methuen, 1979. 162 p. ISBN 0-416-71640-7. $21. ISBN 0-416-71650-4. $10.95. (pbk). LC 74-884413.　　295
A useful well-written text and one of the few to combine an introduction to basic map making techniques with sections on map analysis. Brevity combined with the broad range of topics addressed results in a superficial coverage of some topics and probably confines the text to a supplementary status.

Manual on basic cartography, volume one. Lonnecker, The Netherlands: International Cartographic Association, 1984.　　296
The first of a planned two volume work produced as a joint effort of specialist members of the ICA Commission on Education in Cartography. Topics covered include history, general introduction, mathematical cartography, theory of cartographic expression and design, production techniques, and reproduction.

Robinson, Arthur H.; Sale, Randall D.; Morrison, Joel; and Muehrcke, Phillip C. Elements of cartography. 5th ed. New York: John Wiley and Sons, 1984. 544 p. (1st ed., 1953. 254 p.) ISBN 0-471-09877-9. $27.95. LC 84-11860.　　297
Remains the standard to which all others are compared. Changes from the previous edition include a major reordering of chapters, a new introduction, and integration of computer-assisted cartography into several other chapters. Those who have liked previous editions should find the organization more workable, those who did not will probably not change their mind.

I. SPECIAL TOPICS

Bertin, Jacques. Graphics and graphic information processing. Translated from French by William J. Berg and P. Scott. New York: Walter De Gruyter, 1981. 273 p. ISBN 3-11-008868-1. $29. ISBN 3-11-006901-6. $19. (pbk). LC 81-12610. (La Graphique et le traitement graphique d'information. Paris: Flammarion, 1977. 285 p.).　　298
A sometimes hard to follow presentation of Bertin's concepts concerning processing and display of graphic information. Emphasis is on defining graphics, human processing of graphics, and the role of graphics in information transmission.

Cartography

Bertin, Jacques. Semiology of graphics: diagrams, networks, maps. Translated from French by William J. Berg. Madison, WI: University of Wisconsin Press, 1983. 415 p. ISBN 0-299-09060-4. $75. LC 83-47755. (Sémiologie graphique: les diagrammes, les réseaux, les cartes. 2nd ed. Paris: Mouton, 1973. 431 p.). 299
    Translation of one of the significant contributions by Bertin. Presents an analytical theory for guiding cartographers, graphic designers, and statisticians. Contains more than 1200 illustrations.

Boardman, David. Graphicacy and geography teaching. London; Dover, NH: Croom Helm, 1983. 184 p. ISBN 0-7099-0644-7. $19.50. (pbk). LC 83-140417. 300•
    A stimulating introduction to graphicacy, the communication of spatial information through maps and other forms of illustration.

Burnside, Clifford D. Mapping from aerial photographs. London: Granada Publishing; New York: John Wiley and Sons, 1979. 304 p. ISBN 0-258-97035-9. £18.50. LC 79-11497. 301
    An introduction to the theoretical bases of producing topographical maps from aerial photographs. The main theoretical elements of photogrammetric techniques are presented.

Campbell, James B. Mapping the land: aerial imagery for land use information. (Resource publications in geography). Washington, DC: Association of American Geographers, 1983. 96 p. ISBN 0-89291-167-0. $5. (pbk). LC 83-9936. 302
    A practical presentation of concepts and techniques for land use mapping using aerial imagery. Designed as a supplement to courses in remote sensing. Concepts are presented clearly and concisely with numerous examples, many from actual class projects conducted in conjunction with local governments.

Castner, Henry W. and Robinson, Arthur H. Dot area symbols in cartography: the influence of pattern on their perception. (Technical monograph no. CA-4). Washington, DC: American Congress on Surveying and Mapping, Cartography Division, 1969. 78 p. LC 77-100052. 303
    A presentation of research into the role of pattern in cartographic symbology. Particularly significant as one of the early examples of applying a scientific experimental testing procedure to the investigation of map effectiveness.

Dickinson, Gordon C. Statistical mapping and the presentation of statistics. 2nd ed. London; Baltimore, MD: Edward Arnold, 1973. 195 p. (1st ed., 1963. 160 p.) ISBN 0-7131-5683-X. $9.95. LC 72-91355. 304
    A comprehensive introduction to maps and other graphic techniques for representing statistics. Although concepts presented are generally sound, most North American cartographers will find many objectionable symbolization choices.

Fisher, Howard T. Mapping information: the graphic display of quantitative information. Cambridge, MA: Abt Books, 1982. 384 p. ISBN 0-89011-571-0. $35. LC 82-6858. 305
    This posthumously published work details Fisher's approaches to thematic mapping presenting a variety of controversial approaches to mapping statistical data. Emphasis is on symbolization for quantitative data, value classing, and the use of color.

Freeman, Herbert, and Pieroni, Goffredo G., eds. Map data processing. New York: Academic Press, 1980. 374 p. ISBN 0-12-267180-5. $35. LC 80-23015. 306
    A collection of 18 papers resulting from a NATO Advanced Study Institute meeting held in Maratea, Italy. Contributions represent several disciplines related to the capture, manipulation, and display of spatial information.

Howling, Peter H. and Hunter, Leslie A. Mapping skills and techniques: a quantitative approach. 2nd ed. Edinburgh, U.K.: Oliver and Boyd, 1977. 54 p. (1st ed., 1974. 46 p.) ISBN 0-05-002845-6. ISBN 0-05-003138-4. £2.55. (pbk). LC 77-354116.   307
   A concise easy to follow introduction to the statistical analysis of mapped patterns. Topics include measures of area and shape, sampling techniques, analysis of relief, network analysis, pattern description, and model building.

Hsu, Mei-Ling, and Robinson, Arthur H. The Fidelity of isopleth maps: an experimental study. Minneapolis, MN: University of Minnesota Press, 1970. 92 p. LC 71-99127.   308
   Details the design and results of empirical research into the factors that influence isopleth map accuracy after a brief introduction to isarithmic mapping. Probably the most thorough study of isopleth mapping yet accomplished. It has stimulated a variety of similiar analytic studies of related topics.

Imhof, Eduard. Edited by Steward, H.J. Cartographic relief presentation. Berlin; New York: Walter De Gruyter, 1982. 389 p. ISBN 3-11-006711-0. $79.95. LC 82-5001. (Gelände und Karte. Zürich: Rentsch, 1968. 250 p.).   309
   The premier work on cartographic relief presentation. The concepts presented have stood the test of time. This new edition incorporates a few more recent developments including brief references to computer techniques.

International Cartographic Association. Technical papers of the 12th ICA Conference. Perth, Australia, 1984. 2 v.   310
   Contains 146 papers and/or abstracts presented at the 1984 meeting. A broad range of topics are covered.

Keates, John S. Cartographic design and production. London: Longman; New York: Halsted Press, John Wiley and Sons, 1976. 240 p. ISBN 0-470-15106-4. $39.95. (pbk). LC 72-9251.   311
   One of the most thorough practical introductions to map production and reproduction available. References to materials are becoming dated and the initial section on the graphical basis of cartography is rather superficial. The reference to map design in the title is misleading with only a five page chapter on design.

Keates, John S. Understanding maps. London: Longman; New York: Halsted Press, John Wiley and Sons, 1982. 139 p. ISBN 0-470-27271-6. $15.95. (pbk). LC 81-6921.   312
   A well-organized presentation of perception and cognition as applied to map making and use. While some concepts presented are controversial with little empirical evidence presented to support them, numerous hypotheses and topics for discussion are presented. Useful as a point of departure for advanced courses.

Kidwell, Ann Middleton, and Greer, Peter Swartz. Sites perception and the nonvisual experience: making and designing mobility maps. New York: American Foundation for the Blind, 1973. 192 p. ISBN 0-89128-055-3. $7. (pbk).   313
   A presentation of research on tactual mapping. An overview of nonvisual perception is included followed by discussion of research, design, and preliminary testing of a detailed mobility map for the blind.

Kjellström, Björn. Be expert with map and compass: the complete "orienteering" handbook. 2nd ed. New York: Charles Scribner's Sons, 1976. 214 p. (1st ed., 1967. 136 p.) ISBN 0-684-14270-8. LC 76-12550.   314•
   The best of a number of how-to books on the use of topographic maps for traveling by foot in off-road or wilderness areas. Emphasis is on the growing sport of orienteering.

Lobeck, Armin K. Block diagrams: and other graphic methods used in geology and geography. 2nd ed. Amherst, MA: Emerson-Trussell, 1958. 212 p. (1st ed., 1924. 206 p.) LC 58-1245.   315
   The classic work on techniques and methods for mapping terrain features. The majority of the book is devoted to block diagram maps with some attention to physiographic mapping.

Monmonier, Mark S. Maps, distortion, and meaning. (Resource paper no. 75-4). Washington, DC: Association of American Geographers, 1977. 51 p. ISBN 0-89291-120-4. $4. (pbk). LC 76-44640.   316•

A well-written and illustrated introduction to map uses and limitations. The focus is on the interface between map author and map user. Concepts presented are basic ones not dependent on technology, therefore, the volume should continue to remain a useful supplement to map making and map use courses.

Morrison, Joel L. Method-produced error in isarithmic mapping. (Technical monograph no. CA-5). Washington, DC: American Congress on Surveying and Mapping, 1971. 76 p. LC 76-156328.   317

This groundbreaking work presents results of empirical research on isarithmic map accuracy. Implications for accuracy of surface variability, number of sample values, sample spacing, and interpolation are all considered.

Robinson, Arthur H. The Look of maps: an examination of cartographic design. Madison, WI: University of Wisconsin Press, 1952. 105 p. ISBN 0-299-00950-5. LC 52-4466.   318

Recognized as the beginning of modern cartographic research into visual aspects of maps and their effectiveness. Ten readable essays identify key components of cartographic design and symbolization and suggest numerous avenues for research. Ideas presented and efforts of the author have been a driving force in cartographic research for three decades.

Robinson, Arthur H. and Petchenik, Barbara Bartz. The Nature of maps: essays toward understanding maps and mapping. Chicago, IL: University of Chicago Press, 1976. 138 p. ISBN 0-226-72281-3. $10. LC 75-36401.   319

A collection of theoretical essays on maps and their role in knowledge acquisition. Completed over a five-year period, the connection between essays often seems forced. Each however, is significant in its own right as a synthesis of concepts and research from psychology and cartography that leads to understanding cartography's perceptual/cognitive foundations.

Southworth, Michael, and Southworth, Susan. Maps: a visual survey and design guide. Boston, MA: Little, Brown, and Co. for the New York Graphic Society, 1982. 223 p. ISBN 0-8212-1503-5. $39.95. LC 82-3403.   320

A fascinating collection of published maps from ancient to computer generated. Common or popular maps are included as are analytical maps of geographic relationships. A brief textual description accompanies each map and introduces each of ten chapters, two devoted to maps, humans, and space, with eight on techniques.

Taylor, D.R. Fraser (David), ed. Graphic communication and design in contemporary cartography. (Progress in Contemporary Cartography, v. 2). Chichester, U.K.; New York: John Wiley and Sons, 1983. 314 p. ISBN 0-471-10316-0. $59. LC 82-2843.   321

A stimulating collection of essays on cartographic design and communication. The papers provide a synthesis of concepts presented elsewhere, but also offer many new insights. Significant in bringing together ideas of an international group of scholars whose other work may be less accessible to the U.S. reader.

Thompson, Morris M. Maps for America: cartographic products of the U.S. Geological Survey and others. 2nd ed. Washington, DC: Government Printing Office; Reston, VA: U.S. Geological Survey, 1981. 265 p. (1st ed., 1979. 265 p.) $15. LC 81-607878.   322•

A detailed account of mapping practices by federal government agencies with heavy emphasis on U.S. Geological Survey topographic mapping. This well conceived and written volume explains symbolization, map categories, map sources, and reliability of government produced maps available to the public. A suitable supplement to map reading courses.

Wiedel, Joseph, ed. Proceedings of the First International Symposium on Maps and graphics for the visually handicapped. Washington, DC: Association of American Geographers, 1983. 185 p. ISBN 0-89291-178-6. $9. (pbk). LC 83-073096.   323

The papers represent the state of the art in design, production, and reproduction of maps and graphics for the visually handicapped. Included are papers concerning both technical problems of production and reproduction as well as results of recent research.

Worthington, Basil D.R. and Gant, Robert. Techniques in map analysis. London: Macmillan Education, Ltd., 1975. 104 p. ISBN 0-333-14651-4. £4.25 (pbk). LC 78-303422.

324

Exercises in map analysis. Elementary statistical and diagrammatic techniques are employed to summarize and describe specific distributions of map elements. Techniques presented include grid analysis, transects, network analysis, and land use evaluation methods.

6. HISTORY OF CARTOGRAPHY

Richard W. Stephenson

Bagrow, Leo. History of cartography. Revised and enlarged by R.A. Skelton. Translated from German by D.L. Paisey. Cambridge, MA: Harvard University Press, 1964. 312 p. LC 64-55230. (Die Geschichte der Kartographie. Berlin: Safari-Verlag, 1951. 383 p.).

325

Outstanding general history of cartography. Revised and supplemented by R.A. Skelton. Includes an extensive list of cartographers to 1750 (pages 227-280) and a select bibliography of the history of cartography (pages 283-300). The mapping of the New World is largely ignored in this work.

Brown, Lloyd A. The Story of maps. Boston, MA: Little Brown, 1949. 397 p. Reprinted: New York: Dover Publications, 1979. ISBN 0-486-23873-3. $7.95. (pbk). LC 79-52395.

326

A well-written survey of the history of maps and map makers. Includes a large bibliography and useful notes.

Crone, G.R. (Gerald Roe.) Maps and their makers: an introduction to the history of cartography. 5th ed. Folkestone, U.K.: William Dawson; Hamden, CT: Archon Books, 1978. 152 p. (1st ed., 1953. 181 p.) ISBN 0-7129-0756-4. £12.50. LC 78-40892.

327

Revised and illustrated edition. A brief tracing of map-making, principally in Europe from classical beginnings to modern cartography. Bibliography.

Harvey, P.D.A. (Paul). The History of topographical maps: symbols, pictures, and surveys. London: Thames and Hudson, 1980. 199 p. ISBN 0-500-24105-8. $29.95. (Distributed in the U.S. by W.W. Norton, New York). LC 80-505257.

328

Important work tracing the development of topographical mapping. Includes bibliography.

Ristow, Walter W. American maps and mapmakers: commercial cartography in the nineteenth century. Detroit, MI: Wayne State University Press, 1985. 536 p. ISBN 0-8143-1768-5. $60. LC 84-25798.

329 •

Important book tracing the development of American commercial cartography in the previous century. Contains 212 illustrations.

Ristow, Walter W. Guide to the history of cartography: an annotated list of references on the history of maps and mapmaking. Washington, DC: Library of Congress, Geography and Map Division, 1973. 96 p. LC 73-9776.

330

Includes 398 annotated citations to noteworthy general works as well as references relating to individual countries and to specialized aspects of cartography.

History of Cartography

Robinson, Arthur H. Early thematic mapping in the history of cartography. Chicago, IL; London: University of Chicago Press, 1982. 266 p. ISBN 0-226-72285-6. $35. LC 81-11516.   331
Essential guide to the history of thematic mapping with special emphasis on the first half of the 19th century. The book is concerned mostly with the growth of thematic mapping in northwestern Europe, particularly the British Isles, France, Germany, and the Low Countries. Bibliography.

Schwartz, Seymour I. and Ehrenberg, Ralph E. The Mapping of America. New York: Harry N. Abrams, 1980. 363 p. ISBN 0-8109-1307-0. LC 79-24113.   332 •
Beautifully illustrated book written with the general reader in mind, tracing the history of the mapping of North America from 1500 to the present. Part one, "1500-1800," was written by Seymour I. Schwartz and part two, "since 1800," by Ralph E. Ehrenberg. Bibliography.

Skelton, Raleigh A. Ed. by Woodward, David. Maps: a historical survey of their study and collecting. (The Kenneth Nebenzahl, Jr., lectures in the history of cartography at the Newberry Library). Chicago, IL: University of Chicago Press, 1975. 138 p. ISBN 0-226-76166-5. $10. ISBN 0-226-76165-7. $2.75. (pbk). ISBN 0-226-76164-9. (cloth). LC 74-21344.   333
Based on a series of lectures presented by R.A. Skelton at the Newberry Library on Oct. 27-28 and Nov. 10-11, 1966. The study and collecting of maps is emphasized rather than simply the maps themselves. Includes a bibliography of the author's published works, compiled by Robert W. Karrow, Jr.

Stevenson, Edward L. Terrestrial and celestial globes: their history and construction, including a consideration of their value as aids in the study of geography and astronomy. New Haven, CT: Yale University Press for the Hispanic Society of America, 1921. 2 v. Reprinted: New York: Johnson Reprint Co. ISBN 0-384-58127-7. $95. (set). LC 21-18954.   334
A classic.

Thrower, Norman J.W. Maps and man: an examination of cartography in relation to culture and civilization. Englewood Cliffs, NJ: Prentice-Hall, 1972. 184 p. ISBN 0-13-555961-8. ISBN 0-13-555953-7. $16.95. (pbk). LC 70-166141.   335
Succinctly written book tracing the history of maps from antiquity to modern times. Includes information about nineteenth century American mapmaking not found in other general works.

Tooley, Ronald V. Maps and map-makers. 6th ed. London: B.T. Batsford, 1978. 140 p. (1st ed., 1949. 128 p.) ISBN 0-7134-1395-6. £20. LC 78-318874.   336
Popular introduction to the history of cartography written for the student as well as the collector. The book is arranged in schools of geography, the classics, Italian, Dutch, French, and English, with brief notes on the principal map-makers in each group and a list of their main productions.

Tooley, Ronald V. Tooley's dictionary of mapmakers. New York: Alan R. Liss; Amsterdam, The Netherlands: Meridian, 1979. 684 p. ISBN 0-8451-1701-7. $120. ISBN 0-8451-1702-5. $40. (pbk). LC 79-1936.   337
Brief biographical descriptions of persons associated with the production of maps from the earliest times to 1900. Originally, the first half of the work appeared in parts in Map Collector's Circle *which is now discontinued.*

Wilford, John Noble, Jr. The Mapmakers. New York: Alfred A. Knopf, 1981. 414 p. ISBN 0-394-46194-0. LC 80-2716.   338 •
Popular book on selected aspects of the history of cartography, with special emphasis on mapping in the twentieth century. Bibliographical notes.

Woodward, David, ed. Five centuries of map printing. (The Kenneth Nebenzahl Jr., lectures in the history of cartography at the Newberry Library). Chicago, IL: University of Chicago Press, 1975. 177 p. ISBN 0-226-90724-4. $7.95. (pbk). LC 74-11635.   339

Six invited papers presented on November 2-4, 1972, focusing on the printing processes used in the making of maps. Included are Arthur H. Robinson's Mapmaking and Map Printing: The Evolution of a Working Relationship; *David Woodward's* The Woodcut Technique; *Coolie Verner's* Copperplate Printing; *Walter W. Ristow's* Lithography and Maps, 1796-1850; *Elizabeth M. Harris'* Miscellaneous Map Printing Processes in the Nineteenth Century; *and C. Koeman's* The Application of Photography to Map Printing and the Transition to Offset Lithography. *Selected bibliography.*

7. REMOTE SENSING, AIR PHOTO INTERPRETATION,

AND PHOTOGRAMMETRY

John R. Jensen

A. SERIALS

American Society for Photogrammetry and Remote Sensing. (Formerly American Society of Photogrammetry). Technical Papers. 1- (1965-). Semi-annual. American Society for Photogrammetry and Remote Sensing, 210 Little Falls Street, Falls Church, Virginia 22046. ISSN 0277-2094.   340

Short papers on new research in remote sensing, photogrammetry, and surveying applications, primarily by scientists in the United States.

Geo abstracts. Series G. Remote sensing, photogrammetry, and cartography. 1- (1974-). Bimonthly. Geo Abstracts, Ltd., Regency House, 34 Duke Street, Norwich NR3 3AP, England. ISSN 0305-1951.   341

Abstracts organized by general remote sensing, data acquisition, data processing, interpretation, and photogrammetry as well as by subdivisions of cartography.

International journal of remote sensing. 1- (1980-). Bimonthly. Taylor and Francis, Ltd., Rankine Road, Basingstoke, Hants RG24 0PR, England. ISSN 0143-1161.   342

A multidisciplinary journal that reports on recently developed sensor systems and on remote sensing and photo-interpretation techniques. Includes numerous papers from European scientists.

Photogrammetria. International Society of Photogrammetry. 1- (1949-). Bimonthly. Elsevier Scientific Publishing Co., Box 211, 1000 AE Amsterdam, The Netherlands. ISSN 0031-8663. LC 48-4262 rev.   343

A periodical with international scope that focuses on recent developments in air photo interpretation and photogrammetry.

Photogrammetric engineering and remote sensing. American Society for Photogrammetry and Remote Sensing. Formerly (until 1975): Photogrammetric Engineering (ISSN 0031-8671). 1- (1934-). Monthly. American Society for Photogrammetry and Remote Sensing, 210 Little Falls Street, Falls Church, VA 22046. ISSN 0099-1112. LC 38-20703.   344

A professional journal containing articles, abstracts, notes, and book reviews on a full range of remote sensing topics, from the most recent research in applied digital techniques to the conventional methods of air photo interpretation and photogrammetry. The most important remote sensing journal published in the United States.

Remote sensing of environment: an interdisciplinary journal. 9 per annum. 1- (1969-). Elsevier Scientific Publishing Co., 52 Vanderbilt Avenue, New York, New York 10017. ISSN 0034-4257.   345

An interdisciplinary journal containing research papers on the uses of remotely sensed data and imagery in analysis of the landscape and the environment.

Remote Sensing

Proceedings. Machine processing of remotely sensed data. West Lafayette, IN: Purdue University, Laboratory for the Application of Remote Sensing, 1973- (except for 1976). LC 79-93130.

The most widely respected forum for presenting important breakthroughs in digital image processing of remote sensor data.

346

Proceedings of the series of international symposium on Remote sensing of the environment. 1- (1962-). Annual. Environmental Research Institute of Michigan, P.O. Box 8618, Ann Arbor, Michigan 48107.

Results of the international symposium conducted each year. The definitive summary of international progress in remote sensing.

347

Proceedings Pecora symposium. Remote sensing: an input to geographic information systems in the 1980's. Falls Church, VA: American Society for Photogrammetry and Remote Sensing, 1977- (1982. 619 p. ISBN 0-937294-36-5).

One of the editions of the proceedings especially devoted to the concept of incorporating remotely sensed thematic information into Geographic Information Systems.

348

Technical Papers of the American Society for Photogrammetry and Remote Sensing. (Formerly American Society of Photogrammetry). 1- (1965-). Semi-annual. American Society for Photogrammetry and Remote Sensing, 210 Little Falls Street, Falls Church, Virginia 22046. ISSN 0277-2094.

Short papers on new research in remote sensing, photogrammetry, and surveying applications, primarily by scientists in the United States.

349

Avery, Thomas Eugene, and Berlin, Graydon Lennis. Interpretation of aerial photographs. 4th ed. Minneapolis, MN: Burgess, 1985. 470 p. ISBN 0-8087-0096-0. (Price not set.) LC 84-23249.

An introductory text describing the fundamental principles of aerial photography data collection and analysis.

350•

Barrett, Eric Charles, and Curtis, Leonard F. Introduction to environmental remote sensing. 2nd ed. London: Chapman and Hall; New York: Methuen, 1982. 352 p. (1st ed., 1976. 336 p.) ISBN 0-412-23080-1. $60. ISBN 0-412-23090-9. $31. (pbk). LC 83-138640.

Introductory textbook on remote sensing with an environmental emphasis. Contains applications sections on weather analysis, global resources, crops and land use, and urban studies.

351•

Branch, Melville C. City planning and aerial information. (Harvard city planning studies no. 17). Cambridge, MA: Harvard University Press, 1971. 283 p. ISBN 0-674-13225-4. LC 76-158428.

One of the few texts devoted to describing the use of aerial photography and photogrammetry in 'master' city planning. The excellent annotated bibliography is an important source of information for historical studies in remote sensing.

352•

Burnside, Clifford Donald. Mapping from aerial photographs. New York: John Wiley and Sons, 1979. 304 p. ISBN 0-470-26690-2. LC 79-11497.

An excellent text emphasizing fundamentals of aerial photography data collection and analytical photogrammetry especially in the preparation of planimetric and topographic maps.

353•

Campbell, James B. Ed. by Knight, C. Gregory. Mapping the land: aerial imagery for land use information. (Resource publications in geography). Washington, DC: Association of American Geographers, 1983. 96 p. ISBN 0-89291-167-0. $5. (pbk). LC 83-9936.   354

Describes the theory of land use classification based on both manual and digital analysis of remotely sensed data. An excellent tutorial for those not acquainted with land use mapping principles and strategies.

Castelman, Kenneth R. Digital image processing. Englewood Cliffs, NJ: Prentice-Hall, 1979. 429 p. ISBN 0-13-212365-7. $41.95. LC 78-27578.   355

An advanced treatment of the principles of digital image processing. Examples are based on the use of the VICAR image processing system of the California Institute of Technology Jet Propulsion Laboratory, considered by many to be one of the best image processing systems in the world.

Colwell, Robert N., ed. Manual of remote sensing. 2nd ed. Falls Church, VA: American Society for Photogrammetry and Remote Sensing, 1983. 2 v. 2440 p. (1st ed., 1975. 2144 p.) ISBN 0-937294-52-7. $132. (set). Vol 1. ISBN 0-937294-41-1. Vol 2. ISBN 0-937294-42-X. LC 83-6055.   356

The most comprehensive work on remote sensing to date. With 279 color plates and numerous contributions by leaders in the field, this is an essential text for the serious scholar. The first volume covers topics general to all remote sensing applications such as optics, atmospherics, and properties of electromagnetic radiation. The second volume reviews specific applications such as geology, archaeology, urban/suburban land use, and agriculture. Each chapter contains an extensive bibliography.

Dickinson, G.C. Maps and air photographs: images of the earth. 2nd ed. New York: John Wiley and Sons; London: Edward Arnold, 1979. 348 p. ISBN 0-470-26641-4. $24.95. LC 78-31287.   357 •

A cartographic text that emphasizes how to convert data on an aerial photograph into cartographic information. Emphasis is placed on map projections, scale changes, measurement of areas, and measuring the third dimension (elevation) on aerial photographs.

Eastman Kodak Company. Applied infrared photography. (Kodak publication no. M-28). Rochester, NY: Eastman Kodak Company, 1981. 84 p. ISBN 0-87985-288-7. $6. (pbk). LC 81-65754.   358

Technical manual describing the films, filters, and applications of using black and white and color-infrared ground and aerial photography.

Ford, Kristina, ed. Remote sensing for planners. New Brunswick, NJ: Rutgers University, Center for Urban Policy Research, 1979. 219 p. ISBN 0-88285-058-X. $25. LC 78-31594.   359

The definitive work on the use of remotely sensed data for city planning, site selection for engineering projects, transportation planning, and housing and population analysis. Chapters are written by experts in the fields.

Holz, Robert K., ed. The Surveillant science: remote sensing of the environment. 2nd ed. New York: John Wiley and Sons, 1985. 413 p. (1st ed., Boston, MA: Houghton Mifflin, 1973. 390 p.) ISBN 0-471-08638-X. $18.95. (pbk). LC 84-7508.   360 •

A collection of new and reprinted articles that summarize fundamental principles of remote sensing and applications.

Hord, R. Michael. Digital image processing of remotely sensed data. New York: Academic Press, 1982. 256 p. ISBN 0-12-355620-1. $27.50. LC 82-16267.   361

Mathematical discussion of most of the quantitative techniques that can be applied to digital remotely sensed data. Includes information on data sources, computer equipment, algorithms, and selected case studies.

Remote Sensing

Johannsen, Chris J. and Sanders, James L. Remote sensing for resource management. Ankeny, IA: Soil Conservation Society of America, 1982. 665 p. ISBN 0-935734-08-2. $45. LC 82-16740.   362 •

A collection of articles dealing with the use of remote sensing for earth resources management. Good source of information on the relationship between geographic information systems and remote sensing.

Lillesand, Thomas M. and Kiefer, Ralph W. Remote sensing and image interpretation. New York: John Wiley and Sons, 1979. 612 p. ISBN 0-471-02609-3. $34.95. LC 78-27846.   363 •

The most widely used book in both airphoto interpretation and introductory remote sensing courses in the United States. Carefully reviews principles of data collection and analysis of data by remote sensor systems operating in the visible, near-infrared, thermal infrared, and microwave regions of the spectrum.

Lintz, Joseph, Jr. and Simonett, David S., eds. Remote sensing of environment. Reading, MA: Addison-Wesley, 1976. 694 p. ISBN 0-201-04245-2. $39.50. LC 76-47661.   364

An advanced treatment by selected authors of remote sensing principles, atmospheric effects, instruments, and data processing alternatives. The section on "Principles, Concepts, and Philosophical Problems in Remote Sensing" is thought provoking.

Lo, C.P. Geographical applications of aerial photography. New York: Crane, Russak, and Co., 1976. 304 p. ISBN 0-844808-72-5. $27.50. LC 75-37401.   365

Written by a geographer for geographers who have an interest in the application of aerial photography to their problems. Presents fundamental principles of data collection and photogrammetry.

Lyons, Thomas R. and Avery, Thomas E. Remote sensing: a handbook for archaeologists and cultural resource managers. Washington, DC: U.S. Department of the Interior, National Park Service, Cultural Resources Management Division, 1977. 109 p. LC 77-608037.   366 •

Discussion of photo scale, principles of stereoscopy, simple mapping techniques, and sources of information for the layperson interested in using aerial photography for either archaeological or cultural resource management studies.

National Academy of Sciences. National Resource Council. Resource sensing from space: prospects for developing countries. Washington, DC: National Academy of Sciences, 1977. 201 p. LC 78-109033.   367 •

Covers many aspects of remote sensing from an international perspective and evaluates the remote sensing inventory capabilities of Third World countries. A basic introduction to remote sensing is also provided. The question of industralized nations using remote sensing to monitor the resources of developing countries is examined through critical debate.

National Academy of Sciences. National Resource Council. Remote sensing with special reference to agriculture and forestry. Washington, DC: National Academy of Sciences, 1970. 424 p. ISBN 0-309-01723-8. LC 77-600961.   368 •

Although this work is dated, it should be read by those interested in either of these two fields of application. The principles and problems specific to the disciplines are covered with simplicity and insight. The blueprint for a global information system remains significant today.

Richason, Benjamin F., Jr. Introduction to remote sensing of the environment. 2nd ed. Dubuque, IA: Kendall/Hunt, 1983. 582 p. (1st ed., 1978. 496 p.) ISBN 0-8403-2834-6. $32.95. LC 82-83014.   369 •

An introductory text.

Sabins, Floyd F. Remote sensing: principles and interpretations. San Francisco, CA: W.H. Freeman, 1978. 426 p. ISBN 0-7167-0023-9. $38.95. LC 77-27595.   370
   An introductory text from a geologist's perspective. Especially useful for remote sensing courses that stress geomorphic applications.

Schowengerdt, Robert A. Techniques for image processing and classification in remote sensing. New York: Academic Press, 1983. 249 p. ISBN 0-12-628980-8. $25. LC 83-11769.   371
   Mathematic presentation of the fundamental algorithms associated with remote sensing image processing, enhancement, and classification.

Short, Nicholas M. The LANDSAT tutorial workbook: basics of satellite remote sensing. Washington, DC: National Aeronautics and Space Administration, Scientific and Technical Information Branch, 1982. 553 p. $55. LC 81-600117.   372 •
   Could be titled "Everything you want to know about LANDSAT." Describes the LANDSAT sensor system and presents numerous applications of its data, often in color.

Short, Nicholas M.; Lowman, Paul D., Jr.; Freden, Stanley C.; and Finch, William A., Jr. Mission to earth: LANDSAT views the world. (NASA publication no. SP360). Washington, DC: National Aeronautics and Space Administration, 1976. 459 p. LC 76-608116.   373 •
   Provides 400 full color plates of LANDSAT images throughout the world. Each is accompanied by a detailed description of the imagery. Serves as an excellent source of geomorphic, hydrologic, oceanographic, and urban LANDSAT images.

Siegal, Barry S. and Gillespie, Alan R. Remote sensing in geology. New York: John Wiley and Sons, 1980. 702 p. ISBN 0-471-79052-4. $59.95. LC 79-17967.   374
   A well-written textbook from a geologist's perspective. Results from work performed by the geology group at the California Institute of Technology Jet Propulsion Laboratory.

Slama, Chester C., ed. Manual of photogrammetry. 4th ed. Falls Church, VA: American Society for Photogrammetry and Remote Sensing, 1980. 1056 p. ISBN 0-937294-01-2. $58. LC 80-21514.   375
   The definitive work on the use of photogrammetric techniques to extract quantitative information from vertical and oblique aerial photography. Includes sections on the mathematics of photogrammetry, camera systems, films, filters, principles of stereoscopy, plotting instruments, and photogrammetric principles for planimetric and topographic mapping.

Slater, Philip N. Remote sensing: optics and optical systems. Reading, MA: Addison-Wesley, 1980. 575 p. ISBN 0-201-07250-5. $49.95. LC 80-13224.   376
   A detailed, quantitative description of remote sensor systems, electromagnetic theory, and spectroradiometry. The appendix on "Radiance" is especially useful for quantitative studies in remote sensing.

Swain, Philip H. and Davis, Shirley M. Remote sensing: the quantitative approach. London; New York: McGraw-Hill, 1978. 396 p. ISBN 0-07-062576-X. $48.50. LC 77-30577.   377
   The best single source on the use of quantitative techniques in remote sensing image interpretation. In addition, it provides information on radiation principles and instrumentation. The chapter titled "Biological and Physical Considerations in Applying Computer Aided Analysis Techniques to Remote Sensor Data" is essential for anyone contemplating the use of digital image processing.

Quantitative Methods

Townshend, John R.G., ed. Terrain analysis and remote sensing. London; Boston, MA: George Allen and Unwin, 1981. 232 p. ISBN 0-04-551036-9. ISBN 0-04-551037-7. $24.95 (pbk).
The first book by a geographer that deals directly with the use of remote sensor data for terrain analysis. In addition to reviewing sensor systems and analysis techniques, it discusses the concept of terrain analysis and what remote sensing can contribute. Case studies on gully erosion, arid land analysis, soil surveys, and mapping of surficial deposits are provided.

378

Williams, Richard S., Jr. and Carter, William D., eds. ERTS 1: a new window on our planet. Washington, DC: Government Printing Office, 1976. 362 p. $15. LC 75-37451.
Reports on the use of LANDSAT data for applications in cartography, geology and geophysics, water resources, land use mapping and planning, environmental monitoring, conservation, and oceanography. Each section is supported by case studies in full color and authored by project scientists. Provides an excellent introduction to the utility of LANDSAT MSS data.

379•

Wolf, Paul R. Elements of photogrammetry with air photo interpretation and remote sensing. 2nd ed. New York: McGraw-Hill, 1983. 628 p. (1st ed., 1974. 562 p.) ISBN 0-07-071345-6. $37.50. LC 82-4700.
Describes fundamental principles of photogrammetry using aerial photography. A good reference.

380•

8. QUANTITATIVE METHODS

Grant Ian Thrall

Bartels, Cornelius P.A. and Ketellapper, Ronald H., eds. Exploratory and explanatory statistical analysis of spatial data. The Hague, The Netherlands: Martinus Nijhoff; London; Boston, MA: Kluwer Academic, 1979. 268 p. ISBN 0-89838-004-9. $34.40. LC 79-13142.
Collection of papers on spatial statistics from the 1977 "Regional Science Symposium" at the University of Groningen, The Netherlands. Topics include: methods for analyzing spatial data; analysis of maps; input-output table evaluation; spatial autocorrelation; multivariate models of dependent spatial data, Bayesian analysis of linear model with spatial dependence, and spatial interaction models.

381

Batty, Michael. Urban modelling: algorithms, calibrations, predictions. (Cambridge urban and architectural studies no. 3). Cambridge, U.K.; New York: Cambridge University Press, 1976. 381 p. ISBN 0-521-20811-4. $75. LC 75-41592.
The author takes the position that modelling is an art, then formalizes the process of the art. Topics include: urban models as systems of equations; calibration techniques for models; and urban dynamics. Batty presents an exhaustive and readable treatment of a complex subject.

382

Bennett, Robert J., ed. European progress in spatial analysis. London: Pion; New York: Methuen, 1981. 305 p. ISBN 0-85086-091-1. $26. LC 82-110511.
Collection of papers presented at the Second European Colloquium on Quanititative and Theoretical Geography, 11-14 September, 1980, in Cambridge. The papers focus on the state of development of quantitative geography in various European countries. Part of the book is devoted to applications of quantitative techniques.

383

Bennett, Robert J. Spatial time series: analysis forecasting control. London: Pion; New York: Methuen, 1979. 674 p. ISBN 0-85086-069-5. $87.50. LC 79-319356.
Authoritative work on analysis of space-time series; excellent review of time series methods and forecasting. Very comprehensive. Good for advanced and intermediate level graduate students and for reference.

384

Bennett, Robert J. and Chorley, Richard J. Environmental systems: philosophy, analysis, and control. London: Methuen; Princeton, NJ: Princeton University Press, 1978. 624 p. ISBN 0-691-08217-0. $100. LC 79-100507.   385

The aim of the authors is two fold: to explore to what extent systems theory can provide an interdisciplinary focus for those concerned with issues of the environment; and to examine the manner in which systems approaches aid in the development of an integrated theory on the one hand and in physical and biological theory on the other. Topics include: systems methods; psychological systems; interfacing physico-ecological and socio-economic systems. This is a brilliant and comprehensive work.

Berry, Brian J.L. and Teicholz, Eric, eds. Computer graphics and environmental planning. Englewood Cliffs, NJ: Prentice-Hall, 1983. 250 p. ISBN 0-13-164830-6. $35. LC 82-12203.   386

Inferential and descriptive statistics are considered principal manipulative and analytic operative components of the geographic information systems data environment. A collection of fourteen essays on the general role of statistics and geocoded data in an applied environment. Also interesting from the perspective of the transition of Brian Berry, one of the major forces in the quantitative movement in geography, into applying geographic analysis to planning.

Carlstein, Tommy; Parks, Don; and Thrift, Nigel J., eds. Time and regional dynamics. (Timing Space and Spacing Time, v. 3). New York: Halsted Press, John Wiley and Sons; London: Edward Arnold, 1978. 120 p. ISBN 0-470-26512-4. $29.95. LC 79-302717.   387

Essays on the role of time and space in society and the environment. Topics include: scale; contagious processes; flows; mathematical programming; and calibration dynamics. Extensive bibliography on related work for the researcher.

Chapman, Geoffrey P. Human and environmental systems: a geographer's appraisal. London; New York: Academic Press, 1977. 421 p. ISBN 0-12-168650-7. $67. LC 77-74364.   388

The author formulates geographic concepts and methods as systems analysis. A review of systems analysis and its terminology is presented, including an extensive discussion of entropy and information theory.

Chorley, Richard J. and Haggett, Peter, eds. Models in geography. London; New York: Methuen, 1967. 816 p. ISBN 0-416-29020-5. $82. LC 68-71825.   389

A classic and comprehensive analysis of geographical models.

Cliff, Andrew D. and Ord, J.K. Spatial processes: models and applications. 2nd ed. London: Pion; New York: Methuen, 1981. 266 p. (1st ed., Spatial Autocorrelation, 1973. 178 p.) ISBN 0-85086-081-4. $30. LC 81-132089.   390

Reviews the methods available for analyzing patterns formed on a map. Topics include: autoregressive models; regressive models; moving-average frameworks; and Poisson point patterns. To this seminal book can be attributed both increasing the recognition that geocoded data analysis requires new spatial statistics, and developing the foundations of the spatial statistic.

Cliff, Andrew D.; Haggett, Peter; Ord, J.K.; Bassett, Keith; and Davis, R.B. Elements of spatial structure: a quantitative approach. (Geographical studies series no. 6). London; New York: Cambridge University Press, 1975. 258 p. ISBN 0-521-20689-8. $42.50. LC 74-12973.   391

Illustrates the use of combinatorial structures, trend surfaces, lagged regressions, and nearest neighbor methods in analyzing spatial data.

Davidson, Donald A. Science for physical geographers. London: Edward Arnold; New York: John Wiley and Sons, 1978. 187 p. ISBN 0-7131-6123-X. £9.95. (pbk). LC 78-11957.   392

Presents the fundamental principles of physics and chemistry relevant to the physical geography student. Topics include: measurement scale; atomic structure; foundations of physics, i.e., acceleration and work; principles of energy; chemical reactions and equilibrium; heat and phase change; solids, liquids, and gases; and types of errors.

Quantitative Methods

Dawson, John A. and Unwin, David J. Computing for geographers. Newton Abbot, U.K.: David and Charles; New York: Crane, Russak and Co., 1976. 362 p. ISBN 0-8448-0571-7. $16.95. LC 75-37400.

393

A textbook introduction to programming in FORTRAN with exercises and examples directed towards geographic problems. Text includes many programs, for example, using a line printer to construct choropleth maps.

Gaile, Gary L. and Willmott, Cort J., eds. Spatial statistics and models. (Theory and Decision Library, v. 40). Dordrecht, Holland: D. Reidel and Co.; Hingham, MA: Kluwer Academic, 1984. 482 p. ISBN 9-02-771618-8. $69. LC 83-22984.

394

A collection of twenty-three papers on spatial processes including: spatial statistics; algebraic topology; geocoding; geosampling; classification; computer cartography; trend surfaces; nonlinear models; regression; spatial autocorrelation; equality; spatial time series; directional statistics; topology; shape; and general spatial models.

Gatrell, Anthony C. Distance and space: a geographic perspective. Oxford, U.K.: Clarendon Press; New York: Oxford University Press, 1983. ISBN 0-19-874128-6. $24.95. ISBN 0-19-874129-4. $12.95. (pbk). LC 83-7829.

395

Textbook survey of the different concepts of distance and space.

Getis, Arthur, and Boots, Barry N. Models and spatial processes: an approach to the study of point, line, and area patterns. Cambridge, U.K.; New York: Cambridge University Press, 1978. 198 p. ISBN 0-521-20983-8. $39.50. LC 75-17118.

396

Integrates literature on point patterns from ecology, biology, geology, forestry, astronomy, and statistics. Topics include: the spatial process; Poisson process models; mixed process models; contagious models; line patterns; and area patterns.

Golledge, Reginald G. and Rayner, John N. Proximity and preference: problems in the multidimensional analysis of large data sets. Minneapolis, MN: University of Minnesota Press, 1982. 310 p. ISBN 0-8166-1042-8. $35. LC 81-14634.

397

Directed to the problem of collecting and handling large data sets, what data should be collected, and how to collect it. The emphasis is upon designing experiments for collecting large volumes of subjective data. The data discussed are primarily proximity and preference type within spatial and psychospatial contexts.

Haggett, Peter; Cliff, Andrew D. and Frey, Allan. Locational analysis in human geography. 2nd ed. London: Edward Arnold; New York: John Wiley, 1977. 2 v. 605 p. (1st ed., 1965. 339 p.) ISBN 0-470-99207-7. ISBN 0-470-99208-5. $24.95. (pbk). LC 77-8967.

398

Outlines techniques for gathering, measuring, classifying and describing information that can be used to test models. The example models focus upon human settlements. The authors emphasize how the material can be used in practical applications to social, economic, and regional planning.

Hammond, Robert, and McCullagh, Patrick S. Quantitative techniques in geography: an introduction. 2nd ed. Oxford, U.K.: Clarendon Press; New York: Oxford University Press, 1978. 364 p. (1st ed., 1974. 318 p.) ISBN 0-19-874066-2. $21.95. ISBN 0-19-874067-0. $14.95. (pbk). LC 79-308406.

399 •

A useful introduction to elementary statistics for geographers. Topics include: summarizing data; measures of spatial distribution; time series; probability distributions; samples and estimates; hypothesis testing; correlation, regression, tests for distributions in space; and models as techniques (simulation, computers).

Haynes, Kingsley E. and Fotheringham, A. Stewart. Gravity and spatial interaction models. (Scientific geography series vol. 2). Beverly Hills, CA: Sage Publications, 1984. V. 2. 88 p. ISBN 0-8039-2326-0. $6.50. (pbk).

400 •

Provides a clear and comprehensive introduction to the gravity and spatial interaction models considered by many to be the most useful in all of geographical analysis. Presents many "real world" applications of the model including: planning a new service; defining retail shopping boundaries; forecasting migration and voting patterns; analyzing university enrollments; determining the optimal size of a shopping center; and locating a facility for maximum patronage. The best work of its type so far.

Isard, Walter. Introduction to regional science. Englewood Cliffs, NJ: Prentice-Hall, 1975. 506 p. ISBN 0-13-493841-0. $29.95. LC 74-22031.
401
A classic work at the entry level explaining concepts for analyzing urban and regional economic problems. Topics include: regional science and geography; the city and region; spatial distributions and regional differences; the spatial market system; cost benefit analysis; industrial location; economic base, and trade.

Johnston, Ronald J. Multivariate statistical analysis in geography: a primer on the general linear model. London; New York: Longman, 1978. 280 p. ISBN 0-582-48677-7. ISBN 0-582-30034-7. $13.95. (pbk). LC 77-22424.
402
Intended for reference and for a second course in statistics for geographers. Topics reviewed include: bivariate analysis; multiple correlation and regression; multivariate extensions of analysis of variance; factor analysis; canonical correlation; and discriminant analysis data.

Killen, James E. Mathematical programming methods for geographers and planners. London: Croom Helm; New York: St. Martin's Press, 1983. 363 p. ISBN 0-312-50133-1. $35. LC 82-42839.
403
Presents a comprehensive review of mathematical programming along with an extensive collection of examples illustrating the use of mathematical programming in geography. Topics include: network design; shortest route models; the transportation problem; sensitivity analysis; nonlinear programming, integer programming, dynamic programming, and goal programming.

King, Leslie J. Central place theory. (Scientific geography series vol. 1). Beverly Hills, CA: Sage Publications, 1984. V. 1. 96 p. ISBN 0-8039-23244. $6.50. (pbk).
404
Seeks to explain the numbers, sizes, and locations of urban settlements. Provides a review of central place theory and its antecedents. Directed towards entry level students yet can be a valuable reference tool for advanced students and researchers. Includes applications to planning as well as the use of the theory in disciplines other than geography.

MacDougall, E. Bruce. Computer programming for spatial problems. London: Edward Arnold; New York: John Wiley and Sons, 1976. 158 p. ISBN 0-7131-5865-4. £8.95. LC 76-46976.
405
An introduction to computer programming and its applications to geography. Provides many operational programs at the introductory level, such as solving for great circle distances.

MacDougall, E. Bruce. Microcomputers in landscape architecture. Amsterdam, The Netherlands; New York; Oxford, U.K.: Elsevier Scientific, 1983. 271 p. ISBN 0-444-00771-3. ISBN 0-444-00771-7. $34.50. (pbk). LC 83-5649.
406•
An introduction to the use of microcomputers for handling spatial data. Topics include: hardware; applications (text editing, regional analysis); the BASIC language; digital terrain models; shape and runoff; line plotting; perspectives; sun and shadow calculations; and data base management. A practical and useful book for the student and researcher becoming interested in geographic data management using microcomputers. Many program listings in the BASIC language are included that operationalize the topics discussed.

Quantitative Methods

Martin, R.L.; Thrift, Nigel J.; and Bennett, Robert J. Towards the dynamic analysis of spatial systems. London: Pion; New York: Methuen, 1978. 210 p. ISBN 0-85086-072-5. $24.50. LC 79-317993.

A collection of review papers on methods of dynamic analysis and modelling issues.

407

Mather, Paul M. Computational methods of multivariate analysis in physical geography. London; New York: John Wiley and Sons, 1976. 532 p. ISBN 0-471-57626-3. $79.95. LC 75-23376.

Text on multivariate statistical analysis directed to the geography audience. Each statistical method is accompanied by an operational computer program in FORTRAN. The examples are primarily from physical geography, but are useful to any branch of geography.

408

Mather, Paul M. Computers in geography: a practical approach. Oxford, U.K.: Basil Blackwell, 1976. 125 p. ISBN 0-631-16870-2. LC 76-382907.

A series of examples in geography with short FORTRAN programs that operationalize the example. Topics include: the architecture of the computer; an introduction to FORTRAN; and applications. The programming examples include: nearest-neighbor; Spearman's rank correlation coefficient; simulation of drainage networks; the linear programming simplex method; and linear regression.

409

Putman, Stephen H. Integrated urban models. London: Pion; New York: Methuen, 1983. 330 p. ISBN 0-85086-098-9. $30.

A comprehensive discussion of transportation and location models. Presents and discusses the algorithms necessary for operationalizing the models. The author includes many real world examples.

410

Rayner, John N. Introduction to spectral analysis. (Monographs in spatial and environmental systems analysis no. 2). London: Pion; New York: Methuen, 1971. 174 p. ISBN 0-85086-026-1. $17.95. LC 72-177346.

An introduction for graduate students to spectral analysis. Provides many examples of the application of spectral analysis to spatial systems; most are from physical geography. Topics include: the history of the subject; distinction between discrete and continuous spectrum; explanations of trigonometric functions and Fourier analysis; estimation of spectral densities for nonperiodic data; filtering and aliasing; and cross-spectral analysis.

411

Ripley, Brian D. Spatial statistics. (Probability and mathematical series: applied probability and statistics). New York; Toronto: John Wiley and Sons, 1981. 252 p. ISBN 0-471-08367-4. $33.95. LC 80-26104.

Summarizes techniques for the analysis of spatial data. Two of the most important examples are smoothing and interpolation for producing contour maps (estimating rainfall or petroleum reserves), and mapped point patterns (trees, towns, and galaxies). Other chapters cover: regional variables in human geography; spatially arranged experiments; quadrat counts; sampling a spatially correlated variable; and testing spatial patterns using image analyzers.

412

Rogers, Andrei. Migration, urbanization, and spatial population dynamics. Boulder, CO: Westview Press, 1984. 350 p. ISBN 0-86531-896-4. $25. LC 84-14852.

The first part of the book focuses upon applications of mathematical techniques to forecast migration and population growth. The latter part focuses upon how the mathematical apparatus developed primarily for migration models can also be used for modeling marital and employment status.

413

Rogers, Andrei. Statistical analysis of spatial dispersion: the quadrat method. (Monographs in spatial and environmental systems analysis no. 6). London: Pion; New York: Academic Press, 1974. 164 p. ISBN 0-85086-045-8. £9.50. LC 75-327493.

414

Directed toward geographers on the explanation of quadrat analysis with applications focusing upon retail spatial structure. Topics include: quadrat analysis; nearest neighbor analysis; random, regular, and clustered spatial dispersion; compound and generalized distributions; parameter estimation; hypothesis testing; structure of retail trade; bivariate distributions; and spatial sampling.

Selkirk, Keith E. Pattern and place: an introduction to the mathematics of geography. London; New York: Cambridge University Press, 1982. 203 p. ISBN 0-521-28208-X. $16.95. (pbk). LC 81-3847.   415 •

An entry level book on the basic mathematical components of geography. Appropriate for reference and for the student with high school algebra and geometry. A practical and useful text. Topics include: sets; coordinates; distance; density; area; networks; point pattern analysis; and simulation.

Silk, John. Statistical concepts in geography. London; Boston, MA: George Allen and Unwin, 1979. 276 p. ISBN 0-04-910065-3. $25. LC 78-40957.   416

Entry level textbook on statistics for geographers.

Sumner, Grahm N. Mathematics for physical geographers. New York: John Wiley and Sons, 1978. 236 p. ISBN 0-470-26557-4. $21.95. LC 78-12156.   417

The logic of mathematics can conceptualize many aspects of physical geography. An introduction to mathematics with many examples from physical geography. Topics include: trigonometry; linear functions; matrices; nonlinear functions; series and progressions; probability; integration; and elementary and differential equations.

Tan, K.C. and Bennett, Robert J. Optimal control of spatial systems. (London research series in geography no. 6). London; Boston, MA: Allen and Unwin, 1984. 172 p. ISBN 0-04-519018-6. $29.95. LC 83-25701.   418

The authors use a systematic approach to the application of control theory to spatial systems; the focus is upon dynamic linear programming for the upper-division undergraduate or entry level graduate student.

Thomas, Reginald W. and Huggett, R.J. Modelling in geography: a mathematical approach. New York: Harper and Row, 1980. 338 p. ISBN 0-06-318060-X. $21.50. ISBN 0-06-318171-1. $17.25. (pbk). LC 80-138951.   419

Geographical research is reviewed from the perspective of its mathematical logic; though the mathematical technique is explained, the emphasis remains upon the concept of "model structures." Topics include: mathematical review; deterministic models (space-time, cascading, spatial interaction, spatial allocation) and probability models (quadrat, spatial autocorrelation, decision and strategy models).

Thrall, Grant I., series ed. Scientific geography series. Beverly Hills, CA; London: Sage Publications, 1984- . Each book approximately 100 pages.   420 •

Presents the contributions to scientific geography in a unique manner. Each topic is explained in a small inexpensive book or module. The books are designed to reduce the barriers to learning; successive books at the more advanced level follow the entry level books. The series begins with several important topics in human geography, to be followed by topics in other branches of scientific geography. The modules are written by the leading scholars in each area. Designed for entry level students, advanced students, and reference. Topics include: industrial location; central place theory; gravity and spatial interaction models; urban problems; population projections; land use; input-output modeling; migration; etc. The first three volumes are cited elsewhere in this section. Several other volumes in press at this time with publication of additional volumes planned for the future.

Quantitative Methods

Unwin, David J. Introductory spatial analysis. London; New York: Methuen, 1981. 212 p. ISBN 0-416-72190-7. $27. ISBN 0-416-72200-8. $12.95. (pbk). LC 81-197262.   421
  A textbook on stochastic processes. The author first considers the type of data used and the varieties of maps drawn; then reviews the methods for evaluating the resulting point patterns. The emphasis is upon the geographic theme and their use of statistics rather than on the mechanics of statistics.

Webber, Michael J. Industrial location. (Scientific geography series vol. 3). Beverly Hills, CA: Sage Publications, 1984. 96 p. ISBN 0-8039-23252. $6.50. (pbk).   422
  The author presents a review, for entry level students, of the literature on industrial location, techniques used in industrial location analysis, and a comprehensive and detailed discussion of data sources on manufacturing.

Webber, Michael J. Information theory and urban spatial structure. London: Croom Helm; New York: St. Martin's Press, 1979. 394 p. ISBN 0-85664-665-2. £33.50. LC 79-16812.   423
  Advanced research book on information theoretic models related to the effect of social and economic constraints upon location and land use.

Wilson, Alan G. Catastrophe theory and bifurcation: applications to urban and regional systems. London: Croom Helm; Berkeley, CA: University of California Press, 1981. 331 p. ISBN 0-520-04370-7. $41. LC 80-6059.   424
  An advanced level text on modeling discontinuous change in geographic systems using the catastrophe theory paradigm. The book includes a guide to catastrophe theory and differential equations for the layman. Many examples of the use of catastrophe theory at various scales of analysis are provided.

Wilson, Alan G. Geography and the environment: systems analytical methods. Chichester, U.K.; New York: John Wiley and Sons, 1981. 297 p. ISBN 0-471-27956-0. $44.95. ISBN 0-471-27957-9. $21.95. (pbk). LC 80-41696.   425
  The author outlines the utility of the systems analysis approach to geography. The topics include: optimization, network analysis, entropy maximization, and several approaches to dynamics.

Wilson, Alan G. Urban and regional models in geography and planning. New York; London: John Wiley and Sons, 1974. 418 p. ISBN 0-471-95192-7. £11.85. (pbk). LC 73-8200.   426
  A very comprehensive work on integration of the geographic theory of cities and regions. Topics include: planning; models and urban problems; how to construct a model; mathematical prerequisites; data calibration and testing; and examples.

Wilson, Alan G.; Coelho, J.D.; MacGill, S.M.; and Williams, H.C.W.L. Optimization in locational and transport analysis. Chichester, U.K.; New York: John Wiley and Sons, 1981. 283 p. ISBN 0-471-28005-4. $53.95. LC 80-42068.   427
  Provides a comprehensive review of optimization techniques and models as used in urban analysis. Topics include: mathematical programming methods; random utility theory; entropy models; and catastrophe theory.

Wilson, Alan G. and Kirkby, Michael J. Mathematics for geographers and planners. (Contemporary problems in geography). 2nd ed. Oxford, U.K.: Clarendon Press; New York: Oxford University Press, 1980. 408 p. (1st ed., 1975. 325 p.) ISBN 0-19-874114-6. $36. ISBN 0-19-874115-4. $19.95. (pbk). LC 79-41132.   428
  Provides the basic mathematical tools to study the major problems of spatial analysis in geography. Presents a solid discussion of the fundamental mathematical concepts with applications to geography. Topics include: algebraic operations; calculus; mathematical functions; matrix algebra; differential and difference equations; and probability. The last part of the book includes a discussion of stochastic processes, and optimization and catastrophe theory.

Wrigley, Neil. Statistical applications in the spatial sciences. London: Pion; New York: Methuen, 1979. 310 p. ISBN 0-85086-075-X. $32. LC 80-151661.　　　429

A collection of papers presented at a joint conference of the Institute of British Geographers and the Royal Statistical Society. Topics include the use of spatial statistics in archaeology, climatology, demography, economics, epidemiology, entomology, hydrology, and soil science.

Wrigley, Neil, and Bennett, Robert J., eds. Quantitative geography: a British view. London; Boston, MA: Routledge and Kegan Paul, 1981. 419 p. ISBN 0-7100-0731-0. $65. LC 82-172048.　　　430

A collection of research papers by British geographers. The topics are both statistical and mathematical. The papers are diverse and cover climatology, hydrology, and eight human geography specializations. The book presents the evolution of quantitative methods in all fields of geography.

Yeates, Maurice H., ed. Proceedings of the 1972 meetings of the IGU commission on quantitative geography. Montreal, Canada: McGill-Queen's University Press, 1974. 182 p. ISBN 0-7735-0168-1. LC 74-75799.　　　431

A collection of articles dealing with research methods, probability models, multiple stage regression, and density models.

9. FIELD METHODS

Association of American Geographers, Commission on College Geography. Field training in geography. (Technical paper no. 1). Washington, DC: Association of American Geographers, 1968. 69 p. LC 68-31526.　　　432

A guide for college instructors.

Association of American Geographers. High School Geography Project. Committee on Local Geography. The Local community: a handbook for teachers. New York: Macmillan, 1971. 255 p.　　　433

A guide for high school teachers.

Lounsbury, John F. and Aldrich, Frank T. Introduction to geographic field methods and techniques. Columbus, OH: Merrill, 1979. 181 p. ISBN 0-675-08304-4. $14.50. (pbk). LC 78-61610.　　　434

A general introduction.

PART III. SYSTEMATIC FIELDS OF PHYSICAL GEOGRAPHY

1. GENERAL PHYSICAL GEOGRAPHY

Harold A. Winters

A. SERIALS

Catena: an interdisciplinary journal of soil science, hydrology, and geomorphology focusing on geoecology and landscape evolution. 1- (1973-). Quarterly. Catena Verlag, Margot Rohdenburg, Brockenblick 8, D-3302 Cremlingen-Destedt, Federal Republic of Germany. ISSN 0341-8162.   435
> *Articles, reviews of research fields, commentaries, and book reviews with emphasis on interdisciplinary connections among pedology, hydrology, geomorphology, and related aspects of other disciplines. Mainly in English, some papers in German or French. Abstracts in English. Cumulative index v. 1-5 (1973-1978).*

Geografiska annaler. Series A. Physical geography (Svenska Sällskapet för Antropologi och Geografi). 47a- (1963-). Quarterly. Almqvist & Wiksell Periodical Co., Box 62, S-101 20 Stockholm, Sweden. ISSN 0435-3676.   436
> *A leading international scientific journal in physical geography. Particularly valuable in the fields of geomorphology, glaciology, and climatology and in studies of Arctic areas. Authors from many countries. In English.*

Géographie physique et quaternaire. 1- (1947-). Quarterly. Les Presses de l'Université de Montréal, B.P. 6128, stat. 'a,' Montréal, P.Q. H3C 3J7 Canada. ISSN 0705-7199.   437
> *Devoted to physical geography and the Quaternary: geomorphology (especially glacial and periglacial processes and forms), climatology, hydrology, pedology, and biogeography. In French or English. Abstracts in English, French, and German. v. 1-17 (1947-1963) as Revue Canadienne de géographie, v. 18-30 (1964-1976) as Revue de géographie de Montréal. Cumulative index v. 1-17 (1947-1963).*

Physical geography. 1- (1980-). 3 per annum. V.H. Winston and Sons, 7961 Eastern Avenue, Silver Spring, MD 20910. ISSN 0272-3646.   438
> *Original research articles on general physical geography, geomorphology, climatology, hydrology, soil science, and biogeography.*

Progress in physical geography: an international review of geographical work in the natural and environmental sciences. 1- (1977-). Quarterly. Edward Arnold, 41 Bedford Square, London WC1B 3OQ, U.K., or Cambridge University Press, 32 East 57th Street, New York, NY 10022 U.S.A. ISSN 0309-1333.   439
> *Articles and progress reports on major fields in physical geography. Book reviews. Invaluable for following the expanding frontiers and research thrusts in physical geography.*

Quaternary research: an interdisciplinary journal. 1- (1970-). Bimonthly. Academic Press, 111 Fifth Avenue, New York, NY 10003. ISSN 0033-5894.   440
> *Articles on all aspects of Quaternary research from geography and geomorphology to anthropology, archeology, botany and palynology, geochemistry and geophysics, geochronology, glaciology, paleoclimatology, paleoecology and zoogeography, paleontology, geology, oceanography, and soils.*

B. DICTIONARIES

Bates, Robert L. and Jackson, Julia A., eds. Glossary of geology. 2nd ed. Falls Church, VA: American Geological Institute, 1980. 749 p. (1st ed., 1972. 805, 52 p. ed. by Margaret Gary, Robert McAfee, Jr., and Carol L. Wolf.) ISBN 0-913312-15-0. $60. LC 79-57360.   441•
> *A thorough list of geologic and related terms with their definition(s) and usage. Original sources for the entries are often cited. This is a very useful basic reference book.*

General Physical Geography

Goudie, Andrew S., ed. The Encyclopaedic dictionary of physical geography. Oxford, U.K.: Basil Blackwell, July 1985 (Scheduled date of publication). ISBN 0-631-13292-9. £30.

    A comprehensive, detailed dictionary of more than 2000 terms and essays on areas of study.

442 •

C. GENERAL WORKS

Bennett, Charles F. Man and Earth's ecosystem. New York: John Wiley and Sons, 1975. 331 p. ISBN 0-471-06638-9. $30.95. LC 75-22330.

    Not a traditional physical geography text, but instead uses a regional approach to various ecosystems. The book is useful to those who treat introductory physical geography from a regional ecosystem-human oriented approach.

443 •

Bunnett, R.B. Physical geography in diagrams. London: Longman; New York: Praeger, 1973. 180 p. ISBN 0-582-69906-1. LC 74-183167. Metric 3rd ed., London: Longman, 1976. 196 p. (1st ed. reissued London: Longman, 1966; New York: Praeger, 1968. 172 p.). ISBN 0-582-34122-1. £3.50. (pbk).

    A descriptive non-traditional book that includes an unusually large number of line drawings and diagrams. The text is limited but the illustrations may be quite useful.

444 •

Chorley, Richard J. and Kennedy, Barbara A. Physical geography: a systems approach. London: Prentice-Hall, 1971, 370 p. ISBN 0-13-669036-X. ISBN 0-13-669028-9 (pbk). LC 75-325275.

    A good treatment of physical geography which emphasizes relationships and interactions between variables rather than examining each separately in detail. Heavy emphasis on systems.

445

Davis, William Morris. Geographical essays. Edited by Douglas Wilson Johnson. Boston, MA: Ginn and Co., 1909. 777 p. Reprinted: New York: Dover, 1957. 777 p. LC 54-11800.

    Includes many interesting and classic works by this famous geographer. Especially useful for historical perspectives.

446

Dury, George H. An Introduction to environmental systems. London; Portsmouth, NH: Heinemann Educational, 1981. 366 p. ISBN 0-435-08001-6. $19.95. ISBN 0-435-08002-4. (pbk). LC 80-29151.

    An excellent introduction to the use of General Systems Theory in the study of physical geography. Best used as a second-year college course because the text emphasizes systems analysis of the elements of physical geography; hence students need prior knowledge of those subjects.

447

Gabler, Robert E.; Sager, Robert J.; Brazier, Sheila; and Wise, Daniel. Essentials of physical geography. 2nd ed. Philadelphia, PA: Saunders College, 1982. 568 p. (1st ed., New York: Holt, Rinehart, and Winston, 1977. 496 p.). ISBN 0-03-058551-1. $34.95. LC 81-53083.

    A comprehensive non-technical introductory college textbook that uses a descriptive and traditional approach to the subject.

448 •

Gersmehl, Philip; Kammrath, William; and Gross, Herbert. Physical geography. Philadelphia, PA: Saunders College, 1980. 415 p. ISBN 0-03-014476-0. $26.95. LC 78-12212.

    A good introductory text that seeks to combine environmental science and physical geography. Topical coverage is complete and well keyed to matters of environmental concern.

449 •

King, Cuchlaine A. M. Physical geography. Oxford, U.K.: Blackwell; Totowa, NJ: Barnes and Noble Books, 1980. 332 p. ISBN 0-389-20089-1. $34.50. LC 80-134479.   450
One of the best introductory physical geography texts now available. Not only are the topical discussions clear and up-to-date but each is accompanied by an appraisal of how field studies of the topic may be carried out. Topics are covered at local, regional, and continental scales.

Marsh, William M. and Dozier, Jeff. Landscape: an introduction to physical geography. Reading, MA: Addison-Wesley Publishing Co., 1981. 637 p. ISBN 0-201-04101-4. $29.95. LC 80-68120.   451
A high quality, well-written introductory college textbook with numerous innovative illustrations.

McKnight, Tom L. Physical geography: a landscape appreciation. Englewood Cliffs, NJ: Prentice-Hall, 1984. 488 p. ISBN 0-13-669101-3. $32.95. LC 83-11199.   452•
A traditional introductory college physical geography textbook with numerous helpful illustrations and a good index. Quite useful for a succinct general subject review.

Miller, G. Tyler, Jr. (George Tyler). Living in the Environment: an introduction to environmental science. 4th ed. Belmont, CA: Wadsworth, 1985. (1st ed., 1974. 379 p.). ISBN 0-534-84332-1. $24. LC 84-15297.   453
An environmental studies approach pertinent to physical geography.

Muller, Robert A. and Oberlander, Theodore M. Physical geography today: a portrait of a planet. 3rd ed. New York: Random House, 1984. 591 p. ISBN 0-394-33264-4. $30.00. LC 83-23122.   454•
A well-written and nicely illustrated general introductory college physical geography text. Subject matter is treated according to the process approach.

Navarra, John G. Contemporary physical geography. Philadelphia, PA: Saunders College, 1981. 523 p. ISBN 0-03-057859-0. $31.95. LC 80-53922.   455•
A largely descriptive introductory text designed for non-technical junior college or university freshman students.

Oliver, John E. Perspectives on applied physical geography. North Scituate, MA: Duxbury Press, 1977. 315 p. ISBN 0-87872-131-2. LC 77-4194.   456
Approach emphasizing problem solving and applied aspects of physical geography.

Oliver, John E. Physical geography: principles and applications. North Scituate, MA: Duxbury Press, 1979. 786 p. ISBN 0-87872-185-1. LC 78-10642.   457
A general college physical geography text that incorporates case studies involving human implications with some of the subject matter in an interesting fashion.

Smythe, James M.; Brown, Charles G.; Fors, Erich H.; and Lord, R.C. Physical geography: metric edition. Toronto: Macmillan of Canada, 1978. 342 p. ISBN 0-77051-689-0. LC 79-321553.   458•
Well written at a level appropriate for high school students who have had no prior physical geography. Many useful diagrams and maps.

Stanford, Quentin H. and Moran, Warren. Geography: a study of its physical elements. Toronto; New York: Oxford University Press, 1978. 308 p. ISBN 0-19-540282-0. Can $15.50. LC 81-458655.   459•
Standard approach, well written, with good maps and illustrations. Many topics treated briefly.

Strahler, Arthur N. and Strahler, Alan H. Elements of physical geography. 3rd ed. New York: John Wiley and Sons, 1984. 538 p. (1st ed., 1976. 469 p.) ISBN 0-471-88973-3. $29.95. LC 83-7005.   460
A widely used, current, concisely written, somewhat encyclopedic, comprehensive, well illustrated, and nicely presented introductory college textbook.

2. GEOMORPHOLOGY

William L. Graf and Jeffrey A. Lee

A. SERIALS

Earth surface processes and landforms: a journal of geomorphology. (British Geomorphological Research Group). 1- (1976-). Bimonthly. John Wiley and Sons, Baffins Lane, Chichester P019 10D, England. ISSN 0197-9337.   461
Articles, short communications, and book reviews in geomorphology and related fields.

Environmental geology and water sciences. 1- (1975-). Quarterly. Springer Verlag, 175 Fifth Avenue, New York, NY 10010. 1975-1983 as Environmental Geology. ISSN 0099-0094.   462

Geo abstracts. Part A. Landforms and the Quaternary. 1- (1960-). Bimonthly. Geo Abstracts, Ltd., Regency House, 34 Duke Street, Norwich NR3 3AP, England. ISSN 0305-1897.   463

Geo abstracts. Part E. Sedimentology. 1- (1972-). Bimonthly. Geo Abstracts, Ltd., Regency House, 34 Duke Street, Norwich NR3 3AP, England. ISSN 0305-1935.   464

Geological society of America. Bulletin. 1- (1888-). Monthly. Geological Society of America, Box 9140, Boulder, CO 80301. ISSN 0016-7606.   465

Zeitschrift für Geomorphologie. Annals of geomorphology. 1-11 (1925-1940); New series 1- (1957-). Quarterly. Gebrüder Borntraeger, Johannesstrasse 3A, D-7000 Stuttgart, Federal Republic of Germany. ISSN 0372-8854.   466
High-quality scientific journal for geomorphology with international board of editors and world-wide distribution of authors. Articles. Notes and news. Letters. Extensive book reviews. In English, German, or French with well-written abstracts in English, German, and French. A special series, "Supplementband," 1- (1960-) contains articles on special themes in geomorphology. Cumulative index: 1925-1976.

See also: Catena (435); Geografiska annaler. Series A. (436); Géographie physique et quaternaire (437); Physical geography (438); Progress in physical geography (439); Quaternary research (440); Arctic and Alpine research (2798).

B. ATLAS

Snead, Rodman E. World atlas of geomorphic features. Huntington, NY: R.E. Krieger, 1980. 301 p. ISBN 0-88275-272-3. $39.50. LC 77-28009.   467 •
A largely exemplary work including maps and satellite photographs.

C. GENERAL WORKS

Bloom, Arthur L. Geomorphology: a systematic analysis of late Cenozoic landforms. Englewood Cliffs, NJ: Prentice-Hall, 1978. 510 p. ISBN 0-13-353086-8. $37.95. LC 77-25816.   468
A thorough introduction to the field.

Butzer, Karl W. Geomorphology from the earth. New York: Harper and Row, 1976. 463 p. ISBN 0-06-041097-3. $30.95. LC 76-10863.   469 •
  A reasonably non-technical introduction to the landforms of the earth.

Ritter, Dale F. Process geomorphology. Dubuque, IA: Wm. C. Brown Co., 1978. 603 p. ISBN 0-697-05035-1. LC 77-80412.   470
  A well-written somewhat advanced text emphasizing the processes of geomorphology.

D. FLUVIAL

Dingman, Stanley L. Fluvial hydrology. New York: W.H. Freeman, 1983. 383 p. ISBN 0-7167-1452-3. $31.95. LC 83-20763.   471
  A quantitatively oriented, basic-level approach to hydrologic processes related to rivers and slopes.

Gregory, Kenneth J. ed. River channel changes. Chichester, U.K.; New York: John Wiley and Sons, 1977. 448 p. ISBN 0-471-99524-X. £40.75. LC 77-4342.   472
  A publication of the British Geomorphological Research Group featuring a collection of papers addressing mechanics, geometry, patterns, and theory of river channel change.

Leopold, Luna B.; Wolman, M. Gordon; and Miller, J.P. Fluvial processes in geomorphology. San Francisco, CA: W.H. Freeman, 1964. 522 p. ISBN 0-7167-0221-5. $35.95. LC 64-10919.   473
  A major work in fluvial geomorphology that is partially outdated but still useful.

Richards, Keith S. Rivers: form and process in alluvial channels. London; New York: Methuen, 1982. 358 p. ISBN 0-416-74900-3. $35. LC 82-8133.   474
  A complete quantitatively-oriented review with examples.

Schumm, Stanley A. The Fluvial system. New York: John Wiley and Sons, 1977. 338 p. ISBN 0-471-01901-1. $42.50. LC 77-9333.   475
  An integrated general systems view of fluvial geomorphology.

E. SLOPES

Carson, Michael A. and Kirkby, Michael J. Hillslope form and process. London: Cambridge University Press, 1972. 475 p. ISBN 0-521-08234-X. $65.00. LC 74-163061.   476
  A major reference work on slope processes and morphology directed at an advanced audience.

Schumm, Stanley A. and Mosley, M. Paul., eds. Slope morphology. New York: Van Nostrand Reinhold, 1973. 454 p. ISBN 0-87933-024-4. $55. LC 72-95135.   477
  A collection of 32 previously published papers dealing with various aspects of slope morphology. The reader is led through the history of theories of slope development and introduced to the processes and environmental problems associated with slopes.

Selby, Michael J. Hillslope materials and processes. New York: Oxford University Press, 1982. 264 p. ISBN 0-19-874126-X. $46. LC 81-22458.   478
  Emphasizes rock and soil resistance to erosion in the context of slope development more than other books on the topic.

Young, Anthony. Slopes. 2nd ed. London; New York: Longman, 1975. 288 p. (1st ed., Edinburgh: Oliver and Boyd, 1972. 288 p.). ISBN 0-582-48433-2. LC 75-25521.   479
  A useful and less technical text emphasizing processes and slope form.

Geomorphology

F. EOLIAN

Bagnold, Ralph A. The Physics of blown sand and desert dunes. 2nd ed. London: Chapman and Hall, 1954. 265 p. (1st ed., 1941. 265 p.). ISBN 0-412-10270-6. £19.00 LC 54-34292.   480
  A major reference work presenting major theories of aeolian geomorphology. Emphasizes the movement of sand sized particles, wind ripples, and dunes.

Greeley, Ronald, and Iverson, James D. Wind as a geological process: Earth, Mars, Venus, and Titan. (Planetary Science Series). New York, London: Cambridge University Press. 1984. 335 p. ISBN 0-521-24385-8. LC 83-18878.   481
  The most recent review of aeolian processes. Our understanding of wind-related geomorphic activity on Earth is used to compare aeolian processes and features on Mars and Venus (with speculations for Titan) using results from U.S. and Soviet planetary missions. A highly useful reference for Earth-bound research as well.

Péwé, Troy L. Desert dust: origin, characteristics, and effect on man. (GSA special paper no. 186). Boulder, CO: Geological Society of America, 1981. 303 p. ISBN 0-8137-2186-5. $30. LC 81-7031.   482
  A volume of 21 papers on blowing dust as a natural process and environmental problem. The papers (and their references) serve as useful sources of information on numerous aspects of blowing dust and desertification.

G. GLACIAL

Embleton, Clifford, and King, Cuchlaine, A.M. Glacial and periglacial geomorphology. 2nd ed. London: Edward Arnold; New York: Halsted Press of John Wiley, 1975. (1st ed., London: Edward Arnold; New York: St. Martin's Press, 1968. 608 p.).
  Vol. 1: **Glacial geomorphology.** 573 p. ISBN 0-470-23892-5. ISBN 0-470-23893-3. (pbk). LC 75-14188.
  Vol. 2: **Periglacial geomorphology.** 203 p. ISBN 0-470-23894-1. ISBN 0-470-23895-X. (pbk). LC 75-14187.   483
  A competent and comprehensive text. Well illustrated with an extensive bibliography.

Flint, Richard F. Glacial and Quaternary geology. New York: John Wiley and Sons, 1971. 892 p. ISBN 0-471-26435-0. $47.95. LC 74-141198.   484
  One of the most complete volumes available on the subject. Expertly done with a large bibliography.

Paterson, W.S.B. The Physics of glaciers. 2nd ed. Oxford, U.K.; New York: Pergamon, 1981. 380 p. (1st ed., 1969. 250 p.) ISBN 0-08-024004-6. $53. ISBN 0-08-024005-4. $17.50. (pbk). LC 81-181869.   485
  A quantitative review of the basics of glacial processes.

Sugden, D.E. and John, B.S. Glaciers and landscape: a geomorphological approach. New York: John Wiley and Sons, 1976. 376 p. ISBN 0-470-15113-7. $24.95. LC 76-11014.   486
  Description of processes and forms associated with Pleistocene glaciation.

H. COASTS

Bascom, Willard. Waves and beaches: the dynamics of the ocean surface. New York: Doubleday (Anchor Press), 1980. 267 p. (1st ed., 1964. 267 p.). ISBN 0-385-14844-5. $8.95. (pbk). LC 79-7038.   487•
  A non-technical book dealing with coastal geomorphology. An easily read introduction to the topic.

Davies, John L. Geographical variation in coastal development. 2nd ed. London; New York: Longman, 1980. 204 p. (1st ed., Edinburgh: Oliver and Boyd; New York: Hafner, 1972. 204 p.). ISBN 0-582-49006-5. $16.95. LC 76-30821.   488•

Coastal processes are presented within the context of explaining why coasts vary morphologically over the earth. A relatively non-technical text on coastal processes.

King, Cuchlaine A.M. Beaches and coasts. 2nd ed. London: Edward Arnold, New York: St. Martin's Press, 1972. 570 p. (1st ed., 1960. 403 p.). ISBN 0-312-07035-7. $45. LC 74-187106.   489

A thorough text on coastal geomorphology.

Komar, Paul D. Beach processes and sedimentation. Englewood Cliffs, NJ: Prentice-Hall, 1976. 464 p. ISBN 0-13-072595-1. $39.95. LC 75-44005.   490

A reasonably technical text on coastal geomorphology.

Komar, Paul D., ed. CRC handbook of coastal processes and erosion. Boca Raton, FL: CRC Press, 1983. 305 p. ISBN 0-8493-0208-0. $70. LC 83-3773.   491

A collection of papers that provides a useful introduction to various aspects of coastal geomorphology. Includes basin processes, research techniques, and environmental problems associated with coastal processes. Also lists a large bibliography. As a handbook, it serves as a useful reference for professionals.

I. SOIL

Hunt, Charles B. Geology of soils: their evolution, classification, and uses. San Francisco, CA: W.H. Freeman, 1972. 344 p. ISBN 0-7167-0253-3. $34.95. LC 71-158739.   492•

A spatial perspective and systematic approach to soils.

J. THE FIELD AND ITS HISTORY

Chorley, Richard J.; Dunn, Anthony J.; and Beckinsale, Robert P. The History of the study of landforms; geomorphology: v. 1. Geomorphology before Davis. London: Methuen; New York: John Wiley and Sons, 1964. 678 p. LC 64-6231.   493

A comprehensive review of the early development of geomorphology.

Chorley, Richard J.; Dunn, Anthony J.; and Beckinsale, Robert P. The History of the study of landforms; or the development of geomorphology, v. 2. The life and work of William Morris Davis. London: Methuen; New York: John Wiley and Sons, 1973. 864 p. ISBN 0-416-26890-0. $75. LC 64-6231 rev.   494

A biography of Davis that links his career with the development of geomorphology.

Gregory, Kenneth J. The Nature of physical geography. London: Edward Arnold, June, 1985 (Scheduled date of publication). 272 p. ISBN 0-7131-6431-X.   495

A treatment of the recent history of physical geography, paralleling Johnston's treatment in Geography and Geographers *of human geography.*

King, P.B. and Schumm, Stanley A. The Physical geography (geomorphology) of W.M. Davis. Norwich, U.K.: Geo Books, 1980. 217 p. ISBN 0-86094-046-2. $37.   496

A collection of Davis' lecture notes revealing a thought-provoking, modern perspective. Includes a complete bibliography.

Pitty, Alistar F. The Nature of geomorphology. London: Methuen, 1982. 161 p. ISBN 0-416-32110-0. $8.95. LC 82-7921.   497•

Brief review of major themes in the field and an analysis of its connection with geography and geology.

Climatology

Pyne, S.J. Grove Karl Gilbert: a great engine of research. (History of science, no. 2). Austin, TX: University of Texas Press, 1980. 312 p. ISBN 0-292-72719-4. $22.50. LC 80-13881.

498

Biography of a distinguished American geomorphologist who became a legend in the field.

Dackombe, Roger V. and Gardiner, V. Geomorphological field manual. London; Boston, MA: George Allen and Unwin, 1983. 254 p. ISBN 0-04-551061-X. ISBN 0-04-55102-8 $14.95. (pbk). LC 82-6767.

499

Describes basic sampling and measurement techniques. Useful in field research.

Goudie, Andrew, ed. Geomorphological techniques. (British Geomorphological Research Group.) London; Winchester, MA: George Allen and Unwin, 1981. 395 p. ISBN 0-04-551042-3. $60. ISBN 0-04-551043-1. $29.95. (pbk). LC 80-41361.

500

Wide-ranging review that is most useful as a guide to the literature rather than providing the specifics of particular techniques.

Wright, John. Ground and air survey for field scientists. Oxford, U.K.: Clarendon Press; New York: Oxford University Press, 1982. 327 p. ISBN 0-19-857560-2. $69. ISBN 0-19-857601-3. $26.95. (pbk). LC 82-4155.

501•

Provides basic instructions for a physical survey of the land surface.

3. CLIMATOLOGY
John E. Oliver

American Meteorological Society bulletin. 1-(1920-). Monthly. American Meteorological Society, 45 Beacon St., Boston, MA 02108. ISSN 0003-0007.

502

Official organ of the Society, devoted to editorials, survey articles, professional and membership news, announcements, and society activities.

Archives for meteorology, geophysics, and bioclimatology. (Archiv für Meteorologie, Geophysik, und Bioklimatologie). Series B: Climatology, environmental meteorology, radiation research. 1-(1948-). Quarterly. Springer Verlag, Mölkerbastei 5, P.O. Box 367 A-1011. Wien, Austria, or 175 Fifth Avenue, New York, NY 10010. ISSN 0066-6424.

503

One of the most important journals for modern energy balance climatology. Most articles are in English.

Climatic change: an interdisciplinary, international journal devoted to the description, causes, and implications of climatic change. 1-(1977-). Quarterly. D. Reidel Publishing Co., Box 17, 3300AA Dordrecht, The Netherlands. ISSN 0165-0009.

504

An interdisciplinary, international journal devoted to the description, causes, and implications of climatic change.

Geo Abstracts. Part B. Climatology and hydrology. 1- (1966-). Bimonthly. Regency House, 34 Duke Street, Norwich NR3 3AP, U.K. ISSN 0-305-1897.

505

Journal of climate and applied meteorology. 1-(1962-). American Meteorological Society, 45 Beacon St., Boston, MA 02108. Formerly as Journal of Applied Meteorology, 1962-1982.ISSN 0733-3021.

506

A medium for the publication of research concerned with the application of the atmospheric sciences to the health, economy, and general well-being of the human community, e.g., air pollution meteorology, weather modifications, agricultural meteorology, climate and its impact, paleoclimatology, energy resources, etc.

Journal of climatology. (Royal Meteorological Society). 1-(1981-). Bimonthly. John
Wiley and Sons, Baffins Lane, Chichester, Sussex P019 1UD, England. ISSN 0196-
1748.   507
   An international journal providing a wide range of climatic research papers.

Meteorological and geoastrophysical abstracts. 1-(1950-). Monthly. American Meteorological Society, 45 Beacon St., Boston, MA 02108. ISSN 0026-1130.   508
   *Surveys a wide range of publications on meteorology and others which include me-
teorological and climatological studies, both English and foreign language publications. A basic research tool, it will have greatest value to the more sophisticated
student.*

National weather digest. 1-(1976-). Quarterly. National Weather Association, 4400
Stamp Road, Room 404, Temple Hills, MD 20748. ISSN 0271-1052.   509•
   *The official publication of the National Weather Association, devoted to articles,
studies, correspondence, and official news of the Association. Especially intended for
the operational meteorologist.*

Weatherwise: popular weather magazine. 1-(1948-). Bimonthly. Heldref Publications, 4000 Albemarle Street, N.W., Washington, DC 20016. ISSN 0043-1672.   510•
   *A magazine about weather written for students and interested laymen. Reports em-
phasize storms, weather phenomena, seasons of heavy snowfall, and other ususual
conditions.*

B. ATLASES

Landsberg, Helmut E., et al. World maps of climatology. 2nd ed. Berlin; New York:
Springer, 1965. 28 p. 5 maps. LC Map 65-10.   511
   *Particularly valuable for maps of solar radiation and sunshine. (Edited under the
sponsorship of the Heidelberger Akademie der Wissenschaften by Ernst Rodenwaldt
and H.J. Jusatz).*

U.S. Environmental Data Service. Weather atlas of the United States. 2nd ed. Detroit, MI: Gale Research, 1975. 262 p. (1st ed., Climatic atlas of the United States. Washington, DC: Environmental Science Services Administration, Environmental Data Service, 1968. 80 p.). ISBN 0-81031-10481. $76.00. LC 74-11931.   512•
   *Contains 271 maps and 15 tables depicting the climate of the U.S. in terms of the
distribution of climatic measures such as precipitation, wind, barometric pressure,
relative humidity, mean, normal, or extreme values of temperature, etc.*

U.S. Quartermaster Research and Engineering Center (U.S. Army Natick Laboratories). Natick, MA., Earth Sciences Division. Atlas of mean monthly temperatures.
(Technical paper ES-10). Natick, MA, 1964. 72 sheets of colored maps. LC 66-65797.   513
   *Excellently prepared and printed maps, in color, on a good atlas scale of the conti-
nents. Isotherms at intervals of 10°C (18°F).*

C. DATA SOURCES

U.S. National Oceanic and Atmospheric Administration. Climates of the states: a
practical reference containing basic climatological data of the U.S. Port Washington,
NY: Water Information Center, 1974. 2 v. 1004 p. ISBN 0-912394-099. $45. LC 73-
93482.   514
   *A convenient reference of general climatic conditions within individual states,
culled from the monthly government publication Climatological data. Includes specific tables of temperature, precipitation, and miscellaneous information. Useful to
the layman as well as the researcher.*

Climatology

Butson, Keith D. and Hatch, Warren L. A Selective guide to climatic data sources. (Key to meteorological records documentation, no. 4.11). Washington, D.C.: National Oceanic and Atmospheric Administration, Environmental and Information Service. 1979. 142 p. LC 80-602294.   515

A basic guide to available climatic data.

Rudloff, Willy B. World climates: with tables of climatic data and practical suggestions. Stuttgart, West Germany: Wissenschaftliche Verlagsgesellschaft, 1981. ISBN 3-8047-0509-X. ISBN 0-9960099-X. $96. Distributed in U.S. by Heyden and Son, Inc., 247 S. 41st Street, Philadelphia, PA 19104. ISSN 0077-6157.   516•

A most useful data source with listings for some 1474 stations. Data provided include guides to clothing requirements and climatic stress.

D. SERIES

Landsberg, Helmut Erich, editor-in-chief. World survey of climatology. New York: Elsevier, 1969- . 15 v. ISBN 0-444-40734-0. LC 78-477739.
- Vol. 1: **General climatology.** Flohn, H., ed.
- Vol. 2: **General climatology.** Flohn, H., ed. 1969. 266 p. ISBN 0-444-40704-9. $102.25. LC 68-12480.
- Vol. 3: **General climatology.** Landsburg, H.E., ed. 1981. 408 p. ISBN 0-444-41776-1. $117. LC 78-103353.
- Vol. 4: **Climate of the free atmosphere.** Rex, D.F., ed. 1969. 450 p. ISBN 0-444-40703-0. $127.75. LC 68-12478.
- Vol. 5: **Northern and Western Europe.** Wallén, C.C., ed. 1970. 253 p. ISBN 0-444-40705-7. $106.50. LC 68-12479.
- Vol. 6: **Central and Southern Europe.** Wallén, C.C., ed. 1977. 248 p. ISBN 0-444-40734-0. LC 76-46572.
- Vol. 7: **Soviet Union.** Lydolph, P.E., ed. 1977. 443 p. ISBN 0-444-41516-5. $121.25. LC 76-46298.
- Vol. 8: **Northern and Eastern Asia.** Arakawa, H., ed. 1969. 248 p. ISBN 0-40704-9. $102.25. LC 68-12480.
- Vol. 9: **Southern and Western Asia.** Arakawa, H. and Takahashi, K., eds. 1981. 333 p. ISBN 0-444-41861-X. $117. LC 81-1046.
- Vol. 10: **Africa.** Griffiths, J.F., ed. 1972. 604 p. ISBN 0-444-40893-2. $157.50. LC 72-135485.
- Vol. 11: **North America.** Bryson, R.A. and Hare, F.K., eds. 1974. 420 p. ISBN 0-41062-7. $112.75. LC 78-477739.
- Vol. 12: **Central and South America.** Schwerdtfegar, W., ed. 1976. 532 p. ISBN 0-444-41271-9. $127.75. LC 78-103353.
- Vol. 13: **Australia and New Zealand.** Gentilli, J., ed. 1971. 405 p. ISBN 0-444-40827-4. $134 p. LC 71-103354.
- Vol. 14: **Polar regions.** Orvig, S., ed. 1970. 370 p. ISBN 0-444-40828-2. $127.75. LC 79-103355.
- Vol. 15: **Oceans.** Van Loon, H., ed. 1984. 716 p. ISBN 0-444-41337-5. $175.   517

This multivolume work comprises volumes on climatic processes and on world regions. Each volume has a different editor and consists of sections written by leading authorities.

E. GENERAL WORKS

Barry, Roger G., and Chorley, Richard J. Atmosphere, weather, and climate. 4th ed. London; New York: Methuen, 1982. 407 p. (1st ed., 1968. 319 p.) ISBN 0-416-33690-6. $33. LC 82-7873.   518

A popular book that deals with principles of both meteorology and climatology. Useful chapters on regional climates and climatic variability.

Boucher, Keith. Global climates. London: English Universities Press; New York: John Wiley, 1975. 326 p. ISBN 0-340-15493-4. LC 75-312416. LC 75-29854.   519
  A well-illustrated, descriptive account of world climates. Subsequent to an overview of processes, treatment is in terms of tropical, mid-latitude, and polar regions.

Critchfield, Howard J. General climatology. 4th ed. Englewood Cliffs, NJ: Prentice-Hall, 1983. 453 p. (1st ed., 1960. 465 p.) ISBN 0-13-349217-6. $31.95. LC 82-7536.   520•
  A standard introductory text in non-technical terms. The three-part approach deals with physical, regional, and applied climatology.

Griffiths, John F., and Driscoll, Dennis M. Survey of climatology. Columbus, OH: Charles E. Merrill, 1982. 358 p. ISBN 0-675-09994-3. $27.95. LC 81-81252.   521
  A basic climatology text that includes good chapters on analytic methods and applied climatology.

Hobbs, John E. Applied climatology: a study of atmospheric resources. Boulder, CO: Westview Press, 1980. 224 p. ISBN 0-89158-697-0. $34. LC 79-5287.   522
  Content is based upon the concept of the atmosphere as a resource. Deals with impacts upon human activities and the ways in which the use and knowledge of the resource can be improved.

Mather, John R. Climatology: fundamentals and applications. New York: McGraw-Hill, 1974. 412 p. ISBN 0-07-040891-2. LC 73-23082.   523
  An overview of climatic elements is preface to many aspects of applied climatology including water resources, agriculture, and health.

Oliver, John E. Climate and man's environment: an introduction to applied climatology. New York: John Wiley, 1973. 517 p. ISBN 0-471-65338-1. $38.95. LC 73-5707.   524
  A broad treatment of climate as it relates to environment, human activities, and climatic change.

Oliver, John E. and Hidore, John J. Climatology: an introduction. Columbus, OH: Charles E. Merrill, 1984. 381 p. ISBN 0-675-20144-6. $24.95. LC 83-63079.   525•
  An introductory text with three parts: physical-dynamic, regional, and past and future climates. Some interesting applied studies are included.

Roberts, Walter Orr, and Lansford, Henry. The Climate mandate. San Francisco, CA: W.H. Freeman, 1979. 197 p. ISBN 0-7167-1054-4. $20.95. LC 78-25677.   526•
  Written in popular style, this book relates climatic variations to food production and population growth.

Sellers, William D. Physical climatology. Chicago, IL: University of Chicago Press, 1965. 272 p. ISBN 0-226-74699-2. $20. LC 65-24983.   527
  A classic treatment of energy-mass budgets and physical atmospheric processes. For advanced students.

Smith, Keith. Principles of applied climatology. New York: John Wiley, 1975. 233 p. ISBN 0-470-80169-7. $19.95. LC 74-10976.   528
  A systematic study of weather and climate as related to human activities.

Trewartha, Glenn T. and Horn, Lyle H. An Introduction to climate. 5th ed. New York: McGraw-Hill, 1980. 416 p. (1st ed., 1937. 373 p.) ISBN 0-07-065152-3. $29.95. LC 79-14203.   529•
  An introduction to climate that provides a good overview of the field. The non-technical treatment permits easy understanding.

Wallace, John M. and Hobb, Peter V. Atmospheric sciences: an introductory survey. New York: Academic Press, 1977. 467 p. ISBN 0-12-732950-1. $28. LC 76-19493.   530
  Somewhat advanced introduction to physical and dynamic meteorology.

Climatology

Barrett, Eric C. Climatology from satellites. London; New York: Methuen, 1974. 418 p. ISBN 0-416-72150-8. $17. (pbk). LC 74-171320.   531
  A standard text on the subject.

Barry, Roger G. and Perry, A.H. Synoptic climatology: methods and applications. London; New York: Methuen, 1973. 555 p. ISBN 0-416-08500-8. $58. LC 73-178642.   532
  A definitive systematic treatment of atmospheric circulation patterns. Covers the synoptic climatology at global, regional, and local scales.

Budyko, Mikhail Ivanovich. The Earth's climate: past and future. (International geophysics series, vol. 29). New York; London; Toronto: Academic Press, 1982. 307 p. ISBN 0-12-139460-3. $39.50. LC 81-17673.   533
  A treatment of energy and mass exchanges presented in the framework of past and future climates. Somewhat advanced.

Chang, Jen Hu (Ching-hu). Climate and agriculture: an ecological survey. Chicago, IL: Aldine, 1968. 304 p. LC 67-27389.   534
  An extensive but somewhat dated study of microclimates dealing largely with respect to plants. Draws upon basic concepts of climatology and agronomy.

Geiger, Rudolph. Climate near the ground. Translated from German by Scripta Technica, Inc. Cambridge, MA: Harvard University Press, 1965. 611 p. ISBN 0-674-13500-8. $35. LC 64-23191. (Das Klima der bodennahen Luftschicht: ein Lehrbuch der Mikroklimatologie, 4th ed. Braunschweig: F. Vieweg, 1961. 646 p. 1st ed., 1927. 246 p).   535
  A study of microclimatology that is now considered a classic on the subject.

Lamb, Hubert H. Climate: present, past, and future. London; New York: Methuen, 1972. V.1. Fundamentals and climate now, 1972. 613 p. ISBN 0-416-11530-6. $110. LC 72-183843. V.2. Climatic history and the future, 1977. 835 p. ISBN 0-416-11540-3. $140.   536
  A valuable book that covers many aspects of physical and dynamic climatology. Extensive bibliography.

Landsberg, Helmut E. The Urban Climate. (International geophysics series vol. 28). New York; London; Toronto: Academic Press, 1981. 275 p. ISBN 0-12-435960-4. $32.50. LC 80-2766.   537
  A survey of the climate of cities. While some mathematics are used, there is much of interest for the non-specialist exploring this subject.

Lockwood, John G. Causes of climate. New York: Wiley, 1979. 260 p. ISBN 0-470-26657-0. ISBN 0-470-26658-9. (pbk). LC 78-31907.   538
  An introduction to the dynamics of climate. Provides good coverage of energy exchanges and circulation patterns.

Miller, David H. Energy at the surface of the Earth: an introduction to the energetics of ecosystems. New York; London: Academic Press, 1981. 516 p. ISBN 0-12-497150-4. $54.50. student ed. $27.50. LC 80-989.   539
  A comprehensive analysis of radiant energy fluxes and energy transfer in crop, forest, and urban ecosystems. Case studies demonstrate interactions.

Nieuwolt, Simon. Tropical climatology: an introduction to the climates of low latitudes. London; New York: John Wiley and Sons, 1977. 207 p. ISBN 0-471-99406-5. £13.05. LC 76-13454.   540
  Descriptive climatology of low latitudes. A systematic approach with several chapters on radiation-temperature, circulation, and moisture.

Oke, T.R. Boundary layer climates. New York: John Wiley and Sons, 1978. 372 p. ISBN 0-470-99364-2. $37. LC 77-25266.   541
  An advanced text that provides a rigorous examination of climate near the ground. Contains a useful appendix on microclimatological field observations.

Trewartha, Glenn T. The Earth's problem climates. 2nd ed. Madison, WI: University of Wisconsin Press, 1981. 340 p. (1st ed., 1961. 334 p.) ISBN 0-299-08230-X. $25. LC 80-5120.   542
  An analysis of the causes of unusual climates throughout the world. Organized by continents.

4. BIOGEOGRAPHY*

Kathleen C. Parker and Albert J. Parker

* In this section the editorial committee reduced the number of entries recommended by the contributors.

A. SERIAL

Journal of biogeography. 1- (1974-). Bimonthly. Blackwell Scientific Publications, Osney Mead, Oxford, OX2 0EL, U.K. ISSN 0305-0270.　　　543
Articles on all aspects of biogeography. Book reviews.

B. ENCYCLOPEDIAS

Nowak, Ronald M., and Paradiso, John L. Walker's mammals of the world. 4th ed., Baltimore, MD: Johns Hopkins University Press, 1983. 2 v. 1472 p. (1st ed. Walker, Ernest P., and others. Mammals of the World., 1964. 3 v.). ISBN 0-8018-2525-3. $65. LC 82-49056.　　　544•
Authority on mammalian taxonomy with life history and range information for mammal species throughout the world. Many species shown in black-and-white photographs. Arranged taxonomically with index at the end. Up-to-date bibliography includes more than 3000 entries.

Terres, John K. The Audubon Society encyclopedia of North American birds. New York: Alfred A. Knopf, 1980. 1280 p. ISBN 0-394-46651-9. $60. LC 80-7616.　　　545•
Up-to-date encyclopedia that presents excellent coverage of 625 topics pertaining to birds (e.g., flight; egg-laying; censuses) and life history information about 847 bird species seen in North America. More than 5500 alphabetically arranged entries, 875 color photographs, 800 drawings and diagrams, and approximately 4,000 references that span 300 years of publications.

C. STUDIES OF SPECIFIC AREAS

Barbour, Michael G. and Major, Jack, eds. Terrestrial vegetation of California. New York: Wiley-Interscience, 1977. 1002 p. ISBN 0-471-56536-9. $96.50. LC 76-53769.　　　546
A rigorously edited technical monograph comprising several introductory chapters on such topics as climate, flora, and the vegetational history of California, followed by twenty chapters treating the principal vegetation types of California, all written by noted authorities and regional specialists. Numbers of citations vary considerably across the topics. Virtually every work dealing with California vegetation up to the time of publication is cited. Küchler's map of the natural vegetation of California, which accompanies the monograph, is especially valuable.

Barrett, John W., ed. Regional silviculture of the United States. 2nd ed. New York: John Wiley and Sons, 1980. 551 p. (1st ed., 1962. 610 p.) ISBN 0-471-05645-6. $42.50. LC 80-17129.　　　547
An assessment of biological, physical, and economic qualities of the forested regions of the U.S. After introductory material on general patterns of forest occurrence in the U.S. and their relationship to climate and geomorphology, each region is treated by a contributor familiar with the silviculture of that area. Many photographs, maps, and other figures are included. References at ends of the regional sections.

Bormann, F. Herbert, and Likens, Gene E. Pattern and process in a forested ecosystem: disturbance, development, and the steady state based on the Hubbard Brook ecosystem study. New York: Springer-Verlag, 1979. 253 p. ISBN 0-387-90321-6. $20.80. LC 78-6015.　　　548

A unique treatise on forest ecosystem processes based on data from an intensively studied watershed within the White Mountains of New Hampshire. Topics covered include energetics, hydrology, biogeochemistry, effects of clear-cutting, landscape management, and steady-state and vegetation dynamics. Contains more than 350 references.

Curtis, John T. The Vegetation of Wisconsin: an ordination of plant communities. Madison, WI: University of Wisconsin Press, 1959. 657 p. ISBN 0-299-01940-3. $27. LC 72-183519.

An enduring contribution. Both a monograph on Wisconsin's vegetation patterns (serving admirably as a benchmark for other regional vegetation treatments) and a catalyst for the emergence of ordination techniques and other multivariate statistical approaches to the examination and interpretation of vegetation data. Nicely illustrated with 66 well-chosen photographic plates depicting vegetation types, 53 figures and diagrams, 30 tables in the text, and 82 pages of appendices with data and maps. More than 500 bibliographic entries.

549

Di Castri, Francesco, and Mooney, Harold A., eds. Mediterranean-type ecosystems: origin and structure. (Ecological studies: analysis and synthesis). New York; Berlin, West Germany: Springer-Verlag, 1973. 405 p. ISBN 0-387-06106-1. $52. LC 72-95688.

Collection of research papers by 24 noted international scholars that focuses on specific themes including the climatology, geomorphology, soils, vegetation patterns, animal ecology, and influences of human activities on the biota of Mediterranean-type regions. Bibliographies at the ends of the papers are generally rather detailed and the book is well-indexed. Eighty-eight maps and other figures are included.

550

Forman, Richard T. T., ed. Pine barrens: ecosystem and landscape. New York: Academic Press, 1979. 601 p. ISBN 0-12-263450-0. $55. LC 79-9849.

An in-depth synthesis of the various components and interactions within the New Jersey pine barrens ecosystem including the physical setting (climate, hydrology, soils, geology), vegetation patterns (communities and some species), animal communities, and human populations and their use of resources. Many figures, plates, maps, and tables are included as are references at the ends of the 33 chapters.

551

Hastings, James Rodney, and Turner, Raymond M. The Changing mile: an ecological study of vegetation change with time in the lower mile of an arid and semiarid region. Tucson, AZ: University of Arizona Press, 1965. 316 p. ISBN 0-8165-0014-2. $25. LC 65-25019.

Study of vegetation change in selected oak woodlands, desert grasslands, and desert shrublands of Arizona and Mexico through the use of photographic comparisons. Most original photos date from the late 1800's or early 1900's; most repeats were taken in the late 1960's. In addition to the 97 photo pairs and 14 maps and figures, a thorough discussion of factors affecting vegetation change is presented. A dated bibliography is included.

552

Ives, Jack D. and Barry, Roger G., eds. Arctic and alpine environments. London; New York: Methuen, 1974. 980 p. ISBN 0-416-65980-2. $134. LC 74-2673.

A detailed, comprehensive treatment of the world's arctic and alpine environments including climate, hydrology, ice cover and permafrost extent, paleoenvironments, present and past biota, geomorphology, soils, and human interactions with these environments. An interdisciplinary project including 31 specialists, it is richly illustrated with numerous thematic maps and diagrams. Includes a detailed index and glossary with more than 2500 bibliographical entries arranged at the ends of the 18 chapters.

553

Janzen, Daniel H. Costa Rican natural history. Chicago, IL: University of Chicago Press, 1983. 832 p. ISBN 0-226-39332-1. $50. LC 82-17625.   554

A monumental storehouse of information about the natural environment of Costa Rica that discusses climatology, soils, geology, plants (native and agricultural) and animals (including checklists of each), paleogeography, biotic history, and historical field biology. An interdisciplinary work by noted authorities in their respective fields. References are arranged at the ends of topical chapters and many figures, maps, and photographs are included.

Leigh, Egbert G., Jr.; Rand, A. Stanley; and Windsor, Donald M., eds. The Ecology of a tropical forest: seasonal rhythms and long-term changes. Washington, DC: Smithsonian Institution Press, 1983. 468 p. ISBN 0-87474-601-9. $25. LC 82-600181.   555

A detailed analysis of the relationships between plants and animals within tropical forest ecosystems on Barro Colorado Island, Panama, with focus on seasonal changes in the physical environment and the associated influences on plants and animals. Numerous figures and maps, up-to-date references topically arranged, and much original data in tabular form. There are 37 contributors to the volume.

Mani, M.S., ed. Ecology and biogeography in India. (Monographiae biologicae series vol. 23). The Hague, Netherlands: Dr. W. Junk, 1974. 773 p. ISBN 9-06-193075-8. $109. LC 74-163908.   556

A detailed consideration of vegetation patterns in India with opening chapters considering the physical geography and flora of the subcontinent, followed by a regional treatment of vegetation features and faunal patterns. References are arranged by chapter; richly illustrated with more than 160 figures; and are well indexed, especially floristically.

Sauer, Jonathan D. Cayman Islands seashore vegetation: a study in comparative biogeography. (University of California publications in geography series vol. 25). Berkeley, CA: University of California Press, 1983. 166 p. ISBN 0-520-09656-8. $14.25. LC 82-2608.   557

More than a detailed investigation of shoreline vegetation patterns and processes on a specific island, this monograph reflects years of experience in tropical strand ecosystems and integrates physical, biological, and cultural influences in a compelling manner. Includes 200 literature citations, 13 tables of original data, 25 vegetation profiles, and 39 photographic plates.

Walter, Heinrich. Ecology of tropical and subtropical vegetation. Translated from German by D. Mueller-Dombois. Edinburgh, U.K.: Oliver and Boyd, 1971. 539 p. ISBN 0-05-002130-3. LC 72-189214. (Die tropischen und subtropischen Zonen. Stuttgart: G. Fischer, 1964).   558

A treatment of tropical and subtropical vegetation with emphasis on functional and ecological relationships. General plant features characteristic throughout each major vegetation type are stressed although a review of the world's major examples is included within each section. Original experimental data and numerous maps and figures are presented. References are arranged at the ends of the chapters. This is the first book of a two-volume set, "Die Vegetation der Erde," authored by Walter. The second volume, entitled "Die Gemässigten und Arktischen Zonen," (1966) has not been translated into English.

D. SERIES

Golley, Frank B., ed. Benchmark papers in ecology. Stroudsburg, PA: Dowden, Hutchinson, and Ross, 1974-1983. 13 v.
>   Vol. 1: **Cycles of essential elements.** Pomeroy, Lawrence R., ed. 1974. 373 p. ISBN 0-87933-129-1. $51. LC 74-4252.
>   Vol. 2: **Behavior as an ecological factor.** Davis, David E., ed. 1974. 390 p. ISBN 0-87933-132-1. $51. LC 74-3006.
>   Vol. 3: **Niche: theory and application.** Whittaker, Robert H. and Levin, Simon A., eds. 1975. LC 74-23328.
>   Vol. 4: **Ecological energetics.** Wiegert, Richard G., ed. 1976. 457 p. ISBN 0-470-94361-0. $63. LC 75-30762.
>   Vol. 5: **Ecological succession.** Golley, Frank B., ed. 1977. 373 p. ISBN 0-87933-256-5. $44.50. LC 76-52930.
>   Vol. 6: **Phytosociology.** McIntosh, Robert P., ed. 1978. 388 p. ISBN 0-87933-312-X $41.50. LC 77-20258.
>   Vol. 7: **Population regulation.** Tamarin, Robert H., ed. 1978. 391 p. ISBN 0-87933-324-3. $46. LC 77-16178.
>   Vol. 8: **Patterns of primary production in the biosphere.** Lieth, Helmut F.H., ed. 1978. 342 p. ISBN 0-87933-327-8. $46. LC 78-18691.
>   Vol. 9: **Systems ecology.** Shugart, H.H. and O'Neill, R.V., eds. 1979. 368 p. ISBN 0-87933-347-2. $43. LC 79-970.
>   Vol. 10: **Tropical ecology.** Jordon, Carl F., ed. 1981. 356 p. ISBN 0-87933-398-7. $45. LC 81-4260.
>   Vol. 11: **Dispersal and migration.** Lidicker, William Z., Jr. and Caldwell, Roy L., eds. 1982. 311 p. ISBN 0-87933-435-5. $39.50. LC 82-9326.
>   Vol. 12: **Origins of human ecology.** Young, Gerald L., ed. 1983. 415 p. ISBN 0-87933-104-6. $48. LC 83-39.
>   Vol. 13: **Diversity.** Patrick, Ruth, ed. 1983. 413 p. ISBN 0-87933-420-7. $42.50. LC 83-4365.   559

Each volume in this series seeks to select and reprint the key works published on a particular theme. The papers are organized topically with editorial comments included for groups of papers sharing a common focus. Inclusion of older classic papers facilitates an appreciation of the historical perspective and context from which current endeavors evolved.

Goodall, David W. ed. Ecosystems of the world. New York: Elsevier Scientific, 1977- . To include 29 volumes. Volumes published to 1984 include:
>   Vol. 1: **Wet coastal ecosystems.** Chapman, V.J., ed. 1977. 428 p. ISBN 0-444-41560-2. $80.75. LC 77-342.
>   Vol. 5: **Temperate deserts and semi-deserts.** West, N.E., ed. 1983. 522 p. ISBN 0-444-41931-4. $170.25. LC 80-128760.
>   Vol. 9: **Heathlands and related shrublands.** Specht, R.L., ed. 1980. 498 p. ISBN 0-444-41810-5. $85. LC 79-13938.
>   Vol. 10: **Temperate broad-leaved evergreen forests.** Ovington, J.D., ed. 1983. 241 p. ISBN 0-444-42091-6. $83. LC 82-8891.
>   Vol. 11: **Mediterranean-type shrublands.** DiCastri, F.; Goodall, D.W.; and Specht, R.L., eds. 1981. 643 p. ISBN 0-444-41858-X. $127.75. LC 80-15879.
>   Vol. 12: **Hot desert and arid shrublands.** Evenari, M.; Noy-Meir, I.; and Goodall, D.W., eds. 1984. ISBN 0-444-42296-X. LC 84-1486.
>   Vol. 13: **Tropical savannas.** Bourliere, F., ed. 1983. 730 p. ISBN 0-444-42035-5. $197.75. LC 81-19415.
>   Vol. 14: **Tropical rainforest ecosystems: structure and evaluation.** Golley, Frank B., ed. 1982. 381 p. ISBN 0-444-41986-1. $112.75. LC 81-7861.
>   Vol. 23: **Lakes and reservoirs.** Taub, F.B., ed. 1983. 643 p. ISBN 0-444-42059-2. $181. LC 82-4958.
>   Vol. 26: **Estuaries and enclosed seas.** Ketchum, B.H., ed. 1983. 500 p. ISBN 0-444-41921-7. LC 81-9853.   560

Monumental compendium of information about processes in and characteristics of the major biomes and selected localized types of ecosystems of the world. Each volume represents collective efforts of international experts in their respective areas. Each includes extensive bibliographies, original data arranged tabularly and graphically, many maps, photographs, and other illustrations. The most thorough treatment of its kind to date.

United States/International Biological Program. Biological program synthesis series. Stroudsburg, PA: Dowden, Hutchinson, and Ross, 1976-1981. 15 v.

Vol. 1: **Man in the Andes: a multidisciplinary study of high-altitude Quechua.** Baker, Paul T. and Little, Michael A., eds. 1976. 482 p. ISBN 0-470-15153-6. $59.60. LC 76-17025.

Vol. 2: **Chile-California Mediterranean scrub atlas: a comparative analysis.** Thrower, Norman J.W. and Bradbury, David E., eds. 1979. 237 p. ISBN 0-87933-211-9. $59.50. LC 76-56094.

Vol. 3: **Convergent evolution in warm deserts: an examination of strategies and patterns in deserts of Argentina and the United States.** Orians, Gordon H. and Solbrig, Otto T., eds. 1977. 333 p. ISBN 0-87933-276-X. $55.50. LC 76-56261.

Vol. 4: **Mesquite: its biology in two desert scrub ecosystems.** Simpson, B.B., ed. 1977. 250 p. ISBN 0-87933-278-6. $49. LC 76-58889.

Vol. 5: **Convergent evolution in Chile and California: Mediterranean climate ecosystems.** Mooney, Harold A., ed. 1977. 224 p. ISBN 0-87933-279-4. $45. LC 77-1884.

Vol. 6: **Creosote bush: biology and chemistry of Larrea in New World deserts.** Mabry, T.J.; Hunziker, J.H.; and DiFeo, D.R., Jr., eds. 1977. 284 p. ISBN 0-87933-282-4. $55.50. LC 76-58381.

Vol. 7: **Big biology: the US/IBP/W.** Blair, Frank, ed. 1977. 261 p. ISBN 0-87933-305-7. LC 77-8512.

Vol. 8: **Eskimos of Northwestern Alaska: a biological perspective.** Jamison, Paul L.; Zegura, Stephen L.; and Milan, Frederick A., eds. 1978. 319 p. ISBN 0-87933-319-7. $46. LC 77-18941.

Vol. 9: **Nitrogen in desert ecosystems.** West, N.E. and Skujins, John., eds. 1978. 307 p. ISBN 0-87933-333-2. $31.50. LC 78-17672.

Vol. 10: **Aerobiology: the ecological systems approach.** Edmonds, Robert L., ed. 1979. 386 p. ISBN 0-87933-346-4. $36. LC 78-23769.

Vol. 11: **Water in desert ecosystems.** Evans, Daniel D. and Thames, John L., eds. 1981. 280 p. ISBN 0-87933-365-0. $44.50. LC 79-22432.

Vol. 12: **An Arctic ecosystem: the coastal tundra at Barrow, Alaska.** Brown, Jerry, et al., eds. 1980. 571 p. ISBN 0-87933-370-7. $38.50. LC 79-22901.

Vol. 13: **Limnology of tundra ponds: Barrow, Alaska.** Hobbie, John E., ed. 1980. 514 p. ISBN 0-87933-386-3. $38.50. LC 80-26373.

Vol. 14: **Analysis of coniferous forest ecosystems in the western United States.** Edmonds, Robert L., ed. 1982. 419 p. ISBN 0-87933-382-0. $44. LC 80-26699.

Vol. 15: **Island ecosystems: biological organization in selected Hawaiian communities.** Mueller-Dombois, Dieter; Bridges, Kent W.; and Carson, Hampton L., eds. 1981. 583 p. ISBN 0-87933-381-2. $38.50. LC 80-27650.

561

This series presents detailed summaries of the ecological research conducted at each of the major International Biological Program study sites in the United States, a program sponsored by the National Academy of Sciences and financially supported primarily by the National Science Foundation. Each volume presents considerable original research.

Biogeography

Walter, Heinrich, ed. Vegetationsmonographien der einzelnen Grossräume. Stuttgart, West Germany: Gustav-Fischer-Verlag, 1965-1975. 10 v.

    Vol. 1: **Die Vegetation von Nord und Mittelamerica,** Knapp, R., ed. 1965. 373 p.

    Vol. 2: **Die Wälder Südamerikas,** Hueck, K., ed. 1966. 422 p.

    Vol. 2a: **Vegetationskarte von Südamerika,** Hueck, K. and Seibert, P., eds. 1972. 71 p. LC 73-690899.

    Vol. 3: **Die Vegetation von Afrika,** Knapp, R., ed. 1973. 626 p.

    Vol. 4: **The Vegetation of Australia,** Beadle, N.C.W., ed. 1981. 690 p. ISBN 3-437-30313-9. LC 81-169423.

    Vol. 7: **Die Vegetation Osteuropas, Nord- und Zentralasiens,** Walter, H., ed. 1974. 452 p. ISBN 3-437-20133-6. LC 75-551173.

    Vol. 10: **Klimadiagrammkarten der einzelnen Grossräume,** Walter, H.; Harnickell, E., and Mueller-Dombois, D., eds. 1975. 36 p. 9 maps. LC 75-509942.   562

The most thorough, cohesive treatment of regional vegetation available. In German or English. Each volume in the series provides detailed consideration of plant ecophysiology and vegetation cover within regions by continent. The final volume presents climate diagram maps with comprehensive coverage of the entire areas.

E. TEXTBOOKS

Anderson, J.M. Ecology for environmental sciences: biosphere, ecosystems, and man.(Resource and environmental science series). New York: John Wiley and Sons, 1981. 175 p. ISBN 0-470-27216-3. $19.95. LC 81-3349.   563•

Text that focuses on synthesizing concepts at the ecosystem level. Topics include structure and function of ecological systems, community theory, population dynamics, primary and secondary production, and decomposition. Human interactions with the environment are emphasized.

Barbour, Michael G.; Burk, Jack H.; and Pitts, Wanna D. Terrestrial plant ecology. Menlo Park, CA: Benjamin/Cummings, 1980. 604 p. ISBN 0-8053-0540-8. $28.95. LC 79-23591.   564

Perhaps the most thorough college-level text which treats the fundamental concepts of plant ecology in terrestrial environments of North America. Topics include the history of ecological thought, community dynamics (succession, productivity, mineral cycling), and environmental factors influencing plant distributions (light, temperature, water, soil, fire). Extensive bibliography (over 1000 entries), not arranged topically, except for references on biomes.

Billings, William D. Plants and the ecosystem. 3rd ed. Belmont, CA: Wadsworth, 1978. 177 p. (1st ed., 1964. 154 p.) ISBN 0-534-00571-3. LC 77-15943.   565•

Text that presents basic phytogeographic principles (plant-environment interactions, community structure, ecosystem processes, plant distributions), briefly discusses terrestrial ecosystem types, and examines the human role in the environment (as an element in past natural ecosystems and as a source of change in present and future ones). Glossary and many figures.

Brown, James H. and Gibson, Arthur C. Biogeography. St. Louis, MO: C.V. Mosby, 1983. 653 p. ISBN 0-8016-0824-4. $32.95. LC 82-14124.   566

Up-to-date and well-referenced college-level text offering broad topical coverage of themes in both historical biogeography (speciation and extinction, dispersal, endemism and disjunction, Quaternary studies, cladistics and vicariance studies) and ecological biogeography (population, community, and biome distributions, insular biology, diversity patterns). Includes more than 600 illustrations and maps, references organized topically at the end of the chapters, more than 1800 bibliographic entries, a useful glossary, and a thorough index.

Collinson, A.S. Introduction to world vegetation. London; Boston, MA: George Allen and Unwin, 1977. 201 p. ISBN 0-04-581012-5. $17.95. ISBN 0-04-581013-3. $10.95. (pbk). LC 78-313063.   567•

Text on plant geography presented in two parts: basic principles governing plant distributions (evolution, nutrient cycles, ecosystems, soils, plant communities), and world vegetation patterns. References at ends of the chapters. Numerous figures.

Eyre, S. Robert (Samuel) Vegetation and soils: a world picture. New ed. London: Edward Arnold, 1975. 344 p. (1st ed., 1963. 324 p.) ISBN 0-7131-5838-7. £6.95. (pbk). LC 64-591 rev.   568

Text that examines world patterns of vegetation and soil occurrence. A preliminary section on vegetation and soil development is presented. Includes a more detailed treatment of soils and vegetation of the British Isles. Includes two appendices of detailed maps and climatic data, and a glossary.

Furley, Peter A. and Newey, Walter W. Geography of the biosphere: an introduction to the nature, distribution, and evolution of the world's life zones. London; Stoneham, MA: Butterworth and Co., 1982. 413 p. ISBN 0-408-70801-8. $82.50.   569

Detailed treatment of world biogeographic patterns beginning with the circulation of energy and matter in the biosphere and basic characteristics and the evolution of soil, plant, and animal systems, followed by discussion of the world's major biomes. Includes a brief section on mapping the biosphere. Bibliography includes more than 1,400 references. Contains a detailed index and many figures but no photographs.

Kellman, Martin C. Plant geography. 2nd ed. London: Methuen; New York: St. Martin's Press, 1980, 181 p. (1st ed., 1974. 135 p.) ISBN 0-312-61461-6. $26. LC 80-5079.   570•

Concise text that stresses ecological aspects of vegetation patterns (e.g., demography, resource allocation, interference and coexistence, environmental factors, migration, structure and function of plants, pattern, and human influence). More than 300 references, 33 figures, seven tables, and 16 plates.

MacArthur, Robert H. Geographical ecology: patterns in the distribution of species. Princeton, NJ: Princeton University Press, 1984. 288 p. ISBN 0-691-08353-3. $45. LC 83-24477.   571

An intermediate to advanced text that first reviews fundamental principles of biogeography (e.g., climatic patterns, competition and predation, optimal feeding strategies), then examines patterns of plant and animal occurrences in various contexts (e.g., island biogeography, species distribution, community structure, species diversity, historical patterns, tropical vs. temperate environments). The mathematical foundation of theories presented is inserted in appendices arranged at the chapter ends. Nearly 150 references.

Nelson, Gareth, and Platnick, Norman. Systematics and biogeography: cladistics and vicariance. New York: Columbia University Press, 1981. 592 p. ISBN 0-231-04574-3. $52. LC 80-20828.   572

A thorough treatise on the systematic approach to biogeography that advocates and demonstrates the explanation of biogeographic patterns by integrating morphological, evolutionary, and spatial relationships between organisms. Includes numerous cladograms and other diagrams. Over 200 references.

Pianka, Eric R. Evolutionary ecology. 3rd ed. New York: Harper and Row, 1983. 416 p. (1st. ed., 1973. 356 p.) ISBN 0-06-045232-3. $31.95. LC 82-6118.   573

One of the few texts which is successful in presenting a balanced consideration of plants and animals from an ecological and biogeographic perspective, woven into chapters treating natural selection, the physical environment, population structure and dynamics, competition and predation, niche concepts and community structure, insular biogeography, and human impacts on biotas. Detailed bibliography of some 1000 references, cited by topical readings at the ends of the chapters.

Biogeography

Simmons, Ian G. Biogeography: natural and cultural. London; New York: Edward Arnold, 1979. 400 p. ISBN 0-7131-6245-7. ISBN 0-7131-6246-5. $18.95. (pbk). LC 80-492727.

574•

A text that focuses on human influence on the biosphere. Includes brief treatment of influences on plant and animal distributions at organism, ecosystem, and biome levels before thorough discussion of human attitudes and activities that have influenced the occurrence of plants and animals. Numerous figures, maps, and plates, recommended references with each chapter, and a bibliography at the end with more than 600 sources.

Tivy, Joy. Biogeography: a study of plants in the ecosphere. 2nd ed. London; New York: Longman, 1982. 459 p. (1st ed., 1971. 394 p.) ISBN 0-582-30009-6. $13.95. LC 80-41366.

575•

College text on plant geography. First part is a systematic analysis of environmental and historical influences (climate, soils, competition, animals, people) on plant distributions; second part examines four major ecosystems (marine, forest, grassland, desert) with particular attention to the human use of resources. Major topical references at the end of each chapter.

Walter, Heinrich. Vegetation of the Earth and ecological systems of the geo-biosphere. Translated from the 3rd German edition by Joy Wieser. 2nd ed. New York: Springer-Verlag, 1979, 1984. 274 p. (1st ed., 1973. 237 p.) ISBN 0-387-13748-3. $17. LC 84-14143. (Vegetationszonen und Klima: die ökologische Gliederung der Biogeosphäre. 3rd ed. Stuttgart: Ulmer, 1977. 309 p.)

576

Detailed treatment of world vegetation patterns with emphasis on climate-plant physiology interactions. In nine chapters, general plant responses to climatic conditions of major biomes are addressed, followed by a detailed analysis of representatives of each of these biomes throughout the world. Richly illustrated with maps, line drawings, numerous climate diagrams, and photographs. The bibliography of approximately 150 sources includes many German references.

Whittaker, Robert H. Communities and ecosystems. 2nd ed. New York: Macmillan, 1975. 385 p. ISBN 0-02-427390-2. (pbk). (College text available only through retailers).LC 74-6636.

577

Concisely written intermediate-level college text highlighting basic principles of organism-environment relations (population and community ecology integrating both plants and animals) and ecosystem structure and function (production and nutrient cycling). Nicely illustrated, with well-chosen supplementary exercises and often insightful examples that augment the presentation of concepts. References organized by topic at the end of the chapters are mostly limited to key words.

F. SPECIAL TOPICS

Box, Elgene Owen. Macroclimate and plant forms: an introduction to predictive modeling in phytogeography. (Task for vegetation science series vol. 1). The Hague, Netherlands; Boston, MA: Dr. W. Junk, 1981. 272 p. ISBN 9-06-193941-0. $69.50. LC 81-12368.

578

Innovative treatment of world vegetation patterns that examines the relationship between climatic conditions and both plant form and vegetation structure through ecoclimatic modeling. Climatic variables are used to predict world-scale vegetation patterns which are compared with existing world vegetation to ascertain the role of climate in governing global vegetation patterns. Includes nine global climatic maps, 25 global maps of growth form distributions, and approximately 800 references.

Cody, Martin L. and Diamond, Jared M., eds. Ecology and evolution of communities. Cambridge, MA: Belknap Press of Harvard University Press, 1975. 545 p. ISBN 0-674-22446-9. $15. LC 74-27749.   579

Collection of essays by leading scholars about such aspects of plant and animal communities as species abundance and diversity; community stability; coexistence of species, competition, and natural selection; community structure and the influence of predation and competition. An outgrowth of papers presented at a symposium in memory of renowned ecologist Robert MacArthur. Each paper includes references and, in most cases, original data.

Eisenberg, John F. The Mammalian radiations: an analysis of trends in evolution, adaptation, and behavior. Chicago, IL: University of Chicago Press, 1982. 640 p. ISBN 0-226-19537-6. $45. LC 80-27940.   580

A thorough integrative discussion of mammalian evolution and radiation that reviews taxonomic, morphological, and behavioral relationships within the class Mammalia, in detail. A storehouse of information that presents 157 figures and maps, nearly 50 plates, 61 tables (in addition to six appendices that contain much data), a glossary, and approximately 2000 sources.

Goudie, Andrew S. The Human impact: man's role in environmental change. Oxford, U.K.: B. Blackwell; Cambridge, MA: MIT Press, 1982. 316 p. ISBN 0-262-57086-3. $11.95. LC 81-15643.   581 •

An examination of the many ways in which human activities have altered the physical environment. Human influence on plant and animal occurrences, soils, waterways, climate, and the atmosphere are covered. Within each section the author pinpoints particular problems (e.g., fire suppression, air pollution, etc.) and in some cases regions where human impact is especially intense. More than 350 references. Well illustrated.

Greig-Smith, Peter. Quantitative plant ecology. (Studies in ecology, v.9). 3rd ed. Berkeley, CA: University of California Press, 1983. 359 p. (1st ed., 1957. 198 p.) ISBN 0-520-04989-6. $38.50. ISBN 0-520-05080-0. $22. (pbk). LC 83-1302.   582

Recent update of a standard methodological text treating vegetation description and sampling techniques, including pattern analysis, species associations, statistical approaches to species-environment and vegetation-environment interactions, and classification and ordination methods. Includes 79 figures and 27 data tables in text, eight statistical tables in an appendix, a bibliography with approximately 600 references cited, plus author and subject indices.

Holzner, W.; Werger, M.J.A.; and Ikusima, I., eds. Man's impact on vegetation. (Geobotany series vol. 5). The Hague, The Netherlands; Boston, MA: Dr. W. Junk, 1983. 370 p. ISBN 90-6193-685-3. $98. LC 82-14039.   583

Collection of papers dealing with human influence on vegetation contributed by 33 international scholars. Papers are organized into three sections: general human influences on vegetation, human impact on specific vegetation types throughout the world, and severely influenced areas. Nicely illustrated with photographs, maps, and other figures.

Holzner, W. and Numata, M., eds. Biology and ecology of weeds. (Geobotany series vol. 2). The Hague, The Netherlands; Boston, MA: Dr. W. Junk, 1982. 461 p. ISBN 90-6193-682-9. $99.50. LC 81-20893.   584

An international synthesis of basic and applied research findings on weeds, with an emphasis on those of arable lands. Includes general ecology (e.g. evolution, reproductive strategies, competition with crops, weeds as indicators) and examples of weed vegetation from throughout the world. References are arranged at the at ends of the 38 chapters, with many figures, photographs, maps, and tables.

Biogeography

Kozlowski, Theodore T. and Ahlgren, Clifford E. Fire and ecosystems. New York: Academic Press, 1974. 542 p. ISBN 0-12-424255-3. $75. LC 74-5695.   585
   Comprehensive analysis of fire in ecosystems that examines their beneficial and harmful effects on soils and soil organisms, animals, and plants in general as well as effects of fire in specific types of ecoystems (temperate forests, grasslands, deserts, chaparral, and Mediterranean and African ecosystems). A chapter on fire as a management tool is included. Thirteen contributors and numerous references topically arranged.

Küchler, A. Wilhelm (August). Vegetation mapping. New York: Ronald Press, 1967. 472 p. LC 66-21857.   586
   A comprehensive work reviewing the background of vegetation mapping and classification. Major sections include relevant cartographic techniques, data collection and compilation, and applications of vegetation maps. Significant for its treatment of cartographic symbols as well as its treatment of vegetation.

Kurtén, Björn, and Anderson, Elaine. Pleistocene mammals of North America. New York: Columbia University Press, 1980. 442 p. ISBN 0-231-03733-3. $50. LC 79-26679.   587
   Summary of known information about adaptations and population dynamics of Blancan and Pleistocene mammals of Canada, the U.S., and northern Mexico. A preliminary section is devoted to a listing and description of sites where local fauna were studied. More than 1000 citations in the bibliography.

MacArthur, Robert H. and Wilson, Edward O. The Theory of island biogeography. (Monographs in population biology series vol. 1). Princeton, NJ: Princeton University Press, 1967. 203 p. ISBN 0-691-08049-6. $22. LC 67-24102.   588
   Major treatise on the theory of island biogeography and its mathematical and conceptual underpinnings. Topics covered include species number-area relationships, area-density patterns, colonization, and subsequent evolutionary change. Bibliography comprises approximately 100 references.

West, Darrell C.; Shugart, Herman H.; and Botkin, Daniel B., eds. Forest succession: concepts and applications. New York: Springer-Verlag, 1981. 517 p. ISBN 0-387-90597-9. $38.60. LC 81-16707.   589
   Proceedings of a National Science Foundation sponsored conference on current research in forest succession. Stimulating and up-to-date presentations on successional modeling (including introductions to several computer-based modeling efforts), population and ecosystem dynamics during vegetation recovery sequences, and disturbance ecology. Much previously unpublished data. More than 1000 bibliographic entries, many very recent.

Wright, Henry A. and Bailey, Arthur W. Fire ecology: United States and southern Canada. New York: John Wiley and Sons, 1982. 501 p. ISBN 0-471-09033-6. $49.50. LC 81-14770.   590
   An up-to-date compendium that reviews the present knowledge on fire ecology. The effects of fire on plants, animals, and soils in major ecosystems of the U.S. and southern Canada and the potential role of fire in ecosystem management (including principles and use of prescribed burning) are discussed. Thorough lists of references.

5. SOILS

John C. Dixon

A. TEXTBOOKS

Basile, Robert M. A Geography of soils. Dubuque, IA: Wm C. Brown, 1971. 152 p. LC
70-146327.
    A basic introductory level book designed primarily for students of geography. Addresses the issues of soil classification; morphology; utilization; and global distribution. Contains a glossary; numerous maps; cross section diagrams; and annotated references.
591•

Batten, James W. and Gibson, J. Sullivan. Soils: their nature, classes, distribution, uses, and care. Rev. and enl. ed. University, AL: University of Alabama Press, 1977. 276 p. (1st ed. by Gibson, J. Sullivan, and Battan, James W., 1970. 296 p.). ISBN 0-8173-2876-9. $14.50. LC 76-40302.
    An easily read text written by geographers for students of earth and environmental science. Discussions of soil orders and their relationship to soil morphology are included. The underlying purpose is to show our dependence on soils and to encourage responsible use of this resource. Many maps and photos.
592•

Birkeland, Peter W. Soils and geomorphology. New York: Oxford University Press, 1984. 372 p. (Revised ed. of Pedology, weathering, and geomorphological research, 1974. 285 p.). ISBN 0-19-503398-1. $37.50. LC 83-13345.
    A systematic examination of pedology, this book examines soil classification, soil forming processes and factors, as well as a detailed discussion of weathering processes and products. Application of soils to Quaternary stratigraphic studies is emphasized. Extensive bibliographic citations are provided.
593

Brady, Nyle C. The Nature and properties of soils. 9th ed. (1st ed. by Lyon, T. Lyttleton, and Buckman, Harry O. 1922. 588 p.) New York: Macmillan; London: Collier; Macmillan, 1984. 750 p. ISBN 0-02-313340-6. LC 83-19545.
    The most recent edition of this comprehensive text includes new material keeping it abreast of recent advances in soil science. Soil charcteristics, processes, and classifiction are discussed with special attention given to agricultural and environmental applications of soil science.
594

Donahue, Roy L.; Shickluna, John C.; and Robertson, Lynn S. Soils: an introduction to soils and plant growth. 5th ed. Englewood Cliffs, NJ: Prentice-Hall, 1983. 587 p. (1st ed., 1958. 349 p.) ISBN 0-13-822288-6. $31.95. LC 73-159573.
    An introductory college-level textbook that stresses modern principles and practices of soil science; soil ecology; and plant environments. Special emphasis is given to soil chemistry; agricultural management; soil-plant relationships; and tropical soil problems. Many illustrations and a 75-page glossary are included.
595

FitzPatrick, E.A. Soils: their formation, classification, and distribution. London; New York: Longman, 1983. 353 p. ISBN 0-582-30116-5. $65. LC 82-12669.
    A college-level text that discusses factors of soil formation and physical; chemical; and biological processes of soil formation. A global classification of soils is presented in the second half of the book.
596•

Foth, Henry D. Fundamentals of soil science. 7th ed. New York: John Wiley and Sons, 1984. 436 p. (1st ed., by Millar, Charles E., and Turk, L.M., 1943. 462 p.) ISBN 0-471-88926-1. $32.95. LC 83-23383.
    A comprehensive college-level textbook, this edition has new illustrations and diagrams, and attempts to direct awareness toward regional and international aspects of soil resource conservation and land use planning. Some discussion of fertilizers and soil plant-animal relationships. In addition, factors and processes of soil formation are discussed.
597

Hausenbuiller, Robert L. Soil science: principles and practices. 3rd ed. Dubuque, IA: Wm C. Brown, 1985. (1st ed., 1972. 504 p.) ISBN 0-697-05856-5.   599

A text for first-course pedology at the college level. Emphasis is twofold. The first part examines soil chemistry; physics; micro and macro nutrient components; and soil genesis. The remainder discusses soil fertility and soil management problems and remedial techniques.

Hillel, Daniel. Introduction to soil physics. London; New York: Academic Press, 1982. 346 p. ISBN 0-12-348520-7. $27.50. LC 81-10848.   600

An upper-level undergraduate textbook that examines most of the basic aspects of soil physics. Extensive use of mathematical formulae and simple problems designed to help the student use these for practical application are included. A significant contribution to pedology, this should be of equal value to students of agronomy, engineering, and the environmental sciences. More than 450 bibliographic entries are included.

Jenny, Hans. The Soil resource: origin and behavior. (Ecological studies series v. 37). New York: Springer-Verlag, 1981. 377 p. ISBN 0-387-90543-X. $29.80. LC 80-11785.   601

An advanced book dealing with pedologic principles and their application to ecosystems. The first half of the book examines soil processes on the molecular level. The second half of the book examines spatial and temporal aspects of soil formation. Diagrams and tables are extensively used.

Knapp, Brian. Soil processes. (Processes in physical geography no. 2). London; Boston, MA: George Allen and Unwin, 1979. 72 p. ISBN 0-04-631011-8. $6.95. LC 78-41040.   602 •

An introductory-level soil science textbook that examines soil forming processes; some soil forming factors; and soils as resources of man. Extensive use of diagrams with a limited glossary. Most examples used are British.

Marshall, T.J. and Holmes, J.W. Soil physics. Cambridge, U.K.: Cambridge University Press, 1980. 345 p. ISBN 0-521-22622-8. $85. ISBN 0-521-29579-3. $27.95 (pbk). LC 78-73809.   603

An advanced textbook for scientists or students in the fields of pedology; hydrology; civil engineering; agronomy; and the basic physical sciences. Strong emphasis is given to the quantifiable aspects of soil-water relationships. Most formula units given are in the international system but conversion tables are provided. Extensive bibliography.

Olson, Gerald W. Soils and the environment: a guide to soil surveys and their applications. London: Chapman and Hall; New York: Methuen, 1982. 178 p. ISBN 0-412-23760-1. $17.95. (pbk). LC 81-10067.   604

This text offers a basis for wider application of soil reports and maps. Discussions of specific applications of soil surveys toward problems in engineering; land classification; erosion control; crop yield correlations; archaeology; and future planning are included. Easily interpreted by the non-specialist.

Steila, Donald. The Geography of soils: formation, distribution, and management. Englewood Cliffs, NJ: Prentice-Hall, 1976. 222 p. ISBN 0-13-351734-9. LC 76-230.   605 •

An introductory text in soil science for non-specialists. The book uses the geographic approach to the subject which is to emphasize the spatio-temporal aspects of soil management and classification, and to address certain global problems relating to soil use and abuse. Many maps; diagrams; and a complete glossary are included.

Thompson, Louis M. and Troeh, Frederick, R. Soils and soil fertility. 4th ed. New York: McGraw-Hill, 1978. 516 p. (1st ed., 1952. 339 p.) ISBN 0-07-064411-X. $31.95. LC 77-21404.   606

A college-level text designed to meet requirements for introductory courses in soil science. Extensive discussion of macro and micronutrient soil components. Although directed toward students of agronomy it contains useful chapters on urban soil and water management and soil pollution. Many references cited.

White, R.E. Introduction to the principles and practice of soil science. New York: John Wiley and Sons, 1979. 198 p. ISBN 0-470-26717-8. $25.95. LC 79-14361.   607

A concise, yet advanced introduction to soil science theory and methodology. Emphasis is on the physical-chemical aspects of soil composition; morphology; and utilization. Much attention is given to representative equations and diagrams. Soil analysis techniques are discussed with limited mention of surveys and classification.

Wilding, Lawrence P.; Smeck, Neil E.; and Hall, George F. Pedogenesis and soil taxonomy: I. Concepts and interactions. (Development in soil science no. 11a). Amsterdam, The Netherlands; New York: Elsevier, 1983. 304 p. ISBN 0-444-42100—9. $49. LC 82-24198.   608

Advanced-level text that discusses principal aspects of pedology. Topics covered include: historical background of soil taxonomy; concepts of soils; soil modelling; spatial variability of soils; pedology and geomorphology; soil composition and genesis; soil development in relation to climate and time; and soil biology and hydrology.

Wilding, Lawrence P.; Smeck, Neil E.; and Hall, George F. Pedogenesis and soil taxonomy: II. The soil orders. (Development in soil science no. 11b). Amsterdam, Netherlands; New York: Elsevier, 1983, 410 p. ISBN 0-444-42137-8. $55.50. LC 82-24198.   609

This advanced text examines the relationship between soil taxonomy and pedogenesis. A detailed discussion of characteristics; formation; and influence of soil forming factors on each soil order is presented.

B. SPECIAL TPOPICS

Alexander, Martin. Introduction to soil microbiology. 2nd ed. New York: John Wiley and Sons, 1977. 467 p. (1st ed., 1961. 472 p.) ISBN 0-471-02179-2. $34.95. LC 77-1319.   610

A comprehensive introductory text at the college level. Discussions of the composition; importance; and behavior of microorganisms as they relate to agronomy are included. Interprets microfloral reactions in terms of their biochemistry and relevance to soil ecology.

Australia. Commonwealth Scientific and Industrial Research Organization, Division of Soils. Soils: an Australian viewpoint. Melbourne; New York: Academic Press, 1983. 928 p. ISBN 0-12-654240-6. $89.   611

Fifty-four papers reporting on fifty years of soil research in Australia make up this book which is divided into seven sections dealing with aspects of soil formation. Soil landscapes; classification; physical and chemical characteristics; soil organic matter; soil fauna/microbial/plant relationships; and soils and land management are discussed.

Bohn, Hinrich L.; McNeal, Brian L.; and O'Connor, George H. Soil chemistry. New York: John Wiley and Sons, 1979. 329 p. ISBN 0-471-04082-7. $29.95. LC 79-14515.   612

An advanced text examining aspects of the inorganic chemistry of soils. Topics covered include: chemistry cycles and principles; mineral chemistry; weathering and soil development; cations and cation exchange; oxidation and reduction; soil acidity; and soil salts. Limited use of diagrams and chemical formulae.

Bolt, G.H., ed. Soil chemistry: B. Physico chemical models. 2nd ed. Amsterdam, Netherlands; New York: Elsevier, 1982. 538 p. (1st ed. 1979. 479 p.) ISBN 0-444-42060-6. $76.50. LC 82-1503.   613

A graduate-level text that discusses in depth theoretical developments in soil chemistry. Examines cation exchange and absorption; solute movements; electro chemical phenomena; and clay transformations. Addresses the problems of generalization and determining the validity of theoretical assumptions.

Soils

Brewer, Roy. Fabric and mineral analysis of soils. 2nd ed. Melbourne, FL: Robert E. Kreiger, 1976. 498 p. Reprint of New York: Wiley, 1964. 470 p., with supplementary material. ISBN 0-88275-314-2. $27.50. LC 75-17850.   614

Presents principles and techniques of soil fabric and mineral analysis. The first part of the book discusses concepts of pedogenesis; while the second part examines the application of mineral analysis to soils. The third and major part of the book examines the components of soil micromorphology. Finally, the application of fabric and mineral analysis to soil genesis and classification is discussed.

Bridges, Edwin M. and Davidson, Donald A. Principles and applications of soil geography. London: Longman Group, 1982. 292 p. ISBN 0-582-30014-2. $13.95. LC 80-41509.   615

A collection of essays covering a wide range of topics within pedology that is relevant to students of geography and the environmental sciences. Includes discussions of soil survey and classification systems; modern data systems and applications (computers); and a historical overview of human-soil relationships.

Buol, Stanley W.; Hole, Francis D.; and McCracken, Ralph J. Soil genesis and classification. 2nd ed. Ames, IA: Iowa State University Press, 1980. 404 p. (1st ed., 1973. 360 p.) ISBN 0-8138-1460-X. $19.50. LC 79-15992.   616

A detailed discussion of pedology. The first half of the book examines soil morphology and micromorphology and soil forming processes and factors. The second half of the book discusses distribution; processes of formation; and classification of the ten soil orders. The concept of soilscapes and the interpretation of soil surveys concludes the book.

Davidson, Donald A. Soils and land use planning. London; New York: Longman Group, 1980. 129 p. ISBN 0-582-49895-7. $11.50. (pbk). LC 79-40868.   617

An outline of soil survey procedures and how to determine from those surveys the best land use evaluation methods and capabilities. The application of pedology to land use planning is emphasized and case studies are cited. Valuable to professional planners and to students of geography; the environmental sciences; and landscape architecture.

Dent, David, and Young, Anthony. Soil survey and land evaluation. London; Boston, MA: George Allen and Unwin, 1981. 278 p. ISBN 0-04-631013-4. $35. ISBN 0-04-631014-2. $17.95. (pbk). LC 81-10834.   618

A general review of techniques and procedures for students, professional surveyors, and general survey users such as planners, agronomists, engineers, or economists. An excellent aid for interpreting soil surveys and making predictions or management decisions on the basis of information contained therein.

Dixon, Joe B. and Weed, Sterling B. Minerals in soil environments. Madison, WI: Soil Science Society of America, 1977. 948 p. ISBN 0-89118-765-0. $25. LC 77-80728.   619

Twenty-six technical papers dealing primarily with soil mineralogy and mineral chemistry. All major clay mineral groups are included. A discussion of nature, origin, mode of occurrence, and methods of identification is presented. Extensive references accompany each chapter.

Food and Agriculture Organization of the United Nations. Soil map of the world: 1:500,000. Paris: UNESCO, 1975. 10 v. LC 78-69221.   620 •

A comprehensive survey of the world's soil. Valuable aid for economic development programs and agricultural planning. Many tables of various soil characteristics and discussions of soil groups; physiography; and the vegetation of principal regions of the world.

Gerrard, A. John. Soils and landforms: an integration of geomorphology and pedology. London; Boston, MA: George Allen and Unwin, 1981. 219 p. ISBN 0-04-551048-2. $35. ISBN 0-04-551049-0. $17.95. (pbk). LC 81-10812.

621

This book discusses the interrelationships between soils, landforms, and geomorphological processes. In the first part of the book soil geomorphic principles are discussed; in the second part, application of these principles is applied to selected landform assemblages. Soils as stratigraphic indicators are also discussed.

Kubiena, Walter L. Micromorphological features of soil geography. New Brunswick, NJ: Rutgers University Press, 1970. 254 p. $30. LC 70-133971.

622

Examines the application of field and micromorphological methods to the study of soils. Principles of soil micromorphology are discussed followed by a detailed analysis of soil genesis in different climatic environments. Contains large numbers of color soil photomicrographs.

Pitty, Alistair F. Geography and soil properties. London; New York: Methuen, 1979. 287 p. ISBN 0-416-75380-9. $25.95. ISBN 0-416-71540-0. $15.95. (pbk).

623

Examines the significance of spatial changes in soil distribution and the influence of the environment on soils. A systematic examination of the physical; chemical; biological; mineralogical; and mechanical properties of soils is presented. Extensive use of diagrams is made as well as an exhaustive bibliography.

Soil Science Society of America. Glossary of soil science terms. Madison, WI: Soil Science Society of America, 1984. 28 p. ISBN 0-89118-774-X. $1. LC 84-14173.

624

A regularly updated glossary that clearly designates those terms which are obsolete and defines those that have remained in common American usage. A valuable aid to researchers and students.

Trudgill, Stephen A. Soil vegetation systems. Oxford, U.K.: Clarendon Press; New York: Oxford University Press, 1977. 180 p. ISBN 0-19-874058-1. $33. ISBN 0-19-874059-X. $11.95. (pbk). LC 77-30179.

625

Examines the soil-vegetation system. Discussion is divided into three sections: basic approaches; components of nutrient systems; and modelling nutrient systems. Topics addressed include models of soil-vegetation systems, weathering, atmospheric additions to soils, leaching, nutrient cycling, and stability of soil vegetation systems. An important contribution to understanding soil-vegetation interactions.

U.S. Soil Conservation Service. Soil taxonomy: a basic system of soil classification for making and interpreting soil surveys. (Agriculture handbook 436) Washington, DC: U.S. Government Printing Office, 1975. 754 p. $23.00 in U.S., $28.75 foreign. (Catalog no. A1.76.436. Stock no. 001-000-02597-0)

626

A detailed and advanced discussion of the latest system of soil classification: soil taxonomy. This technical description of soil taxonomy provides general principles of soil horizon classification. The bulk of the book discusses principal classes within each of the major soil orders.

6. OCEANS AND LAKES*

Fillmore C.F. Earney

*In this section the editorial committee reduced the number of entries recommended by the contributor.

A. BIBLIOGRAPHIES

Champ, Michael A. and Park, P. Kilho. Global marine pollution bibliography: ocean dumping of municipal and industrial wastes. New York: Plenum, 1982. 424 p. ISBN 0-306-65205-6. $69.50. LC 82-28060.

    A concise bibliography dealing with the subject of municipal and industrial waste pollution in the world's oceans.

627

Feltner, Charles E. and Feltner, Jeri B. Great Lakes maritime history: bibliography and sources of information. Dearborn, MI: Seajay, 1982. 124 p. ISBN 0-9609014-0-X. $9.95. LC 82-51175.

    Contains more than 1000 citations divided into twelve categories among which are: Great Lakes history; directories and registers of ships; shipbuilding; shipwreck history; United States government agencies dealing with the Great Lakes; maps and charts; newspapers and periodicals; and photographic collections. Also contains a directory of libraries; societies; and museums that deal with the Great Lakes region.

628 •

Oda, Shigeru, ed. The International Law of the ocean development. Leiden, The Netherlands: A.W. Sijthoff, 1972. v.1. 519 p. ISBN 9-02-860122-8. LC 72-76418. Basic Documents. 1975-1979. Dordrecht, The Netherlands: Kluwer Academic Publishers Group, 2 v. 2070 p. ISBN 0-90-286-0177-5. $289.

    A helpful source for detailed information on the contents of published proceedings and documents resulting from a variety of international meetings and agreements dealing with ocean affairs. Includes items as diverse as government commission reports and bilateral territorial sea and fishery zone agreements.

629

Marine affairs bibliography: a comprehensive index to marine law and policy literature. 1- (1980-). Quarterly. Dalhousie Law Library, Dalhousie University, Halifax, Nova Scotia, Canada, B3H 4H9. ISSN 0226-8361.

    Useful for "Law of the Sea" affairs including living and non-living resources, boundary delimitation problems and settlements, preservation of the marine environment, military use, scientific research, and archaeology. Cites recent books, periodical literature, proceedings, and government documents.

630

B. SERIALS

Coastal zone management journal. 1- (1973-). Quarterly. Crane, Russak, and Co., 3 E. 44th St., New York, NY 10017. ISSN 0090-8339.

    Studies the social, political, legal, and technical issues in the use of oceanic coastal zone resources.

631

Limnology and oceanography. 1- (1956-). Bimonthly. American Society of Limnology and Oceanography, 1530 12th Ave., Grafton, WI 53024. ISSN 0024-3590.

    Contains articles useful to biogeographers with published book reviews.

632

Marine mining: the journal of seafloor minerals exploration, assessment, and ore processing. 1- (1977-). Quarterly. Crane, Russak, and Co., 3 E. 44th St., New York, NY 10017. ISSN 0149-0397.

    Contains items important to those interested in ocean mineral extraction, especially in deep seabed areas, along with many book reviews.

633

Marine policy: the international journal of ocean affairs. 1- (1977-). Quarterly. Butterworth Scientific Limited, Journals Division, Box 63, Westbury House, Bury St., Guildford, Surrey, GU2 5BH U.K. ISSN 0308-597X.

    An excellent source for current developments in all types of ocean industries and legal problems. Contains book reviews and useful lists of recent publications dealing with marine policy.

634

Marine pollution bulletin. 1- (1970-). Monthly. Pergamon Press, Inc., Maxwell House, Fairview Park, Elmsford, NY 10523. ISSN 0025-326X.　　　　635
Examines all aspects of conditions existing in lakes, estuaries, oceans, and seas. Publishes book reviews, commentaries, and research reports on the danger of noxious materials to marine life and looks at the management and productivity of the marine environment.

Ocean development and international law: the journal of marine affairs (formerly Ocean development and international law journal). 1- (1973-). Quarterly. Crane, Russak, and Co., 3 E. 44th St., New York, NY 10017. ISSN 0090-8320.　　　636
Considers all types of ocean use and regulation that have international implications. Contains book reviews.

Ocean industry. 1- (1966-). Monthly. Gulf Publishing Co., Box 2608, Houston, TX 77001. ISSN 0029-8026.　　　　637
Focuses especially on offshore petroleum activities throughout the world. Examines new developments in ocean industry technology.

Oceanic abstracts. 1- (1964-). Bimonthly, plus a year-end index. (1-8, 1964-1971 as Oceanic index). Cambridge Scientific Abstracts, 5161 River Road, Bethesda, MD 20816. ISSN 0093-6901.　　　638
A highly useful publication that screens more than 3500 primary sources, foreign and domestic; designed for researchers in academia, industry, and government. Topics include: marine biology; biological oceanography; physical and chemical oceanography; geochemistry; geophysics; geology; meteorology; marine pollution; ships and shipping; as well as other marine related items. Material is indexed by subject, taxonomy, geographic location, and author. Contents are available as a computerized database searchable through online systems provided by Dialog Information Services *and* Information Retrieval Service/European Space Agency.

Oceans. 1- (1969-). Bimonthly. Ocean Society, Fort Mason Center, Building E., San Francisco, CA 94123. ISSN 0029-8174.　　　639•
Beautifully illustrated in color photographs and popularly written; usually based on field work or historical research.

Oceanus: the magazine of marine science and policy. 1- (1952-). Quarterly. Woods Hole Oceanographic Institute, Woods Hole, MA, ISSN 0029-8182.　　　640•
Contains articles and book reviews of interest to the scientist and those with a general interest in the oceans. Also, publishes articles dealing with fresh water resources (such as aquaculture) and the marine coastal zone.

Offshore: international journal of ocean business. 1- (1954-). Monthly, except June and October when two issues are published. PennWell Publishing, Box 1260, Tulsa, OK 74101. ISSN 0030-0608.　　　641
The best journal available for information on rapidly changing technological innovations; economic conditions; and statistical data for the near offshore petroleum industry throughout the world.

C. ATLASES

Bramwell, Martyn, ed. [Rand McNally] Atlas of the oceans. London: Mitchell Beazley; Chicago: Rand McNally, 1977. 208 p. ISBN 0-528-83082-1. $35. LC 77-73772.　　　642•
Forty editorial consultants have helped the staff prepare and combine maps and pictorial diagrams of oceanic technology; life; minerals; geology; pollution; and current circulation in one concise volume.

Oceans and Lakes

Couper, Alastar D., ed. The Times atlas of the oceans. London; New York: Time Books Limited, 1983. 272 p. ISBN 0-7230-0246-0. $79.95. LC 83-135037.   643 •

Twenty-six contributors and a cartographic editor examine in detail (map and descriptive text) a wide range of topics, including: the physical characteristics of the oceanic basins and environment; the current status of marine resources and technology for their development; world trade and shipping; environmental problems; the strategic military use of ocean space; current marine research and management; and implications of the Law of the sea for the future of the oceans. The authors stress the importance of the oceans as the last resource frontier. An excellent atlas.

Food and Agriculture Organization of the United Nations, Department of Fisheries. Atlas of the living resources of the seas. (Fisheries series no. 15). 4th ed. Rome, Italy: FAO; New York: United Nations, 1981. 57 plates and 23 p. (1st ed., 1971. [FAO Fisheries Circular no. 126]) ISBN 9-250-01000-1. $120. LC 83-134750.   644 •

Contains detailed maps of the distribution of phytoplankton, zooplankton, value of the 1968 world catch, and catches of demersal and pelagic fish and crustaceans. Presents maps on the migrations of North America cod, Atlanto-Scandian herring, southwest Atlantic anchovita, north Pacific tuna, and Gulf of Mexico shrimp. Text is in English, French, and Spanish.

Rondy, Donald R. Great Lakes ice atlas. (U.S. Department of Commerce. National Oceanic and Atmospheric Administration. Great Lakes Environmental Research Laboratory). Detroit, MI: Great Lakes Research Center, 1983. 78 p. LC 78-628665.   645

Based on 2800 historical Great Lakes ice charts that span 20 winters (1960-1979). The charts have been digitized and converted to discrete 5 x 5 grid cells. These have been used to produce 46 plates that show maximum, minimum, and normal ice condition patterns beginning in the last half of December and ending in the last half of April.

D. GLOSSARY

Veatch, Jethro, and Humphrys, C.R. Water and water use terminology. Kaukauna, WI: Thomas Printing and Publishing, 1966. 375 p. LC 66-21407.   646 •

Designed as a reference for those especially interested in lakes, standing waters, and wetlands.

E. YEARBOOK

Borgese, Elisabeth Mann, and Ginsburg, Norton, eds. Ocean yearbook, 1978- . Annual. Chicago, IL: University of Chicago Press, V. 1: (1978), 890 p.; V.2: (1980), 713 p.; V.3: (1982), 581 p.; V. 4: (1984), 621 p. ISBN 0-226-06605-3. $49. LC 79-642855.   647

Includes selected reprints and updated articles and monographs, as well as original pieces. Although the same topics are not included in each volume, most contain items dealing with the living and non-living resources, transportation, marine science and technology, and environment.

F. TEXTBOOKS

Anikouchine, William A. and Sternberg, Richard W. The World ocean: an introduction to oceanography. 2nd ed. Englewood Cliffs, NJ: Prentice-Hall, 1981. 513 p. (1st ed., 1973. 338 p.) ISBN 0-13-967778-X. $28.95. LC 80-24473.   648 •

Investigates the origin and history of the ocean basins and continental margins, and explores the physical properties and chemistry of seawater and chemical transport. Also, shows the operation of the oceanic environment in relation to atmospheric circulation, currents, and waves. Looks at the biological and ecological characteristics of the seas. Marine sediments and oceanographic instrumentation are examined.

Borgese, Elisabeth Mann. The Drama of the oceans. New York: H.N. Abrams, 1975.
258 p. ISBN 0-8109-0337-7. LC 74-16165.   649•
*A beautifully illustrated volume presented in a highly readable style. Examines the
oceans geological and biological settings, fishermen and sailors of the seas, oilmen
and energy engineers, mineral miners, shore developers, the death of the oceans, and
the prospects and potential future of the oceans.*

Doumani, George A. Ocean wealth: policy and potential. Rochelle Park, NJ: Hayden
Book, 1973. 285 p. LC 72-87830.   650•
*Examines the oceans configuration and resources and demonstrates how various
advances in technology for exploiting these resources have influenced United States
marine policies and international diplomacy among world states.*

Goldman, Charles R. and Horne, Alexander J. Limnology. New York: McGraw-
Hill, 1983. 464 p. ISBN 0-07-023651-8. $31.95. LC 82-15356.   651
*Examines lake, stream, and estuarine systems and integrates methodologies used
in chemistry, biology, and physics in limnological studies. Employs a mathematical
approach.*

Marx, Wesley. The Oceans: our last resource. San Francisco, CA: Sierra Club, 1981.
332 p. ISBN 0-87156-291-X. $13.95. LC 81-5332.   652•
*Uses a semi-narrative style to present a variety of critical topics such as the prob-
lems of contemporary efforts to implement aquaculture, ocean dumping of sludge
and nuclear materials, and deep seabed mining.*

Ragotzkie, Robert A., ed. Man and the marine environment. Boca Raton, FL: CRC
Press, 1983. 180 p. ISBN 0-8493-5759-4. $66. LC 83-3777.   653•
*An examination by marine and earth scientists of our interactions with the sea, on
both human and technical levels. Topics include transportation, fishing, recreation,
diving, coastal, and beach management. The Great Lakes region is examined within
the context of its shipping industry and as a microcosm of the world ocean.*

Ray, John B., ed. The Oceans and man. Dubuque, IA: Kendall-Hunt, 1975. 318 p. LC
75-16943.   654•
*A helpful general introduction to oceanic resources including biotic sources, ener-
gy potential (petroleum, tidal, waves, currents, and wind), hard minerals, water as a
transport mode, military uses, and sites for waste disposal.*

Wetzel, Robert G. Limnology. 2nd ed. Philadelphia, PA: Saunders College, 1983. 767
p. (1st ed., 1975. 743 p.) ISBN 0-03-057913-9. $35.95. LC 81-53073.   655•
*An outstanding examination of fresh waters, with the main focus generally being
lakes. Looks at the origin of lakes, distribution, and forms; oxygenation; light pene-
tration, heat budgets, and temperature stratification; mineral cycling; and planktonic
and littoral communities.*

G. SPECIAL TOPICS

Alexander, Lewis M., ed. The Law of the sea: offshore boundaries and zones. Columbus, OH: Ohio State University, 1967. 321 p. LC 67-16949.   656
*Useful as a benchmark to compare world oceanic resource use and legal regimes
after two decades.*

Burk, Creighton A. and Drake, Charles L., eds. The Geology of continental margins. New York: Springer-Verlag, 1974. 1009 p. ISBN 0-387-06866-X. $52. LC 74-16250.   657
*This valuable volume contains contributions by more than 100 researchers in gov-
ernment, industry, and academia. An indispensable reference for those specializing
in the geology of continental margins.*

Oceans and Lakes

Buzan, Barry. Seabed politics. New York: Praeger, 1976. 311 p. ISBN 0-275-22850-9. LC 75-36406. 658
Useful for its detailed analysis of the development of the Law of the Sea Conferences and their antecedents.

Constans, Jacques. Marine sources of energy. (Pergamon policy studies on energy and environment no. 53). New York; Oxford, U.K.: Pergamon Press, 1979. 169 p. ISBN 0-08-023897-1. $33. LC 79-15200. 659 •
A useful book that examines the contemporary state of marine energy technology available; those sites with the best physical potential for energy production; and prospects for offshore coastal wind energy, ocean thermal energy conversion, marine currents, salinity gradients, and biomass conversion.

Dunst, Russell C. et al. Survey of lake rehabilitation techniques and experiences: an inland lake renewal and management demonstration project report. (Technical bulletin no. 75). Madison, WI: Department of Natural Resources, 1974. 179 p. LC 74-624124. 660
Examines methods of limiting and controlling sedimentation and nutrient entry into lakes and managing the consequences of lake aging. Lists by name and location lakes in the United States and the world that have problems. Identifies what these problems are.

Earney, Fillmore C.F. Petroleum and hard minerals from the sea. (Scripta series in geography). New York: John Wiley, 1980. 291 p. ISBN 0-470-27009-8. $49.95. LC 80-17653. 661 •
A detailed analysis of present and potential world seabed mining activities on the continental shelves and contemporary deep seabed mining research and economic potential (especially for manganese nodules). Looks at the overriding political issues confronting the international community of states for an equitable system of exploitation.

Gilbert, Grove Karl. Lake Bonneville. U.S. Geological Survey (Monograph no. 1). Washington, DC: Government Printing Office, 1890. 438 p. (G55-789). 662
Although out of print and limited in number, this is a classic publication on the process-response approach to shoreline geomorphology. Gilbert covers theory and field work in establishing the mechanisms to describe and explain the Pleistocene shorelines stranded above the Great Salt Lake.

Goldstein, Joan, ed. The Politics of offshore oil. New York: Praeger, 1982. 208 p. ISBN 0-03-059813-3. $25.95. LC 82-7697. 663 •
Focuses on the environmentalist, state, federal, and industrial perspectives of the politics of offshore oil.

Heezen, Bruce C. and Hollister, Charles D. The Face of the deep. New York: Oxford University Press, 1971. 659 p. ISBN 0-19-501277-1. $18.95. (pbk). LC 77-83038. 664 •
A beautifully illustrated volume containing a total of 563 black-and-white diagrams and photographs that focus on the oceans' abyssal plains.

Kennett, James P. Marine geology. Englewood Cliffs, NJ: Prentice-Hall, 1982. 813 p. ISBN 0-13-556936-2. $40.95. LC 81-10726. 665
Provides a brief history of the field of marine geology, and examines in detail oceanic geophysics and morphology, stratigraphy, plate tectonics, crustal structure, surface and deep water circulation, sea-level history, the continental shelf and its associated geological processes, sedimentation, and calcareous and siliceous microfossils.

Ketchum, Bostwick H., ed. Estuaries and enclosed seas. Amsterdam, The Netherlands; New York: Elsevier Scientific, 1983. 500 p. ISBN 0-444-41921-7. Dfl 400. LC 81-9853. 666 •
Looks at estuaries as nurseries for animals and suppliers of food for humans. Examines in detail the world's major closed seas, including the Mediterranean Sea, Black Sea, Red Sea, and the Gulf of St. Lawrence.

McLusky, Donald S. The Estuarine ecosystem. New York: John Wiley and Sons, 1981. 150 p. ISBN 0-470-27127-2. $29.95. LC 80-28199. 667 •
Uses case studies to focus on problems of estuary pollution. Examines the normal functions of estuarine environments including a look at trophic levels. Considers the impact of humankind on estuaries throughout the world using examples that show varying physical conditions.

Nairn, Alan E.M., ed. and Stehli, Francis G. The Ocean basins and margins. New York: Plenum Press, 1973-1983. 7 v. ISBN 0-306-37776-4. $85 per volume. LC 72-83046.
V. 1: **The South Atlantic.** 1973. 600 p.
V. 2: **The North Atlantic.** 1974. 662 p.
V. 3: **The Gulf of Mexico and the Carribean.** 1975. 722 p.
V. 4a: **The Eastern Mediterranean.** 1977. 520 p.
V. 4b: **The Western Mediterranean.** 1978. 447 p.
V. 5: **The Arctic Ocean.** 1981. 672 p.
V. 6: **The Indian Ocean.** 1982. 750 p.
V. 7: **The Pacific Ocean.** 668
A useful series on oceanic geology.

Neal, James T., ed. Playas and dried lakes: occurrence and development. (Benchmark papers in geology, v. 20). New York: Academic Press, 1975. 411 p. ISBN 0-12-787110-1. $69. LC 74-31134. 669
Thirty-eight specialists examine past climates and antecedent lakes in playa basins; hydrological variations and implications; variations in playa types, origins, and regions; processes and surface features; and investigative methods.

Oxman, Bernard H.; Caron, David D.; and Buderi, Charles L.O., eds. Law of the sea: U.S. policy dilemma. San Francisco, CA: Institute for Contemporary Studies, 1983. 184 p. ISBN 0-917616-59-6. LC 83-10788. 670
A collection of publications (by authors who have been part of the Law of the Sea negotiations) that examines the controversy over the U.S. refusal to sign the Law of the Sea Treaty of 1982.

Prescott, John R. The Political geography of the oceans. New York: John Wiley and Sons, 1975. 247 p. ISBN 0-470-69672-9. $24.95. LC 74-31813. 671 •
Focuses on the concepts of territorial sea measurement and claims to territorial seas, fishing zones, continental shelves, and the high seas.

Reid, George K. and Wood, Richard D. Ecology of inland waters and estuaries. 2nd ed. New York: Van Nostrand, 1976. 485 p. (1st ed., 1961. 375 p.) ISBN 0-442-17605-8. LC 75-18032. 672
Investigates the major principles of aquatic ecology, especially from a quantitative systems approach. Looks at origins and features of lake basins, estuaries, and oceans; environmental variables of natural waters including solar radiation, thermal relationships, motion, gasses, and dissolved solids; and also considers organisms and communities in aquatic environments. Contains a useful reference bibliography of more than 600 articles and books.

Serruya, Colette, and Pollingher, Utsa. Lakes of the warm belt. Cambridge, U.K.; New York: Cambridge University Press, 1983. 568 p. ISBN 0-521-23357-7. $89.50. LC 82-19857. 673
Analyzes the warm belt's geodynamic history; climatology; hydrology; aquatic ecosystems of South America, Central America, Africa, the Middle East, South-east Asia, and Australia; mechanisms and patterns of circulation; water chemistry; biological diversity; and case studies of food webs. The warm belt is defined as an area lying between the northern and southern hemisphere glaciations at approximately 30° N. and 30° S. Contains an exhaustive bibliography.

United Nations. The Law of the sea: official text of the United Nations convention on the Law of the Sea: final act of the United Nations conference on the Law of the Sea. New York: United Nations, 1983. 224 p. ISBN 0-312-47555-1. $12.95. LC 83-40157.   674
  A useful volume that presents the final document which will outline the future use of the oceans by the world's states, once the convention is adopted. Defines various territorial zones and outlines mandates for the use of these zones for scientific research, navigation, fishing, mining, and other activities. Outlines an administrative system for the control of seabed areas (the International Deep Seabed Authority) and (the International Tribunal for the Law of the Sea) for settling disputes.

Wick, Gerald L. and Schmitt, Walter R., eds. with the assistance of Clarke, Robin. Harvesting ocean energy. Paris: UNESCO Press; New York: Unipub, 1981. 171 p. ISBN 9-231-01873-6. $17. (pbk). LC 81-211128.   675•
  Considers the oceans as a source of tidal, current, wave, wind, salinity, and temperature differential energy. The resource potential, distribution, extraction methods, economics, and environmental impacts are examined for each energy type.

Williams, W.D., ed. Salt lakes: proceedings of the International Symposium on Athalassic (Inland) Salt Lakes. (Developments in hydrology no. 5). The Hague, The Netherlands; Boston, MA: W. Junk, 1981. 444 p. ISBN 9-06-193756-6. $95. LC 81-11780.   676
  Explores the plant and animal life of salt lakes with examples taken from Australia, Africa, North America, the Pacific Islands, Antarctica, and Europe. Each of the 30 separately authored chapters contains a useful bibliography.

7. COASTAL AREAS

Norbert P. Psuty

A. BOOKS

Barth, Michael C. and Titus, James G., eds. Greenhouse effect and sea level rise: a challenge for this generation. New York: Van Nostrand Reinhold, 1984. 325 p. ISBN 0-442-20991-6. $24.50. LC 84-7207.   677
  Based on papers presented at the 1983 EPA conference on sea level rise, this book looks at the causes and consequences of the "greenhouse effect" and the related rise in the world's sea level. Coastal flooding, salinity intrusion, implications for hazardous waste siting, and economic planning are among the topics well covered on this important issue for the next century.

Bascom, Willard. Waves and beaches: the dynamics of the ocean surface. Garden City, NJ: Anchor Press/Doubleday, 1980. 366 p. ISBN 0-385-14844-5. $8.95. (pbk). LC 79-7038.   678•
  Provides an introduction of mechanics of waves, currents, and sediment transport in a non-technical fashion. Outstanding for readability and communication of information in colloquial style.

Bird, Eric C.F. Coasts: an introduction to coastal geomorphology. 3rd ed. Oxford, U.K.; New York: Basil Blackwell, 1984. 320 p. (1st ed., Canberra, Australia: Australian National University, 1965. 193 p.) ISBN 0-262-02050-5. $25.   679•
  Non-technical portrayal of coastal landforms with many examples from Australia and the U.K. Little discussion of processes. A very readable introduction to a very complex system.

Clark, John R. Coastal ecosystem management: a technical manual for the conservation of coastal zone resources. New York: John Wiley and Sons, 1977. 928 p. Reprinted: Melbourne, FL: Robert E. Krieger, 1983. 928 p. ISBN 0-89874-456-3. $56.50. LC 81-18650.   680

A technical reference book incorporating several-page descriptions of nearly every variable and management issue found in the coastal zone. The theme is coastal management and the procedures for evaluating and accommodating multiple uses in a very dynamic zone.

Davies, John L. Geographical variation in coastal development. 2nd ed. (Geomorphological text series no. 4). London; New York: Longman, 1980. 212 p. ISBN 0-582-49006-5. (pbk). $16.95. (1st ed., Edinburgh, U.K.: Oliver and Boyd, 1972. 204 p.) LC 76-30821. 681
Evaluates the hypothesis that coastal processes and form have geographical order in their distribution. The spatial organization of biologic and physical processes leads to the eventual conclusion that process-response models are geographically constrained. The text is unusually well illustrated.

Davis, Richard A., Jr., ed. Coastal sedimentary environments. New York: Springer-Verlag, 1978. 420 p. ISBN 0-387-90300-3. $25. LC 77-16182. 682
Excellent review type chapters on the sedimentary characteristics of the major coastal environments. Most presentations follow process-response format. The bibliographic entries following each chapter are extensive.

Ducsik, Dennis W. Shoreline for the public: a handbook on social, economic, and legal considerations regarding public recreational use of the nation's coastal shoreline. Cambridge, MA: MIT Press, 1974. 257 p. ISBN 0-262-04045-X. $20. LC 74-4528. 683
A focus on the social and economic values associated with shoreline recreation. An early, easily read treatment of a very complex topic. Economic variables are brought forward as well as public and private rights. The text includes a discussion of non-traditional strategies and options conducted within coastal zone management programs.

Hildreth, Richard G. and Johnson, Ralph W. Ocean and coastal law. Englewood Cliffs, NJ: Prentice-Hall, 1983. 514 p. ISBN 0-13-629204-6. $39.95. LC 82-10221. 684
A kind of glossary which identifies issues in the coastal zone and then brings in applicable case law. Topics range from oil spill liability to public and private rights to resource management. A very instructive compilation of the several sides of critical issues in the coastal and marine zones.

Johnson, Douglas Wilson. Shore processes and shoreline development. New York: John Wiley, 1919. 584 p. Reprinted: New York: Hafner, 1956. 584 p. LC 19-8228. 685
A true classic in coastal geomorphology. Dated in part, this book has historical value in tracing the development of the scientific inquiry in the full range of coastal geomorphology.

Ketchum, Bostwick H., ed. The Water's edge: critical problems of the coastal zone. (Coastal Zone Workshop, Woods Hole, MA, 1972). Cambridge, MA: MIT Press, 1972. 393 p. ISBN 0-262-11048-2. LC 72-7067. 686
A far-ranging document treating nearly every aspect of coastal issues. This book laid the foundation for the U.S. Coastal Zone Management Program. Each chapter focuses on a particular subset of questions and thoroughly presents background, issues, and management needs. Chapters have excellent bibliographies.

King, Cuchlaine, A.M. Beaches and coasts. 2nd ed. London: Edward Arnold; New York: St. Martin's Press, 1972. 570 p. (1st ed., 1959. 403 p.) ISBN 0-312-07035-7. $45. LC 74-187106. 687
An excellent summation of process-response studies applied to coastal geomorphology. Incorporates quantitative approaches in coastal engineering with explanatory descriptions of coastal landforms and beach features. Provides the best coverage of international literature.

Coastal Areas

Komar, Paul D. Beach processes and sedimentation. Englewood Cliffs, NJ: Prentice-Hall, 1976. 429 p. ISBN 0-13-072595-1. $39.95. LC 75-44005.
688
The best textbook on coastal processes and sediment transport. Coastal landforms and coastal change are not well presented. Extremely mathematical in approach. Excellent diagrams contribute greatly to the communication.

McLachlan, Anton, and Erasmus, Theuns, eds. Sand beaches as ecosystems. (Proceedings of the First International Symposium on Sandy Beaches, Port Elizabeth, South Africa, 1983). The Hague, The Netherlands; Boston, MA: Dr. W. Junk, 1983. 757 p.
689
This proceedings volume incorporates review articles on the physical, chemical, ecological, and management aspects of beaches, as well as original papers. The publication is unique in its multidisciplinary focusing on the dynamic beach zone.

Mitchell, James K. Community response to coastal erosion: individual and collective adjustments to hazard on the Atlantic shore. (Research series no. 156). Chicago, IL: University of Chicago, Department of Geography, 1974. 209 p. ISBN 0-89065-063-2. $10. LC 73-92652.
690
An early publication focusing on natural hazard research in the coastal zone. A well-organized presentation on environmental perception and adjustments at individual and community levels at several sites along the Atlantic coast of the United States.

Pethick, John. An Introduction to coastal geomorphology. London; New York: Edward Arnold, 1984. 260 p. ISBN 0-7131-6391-7. $16.95. (pbk).
691
A very well-done introduction to the process-response approach in coastal geomorphology. Basic equations of wave theory and longshore currents are included and described. The coastal landforms are related to process in classical style. An excellent college level introduction to coastal geomorphology.

Schwartz, Maurice L. The Encyclopedia of beaches and coastal environments. (Encyclopedia of earth science series, v. 15). Stroudsburg, PA: Hutchinson Ross, 1982. 940 p. ISBN 0-87933-213-1. $95. LC 81-7250.
692
The most thorough glossary of its kind. Includes regional descriptions of coastal ecology and coastal geomorphology at the continental scale as well as specific coastal types. Entries are usually accompanied by references. The volume is well illustrated by photos and line drawings.

Shapiro, Sidney, ed. Our changing fisheries. (National Marine Fisheries Service). Washington, DC: Government Printing Office, 1971. 534 p. $9. LC 72-600832.
693 •
A handsomely produced, non-technical account of our national fisheries. Basic background is presented as is a region-by-region account of current and future fisheries activity. Although the catch statistics are dated, the issues remain largely current.

Snead, Rodman E. Coastal landforms and surface features: a photographic atlas and glossary. Stroudsburg, PA: Hutchinson Ross, 1982. 247 p. ISBN 0-87933-052-X, $44. LC 81-2949.
694 •
An aerial and ground photo review of many types of coasts and beaches. Excellent international coverage. The organization of photos based on a coastal classification system is questionable but does not detract from the informative scenes.

B. PROCEEDINGS

Association of American Geographers, Specialty Group on Coastal and Marine Geography: Proceedings. 1970- . Annual. Two years sometimes combined. Available from the Chairman of the AAG Specialty Group on Coastal and Marine Geography. In 1985: Prof. James M. McCloy, Director, Coastal Zone Laboratory, Texas A & M University at Galveston, P.O. Box 1675, Galveston, TX 77553.
695
Contains papers presented at the AAG annual meetings in the coastal and marine geography sponsored sessions. Begun in 1970, the coastal and marine geographers have been faithfully recording the interests and activities of their group. A unique collection of coastal geography as presented by the practitioners.

Coastal Engineering Conferences: Proceedings, 1- (1950-). American Society of Civil Engineers, 345 East 47th Street, New York, NY 10017. 696
The single best source for state-of-the-art information on coastal engineering practices, sedimentation, and geomorphology. Aspects of coastal environmental systems are also included. Tends to incorporate theoretical, laboratory, and field aspects of coastal mechanics, structures, and geomorphology. The 1985 proceedings consisted of three volumes with 2780 pages.

Coastal Society Annual Conferences: Proceedings, 1975- . Annual. The Coastal Society, Suite 150, 5410 Grosvenor Lane, Bethesda, MD 20814. 697
These volumes usually stress coastal management issues. The political and social aspects of the coastal zone are well represented. Geographers contribute to these presentations in considerable number in both physical and cultural topics.

Symposia on Coastal and Ocean Management: Proceedings, Coastal Zone 78, 80, 83. American Society of Civil Engineers, 345 East 47th Street, New York, NY 10017. 698
A grouping of papers presented at the International Coastal Zone Conferences. An eclectic combination. Emphasis is on management of coastal resources. Many presentations are excellent state-of-the-art communications on physical and/or cultural geography of the coastal zone. Coastal zone 83, published in 1983 in four volumes, contained 3232 pages.

8. WATER RESOURCES*

John L. Harper

*In this section the editorial committee reduced the number of entries recommended by the contributor.

A. SERIALS

Water resources bulletin. 1- (1965-). Bimonthly. American Water Resources Association, 4510 Grosvenor Lane, Suite 220, Bethesda, MD 20814. ISSN 0043-1370. 699
Concerned with all aspects of water resources from professional papers to technical notes and items of general interest.

Water resources research. 1- (1965-). Bimonthly. American Geophysical Union, 2000 Florida Ave., N.W., Washington, DC 20009. ISSN 0043-1397. 700
An "interdisciplinary journal integrating research in the social and natural sciences of water." Includes reviews, papers, and technical notes, with contributions in hydrology, the physical, chemical and biological sciences, and the social and policy sciences, including economics, systems analysis, sociology, and law.

B. ATLAS

Geraghty, James J.; Miller, David W.; Van Der Leeden, Frits; and Troise, Fred L. Water atlas of the United States. 3rd ed. Syosset, NY: Water Information Center, 1973. 190 p. (1st ed., by Miller, David W. and others, 1962), ISBN 0-912394-03-X. $45. LC 73-76649. 701 •
An updated, considerably expanded version (with 122 plates) of the 1962 atlas (with 40 plates). Atmospheric, hydrologic, and water-use data for all fifty states.

Water Resources

C. INFORMATION SOURCES

Gieffer, Gerald J. Sources of information in water resources: an annotated guide to printed materials. Port Washington, NY: Water Information Center, 1976. 290 p. ISBN 0-912394-15-3. $30. LC 75-20953. 702 •

Easily accessed guide with a section including general works, maps and atlases, scientific treatises, theses, proceedings of meetings and conferences, and water resources research. A specific subject section follows with source listings arranged according to fifty-five categories.

Todd, David K., ed. The Water encyclopedia: a compendium of useful information on water resources. Port Washington, NY: Water Information Center, 1970. 559 p. LC 76-140311. 703 •

A rich storehouse of facts, figures, agency and organization addresses (including university programs in water resources, current to 1970), constants and conversion factors relevant to all aspects of the realm of water.

Water Resources Scientific Information Center. Selected water resources abstracts. 1- (1968-). Monthly. Water Resources Scientific Information Center, U.S. Geological Survey, 425 National Center, Reston, VA, 22092. ISSN 0037-136X. 704

Monthly publication of the Water Resources Scientific Information Center *of the U.S. Geological Survey. Contains "abstracts of current and earlier pertinent monographs, journal articles, reports, and other publication formats." Entries classified in ten fields and sixty groups similiar to the water resource research categories established by the* Committee on Water Resources Research.

D. GENERAL WORKS

Deming, Horace G. Water: the fountain of opportunity. Original MS rev. and updated by W. Sherman Gillam and W.H. McCoy. New York: Oxford University Press, 1975. 342 p. ISBN 0-19-501841-9. $25. LC 74-16657. 705 •

An excellent comprehensive introduction to Earth's most basic resource: water. Chemist Deming left no aspect of water's influence untouched. The book reviews the molecular structure of water, the fluid's effects on land use, and human abuses of water, among other subjects discussed. Informative illustrations. Well prepared.

Dunne, Thomas, and Leopold, Luna B. Water in environmental planning. San Francisco, CA: W.H. Freeman, 1978. 818 p. ISBN 0-7167-0079-4. $43.95. LC 78-8013. 706

Compendium of information presented in a scholarly fashion for all water resource planners. The twenty-two chapters are arranged in three parts: hydrology; geomorphology; and river quality. Exceptionally thorough coverage, amply illustrated; an all around encyclopedic treatise on water in the environment.

Leopold, Luna B. Water: a primer. San Francisco, CA: W.H. Freeman, 1974. 172 p. ISBN 0-7167-0264-9. $20.95. ISBN 0-7167-0263-0. $11.95. (pbk). LC 73-19844. 707 •

A "must" for any library's water resource collection. Hydrologist Leopold provides a clear and concise introduction (an honest primer) to water as a resource. Particularly helpful to the education of the layperson are Leopold's explanations of the drainage network, the floodplain, load in streams, and the water budget. Excellent illustrations accompany the text.

Miller, David H. Water at the surface of the earth: an introduction to ecosystem hydrodynamics. (International geophysical series v. 21). New York: Academic Press, 1977. 557 p. ISBN 0-12-496750-7. $74.50. LC 76-13947. ISBN 0-12-496752-3. $24. (text ed.). LC 82-13769. 708

General scientific treatment of the entire spectrum of water in ecosystems. Special concentration on the atmospheric portion of the hydrologic cycle, on the critical matter of infiltration at the soil surface, and on the effects of evaporation. Monumental contribution worthy of inclusion in major library collections.

Powledge, Fred. Water: the nature, uses, and future of our most abused resource. New York: Farrar, Straus Giroux, 1982. 423 p. ISBN 0-374-28660-4. $13.95. ISBN 0-374-51798-3. $7.95. (pbk). LC 82-7471.   709 •

A "battle-book," or summons to action, written for the edification of the general public by an independent author with literary skills. The book coverage spans the problems of water quantity and quality that afflict America in the absence of a national water policy.

U.S. Water Resources Council. The Nation's water resources, 1975-2000: second national water assessment. Washington, DC: Government Printing Office, 1978. 4 v. LC 79-601659.   710 •

The most comprehensive appraisal yet assembled by the federal government. An update and expansion of the first assessment completed in 1968. Volume 1 is the summary. Volume 2, in five parts, looks at water quantity and quality. Volume 3, five appendices plus a summary document, compiles statistics. Volume 4 includes 21 regional water resource reports.

E. SPECIAL TOPICS

Anderson, Terry L. Water crisis: ending the policy drought. Baltimore, MD: Johns Hopkins University Press, 1983. 121 p. ISBN 0-8018-3087-7. $15. ISBN 0-8018-3088-5. $7.95. (pbk). LC 83-48046.   711

Anderson perceives the water crisis as a failure in government direction of water resource development and management. A spokesman for the school of "new" economics that would convert water rights into absolute property rights and let market forces play freely, he foresees a convergence of purpose and coalition of environmentalists and fiscal conservatives.

Angelides, Sotirios, and Bardach, Eugene. Water banking: how to stop wasting agricultural water. San Francisco, CA: Institute for Contemporary Studies, 1978. 52 p. ISBN 0-917616-26-X. $2. (pbk). LC 78-50766.   712

An argument for water use efficiency through water banking, the legal transfer of the unused portion of a water allocation from one party to another without disturbing either party's water rights. Market prices prevail; compensation to damaged third parties is implicit. Conservation incentives are automatic, notably in reducing demand for development of new water sources.

Davis, Stanley N. and DeWiest, Roger J.M. Hydrogeology. New York: John Wiley and Sons, 1966. 463 p. ISBN 0-471-19900-1. $36.95. LC 66-14133.   713

An exceptionally readable textbook treatment of groundwater by a geologist and a hydrologist. Though technical, especially in reviewing flow theory, early chapters on the properties of water and concluding chapters on groundwater occurrence and exploration are digestible by anyone with modest scientific training. An abundance of lucid illustrations.

Eckstein, Otto. Water-resource development: the economics of project evaluation. (Economic studies no. 104). Cambridge, MA: Harvard University Press, 1958. 300 p. ISBN 0-674-94785-1. $20. LC 58-7501.   714

One of the classics for any student of water resources despite the age of the book. An examination of cost-benefit analysis in theory and practice, with special reference to flood control, navigation, irrigation, and hydropower generation.

Greeson, Phillip E.; Clark, John R.; and Clark, Judith E., eds. Wetland functions and values: the state of our understanding. (Proceedings of the National Symposium on Wetlands, held in Disneyworld Village, Lake Buena Vista, Florida, November 7-10, 1978). (Technical publication series, TPS 79-2). Minneapolis, MN: American Water Resources Association, 1979. 674 p. NUC 80-372186.   715 •

A grand collection of 57 papers on the often overlooked wetlands environment. Studies are grouped in such categories as wetlands food chains, habitats, hydrology, water quality, human harvests, and conservation. The book provides good foraging for any student of water resources.

Hirshleifer, Jack; DeHaven, James D.; and Milliman, Jerome W. Water supply: economics, technology, and policy. Chicago, IL: University of Chicago Press, 1960. 378 p. LC 60-14355.

716

Reporting the results of a Rand Corporation study, the three authors have produced a timeless analysis of water supply practice. They present a case for the application of principles of efficiency and economy to an industry which overbuilds and underprices. Every major library should have this volume.

Kneese, Allen V. and Bower, Blair T. Managing water quality: economics, technology, institutions. Baltimore, MD: Johns Hopkins Press, 1968. 328 p. LC 68-8290. Available from Books on Demand, Ann Arbor, MI: University Microfilms International. $84.50. (pbk).

717

A first-rate assessment, in textbook format, of the nation's water quality problems and solutions with carefully documented case studies, i.e., the Deleware estuary, ORSANCO in the Ohio basin, and Germany's Ruhr region "Genossenschaften." Applicable economic concepts and management alternatives are reviewed in detail; the authors promote regional agency management.

Sheaffer, John R. and Stevens, Leonard A. Future water: an exciting solution to America's most serious resource crisis. New York: William Morrow and Co., 1983. 269 p. ISBN 0-688-01575-1. $14.95. LC 83-61855.

718

Vigorous promotion by the authors of water reclamation employing a "circular system" (or feedback-loop array) with recovery of pollutants for use or sale. Disavowal of the standard "linear system" approach to clean water--treatment of virgin supply, use, treatment (or not) of waste or discharge.

Ward, James V. and Stanford, Jack A., eds. The Ecology of regulated streams. (Proceedings of the first International Symposium on Regulated Streams, Erie, Pennsylvania, April 1979). New York: Plenum Press, 1979. 398 p. ISBN 0-306-40317-X. $49.50. LC 79-21632.

719

With its perspective being "dammed rivers of the world" and their effects on riverine ecology, the symposium's papers deal with natural stream ecosystem character, phyto- and zoobenthos and fish in regulated streams, chemistry and temperature modification in such waters, and changes in channel morphology.

White, Gilbert F. Strategies of American water management. Ann Arbor, MI: University of Michigan Press, 1969. 155 p. LC 69-15842.

720

Distinguished American geographer Gilbert White analyzes the institutionalized standards by which water resource developments are justified and management of water supply is undertaken. He counters with an alternative strategy, multiple means evaluation, which since 1969 has influenced the thinking of economists, water planners, and environmentalists.

Wollman, Nathaniel, and Bonem, Gilbert W. The Outlook for water: quality, quantity, and national growth. Baltimore, MD: Johns Hopkins Press, 1971. 286 p. ISBN 0-8018-1260-7. $24. LC 75-149243.

721

Nathaniel Wollman is prominent among water resource authorities in America. At the behest of Resources for the Future, *economist Wollman and co-author Bonem prepared a scholarly assessment and warning supported by economic theory and modeling, of the plight of the water resource if current development and use trends continue.*

Zwick, David, and Benstock, Marcy. Water wasteland: Ralph Nader's study group report on water pollution. New York: Grossman, 1971. 494 p. ISBN 0-670-75169-3. LC 77-112516.

722•

This hard-hitting study gets it "out in the open" regarding specific polluting industries, the problem of non-point sources, the federal government itself as a polluter, and the general ineffectiveness of the Federal Water Quality Administration's control program. Typical of Nader group investigations, this report spotlights offenders and makes tough recommendations.

Dunbar, Robert G. Forging new rights in western waters. Lincoln, NE: University of Nebraska Press, 1983. 278 p. ISBN 0-8032-1663-7. $19.95. LC 82-13421.

    723

    Historian Dunbar provides an excellent foundation for understanding the evolution of irrigated agriculture, the West's largest water consumer. From Hohokam, Mormon, and Greeley Colony farmers to federal intervention in reclamation, the author adroitly sets the stage for analysis of the emergence of appropriation rights in western states upon failure of riparian doctrine.

Ferejohn, John A. Pork barrel politics: rivers and harbors legislation, 1947-1968. Stanford, CA: Stanford University Press, 1974. 288 p. ISBN 0-8047-0854-1. $22.50. LC 73-98959.

    724

    Unusual in its probing investigation of the progress of water resource project development in the United States from grass-roots conception of an idea through cost-benefit analysis to the battles in Congress. Especially revealing are sections on the appropriation process for the Army Corps of Engineers *and the distribution of the Corps' construction projects.*

Foreman, Richard L. Indian water rights: a public policy and administrative mess. Danville, IL: Interstate Printers and Publishers, 1981. 233 p. ISBN 0-8134-2160-8. $8.95. (pbk). LC 80-82922.

    725 •

    Foreman, a non-Indian political scientist, introduces the white man's imprint on the American West, particularly his law which conflicts with Indian common law. He covers the evolution of "trust responsibility" embodied in the reservation movement by our federal government and argues for state control of water rights with protection of reserved rights for Native Americans.

Fradkin, Philip L. A River no more: the Colorado River and the West. Tucson, AZ: University of Arizona Press, 1984. 360 p. ISBN 0-8165-0823-2. $10.95. (pbk). LC 83-18053.

    726 •

    With its principal focus being the long history of negotiation, intrigue, and structural development in the Colorado River system, this book by a Pulitzer prize-winning journalist also looks acutely at the landscapes (high country, Indian lands, canyon country, desert) and the people that characterize the water deficient West.

Hargrove, Erwin C. and Conkin, Paul K., eds. TVA: fifty years of grass-roots bureaucracy. Urbana, IL: University of Illinois Press, 1983. 345 p. ISBN 0-252-01086-8. $24.95. LC 83-6475.

    727 •

    An anthology of papers treating the enigmatic Tennessee Valley Authority, *its creation and history, leadership and special privileges, and regional development made possible by the basin wide authority. Reviews how* TVA *is viewed both as monolithic bureaucracy intruding on private enterprise and as cherished government benefactor to a region in need.*

Haveman, Robert H. Water resource investment and the public interest: an analysis of federal expenditures in ten southern states. Nashville, TN: Vanderbilt University Press, 1965. 199 p. ISBN 0-8265-1077-9. $11.95. LC 65-18545.

    728

    Investigation of the inadequacies of the cost-benefit ratio. Under Haveman's scrutiny go matters of regional income distribution and risk and uncertainty in project development, all presenting problems in measurement when the concern is econmic efficiency.

Minerals

Howe, Charles W. and Easter, K. William. Interbasin transfers of water: economic issues and impacts. Baltimore, MD: Johns Hopkins Press, 1971. 196 p. ISBN 0-8018-1206-2. $14.50. LC 78-149241.   729

Interbasin water transfer often is perceived as the ultimate solution to slake a water-deficient region's thirst. Howe and Easter look critically at the economic efficiency, benefits and costs, and notably the social impacts of such transfers. Hypothetical Pacific Northwest to Pacific Southwest transfer is employed as a model.

Maass, Arthur. Muddy waters: the Army Engineers and the nation's rivers. Cambridge, MA: Harvard University Press, 1951. 306 p. Reprinted. New York: Da Capo Press, 1974. 306 p. ISBN 0-306-70607-5. $39.50. LC 73-20238.   730

A clinical probe into the workings of the civilian side of the Army Corps of Engineers, the dam builders. The book reveals the Corps' several allegiances to the engineering profession, group interests, and the executive and legislative branches of government that complicate and taint objectivity in project choice and development.

Meyers, Charles J. and Tarlock, A. Dan. Water resource management: a coursebook in law and public policy. 2nd ed. Mineola, NY: Foundation Press, 1980. 1127 p. (1st ed., 1971. 984 p.) 1983 Supplement by A. Dan Tarlock, 1982. 223 p. ISBN 0-88277-103-5. LC 79-55248.   731

In law book format, the authors review riparian and appropriation law in the United States noting economic aspects of water allocation. State powers and the permit system are analyzed, as are reserved water rights. Interstate and federal-state disputes are assessed. Special cases of groundwater management, water pollution, and water based recreation are addressed.

Warne, William E. The Bureau of Reclamation. New York: Praeger, 1973. 270 p. LC 70-151960.   732

Mastermind of California's water plan and former bureau employee, Warne paints a favorable image of "BuRec" as dam builder and irrigator. He reviews the 160-acre limitation and hydropower's "cash-register effect" that pays for projects. Good analyses of Colorado River, Columbia and Missouri basins, and California Central Valley reclamation developments.

9. MINERALS

E. Willard Miller

See also Part III, Section 10, Energy.

Alexandersson, Gunnar, and Klevebring, Bjorn-Ivar. World resources, energy, and minerals: studies in economic and political geography. Berlin; New York: De Gruyter, 1978. 248 p. ISBN 3-11-006577-0. $18. LC 77-27560.   733 •

A survey of the exploitation and utilization of the products of the mineral realm. Reveals the importance of mineral raw materials in our modern industrial society.

Anders, Gerhard, et al. The Economics of mineral extraction. New York: Praeger, 1980. 316 p. ISBN 0-03-053171-3. $45.95. LC 79-22949.   734

Provides a pragmatic approach to the analysis of public policy issues in mineral exploitation. Analyzes primary determinants such as price, output, and investment in mineral extraction. Provides a methodology for evaluating the impact of governmental policies on the mineral extracting industries.

Banks, Ferdinand E. Bauxite and aluminum: an introduction to the economics of non-fuel minerals. Lexington, MA: Lexington Books, 1979. 189 p. ISBN 0-669-02771-5. $24.50. LC 78-24632.
735•
An introduction to the economics of nonfuel minerals in general and the bauxite-alumina-aluminum industry in particular. Emphasis is placed on the outlook and problems of the industry in Australia which is the largest supplier of bauxite in the world.

Barger, Harold, and Schurr, Sam H. The Mining industries, 1899-1939: a study of output, employment, and productivity. (National Bureau of Economic Research series). Salem, NH: Ayer, 1975. 447 p. ISBN 0-405-07575-8. $32. LC 75-19694.
736•
A historical study of the progress in the physical efficiency of the mining industry in the United States from 1899 to 1939. Examines technological change, new production methods, shortening of the workday, and other factors that influence productivity.

Bosson, Rex, and Varon, Bension. The Mining industry and the developing countries. New York: Oxford University Press, 1977. 292 p. ISBN 0-19-920096-3. $29.50. ISBN 0-19-92099-8. (pbk). $14.95. LC 77-2983.
737•
An overview of the world mining industry: structure; objectives; and operation. Considers such major factors as physical characteristics of mineral resources, economies of scale, capital requirements, economic and political risks, production, consumption, trade, and the industry's impact on economic growth with particular reference to developing countries.

Carlson, Rodger D. The Economics for profitable mining and marketing of gold, silver, copper, lead, and zinc ores. Lanham, MD: University Press of America, 1982. 78 p. ISBN 0-8191-2021-9. $20.75. ISBN 0-8191-2022-7. $7.50. (pbk). LC 81-40818.
738
A brief survey of selected mineral resources and individual mining sites. Provides specific insights to the exploitation of minerals from a geologic, extractive, and economic perspective.

Cobbe, James. Governments and mining companies in developing countries. Boulder, CO: Westview, 1979. 332 p. ISBN 0-89158-562-1. $33. LC 79-4851.
739
A monograph on the role of governmental actions in the establishment of mining companies in the developing countries. Stresses political and economic consequences of mineral development.

Dixon, Colin J. Atlas of economic mineral deposits. Ithaca, NY: Cornell University Press, 1979. 143 p. ISBN 0-8014-1231-5. $115. LC 78-65360.
740•
Depicts the geology of a select number of mineral deposits showing where they are, what they consist of, how large they are, and how they are related to the geological environment in which they occur.

Eckes, Alfred E. The United States and the global struggle for minerals. Austin, TX: University of Texas Press, 1979. 353 p. ISBN 0-292-78506-2. $20. ISBN 0-292-78511-9. $9.95. (pbk). LC 78-11082.
741
A survey from World War I to the present exploring how natural resource considerations have influenced American foreign relations. Evaluates the changing relationships in world mineral availability.

Fischman, Leonard L. World mineral trends and U.S. supply problems. (Resources for the Future research paper no. R-20). Baltimore, MD: Johns Hopkins, 1981. 535 p. ISBN 0-8018-2491-5. $20. (pbk). LC 80-8025.
742
Considers long-term potential supply and price problems for seven major nonfuel minerals (aluminum, chromium, cobalt, copper, lead, manganese, and zinc). Changing world policies affecting mineral accessibility highlights the need for an American policy on mineral stockpiles, tariffs, and substitution.

Minerals

Harris, De Verle P. Mineral resources appraisal: mineral endowment, resources, and potential supply: concepts, methods, and cases. New York: Oxford University Press, 1984. 445 p. ISBN 0-19-854456-1. $69. LC 83-11445.　　　　　743
A quantitative study providing the conceptual framework for appraisal of potential mineral resources of an area. Because of the uncertainty of undiscovered mineral deposits, the model is based on issues of statistical estimation.

Hartley, James N.; Eschbach, E.A.; and Wick, O.J. World mineral and energy resources: some facts and assessments. (Battelle monograph series no. 6). Richland, WA: Battelle, Pacific Northwest Laboratories, 1974. 136 p. LC 78-323226. NUC 78-84268.　　744
An assessment of the basic mineral and energy resources of the world providing a foundation for developing models to predict future mineral sources.

Klass, Michael W.; Burrows, James C.; and Beggs, Stephen D. International minerals cartels and embargoes: policy implications for the United States. New York: Praeger, 1980. 274 p. ISBN 0-03-044-366-0. $39.95. LC 80-11123.　　　　745
Analyzes the development of cartels in the developing countries since 1970 in order to control mineral production and world prices, and the effects on trade with the United States.

Leontief, Wassily, et al. The Future of nonfuel minerals in the U.S. and world economy: input-output projections, 1980-2030. Lexington, MA: Lexington Books, 1983. 454 p. ISBN 0-669-06377-0. $41. LC 82-48956.　　　　746
Development of an input-output model to set projections of production, processing, and demand for nonfuel minerals to 2030 A.D. in the United States and the world economy. Presents a logical progression from a brief survey of resources and the economy through the methodological approach, assumptions, alternative projections, and projections of demand.

McDivitt, James F. and Manners, Gerald. Minerals and men: an exploration of the world of minerals and metals including some of the major problems that are posed. Rev. and enl. ed. Baltimore, MD: Johns Hopkins University Press for Resources for the Future, 1974. 175 p. (1st ed., 1965. 158 p.) ISBN 0-8018-1536-3. $15. ISBN 0-8018-1827-3. $4.95. (pbk). LC 73-8138.　　　　747 •
For interested, but not expert, readers who want to know more about the nonfuel minerals that are so important in our modern economy. Provides a blend of geological and economic information.

Mikesell, Raymond F. The World copper industry: structure and economic analysis. Baltimore, MD: Johns Hopkins University Press, 1979. 393 p. ISBN 0-8018-2270-X. $10.95. (pbk). LC 79-4581.　　　　748
Survey of the world copper industry including physical characteristics, markets, prices and costs, demand, quantitative analysis of supply, investments, economic issues in nationalization, and special problems in the future.

Muller-Ohlsen, Lotte. Non-ferrous metals: their role in industrial development. Cambridge, U.K.: Woodhead-Faulkner in association with Metallgesellschaft AG, 1981. 297 p. ISBN 0-85941-190-7. $50. (New York: State Mutual Book and Periodical Service). LC 82-166537.　　　　749
Provides a broad perspective of the key role non-ferrous metals play in industrial production. Major topics include production structure, origins of the industry, factors determining demand, patterns of consumption, supply situation, pricing, regulation of markets, and trends in world trade.

Roberts, Peter W. and Shaw, Tim. Mineral resources in regional and strategic planning. Aldershot, U.K.; Brookfield, VT: Gower, 1982. 165 p. ISBN 0-566-00395-3. $37.25. LC 82-125939.　　　　750
An evaluation of planning procedures that are needed in order to ensure a supply of essential minerals in both the long and short term. Analysis includes the significance and need for planning, decision making, current failures and weaknesses in planning, and alternative approaches and new policies.

Sideri, S. and Johns, Sheridan., eds. Mining for development in the Third World: multinational, corporations state enterprises and the international economy. Oxford, U.K.; New York: Pergamon, 1980. 360 p. ISBN 0-08-026308-9. $39. LC 80-20930.   751
  Evaluates the mining industry in its global perspective. A spectrum of viewpoints ranging from advocacy of direct foreign investment to vigorous opposition is presented. Concludes with a major discussion of mining policies for economic development. Based on a workshop organized by the Institute of Social Studies at The Hague in 1979.

Spurr, Josiah E., ed. Political and commercial geology and the world's mineral resources. New York: McGraw-Hill, 1920. 562 p. Reprinted: New York: Garland, 1983. 562 p. ISBN 0-8240-5378-8. $60. LC 82-48322.   752
  An analysis of the control of mineral exploitation by governments and multinational companies. The question of domestic and foreign governmental policies of the United States is evaluated by specialists following World War I.

Sutulov, Alexander. Minerals in world affairs. Salt Lake City, UT: University of Utah Printing Services, 1972. 200 p. LC 72-88498.   753 •
  A general treatment of the place of minerals in our present day society. Ore reserves, supply-demand relations, markets, trade, and investments provide the background for understanding the importance of minerals in providing basic human needs.

Tanzer, Michael. The Race for resources: continuing struggles over minerals and fuels. New York: Monthly Review Press, 1980. 285 p. ISBN 0-85345-541-4. $6.50. (pbk). LC 80-18027.   754 •
  An analysis of the widely proclaimed "mineral crisis" within the historical context of the workings of the international capitalist system. The focal point is the many sided and complex struggles for control of mineral resources and profits among companies and countries.

Tilton, John E. The Future of nonfuel minerals. Washington, DC: Brookings Institution, 1977. 113 p. ISBN 0-8157-8460-0. $14.95. LC 77-8186.   755
  A study of the world's economic and political changes in relation to the ability of the multinational mineral corporations to provide adequate mineral supplies, at competitive prices, to the industrial nations.

Tilton, John E., ed. Material substitution: lessons from tin-using industries. Washington, DC: Resources for the Future, 1983. 118 p. ISBN 0-8018-3161-X. LC 83-16164.   756
  A detailed examination of the factors causing substitution in three major tin-using industries: beverage containers; solder; and water and waste pipes.

Vogely, William A., ed. and Risser, Hubert E., assoc. ed. Economics of the mineral industries: a series of articles by specialists. 3rd ed. Completely rev. and rewritten. New York: American Institute of Mining, Metallurgical, and Petroleum Engineers, 1976. 863 p. (1st ed., by Edward H. Robie, 1959. 755 p.; 2nd ed., 1964. 787 p.) ISBN 0-89520-033-3. $30. LC 72-86920.   757
  A substantial survey of the mineral industries considering the economic role of minerals, central problems arising from the analysis of resources, function, structure, and public policy of the mineral industries and minerals and the future. The volume provides a bridge between mineral economics and mineral engineering.

Warren, Kenneth. Mineral resources. Newton Abbot, U.K.: David and Charles; New York: John Wiley and Sons, 1973. 272 p. ISBN 0-7153-5851-0. LC 73-162344.   758
  Evaluates problems of supplying the growing demand for minerals in the world's industrial economies.

Weston, Rae. Strategic materials: a world survey. Totowa, NJ: Rowman and Allanheld, 1984. 189 p. ISBN 0-86598-165-5. $34.50. LC 84-3358.   759 •
  A survey of strategic mineral resources and the possibility of disruption of supply due to political events in major producing countries. Suggests that stockpiles threaten prices of these materials and inhibit supply and should be replaced, whenever possible, by development of potential new sources.

10. ENERGY

Martin J. Pasqualetti

See also Part IV, Section 23, Technological Hazards.

A. SERIALS

Energy journal. 1- (1980-). Quarterly. Oelgeschlager, Gunn and Hain, Inc., 131 Clarendon St., Boston, MA 02116. ISSN 0195-6574. 760
An important eclectic journal. Papers ranging from descriptive to highly technical.

Energy policy. 1- (1973-). Quarterly. Buttersworth Scientific Ltd., Journals Division, Box 63, Westbury House, Bury St., Guildford, Surrey GU2 5BH, U.K. ISSN 0301-4215. 761
An important journal of growing research interest.

Petroleum economist. 1- (1934-). Monthly. Petroleum Press Bureau, Ltd., P.O. Box 105, 107 Charterhouse St., London EC1M 6AA, U.K. ISSN 0306-395X. 762
An indispensable publication on current events and developments in oil.

World energy industry. Quarterly. Business Information Display, Inc., 4202 Sorrento Valley Blvd., Ste. J., San Diego, CA 92121. 763
An excellent compilation of a wide variety of data in common units and clear graphics.

B. ATLAS

Cuff, David J. and Young, William J. The United States energy atlas. 2nd ed. New York: Free Press, 1984. 416 p. (1st ed., 1980. 416 p.) ISBN 0-02-691250-3. $85. LC 80-24942. 764•
The best energy atlas for the U.S. resources, transportation, production, consumption, and more.

C. DICTIONARIES

Couniham, Martin. A Dictionary of energy. London; Boston, MA: Routledge and Kegan Paul, 1981. 157 p. ISBN 0-7100-0847-3. $14.95. LC 81-8703. 765•
Defines terms related to all aspects of energy: nature; farms; conversion and use; sources; and economic and political aspects. Some definitions extend to factual summaries on particular energy-related subjects.

Slesser, Malcolm, general ed. Dictionary of energy. London; New York: Macmillan Press, 1983. 299 p. ISBN 0-8052-3816-6. $29.95. LC 82-10252. 766•
The best one-volume energy dictionary of a relatively non-technical nature. Less all-inclusive than other energy dictionaries, but with more detail per definition. Clearly written.

World Energy Conference. Energy terminology: a multi-lingual glossary. Oxford, U.K.; New York: Pergamon, 1978. 270 p. ISBN 0-08-029314-X. $100. ISBN 0-08-029315-8. $40. (pbk). LC 82-337. 767
A somewhat dated but excellent dictionary giving equivalent definitions in English, French, German, and Spanish.

D. COAL

James, Peter. The Future of coal. 2nd ed. London: Macmillan, 1984. (1st ed., 1982.)
275 p. ISBN 0-333-36567-4. £25. ISBN 0-333-36521-6. £8.95. (pbk).   768 •
An excellent recent overview of the worldwide coal situation. Many maps and tables of data.

Manners, Gerald. Coal in Britain: an uncertain future. London; Boston, MA: Allen
and Unwin, 1981. 108 p. ISBN 0-04-333018-5. $25. ISBN 0-04-333019-3. $11.95. (pbk).
LC 81-189518.   769
The future need for coal is investigated in terms of energy policies, the power station market, the remainder of the home market, and the prospects for international trade. An excellent introduction to coal in Great Britain.

E. PETROLEUM AND NATURAL GAS

Manners, Ian R. North Sea oil and environmental planning: the United Kingdom experience. Austin, TX: University of Texas Press, 1982. 332 p. ISBN 0-292-76475-8.
$37.50. LC 81-16170.   770
An insight appraisal of an increasingly important source of petroleum. Applicability in places other than the U.K. Special emphasis is given to marine pollution and onshore planning.

Odell, Peter R. An Economic geography of oil. London: G. Bell, 1963. Reprinted:
Westport, CT: Greenwood, 1976. 219 p. ISBN 0-8371-8721-4. $19.75. LC 75-43868.   771 •
An analysis of the locational patterns of the world oil industry on a function-by-function basis. A revised edition is in preparation.

Odell, Peter R. Oil and world power. 7th ed. New York: Penguin Books, 1983. 288 p.
(1st ed., 1970. 188 p.) ISBN 0-14-022284-7. (pbk). $5.95. LC 84-673187.   772 •
The best brief book on oil in terms of its international implications. A new edition is in preparation.

Odell, Peter R. and Rosing, Kenneth E. The Future of oil. 2nd ed. Mundelein, IL:
Nichol Publications, 1983, 265 p. (1st ed., London: Kogan Page, 1980. 265 p.) ISBN 0-
89397-146-4. $23.95. (pbk). LC 80-26192.   773
Estimates of oil future in the 1960's and 1970's are taken and used to simulate production profiles to produce a forecast for the future of oil. A cleverly argued book, the conclusions of which have implications on energy policy makers in many countries.

Russell, Jeremy. Geopolitics of natural gas. Cambridge, MA: Ballinger, 1983. 181 p.
ISBN 0-88410-610-1. $24.50. LC 83-2670.   774 •
A report of Harvard's Energy and Environmental Policy Center, this book provides a good introduction to the geopolitical factors likely to influence future developments of the international trade in natural gas through the year 2000.

Tiratsoo, E.N. Oilfields of the world. 3rd ed. Beaconsfield, U.K.: Scientific Press,
1984. 392 p. (1st ed., 1973. 376 p.) ISBN 0-901360-14-7. £40. LC 76-367011.   775 •
A current, comprehensive account, with many maps. The essence of each oil field is given in terms of location and production.

F. NUCLEAR POWER

Glasstone, Samuel, and Jordan, Walter H. Nuclear power and its environmental effects. La Grange Park, IL: American Nuclear Society, 1980. 395 p. ISBN 0-89448-022-7.
$29.95. ISBN 0-89448-024-3. $17.95. (pbk). LC 80-67303.   776
Very good overview of actual, anticipated, and potential environmental impacts. Of particular interest are chapters on siting, transportation, and waste disposal.

Energy

Kasperson, Roger E. and Berberian, Mimi, eds. Equity issues in radioactive waste management. Boston, MA: Oelgeschlager, Gunn and Hain, 1984. 416 p. ISBN 0-89946-055-0. $30. LC 81-18702.   777

A collection of papers on a subject of relatively new geographic interest. Intertemporal and interspatial equity problems are addressed.

Pasqualetti, Martin J. and Pijawka, K. David, eds. Nuclear power: assessing and managing hazardous technology. Boulder, CO: Westview Press, 1984. 423 p. ISBN 0-86531-811-5. $31. LC 83-21797.   778

Fifteen papers on risk acceptability, spatial assessment and impact mitigation, socioeconomic impacts, risk perception and emergency planning, and present issues (i.e. waste, decommissioning). A representative sampling.

Sills, David L.; Wolf, C.P.; and Shelanski, Vivien B., eds. Accident at Three Mile Island: the human dimensions. Boulder, CO: Westview Press, 1982. 258 p. ISBN 0-86531-165-X. $25. ISBN 0-86531-187-0. $12.50. (pbk). LC 81-10413.   779 •

An important book in terms of its treatment of the human aspects of a nuclear accident. It is divided into five parts: public perceptions; local responses; institutional responsibilities; interactions of social and technical systems; and implications for public policy.

G. NONCONVENTIONAL ENERGY SOURCES

Armstead, H. Christopher H., ed. Geothermal energy: its past, present, and future contributions to the energy needs of man. 2nd ed. London; New York: Methuen, 1983. 404 p. (1st ed., London: E. and F.N. Spon; New York: John Wiley and Sons, 1978. 357 p.) ISBN 0-419-12220-6. $53. LC 83-337.   780 •

A very good introductory book on an energy resource with great potential in two-thirds of the countries in the world. Many references, diagrams, and some photographs make it of value to the scientist and layperson alike.

Edmunds, Stahri W. and Rose, Adam L., eds. Geothermal energy and regional development: the case of Imperial County, California. New York: Praeger, 1979. 371 p. ISBN 0-03-053316-3. $44.95. LC 79-19219.   781

A detailed look at the development of the most site specific of all energy resources in one of the most intensive agricultural areas of the world. A micro-regional approach is advanced which may usefully be applied during the development of other energy resources, particularly in the western U.S.

Pryde, Philip R. Nonconventional energy resources. New York: John Wiley and Sons, 1983. 270 p. ISBN 0-471-86807-8. $22.50 LC 82-21827.   782 •

A broad survey of the attributes, distribution, limitations, and environmental effects of solar, wind, geothermal, hydroelectric, biomass, nuclear, electrochemical, and oceanic energy resources.

H. GENERAL WORKS

Anderer, Jeane, with McDonald, Alan, and Nakicenovic, Nebojsa, eds. Energy in a finite world: paths to a sustainable future. Cambridge, MA: Ballinger, 1981. 256 p. ISBN 0-88410-641-1. $25. LC 80-20057.   783 •

A summary of comprehensive works which reported on a study by the Energy Systems Program Group *of the* International Institute for Applied Systems Analysis. *Similiar to* Energy in a Finite World: A Global Systems Analysis *by Wolf Häfele.*

Cook, Earl. Man, energy, society. San Francisco, CA: W.H. Freeman, 1976. 478 p. ISBN 0-7167-0725-X. $25.95. ISBN 0-7167-0724-1. $13.95. (pbk). LC 75-33774.   784 •

A broad review of the role of energy in modern society with an emphasis on the fossil fuels and energy policy questions that face the U.S.

Daneke, G.A., ed. Energy, economics, and the environment: toward a comprehensive perspective. Lexington, MA: Lexington Books, 1982. 283 p. ISBN 0-669-04717-1. $31. LC 81-47690.
785
The sixteen papers of this publication stimulate thinking about alternative modes of analysis and energy perspective affecting U.S. energy policy. Many have geographical implications or authorship.

Häfele, Wolf. Energy in a finite world: a global systems analysis. Cambridge, MA: Ballinger, 1981. 837 p. ISBN 0-88410-642-X. $50. LC 80-20057.
786
A comprehensive report by the Energy Systems Program Group of the International Institute for Applied Systems Analysis with excellent coverage of virtually all fuel types.

Lakshmanan, T.R., and Nijkamp, Peter, eds. Systems and models for energy and environmental analysis. Aldershot, U.K.; Brookfield, VT: Gower, 1983. 233 p. ISBN 0-566-00558-1. $35.50. LC 83-101241.
787
Fourteen papers divided into three sections including a general framework, The Netherlands, and the U.S. An excellent collection which introduces a range of approaches. Includes helpful bibliographies.

Landsberg, Hans H.; Schanz, John T.; Shurr, Sam T.; and Thompson, Grant P. Energy and the social sciences: an examination of research needs. (RFF working paper EN-3). Washington, DC: Resources for the Future, 1974. 778 p. ISBN 0-8018-1684-X. LC 74-16949.
788
Helped point the way to further research, some of which still waits.

McMullan, John T.; Morgan, R.; and Murray, R.B. Energy resources. 2nd ed. London; New York: Edward Arnold, 1983. 177 p. (1st ed., London: Edward Arnold; New York: John Wiley and Sons, 1977. 177 p.) ISBN 0-7131-3486-0. $12.95. (pbk). LC 77-26748.
789 •
A non-technical introduction into basic energy principles, energy resources, and inefficiencies. A good "first" book.

Pasqualetti, Martin J., ed. Energy. (GeoJournal: supplementary issue no. 3). Wiesbaden, West Germany: Akademische Verlagsgesellschaft, 1981. 102 p. ISBN 3-400-00476-6.
790
A collection of eight papers by geographers on a range of topics including fossil and renewable fuels, modeling, and U.S. and European perspectives. Unusual in its variety and indicative of research paths geographers follow.

Pimentel, David, and Pimentel, Marcia. Food, energy, society. New York: Halsted Press, 1979. 165 p. ISBN 0-470-26840-9. $16.95. (pbk). LC 79-9484.
791 •
A very good overview of a subject which has substantial geographic implications.

Ruedisili, Lon C. and Firebaugh, Morris W., eds. Perspectives on energy: issues, ideas, and environmental dilemmas. 3rd ed. New York: Oxford University Press, 1982. 605 p. (1st ed., 1975. 527 p.) ISBN 0-19-503289-6. $24.95. ISBN 0-19-503038-9. (pbk). $16.95. LC 82-12558.
792
A good collection of important short papers on energy.

Stobaugh, Robert, and Yergin, Daniel, eds. Energy future: report of the energy project at the Harvard Business School. 3rd ed. New York: Vintage Books, 1983. 459 p. (1st ed., 1979. 353 p.) ISBN 0-394-74750-X. $2.95. (pbk). LC 81-40078.
793
Important overview of present and future energy conditions in the U.S. Particularly significant is its support for conservation.

Energy

Bach, Wilfred; Pankrath, Jurgen. W.; and Williams, Jill., eds. Interactions of energy and climate: Proceedings of an international workshop, 1980, Münster. Dordrecht, The Netherlands: D. Reidel; Boston, MA: Kluwer, 1980. 569 p. ISBN 9-02-771179-9. $58. ISBN 90-277-1177-1. $26.50 (pbk). LC 80-23220.   794

   A wide-ranging collection of 29 papers which deal with both the effects of energy systems on climate and with the effects of climate on energy use. An excellent introduction to these relationships.

Burchell, R.W. and Listoken, D., eds. Energy and land use. New Brunswick, NJ: Rutgers University, 1982. 601 p. ISBN 0-88285-069-5. LC 80-22134.   795

   A compilation of papers of uneven quality which importantly demonstrate the range of interconnections between the two topics. Useful collections of references.

Conant, Melvin A. and Gold, Fern R. The Geopolitics of energy. Boulder, CO; London: Westview Press, 1978. 224 p. ISBN 0-89158-404-8. $33. LC 77-20668.   796 •

   A good introduction to the international relationship arising from the different locations of energy supply and demand of various types including vulnerability and security.

Cope, David R.; Hills, Peter; and James, Peter, eds. Energy policy and land-use planning: an international perspective. Oxford, U.K.; New York: Pergamon, 1984. 308 p. ISBN 0-08-026757-2. $40. $22.50. (pbk). LC 83-13236.   797

   An excellent group of papers by planners and geographers. The perspective is largely European but many lessons have applicability elsewhere.

Cottrell, W. Frederick (William). Energy and society: the relationship between energy, social change, and economic development. New York: McGraw-Hill, 1955. 330 p. Reprinted: Westport, CT: Greenwood Press, 1970. 330 p. ISBN 0-8371-3679-2. $21.75. LC 75-100152.   798

   Classic work on the profound effect that the availability and quality of energy has had on human societies through the ages with the thesis that energy limits what humans can do and strongly influences what they will do.

Kirk, Geoffrey, ed. Schumacher on energy: speeches and writings of E.F. Schumacher. London, U.K.: Jonathan Cape, 1982. 212 p. ISBN 0-224-01965-1. £7.95. LC 82-101020.   799

   A collection of work by one of the strongest advocates of a discontinuation of growing technical concentration in energy supply.

Lovins, Amory B. and Lovins, L. Hunter. Brittle power: an energy strategy for national security. Andover, MA: Brick House, 1983. 486 p. ISBN 0-931790-28-X. $17.95. LC 82-4159.   800

   A further development of ideas advocating a non-technical solution to energy problems first widely noticed in A.B. Lovin's Soft Energy Paths. *Brittle Power examines the extreme vulnerability of a dependence on a highly technical, capital intensive energy infrastructure.*

Miller, David H. Energy at the surface of the earth: an introduction to the energetics of ecosystems. London; New York: Academic Press, 1981. 516 p. ISBN 0-12-497152-2. $54.50. ISBN 0-12-497152-0. (student ed.). $27.50. LC 80-989.   801

   A technical introduction to the relationship surrounding the pathways of energy flow through biosystems. An important book for those interested in natural resource relationship with the environment.

Van Til, Jon. Living with energy shortfall: a future for American towns and cities. Boulder, CO: Westview Press, 1982. 209 p. ISBN 0-86531-135-8. $27. ISBN 0-86531-136-6. $12. (pbk). LC 82-50611.   802

   An examination of the spatial ramifications on urban, suburban, and rural use of space brought about by changes in the availability of amount and types of energy resources. A valuable book.

Dienes, Leslie, and Shabad, Theodore. The Soviet energy system: resource use and policies. New York: Halsted Press, 1979. 298 p. ISBN 0-470-26629-5. $21.95. LC 78-20814.
803 •
The standard reference work on energy matters in the U.S.S.R.; it details reserves, development, exports, and future considerations relating to Soviet fossil fuels, nuclear, and hydroelectric facilities.

Maull, Hanns. Europe and world energy. London; Boston, MA: Butterworth's, 1980. 342 p. ISBN 0-408-10629-8. $39.95. LC 80-40488.
804
Indepth appraisal of the energy relationships and dependencies of Europe including trade, security, vulnerability, and policy, with particular emphasis on oil.

Smil, Vaclav. China's energy: achievements, problems, prospects. New York: Praeger, 1976. 246 p. ISBN 0-275-23050-3. $36.95. LC 75-44939.
805
A dated but nevertheless useful compendium of data and evaluation from a geographic perspective.

Smil, Vaclav, and Knowland, William F., eds. Energy in the developing world: the real energy crisis. Oxford, U.K.; New York: Oxford University Press, 1980. 386 p. ISBN 0-19-854425-1. $74. ISBN 0-19-854421-9. $34.95. (pbk). LC 79-41127.
806
An important book which summarizes the energy situation in the developing world where three-quarters of the world's population lives. Chapters on individual energy resources and on separate areas such as China, India, Brazil, and Africa.

11. CONSERVATION AND ENVIRONMENTAL MANAGEMENT

George Macinko

Conservation foundation letter. 1- (1966-). Bimonthly. (Formerly monthly; Formerly CF Letter). Conservation Foundation, Inc., 1717 Massachusetts Ave., N.W., Washington, DC 20036. ISSN 0091-536X.
807
In-depth coverage of a single topic in each issue together with socioeconomic and political commentary on a wide range of environmental topics.

Ecologist. 1- (1970-). Bimonthly. Ecosystems Ltd., Worthyvale Manor Farm, Camelford, Cornwall, PL32 9TT, U.K. (Formerly monthly; Formerly New Ecologist, Ecologist Quarterly). ISSN 0261-3131.
808
Emphasizes the social and political dimensions of the human-resource-environment relationship.

Environment. 1- (1958-). Ten Issues. Scientists' Institute for Public Information, Heldref Publications, 4000 Albemarle St., N.W., Washington, DC 20016 (Formerly Scientist and Citizen). ISSN 0013-9157.
809
Factual reviews and articles aimed at providing information on leading environmental problems and issues with minimal polemics.

Natural resources journal. 1- (1961-). Quarterly. University of New Mexico, School of Law, 1117 Stanford, N.E., Albuquerque, NM 87131. ISSN 0028-0739.
810
A key policy journal.

Science. 1- (1880-). Weekly. American Association of the Advancement of Science, 1515 Massachusetts Ave., N.W., Washington, DC 20005. ISSN 0036-8075.
811
Broad coverage ranges across the science field but includes authoritative articles, reports, and commentary on environmental affairs.

Conservation

Brown, Harrison S. The Challenge of man's future. New York: Viking, 1954. 290 p. Reprinted (Encore editions): Boulder, CO: Westview, 1984. 290 p. ISBN 0-8133-0033-9. $24. LC 54-6422.   812•
Though statistics are dated, the clear cut, logical, and well-reasoned exposition presented by Brown makes it perhaps the best introduction to the persistent theme of the adequacy of the human resource base. Thoughtful depiction of the scope and magnitude of the resource and attendant social problems facing modern machine civilization.

Brown, Lester R.; Chandler, William; Flavin, Christopher; Postel, Sandra; Starke, Linda; and Wolf, Edward. State of the world: a Worldwatch Institute report on progress toward a sustainable society. New York: W.W. Norton, 1984. 252 p. ISBN 0-393-0185-0. $15.95. ISBN 0-393-30176-1. $8.95. (pbk). LC 83-25123.   813•
Uses the concept of sustainability to determine the extent to which economic and social systems worldwide are successfully adjusting to changes in the underlying global natural resource base. Intended as an annual report that changes its focus each year and supplements and, in part, integrates more specialized assessments.

Cloud, Preston. Cosmos, Earth, and man: a short history of the universe. New Haven, CT: Yale University Press, 1978. 372 p. ISBN 0-300-02146-1. $30. ISBN 0-300-02594-7. $10.95. (pbk). LC 78-2666.   814
A superbly written tour de force in four parts that moves from cosmos to Earth (origin of seas, atmosphere, plate tectonics) to life (amino acids to ancient humans) to humans today wherein the environment and the human impact is the subject of a brilliant exposition. Concludes with suggested programs for regional, national, and global environmental management.

Cole, H.S.D.; Freeman, Christopher; Jahoda, Marie; and Pavitt, K.L.R. Models of doom: a critique of the limits to growth. New York: Universe Books, 1972. 244 p. ISBN 0-87663-184-7. $10. ISBN 0-87663-905-8. $5. (pbk). LC 72-97037.   815
Perhaps the most extensive critique of Limits to Growth. *Uses the computer and a different set of assumptions focused on the social rather than physical environment to come to a radically different set of conclusions. The antagonism generated by the "Limits" debate is evident in the critique. Includes a response by the* Limits to Growth *authors.*

Conservation foundation staff. State of the environment: an assessment at mid-decade. Washington, DC: The Conservation Foundation, 1984. 586 p. ISBN 0-89164-084-3. $16. (pbk). LC 84-12651.   816•
Environmental assessment intended to furnish insight into the issues that will be important in the 1980's and beyond. Description of basic conditions and underlying trends affecting population growth, the economy, environmental contaminants, and natural resources. Analyses of future environmental problems, the methodology of risk assessment, integrated approaches to pollution control, problems of water quality and quantity, and intergovernmental environmental policy.

Council on Environmental Quality and the Department of State, Gerald O. Barney, study director. The Global 2000 report to the President: entering the twenty-first century. Washington, DC: Government Printing Office, 1980-1981. 3v. 47, 766, 401 p. ISBN 0-08-025990-1. $35. LC 80-602859.   817
A three-year effort of thirteen U.S. government agencies to make long-term projections about a variety of population, resource, and environmental concerns. Concludes that world problems will intensify unless present policies and practices are significantly altered.

Ehrlich, Paul R.; Ehrlich, Anne H.; and Holdren, John P. Ecoscience: population resources and environment. San Francisco, CA: Freeman, 1977. 1051 p. ISBN 0-7167-0029-8. $28.95. (pbk). LC 77-6824.   818

*Overview of the population-resource-environment predicament together with a dis-
cussion of strategies for dealing with it. Basic review of physical and biological earth
systems followed by more extensive treatment of population, renewable resources,
energy and materials, environmental disruption, and social, political, and economic
change. Copious footnotes, annotated chapter bibliographies, and suggestions for
further reading.*

Goudie, Andrew S. The Human impact: man's role in environmental change. Oxford,
U.K.: Basil Blackwell; Cambridge, MA: MIT Press, 1982. 316 p. ISBN 0-262-07086-3.
$11.95. (pbk). LC 81-15643.   819•

*Survey of the literature on interactions between natural and cultural systems. Fo-
cuses on the impact human societies have had on the various components of the physical environment and the problems caused by environmental modification.*

Hammond, Kenneth A.; Macinko, George; and Fairchild, Wilma B., eds.
Sourcebook on the environment: a guide to the literature. Chicago, IL: University of
Chicago Press, 1978. 613 p. ISBN 0-226-31522-3. $25. LC 77-17407.   820•

*Basic guide to a broad spectrum of the environmental literature. Critically annotat-
ed, extended commentaries provide introductions to the basic literature and give directions for examination of more advanced and more specialized works. Tutorial
guidance is given on major elements of the environment; case studies of environmental modification; environmental perspectives and prospects; and environmental research aids. About 3800 references.*

Leopold, Aldo. A Sand County almanac: and sketches here and there. New York: Oxford University Press, 1949. 226 p. ISBN 0-19-500777-8. $6.95. (pbk). LC 49-11164.   821•

*Series of essays that give eloquent expression to humanity's emergent sense of en-
vironmental responsibility. The "Land Ethic" essay is the classic statement of the notion that human environmental actions should be governed by a sense of stewardship
rather than exploitation and his guiding conscience should be broadened to include
non-human as well as human considerations.*

Manners, Ian R. and Mikesell, Marvin W., eds. Perspectives on environment. (Association of American Geographers, Commission on College Geography). Washington,
DC: Association of American Geographers, 1974. 395 p. ISBN 0-89291-044-5. $3.95.
(pbk). LC 73-88849.   822•

*One dozen review essays requested by the Panel on Environmental Education of
the Commission on College Geography that call attention to geography's past and
potential contributions to environmental education and suggest those aspects of geographic training most useful in handling environmental issues. Bibliography at the
end of each essay.*

**Meadows, Donnela H.; Meadows, Dennis L.; Randers, Jorgen; and Behrens,
William W., III.** The Limits to growth: a report for the Club of Rome's project on the
predicament of mankind. 2nd ed. New York: Universe Books, 1973. 205 p. (1st ed.,
1972. 205 p.) ISBN 0-87663-222-3. $10. ISBN 0-87663-918-X. $5. (pbk). LC 73-187907.   823

*The computer based study that, more than any other work, focused attention on the
resource and environmental problems associated with growth. Revered and reviled
by different segments of the academic community, it provoked a discussion that
shows little sign of resolution.*

Miller, G. Tyler, Jr. (George). Living in the environment: an introduction to environmental science. 4th ed. Belmont, CA: Wadsworth, 1984. 460 p. (1st ed., 1974. 379 p.)
ISBN 0-534-04332-1. $24. LC 84-15297.   824•

*Widely-used, comprehensive text with coverage of the rudiments of ecology as
background for an examination of major environmental problems and the range of
possible solutions to them. Fifteen enrichment studies, more that 6000 references,
guest editorials, and annotated suggested readings complement the basic text.*

Conservation

Shepard, Paul, and McKinley, Daniel., eds. The Subversive science: essays toward an ecology of man. Boston, MA: Houghton Mifflin, 1969. 453 p. LC 69-15029.　　825
Wide-ranging collection of carefully chosen essays that are valuable contributions toward developing a philosophy of human ecology. Editorial commentary and abundant references enhance this work.

C. SPECIAL TOPICS

American Chemical Society. Cleaning our environment: a chemical perspective. 2nd ed. Washington, DC: American Chemical Society, 1978. 457 p. ISBN 0-8412-0467-5. $12.95. LC 78-73104.　　826
A detailed and authoritative analysis of the problems of solid-waste disposal.

Black, John N. The Dominion of man: the search for ecological responsibility. Edinburgh, U.K.: Edinburgh University Press, 1970. 169 p. ISBN 0-85-224-186-0. £5. LC 79-116691.　　827
Contends that sound environmental stewardship demands serious attention be paid now to the future productivity of world resources and argues that the failure to do so will result in the continued deterioration of natural environments.

Carrol, John E. Environmental diplomacy: an examination and a prospective of Canadian-U.S. transboundary environmental relations. Ann Arbor, MI: University of Michigan Press, 1983. 382 p. ISBN 0-472-10029-7. $22.50. LC 82-17363.　　828
This look at environmental issues involving Canadian-U.S. relations includes a description of existing institutional arrangements to handle problems; a discussion of eight types of environmental problems, with a focus on acid rain; and concludes with prescriptions for the future.

Carson, Rachel. Silent spring. Boston, MA: Houghton Mifflin, 1962. 368 p. ISBN 0-395-07506-8. $16.95. LC 60-5148.　　829•
Environmental classic that first brought to public attention the many and varied unintended side effects of pesticides. Created a furor among agricultural, pesticide, and environmental interests and led to an upsurge of environmental activism.

Clawson, Marion. The Federal lands revisited. Washington, DC: Resources for the Future, 1983. 302 p. ISBN 0-318-00447-X. $25. ISBN 0-318-00448-8. $8.95. (pbk). LC 83-42904.　　830
The use and management of federal lands and suggestions for future policy by an economist and long-time student of American land use.

Culbertson, John M. Economic development: an ecological approach. New York: Knopf, 1971. 305 p. LC 70-144145.　　831
Culbertson urges that the broader ecological approach be added to narrowly traditional economics in the development of a new theory of economics more philosophical and holistic than those currently in vogue and one perhaps better suited to the emergent resource scene.

Dana, Samuel T. and Fairfax, Sally K. Forest and range policy: its development in the United States. 2nd ed. New York: McGraw-Hill, 1980. 458 p. (1st ed., 1956. 455 p.) ISBN 0-07-015288-8. $31. LC 79-13652.　　832
Standard reference for analyses of federal policies affecting forest and range lands.

Ehrenfeld, David W. Conserving life on earth. New York: Oxford University Press, 1972. 360 p. LC 72-85776.　　833
Depicts the impact that population and technological expansion have had on the natural world, and discusses the alternatives for wildlife preservation under political, moral, economic, and managerial constraints.

Ehrlich, Paul, and Ehrlich, Anne. Extinction: the causes and consequences of the disappearance of species. New York: Random House, 1981. 305 p. ISBN 0-394-51312-6. $16.95. LC 80-6036.

834•

Well-documented exploration of the extinction process. Topics examined include, but are not limited to, the economic and biological costs of extinction, the role of zoos and wildlife refuges, the politics of wildlife trade, and the destruction of wildlife habitats.

Farvar, M.T. and Milton, John P., eds. The Careless technology: ecology and international development. (Conference on the ecological aspects of international development). Garden City, NJ: Natural History Press, 1972. 1030 p. LC 73-150890.

835•

An examination of the dark side of large scale development projects. Case studies, examined in detail, disclose some of the pitfalls of environmental modification and reveal that the unintended consequences of environmental change may sometimes offset the intended favorable consequences.

Flint, Mary Louise, and Van Den Bosch, Robert. Introduction to integrated pest management. New York: Plenum Press, 1981. 240 p. ISBN 0-306-40682-9. $19.95. LC 80-28479.

836•

Wide-ranging discussion of pest control by means as diverse as fly swatters and synthetic chemicals. Informative discussion on what constitutes a pest, the history of pest control, the phenomenon of pesticide resistance, and the role of government and industry in pest control.

Hirsch, Fred. Social limits to growth. Cambridge, MA: Harvard University Press, 1976. 208 p. ISBN 0-674-81365-0. $15. ISBN 0-674-81366-9. $6.95. (pbk). LC 76-18974.

837

An examination of the social problems that economic growth creates, especially for advanced democracies. If correct, this analysis has many important implications for environmental policy that have been largely ignored.

Hoose, Philip M. Building an ark: tools for the preservation of natural diversity through land protection. Covelo, CA: Island Press, 1981. 221 p. ISBN 0-933280-09-2. $12. (pbk). LC 80-28279.

838

Laws, deeds, and agreements serve as Hoose's proxy for Noah's Ark in this pragmatic approach to restrain ecological simplification and species extinction. To this end Hoose discusses overall strategy, means of achieving land agreements, and necessary political action.

Odum, Eugene P. Fundamentals of ecology. 3rd ed. Philadelphia, PA: Saunders College Press, 1971. 574 p. (1st ed., 1953. 384 p.) ISBN 0-7216-6941-7. $35.95. LC 76-81826.

839•

Standard reference by a respected ecologist. Provides the basic grounding in ecology necessary to grapple with environmental issues. Includes sections on applied and human ecology as well as a detailed index and useful bibliography.

Ophuls, William. Ecology and the politics of scarcity. San Francisco, CA: W.H. Freeman, 1977. 303 p. ISBN 0-7167-0481-1. $11.95. (pbk). LC 76-46436.

840

A critique of the suitability of American political values and institutions in an era of increasing ecological scarcity. Nearly a thousand literature citations, many copiously annotated, enrich this effort.

Partridge, Ernest., ed. Responsibilities to future generations: environmental ethics. Buffalo, NY: Prometheus Books, 1981. 319 p. ISBN 0-87975-153-3. $18.95. ISBN 0-87975-142-8. $10.95. (pbk). LC 80-84401.

841•

Thoughtful and balanced discussion of the question, "what do we owe posterity?" that is central to many questions of resource allocation. Includes a discussion of the moral issues encompassed by the concept of environmental stewardship.

Conservation

Popper, Frank J. The Politics of land-use reform. Madison, WI: University of Wisconsin Press, 1981. 321 p. ISBN 0-299-08530-9. $25. ISBN 0-299-08534-1. $8.50. (pbk). LC 80-23255.   842

Description of the general problems of land use that led to its regulation followed by a closer examination of the successes and failures of regulatory programs in six states. Treats both the pros and cons of land regulation.

Rosenbaum, Walter A. The Politics of environmental concern. 2nd ed. New York: Holt, Rinehart, and Winston, 1977. 311 p. (1st ed., New York: Praeger, 1973. 298 p.) ISBN 0-275-64820-6. $13.95. (pbk). LC 76-41964.   843

Survey of a wide range of governmental decisions and public policies. Discusses political maneuvering that forms the backdrop for environmental legislation.

Schumacher, Ernst F. Small is beautiful: economics as if people mattered. London: Blond and Briggs; New York: Harper and Row, 1973. 305 p. ISBN 0-06-013801-7. $12.45. LC 73-12710.   844 •

The book that sparked consideration of "appropriate technology," particularly with regard to third-world development efforts. Calls into question many of the basic tenets of orthodox development thought.

Sears, Paul Bigelow. Deserts on the march. 4th rev. ed. Norman, OK: University of Oklahoma Press, 1980. 231 p. (1st ed., 1935. 231 p.) ISBN 0-8061-1667-6. $14.95. LC 79-6721.   845 •

Examination of the physical conditions and the cultural practices that led to the American Dust Bowl. This classic is the forerunner to contemporary concerns with the process of desertification.

Simon, Julian L. and Kahn, Herman, eds. The Resourceful earth: a response to global 2000. Oxford, U.K.; New York: Basil Blackwell, 1984. 585 p. ISBN 0-631-13467-0. $19.95. LC 84-14613.   846

Direct response to the Global 2000 Report to the President *that contends the earlier work was unduly pessimistic. In general, this response reaches conclusions diametrically opposed to* Global 2000.

Stone, Christopher D. Should trees have standing?: toward legal rights for natural objects. Los Altos, CA: William Kaufmann, 1974. 102 p. ISBN 0-913232-08-4. $5.50. (pbk). LC 73-19535.   847

Pioneering effort most responsible for the notion that serious consideration be given to granting legal rights to other than the human species and thence to the natural environment in its entirety.

Tribe, Lawrence H.; Schelling, Corrine S.; and Voss, John. When values conflict: essays on environmental analysis, discourse, and decision. Cambridge, MA: Ballinger, 1976. 178 p. ISBN 0-88410-431-1. $20. LC 75-45448.   848

A series of seven essays which attempt to weigh the conflicting values facing environmental decision makers. In particular, attention is paid to the "fragile" values such as those of aesthetics, endangered species and wilderness preservation, and the means to treat them with adequate sensitivity.

Wenner, Lettie M. The Environmental decade in court. Bloomington, IN: Indiana University Press, 1982. 211 p. ISBN 0-253-31957-9. $22.50. LC 81-47778.   849

Overview of federal judicial policy on environmental matters since passage of the National Environmental Policy Act of 1970. Analyzes the role of federal courts in implementing and formulating environmental policy.

See also entries 784, 798, and 980.

12. NATURAL HAZARDS

David Morton

A. BIBLIOGRAPHIES

Cochran, Anita. A Selected, annotated bibliography on natural hazards. Boulder, CO: University of Colorado, Institute of Behavioral Science. Program on Technology, Environment, and Man. (Natural Hazards Research Working Paper no. 22). 1972. 86 p. $4.50. (Available from Natural Hazards Center, Box 482, University of Colorado, Boulder, CO 80309).   850

The bibliography contains 317 citations dated from the mid-1950's through 1971. References are grouped under twelve specific natural hazards and a general hazards section. Excluded from the bibliography is the vast amount of technical literature in geology and meteorology on the physical aspects of natural hazards and the technical engineering literature. Biological hazards and hazards which are partly natural, partly human-caused, such as air pollution, are not included. Journal articles, books, bibliographies, and research reports represent most of the cited materials. Indexed by first author and subject.

Torres, Kathleen, and Cochran, Anita. A Selected, partially annotated bibliography of recent (1975-1976) natural hazards. Boulder, CO: University of Colorado, Institute of Behavioral Science, Natural Hazards Research and Applications Information Center. 1977. 75 p. with 16 p. supplement. $2. (Available from Natural Hazards Center, Box 482, University of Colorado, Boulder, CO 80309).   851

A partially annotated bibliography of publications which are concerned in some way with the reduction of losses from natural hazards. It should be considered a working bibliography and while it is necessarily incomplete, it does offer many materials not indexed elsewhere in one place. There are a total of 251 citations indexed by author and subject.

Morton, David. A selected, partially annotated bibliography of recent natural hazards publications, 1976/77- . Annual. Boulder, CO: University of Colorado, Institute of Behavioral Science. Natural Hazards Research and Applications Information Center. Variable length and price. (Available from Natural Hazards Center, Box 482, University of Colorado, Boulder, CO 80309).   852

These annual bibliographies represent a continuation of the work begun by Anita Cochran (1972) and Kathleen Torres and Anita Cochran (1977). The volumes average about 125 pages, contain approximately 250 annotations, and are indexed by author and subject. Included are sections on specific hazards (i.e., earthquakes, floods, etc.) together with multi-hazard sections such as land use, hazards planning, health and medical issues, coastal zone management, and mass earth movements.

B. NEWSLETTER

Natural hazards observer. 1- (1976-). Bimonthly. Natural Hazards Research and Applications Information Center, Institute of Behavioral Science, Box 482, University of Colorado, Boulder, CO 80309. ISSN 0193-8355.   853 •

A newsletter on new research and findings from completed projects; pertinent legislation and regulations; applications of research at federal, state, and local levels and by private agencies; and future conferences. Approximately fifteen to twenty annotations of recent publications are included in each issue. Occasional issues feature specialized topics such as technological hazards and climatology.

Natural Hazards

Burton, Ian; Kates, Robert W.; and White, Gilbert F. The Environment as hazard. New York: Oxford University Press, 1978. 240 p. ISBN 0-19-502222-X. (pbk). $12.95. LC 77-9620.   854

Sums up what has been learned thus far as to how individuals and social groups respond to extreme events in nature. It also contains tentative conclusions that many hazard researchers feel should be re-examined, and uses methods that need to be subjected to future refinement. Topics investigated include individual choice in the response process, collective action, national policy, international action, and social resilience to natural extremes.

Dacy, Douglas C. and Kunreuther, Howard. The Economics of natural disasters: implications for federal policy. New York: Free Press, 1969. 270 p. LC 68-29581.   855

The main objective of this book is to formulate a clear-cut case for the development of a comprehensive system of disaster insurance as an alternative to the current paternalistic federal policy. Conclusions and recommendations are supported with statistics and case studies. Included among the twelve chapters are: The Cost of Natural Disaster in the United States; Economic Theory and Natural Disaster Behavior; Short-run Supply and Demand Problems; Planning for Recovery; The Special Problem of Damage Assessment; Population Migration: The Supply of Labor; Equity of Disaster Relief; The Cost of Federal Relief; *and* The Need for Comprehensive Disaster Insurance.

Fitzsimmons, Ann R. Natural hazards and land use planning: an annotated bibliography. (Topical bibliography no. 11). Boulder, CO: University of Colorado, Institute of Behavioral Science, Natural Hazards Research and Applications Information Center. 1984. 103 p. $5.50. (Available from Natural Hazards Center, Box 482, University of Colorado, Boulder, CO 80309).   856

Presents 283 annotated references on thirteen specific hazards along with materials that deal with land-use planning in a multi-hazard context. Citations that contain case studies have been identified to provide practicing planners with examples of what other persons and jurisdictions have done to cope with natural hazards. The annotations are indexed by subject and author, and a special index is provided to aid planners working in small communities (under 30,000 inhabitants).

Foster, Harold D. Disaster planning: the preservation of life and property. New York; Berlin: Springer-Verlag, 1980. 275 p. ISBN 0-387-90498-0. $29.80. LC 80-18910.   857

Concepts and mechanisms of comprehensive risk management are surveyed in the hope that more local authorities, institutions, and planners will accept the responsibilities for improving safety standards for the protection of human lives and property. Among the topics examined are strategies for achieving mitigative goals; acceptable risks; mapping and microzonation, structural design, and predictive techniques; and warning systems, modeling, and simulation methodology.

Haas, J. Eugene; Kates, Robert W.; and Bowden, Martyn J., eds. Reconstruction following disaster. Cambridge, MA: MIT Press, 1977. 331 p. ISBN 0-262-08094-X. $30. LC 77-23176.   858

Two ongoing cases of natural hazard reconstruction (Managua and Rapid City) and two historical cases (Anchorage and San Francisco) were examined to determine whether identifiable social patterns emerged during the reconstruction period. It was found that the reconstruction process is ordered, knowable and predictable; that the central issues and decisions were based upon value choices; and that over-ambitious post-reconstruction planning to reduce future vulnerability or to improve efficiency or amenity appears to be counterproductive.

Hewitt, Kenneth. ed. Interpretation of calamity. (Risks and Hazards series no. 1). London; Boston, MA: Allen and Unwin, 1983. 304 p. ISBN 0-04-301160-8. $35. ISBN 0-04-301161-6. (pbk). $14.95.   859

Centers around the theme that the traditional or dominant view of natural hazards (i.e. emphasizing geophysical or meteorological events) must give way to a broader view that recognizes the central importance of historically determined social, political, and economic structure in creating hazard risk. The contributing authors examine the concept of disasters, their meaning and causes; examine different social and geographic settings for problems associated with climatic hazards, particularly drought; and attempt to reconceptualize hazards within a broader social framework.

Hewitt, Kenneth, and Burton, Ian. The Hazardousness of a place: a regional ecology of damaging events. (University of Toronto, Department of Geography, Research Publication no. 6). Toronto: University of Toronto Press, 1971. 154 p. ISBN 0-8020-3281-8. $5. LC 72-177223.   860

The prevailing research strategy of investigating single hazards is expanded to a more systematic cross-hazard approach. The need was recognized for a theoretical and conceptual framework to organize empirical data from the cross-hazard approach, and the volume presents the initial steps toward the development of such a framework for hazard studies.

Kates, Robert W. Risk assessment of environmental hazard. (Scientific Committee on Problems of the Environment (SCOPE) publication no. 8). New York: John Wiley and Sons, 1978. 112 p. ISBN 0-471-99582-7. $21.95. LC 77-12909.   861

The proliferation of environmental threats places scientists in the center of concern, controversy, and public policy formation. This monograph serves as an introductory review to environmental risk assessment and is a source for tentative generalizations. Trends and attitudes in assessing environmental threat are examined, as are risk assessment methodology and organized modes of assessment.

Kunreuther, Howard, et al. Disaster insurance protection: public policy lessons. New York: John Wiley and Sons, 1978. 400 p. ISBN 0-471-03259-X. LC 78-179.   862

Provides a greater understanding of the decision processes employed by individuals in dealing with low probability events in nature which cause severe losses to themselves and society. Findings reveal that most homeowners in hazard prone areas possess limited knowledge of mitigation measures and relief programs; that most people consider insurance a form of investment rather than a mechanism for transferring risk from themselves to others; and that the financial incentive is presently too low to encourage insurance agents to actively pursue the market for disaster and hazard insurance.

Mileti, Dennis S.; Drabek, Thomas E.; and Haas, J. Eugene. Human systems in extreme environments: a sociological perspective. (Monograph no. NSF-RA-E-75-021; NTIS no. PB 267 836). Boulder, CO: University of Colorado, Institute of Behavioral Science, Program on Technology, Environment, and Man, 1975. 165 p. $14. LC 74-620148.   863

Attempts to codify what the social science literature reveals about how humans, individually and collectively, adapt and respond to natural hazards and disasters. It covers aspects such as preparedness, warning, organizational and social system mobilization, immediate responses following impact, and restoration and reconstruction.

Platt, Rutherford, et al. Intergovernmental management of floodplains. (Monograph no. 30; NTIS no. PB 194 896). Boulder, CO: University of Colorado, Institute of Behavioral Science, Program on Technology, Environmemt, and Man, 1980. 317 p. $24.50. (Available from Natural Hazards Center, Box 482, University of Colorado, Boulder, CO 80309).   864

Analyzes the types of conflicts that arise in intergovernmental management of a floodplain; details the ways to achieve varying degrees of coordination among governments; and presents seven case studies of regional management successes and failures. Suggests that floodplain management is more successful when integrated with other related regional planning endeavors such as wastewater management, housing development, open space, and coastal zone management.

Natural Hazards

Quarantelli, E.L., ed. Disasters: theory and research. (Sage studies in international sociology, v. 13). Beverly Hills, CA: Sage Publications, 1978. 288 p. ISBN 0-8039-9851-1. $28. LC 76-52786.

    A group of original essays focusing on a specific area of current sociological research and analysis that offers theoretical overviews of research, summaries of work undertaken, and a number of illuminating case studies. The essays explore the organization of disaster response, information sources in catastrophic conditions, human behavior during earthquakes, and other topics of vital concern to disaster specialists.

865

White, Gilbert F., ed. Human adjustment to floods: a geographical approach to the flood problem in the United States. (Research paper no. 29). Chicago, IL: University of Chicago, Department of Geography, 1953. (Reproduced from 1942 dissertation). 225 p.

    A comprehensive approach to the formulation of national policy regarding floodplains in the United States would take into account all relevant factors affecting the use of floodplains, the feasible adjustments to the conditions involved, and the practicality of proposed solutions. This approach demands an integration of engineering, geographic, economic, and related techniques.

866

White, Gilbert F., ed. Natural hazards: local, national, global. New York; London; Toronto: Oxford University Press, 1974. 288 p. ISBN 0-19-501757-9. $11.95. (pbk).

    Presents primary results of comparative field observations, of efforts to make integrated reviews of the hazard situation in selected countries, and of human reponses to particular extreme events on a global scale. Hazards examined include floods, windstorms, frost, drought, earthquakes, hurricanes, urban snow, and avalanches. Several national reviews of natural hazards are offered and global summaries are presented for human response to tropical cyclones, earthquakes, and floods.

867

White, Gilbert F. and Haas, J. Eugene. Assessment of research on natural hazards. Cambridge, MA; London: MIT Press, 1975. 487 p. ISBN 0-262-23071-2. $37.50. LC 75-2058.

    Assesses the status of natural hazard research in the United States and recommends future research directions. Discusses topics that cut across hazard boundaries and the specific major hazards. Subjects discussed include national response to natural hazards, acceptable level of risk, the range of adjustment choices, how research results are applied, and how research benefits can be estimated.

868

Whyte, Anne V. and Burton, Ian., eds. Environmental risk assessment. (Scientific Committee on Problems of the Environment (SCOPE) publication no. 15). New York: John Wiley and Sons, 1980. 157 p. ISBN 0-471-27701-0. $34.95. LC 79-42903.

    Evaluates the status of environmental risk assessment which is considered to be a new method and means for the elaboration of the economic and science-based alternatives of processes applied in protecting human health. The subject is examined in terms such as the identification of environmental risks and their management, uncertainty in monitoring processes, quantitative estimation of risk, risk evaluations and national policy, and international collaboration in risk management.

869

PART IV. SYSTEMATIC FIELDS OF HUMAN GEOGRAPHY

1. GENERAL HUMAN GEOGRAPHY

Wilbur Zelinsky

A. SERIALS

Geo abstracts. Series d. Social and historical geography. 1- (1966-). 6 per annum. Geo Abstracts, Ltd., Regency House, 34 Duke Street, Norwich NR3 3AP, U.K. ISSN 0305-1927.   870

  An overview of all aspects of social and historical geography. A must for the geographical researcher.

Geografiska annaler. Series b. Human geography. (Svenska Sällskapet för Antropologi och Geografi). 48B- (1963-). 2 per annum. Almqvist and Wiksell Periodical Co., Box 62, S-101 20 Stockholm, Sweden. ISSN 0435-3684.   871

  International scholarly journal with wide range of articles in human, economic, social, cultural, and historical geography. Many authors from different countries. In English.

Journal of cultural geography. 1- (1980-). Semi-annual. Bowling Green State University, Popular Culture Center, Bowling Green, Ohio 43403.   872

  Brief articles cover highly varied gamut of subjects that fall within the field of cultural geography. No book reviews.

Progress in human geography: an international review of geographical work in the social sciences and humanities. 1- (1976-). Quarterly. Edward Arnold, 41 Bedford Square, London WC1B 3DQ, U.K. or Cambridge University Press, 32 East 57th Street, New York, NY 10022. ISSN 0309-1325.   873

  Articles, progress reports on major fields in human geography, book review essays, and book reviews. Invaluable for following the expanding frontiers and research thrusts in human geography.

B. ATLAS

Bruk, Solomon Il'ich, and Apenchenko, V.S., eds. Atlas narodov mira (Atlas of the peoples of the world). Moscow: Glavnoe Upravlenie Geodezii i Kartografii (GUGK), 1964. 184 p. LC MAP 65-6.   874

  Includes 106 world and regional color plates showing the location and population densities of 900 distinct ethnic and racial groups. Plates on language and religion are also included. Text and statistical tables supplement the work.

C. DICTIONARY

Johnston, Ronald J.; Derek, Gregory; Haggett, Peter; Smith, David; and Stoddart, David R., eds. The Dictionary of human geography. Oxford, U.K.: Basil Blackwell; New York: Free Press, 1981. 411 p. ISBN 0-02-916510-5. $40. LC 81-15084.   875 •

  Major comprehensive reference work by eighteen British geographers providing succinct definitions of terms and concepts used in human geography. The terms and concepts cover philosophical, methodological, and substantive aspects of all areas of human geography. Includes essays defining and reviewing the major sub-fields of the discipline. Selected readings suggested.

General Human Geography

D. TEXTS AND GENERAL STUDIES

Broek, Jan O.M. and Webb, John W. A Geography of mankind. 3rd ed. New York: McGraw-Hill, 1978. 571 p. (1st ed., 1968. 527 p.) ISBN 0-07-008012-7. $32.50. LC 77-12586.

876 •

An excellent introductory text that describes the geographer's approach, discusses cultural, religious, and economic diversity, and proceeds to urbanization and population phenomena. Excellent maps and graphs.

Brown, Lawrence A. Innovation diffusion: a new perspective. London; New York: Methuen, 1981. 345 p. ISBN 0-416-74270-X. $19.95. LC 80-49706.

877

Diverse perspectives on innovation: innovation agency; the firm; applicability to Third World settings; interrelationship with development processes; and public policy context. Extensive bibliography.

Chisholm, Michael. Human geography: evolution or revolution? Hammondsworth, U.K.; Baltimore, MD: Penguin, 1975. 207 p. ISBN 0-14-021883-1. £1.95. (pbk). LC 76-357707.

878

Readable personal essays on recent developments in human geography up to the mid-1970s, particularly of origins, static patterns, dynamic patterns, and theories of spatial structure and process.

Cox, Kevin R. Man, location, and behavior: an introduction to human geography. New York; London: John Wiley, 1972. 399 p. (Reprinted: Melbourne, FL: Krieger). ISBN 0-471-18150-1. $21.50. LC 72-4790.

879

de Blij, Harm J. Human geography: culture, society, and space. 2nd ed. New York: Wiley, 1982. 656 p. (1st ed., 1977. 444 p.). ISBN 0-471-08557-X. $29.95. LC 81-7506.

880 •

Introduction to human geography with emphasis on demographic, economic, and political issues.

Forde, C. Daryll (Cyril). Habitat, economy, and society: a geographical introduction to ethnology. London: Methuen; New York: Harcourt Brace, 1934. 509 p. Reprint of 8th ed., 1950. New York: Dutton, 1963. 500 p. ISBN 0-525-47110-3.

881

Illustrates the interactions among human economies, social structures, and the physical habitat by a series of representative communities. Final section of the volume summarizes the evolution of various socio-economic forms and complexes.

Hägerstrand, Torsten. Innovation diffusion as a spatial process. Translated from Swedish by Allan Pred and Greta Haag. Chicago, IL: University of Chicago Press, 1967. 334 p. (Innovationsförloppet ur korologisk synpunkt, Lund, Sweden: C.W.K. Gleerup, 1953. 304 p.) ISBN 0-226-31261-5. $22.50. (pbk). LC 67-26091.

882

A seminal work applying quantifications to the measurement and modeling of the geographical spread of innovations with particular attention to the diffusion of information and decisions.

Haggett, Peter. Geography: a modern synthesis. 3rd ed. London; New York: Harper and Row, 1983. 644 p. ISBN 0-06-042579-2. $32.50. LC 82-23282.

883 •

Possibly the best single-author introductory text currently available. Covers the whole spectrum of geography in a modern context. The two overarching themes are the role of human beings in the ecological system and the spatial systems of human societies. Well-written and intelligently illustrated.

Jackson, W.A. Douglas (William). The Shaping of our world: a human and cultural geography. New York; Chichester, U.K.; and Toronto: John Wiley, 1985. 622 p. ISBN 0-471-88031-0. $29.95. LC 84-7518.

884 •

Introductory text on human geography with emphasis on cultural geography.

Jakle, John A.; Brunn, Stanley; and Roseman, Curtis C. Human spatial behavior: a social geography. North Scituate, MA: Duxbury Press, 1976. 315 p. ISBN 0-87872-098-7. $13.95. LC 75-11278. 885

A pioneering text for the relatively young field of social geography. Among the major topics covered in this useful work are: territoriality, environmental cognition, social interaction over space, communications, migration, territorial sorting out of social groups, and the spatial dimensions of social problems.

Jordan, Terry G. and Rowntree, Lester. The Human mosaic: a thematic introduction to cultural geography. 3rd ed. New York: Harper and Row, 1982. 444 p. (1st ed. San Francisco, CA: Canfield Press, 1976. 430 p.) ISBN 0-06-043461-9. $32.50. LC 81-6968. 886•

Probably the most attractive and effective textbook devoted to cultural geography. Text is organized around five concepts or themes: culture regions; cultural diffusion; cultural ecology; cultural integration; and cultural landscapes. The illustrations (maps, charts, and photographs) are exceptional in both quantity and quality.

Marsh, George Perkins. Man and nature. **David, Lowenthal, ed.** Cambridge, MA: Harvard University Press, 1965. (First published as Man and Nature; or Physical Geography as Modified by Human Action. New York: C. Scribner, 1864. 560 p.). 472 p. ISBN 0-674-54400-5. $8.95. (pbk). LC 65-11591. 887

An epochal work by the first great figure in the field now known as conservation, or human ecology, that deals skillfully with the human impact upon the physical and biological world. A significant contribution to the development of geographic thought in the United States.

Morrill, Richard L. and Dormitzer, Jacqueline M. The Spatial order: an introduction to modern geography. North Scituate, MA: Duxbury, 1979. 483 p. ISBN 0-87872-180-0. LC 78-32061. 888•

Focus is on how human societies organize their space, why certain patterns emerge, and the human use and modification of the landscape particularly in high technology market economies. Using simple location theory, it seeks to reveal basic spatial order that underlies great diversity in landscapes.

Murphey, Rhoads. Patterns on the earth: an introduction to geography. 4th ed. Chicago, IL: Rand McNally, 1978. 544 p. [Now published Boston, MA: Houghton Mifflin]. (1st ed. Introduction to Geography. Chicago, IL: Rand McNally, 1961. 699 p. ISBN 0-528-63000-8; ISBN 0-395-30827-5. $28.95. LC 77-77728. 889

Although the opening chapters deal with geographical aims, methods, and physical elements, the bulk of this work is historical and cultural, focusing on appropriate local topics within a regional framework.

Pocock, Douglas C.D., ed. Humanistic geography and literature: essays on the experience of place. London: Croom Helm; Totowa, NJ: Barnes and Noble; 1981. 224 p. ISBN 0-389-20158-8. $28.50. LC 81-102027. 890

A collection of essays dealing with place theory as it relates to the human element of geography.

Relph, Edward C. Place and placelessness. (Research in planning and design series no. 1). London: Pion; New York: Methuen, 1976. 156 p. ISBN 0-85086-055-5. $15.50. LC 76-383276. 891

An influential essay that attempts to elucidate the diversity and intensity of our experiences of place. Among the major topics are the relationships between space and place, and how attachment to place manifests itself in the making of landscapes.

Salter, Christopher L. The Cultural landscape. Belmont, CA: Duxbury Press, 1971. 281 p. LC 75-144141. 892•

Collection of 48 relatively brief pieces by a highly varied crew of writers. Intent is to give some structure to the study of human reshaping of the environment. Salter's chief concerns are human mobility, human ecology, the organization of space, and our contemporary cultural landscape. Ethical issues figure prominently in this anthology.

Historical Geography

Stewart, George R. Names on the globe. New York: Oxford University Press, 1975. 411 p. LC 74-21829.

Wide-ranging study of the process of place-naming. Emphasis is on the human dimension, the namer, and what their choices and the subsequent alterations tell us about patterns of human activities and perceptions. This sweeping survey, with its many philosophical insights, is the culmination of a lifetime of productive scholarship.

893 •

Taylor, Isaac. Words and places: illustrations of history, ethnology, and geography. 4th ed. New York: Dutton; London: J.M. Dent, 1911. 467 p. (1st ed. London: Macmillan, 1864. 578 p.) Frequently and widely reprinted.

A full century has passed since the first edition of this charming work, but it still remains without a rival as a highly readable and stimulating introduction to the study of place-names throughout the world.

894

Tuan, Yi-Fu. Space and place: the perspective of experience. Minneapolis, MN: University of Minnesota Press, 1977. 235 p. ISBN 0-8166-0808-3. $17.50 ISBN 0-8166-0884-9. $8.95. (pbk). LC 77-072910.

A classic of twelve relatively self-contained essays dealing with the human encounter and experience of the environment at a range of scales. Provides a perceptive, if personal, reflection on the geographies of everyday life from a humanistic perspective.

895

Tuan, Yi-Fu. Topophilia: a study of environmental perception, attitudes, and values. Englewood Cliffs, NJ: Prentice-Hall, 1974. 260 p. ISBN 0-13-925230-4. $16.95. (pbk). LC 73-8974.

An exhilarating and widely admired volume that has added "topophilia," the affective bond between people and place, to our geographic vocabulary. Tuan's elegant prose explores the human universals and variables that affect our environmental perception and behavior. A rich medley of ideas, observations, and insights that draws extensively on the worlds of literature, art, and architecture, as well as the standard geographical sources.

896

Zelinsky Wilbur. Human geography: coming of age. (American Behavioral Scientist, vol. 22, no. 1). Beverly Hills, CA: Sage Publications, 1978. 167 p.

A set of eight essays on the evolution and status of major foci within human geography. Written with the lay public in mind. Human ecology, social geography, the spatial economy, urban geography, the historical geography of North American regions, perceptual geography, and First and Third World relations are the principal subjects.

897

2. HISTORICAL GEOGRAPHY

Robert D. Mitchell

See also the following sections: 3. Old World Historical Cultural Geography; 4. New World Prehistory; Part VI, Section 2, E. North America, Historical Geography, Section 3, G. United States, Historical Geography, and the following entries in other regional sections: 2072; 2081; 2082; 2083; 2119; 2120; 2122; 2130; 2144; 2152; 2160; 2198; 2263-2269; 2303; 2393; 2521; 2522; 2528; 2529; 2549; 2559; 2562; 2593; 2630; 2646; 2680; 2682; 2694; 2697; 2699; and 2706.

A. SERIAL

Journal of historical geography. 1- (1975-). Quarterly. Academic Press, Ltd., 24-28 Oval Road, London NW1 7DX, U.K. or 111 Fifth Avenue, New York, NY 10003. ISSN 0305-7488.

Scholarly articles on historical geography with a wide range of areas and periods studied. Debates. Review articles. Extensive reviews. Short notices.

898

Baker, Alan R.H., ed. Progress in historical geography. Newton Abott, U.K.: David and Charles; New York: Wiley-Interscience, 1972. 311 p. LC 72-188291.　　　　899
An anthology of ten essays on the progress of historical geography since 1945 in North America, Europe, Australia, Latin America, and Africa. Introductory essay by the editor on the new historical geography, more theoretical, quantitative, and behavioral than before.

Baker, Alan R.H. and Billinge, Mark, eds. Period and place: research methods in historical geography. (Cambridge studies in historical geography no. 1). Cambridge, U.K.; New York: Cambridge University Press, 1982. 377 p. ISBN 0-521-24272-X. $54.50. LC 81-12266.　　　　900
A collection of more than 30 articles on research strategies and methods in historical geography. Divided into six sections: developments in historical geography; reconstructing past geographies; identification and interpretation of geographical change; behavioral approaches; theoretical approaches; and historical sources and techniques.

Baker, Alan R.H. and Gregory, Derek, eds. Explorations in historical geography: interpretative essays. (Cambridge studies in historical geography no. 5). Cambridge, U.K.; New York: Cambridge University Press, 1984. 252 p. ISBN 0-521-24968-6. $49.50. LC 83-19003.　　　　901
A collection of six essays on interpretations in historical geography. Topics include historical geography and history, social class, agricultural revolution, modernization, and new trends in historical-geographical interpretation.

Baker, Alan R.H.; Hamshere, John D.; and Langton, John, eds. Geographical interpretations of historical sources: readings in historical geography. Newton Abbot, U.K.: David and Charles; New York: Barnes and Noble, 1970. 452 p. LC 79-17258. [New ed., Folkestone, Kent, U.K.: William Dawson, 1976. 458 p. ISBN 0-7129-0741-6. £15.]　　　　902
A collection of 20 articles emphasizing geographical interpretation and analysis of particular source materials ranging from the eleventh to the nineteenth centuries. Articles arranged chronologically by date of source material.

Ehrenberg, Ralph E., ed. Pattern and process: research in historical geography. (National Archives Conference, v. 9). Washington, D.C.: Howard University Press, 1975. 360 p. ISBN 0-88258-050-7. $15. LC 74-23617.　　　　903
Proceedings of a National Archives conference on archival sources and historical geography. Emphasis on four principal themes: Afro-American population; exploration, surveying, and mapping; transportation, commerce, and industry; and rural and urban settlement.

Guelke, Leonard. Historical understanding in geography: an idealistic approach. (Cambridge studies in historical geography no. 3). Cambridge, U.K.; New York: Cambridge University Press, 1982. 109 p. ISBN 0-521-24678-4. $27.95. LC 82-4356.　　　　904
A work concerned with establishing historical geography as a legitimate field of historical knowledge based on the idea that historical inquiry is a form of understanding the past independent of approaches derived from the natural and social sciences.

Lemon, Anthony, and Pollock, Norman E., eds. Studies in overseas settlement and population. London; New York: Longman, 1980. 337 p. ISBN 0-582-48567-3. $50. LC 79-42738.　　　　905
A series of thirteen essays providing a broad overview of the historical geography of colonization and settlement. Focuses on themes of pioneering, metropolitan and colonial economies, social and political development, settler-native contacts, and migragrant labor. Case studies of Western Europe, the Americas, Australasia, East Africa, and Palestine.

Historical Geography

Lowenthal, David, and Bowden, Martyn J., eds. Geographies of the mind: essays in historical geography in honor of John Kirtland Wright. (American Geographical Society, special publication no. 40). New York: Oxford University Press, 1976. 263 p. LC 75-10182.

906

Series of essays identifying Wright's broad interest in the study of geographical knowledge. Themes include nature and place, environmental perception and behavior, landscape and the past, the sequent occupance concept, and imagination and exploration.

Morgan, Michael Alan. Historical sources in geography. London; Boston, MA: Butterworth Scientific, 1979. 153 p. ISBN 0-408-10609-3. LC 78-40365.

907

Although based on British sources, this work examines the source materials for historical geography and methods of data extraction and manipulation. Emphasis on the Middle Ages, agriculture, population, transport, and urban sources.

Newcomb, Robert M. Planning the past: historical landscape resources and recreation. Folkestone, U.K.: William Dawson and Sons; Hamden, CT: Shoestring Press, 1979. 255 p. ISBN 0-208-01728-3. $19.50. LC 78-40896.

908

Concerned with the historical geographer's interest in the interpretaion and preservation of old landscapes and their significance in present education. Two parts: a survey of recreational uses of past landscapes and the planning process; and examples of landscapes that reflect rural, urban, industrial, and political dimensions of settlement.

Norton, William. Historical analysis in geography. London; New York: Longman, 1984. 231 p. ISBN 0-582-30104-1. $15.95. (pbk). LC 83-747.

909

An examination of historical geography from the perspectives of its methods and modes of analysis relative to related fields and actual practices in current research. Historical geography presented as concerned primarily with geographical change through time, the development of landscape, and the evolution of spatial form.

Powell, Joseph M. Mirrors of the New World: images and image-makers in the settlement process. Canberra, Australia: Australian National University Press; Hamden, CT: Shoe String Press (Archon), 1978. 207 p. ISBN 0-7081-0580-7. ISBN 0-208-01654-6. $17.50. LC 79-338303.

910

Examines the process of image-making in influencing the evolution of settlement in Australia, New Zealand, and North America. Evaluates the social and intellectual context of eighteenth and nineteenth century Europe and emphasizes the images of the yeoman farmer, the desire for a healthy environment, and the search for utopia.

Vance, James E., Jr. This scene of man: the role and structure of the city in the geography of western civilization. New York: Harper's College Press, 1977. 437 p. ISBN 0-06-167407-9. LC 77-852.

911

A broad survey of western urbanism and urbanization. Emphasis is on two themes: the significance of the city in western society; and the processes involved in the changing internal structure of cities. Focuses on classical, medieval, mercantile, and industrial cities in Western Europe and North America.

3. HISTORICAL CULTURAL GEOGRAPHY (OLD WORLD)

Karl W. Butzer

Bender, Barbara. Farming in prehistory: from hunter-gatherer to food-producer. London: J. Baker; New York: St. Martin's Press, 1975. 268 p. ISBN 0-212-97003-8. LC 76-353698.

912•

A wide-ranging study of Old and New World agricultural origins based on the archaeological data. The cumulative nature of information derived from long term experimentation and manipulation of a broad spectrum of plants and animals are emphasized as a necessary prelude to domestication. The criteria for recognizing domestication are discussed, as are the distributions of potential domesticates.

Butzer, Karl W. Environment and archaeology: an ecological approach to prehistory. 2nd ed. Hawthorne, NY: Aldine-Atherton, 1971. 703 p. (1st ed., 1964. 524 p.). ISBN 0-202-33023-0. $35. LC 74-115938.

913

Presents earth-science and biological criteria for reconstructing prehistoric landscapes and the settings of prehistoric sites. Glacial-age environments of Europe, the Mediterranean Basin, Africa, and North America are synthesized. Human-land relationships of prehistoric hunter-gatherers and agriculturalists are examined, as well as Old World agricultural dispersal.

Butzer, Karl W. Recent history of an Ethiopian delta: the Omo River and the level of Lake Rudolf. (Research papers series no. 136). Chicago, IL: University of Chicago, Department of Geography, 1971. 184 p. ISBN 0-89065-043-8. $10. (pbk). LC 70-184080.

914

A case study of historical ecology in the Omo Delta and floodplain of southwestern Ethiopia, primarily affected by rapid and repeated fluctuations in the level of non-outlet Lake Rudolf (Turkana). The geomorphological and historical records provide a 100-year chronology that is evaluated in terms of changing settlement response and broader climatic implications in the eastern Sahel.

Butzer, Karl W. Early hydraulic civilization in Egypt: a case study in cultural ecology. Chicago, IL; University of Chicago Press, 1976. 134 p. ISBN 0-226-08634-8. $5. LC 75-36398.

915

Examines environmental, technological, settlement, and demographic change in the Nile Valley irrigation system from the 4th to the 1st millennium B.C. The Wittfogel and Carneiro hypotheses are critically examined against the historical data. Greater weight is given to differential regional development reflecting environmental constraints, local particularism, and evolution of provincial central places.

Butzer, Karl W. Archaeology as human ecology: method and theory for a contextual approach. Cambridge, U.K.; New York: Cambridge University Press, 1982. 364 p. ISBN 0-521-24652-0. $29.50. ISBN 0-521-28877-0. $13.95. (pbk). LC 81-21576.

916

A systematic application of human geography principles to archaeology. Spatial and temporal variability of the environmental matrix, at several scales, provide the context in which human communities interacted spatially, economically and socially, including modification of the physical and biotic landscape. The synthetic section deals with quantitative models for spatial organization, socio-ecological models for settlement, reconstruction of settlement systems, and cultural adaptation.

Dennell, Robin. European economic prehistory: a new approach. New York; London: Academic Press, 1983. 217 p. ISBN 0-12-209180-9. $25. LC 82-72666.

917

An interpretation of Europe's Stone Age past in terms of environmental opportunities or constraints, an increasingly complex technology and social repertoire, and the exploitation of new resources, ultimately including introduced domesticates. The economic and ecological perspectives highlight the development of different, postglacial subsistence systems in southern and northern Europe as well as contrasts between intrusive, agricultural colonization and the acculturation of indigenous hunter-gatherers to a farming way of life.

Hassan, Fekri A. Demographic archaeology. New York; London: Academic Press, 1981. 298 p. ISBN 0-12-331350-3. $35. LC 80-1677.   918

Includes ecological methods of estimating population size from archaeological data, together with discussions of mortality, fertility regulation, population growth, theoretical carrying capacity, and population pressure. Complex demographic models are developed for prehistoric hunter-gatherers and agriculturalists, illustrated by Old World examples.

Hodder, Ian; Isaac, Glynn; and Hammond, Norman, eds. Pattern of the past: studies in honour of David Clarke. Cambridge, U.K.; New York: Cambridge University Press, 1981. 443 p. ISBN 0-521-22763-1. $59.50. LC 79-41379.   919

Fifteen original chapters that exemplify how archaeologists have turned to geographical concepts in interpreting the record of land-use, site function, social organization using people/area ratios, colonization processes, settlement distributions, and agricultural intensification.

Klein, Richard G. Ice-Age hunters of the Ukraine. Chicago, IL: University of Chicago Press, 1973. 140 p. ISBN 0-226-43945-3. $12.50. ISBN 0-226-43946-1. $4.50. (pbk). LC 73-77443.   920

A case study of glacial-age settlement and subsistence on the periglacial steppes of southern Russia. The chronology, regional and site-specific environments, and economic activities are described, as are the differences in social organization between Neanderthal and modern human groups.

Larsen, Curtis E. Life and land use on the Bahrain Islands: the geoarchaeology of an ancient society. Chicago, IL: University of Chicago Press, 1983. 339 p. ISBN 0-226-46905-0. $20. ISBN 0-226-46906-9. $9. (pbk). LC 83-5085.   921

Bahrain represents a circumscribed environmental system with productivity depending on a finite, artesian water supply, with minimal recharge. This interdisciplinary case study illustrates the delicate balance between water, agriculture, settlement, demography, and external trade across 7000 years. Periodic land-use expansion is examined in the light of socioeconomic and climatic factors, and interpreted in a modern context of rapid change.

Martin, Paul S. and Klein, Richard G., eds. Quaternary extinctions: a prehistoric revolution. Tucson, AZ: University of Arizona Press, 1984. 892 p. ISBN 0-8165-0812-7. $65. LC 83-18053.   922

During the recent geological or historical past, scores of large animals such as mammoths, mastodons, giant sloths, and kangaroos became extinct on most world continents. These thirty-eight papers explore potential explanations for this disappearance, suggesting complex environmental changes and overkill by new human colonists. Regional empirical studies are complemented by environmental and cultural models.

Reed, Charles A., ed. Origins of agriculture. The Hague, Netherlands: Mouton; Hawthorne, NY: Walter de Gruyter, 1977. 1017 p. ISBN 0-202-90043-6. DM 148. LC 79-300652.   923

Twenty-nine original chapters on all aspects of early plant and animal domestication in the Old and New Worlds. Deals with general principles, chronology, questions of multiple, independent invention, the potential role of climatic change, the mechanics of early agrosystems, cultural idiosyncrasies, and agricultural centers vs. noncenters.

Simmons, Ian, and Tooley, Michael, eds. The Environment in British prehistory. Ithaca, NY: Cornell University Press, 1981. 334 p. ISBN 0-8014-1397-4. $39.50. LC 80-69818.   924

Seven original studies evaluate the environmental factors that affected early people in Britain from the Old Stone Age to the Roman invasion. Settlement and subsistence patterns are discussed for five major phases, together with information on changing landscapes, the contrasting exploitation of the highland and lowland zones, and the relative human impact of the various cultural groups on the environment.

Ucko, Peter J.; Tringham, Ruth; and Dimbleby, G.W., eds. Man, settlement, and urbanism: proceedings of a meeting of the research seminar in archaeology and related subjects held at the Institute of Archaeology, London University. Cambridge, MA: Schenkman; London: Duckworth, 1972. 979 p. ISBN 0-7156-0589-5. $65. LC 72-186318.   925

Eighty-nine papers provide a broad and valuable range of ethnographic, historical, and prehistoric studies dealing with urban and non-urban settlement patterning in the Old and New Worlds. Factors such as resources, mobility, demography, territoriality, and culture are considered, as are determinants for urban growth in pre-industrial societies.

4. NEW WORLD PREHISTORY

B.L. Turner II

Denevan, William M., ed. The Native population of the Americas in 1492. Madison, WI: University of Wisconsin Press, 1976. 353 p. ISBN 0-299-07050-6. $29.50. LC 75-32071.   926•

Compendium of eight classic works on pre-Hispanic populations covering the Caribbean and Central America, Mexico, South America, and North America. Includes well-illustrated controversies about the maximum sizes of these populations and the scale of depopulation after the conquest. Outstanding 32-page bibliography.

Donkin, Robin A. Agricultural terracing in the aboriginal New World. (Viking Fund Publications in anthropology no. 56). Tucson, AZ: University of Arizona Press, 1979. 196 p. ISBN 0-8165-0453-9. $8.50. (pbk). LC 77-15120.   927

Compendium of data and technical interpretations of pre-Hispanic terracing in the New World. Information is ordered by topic and by regions. Broader interpretations include the dominance of terraces in arid and semi-arid areas but with origins in wetter locales, and primary functions to increase soil depth and moisture. Appendix listing site, form, use, and authority of reported terraces. Excellent 33-page bibliography.

Eidt, Robert C. Advances in abandoned settlement analysis: application to prehistoric anthrosols in Colombia, South America. Milwaukee, WI: University of Wisconsin-Milwaukee, Center for Latin America, 1984. 159 p. ISBN 0-930450-53-1. $12.50. (pbk). LC 84-17502.   928

Demonstrates the use of data on phosphate fractions and other soil indicators to analyze ancient residential and agricultural features in several environments of Colombia. Interpretations include a sequence of land use from drier valley floors to upslope settlement and abandonment of valleys and ultimate resettlement of valley floors with the use of raised fields. Hailed as a landmark.

Flannery, Kent V. The Early Mesoamerican village. New York; London: Academic Press, 1982. 377 p. ISBN 0-12-259852-0. $17.50. (pbk). LC 82-6737.   929

Based primarily on long-term archaeological field work in Oaxaca, this book explores the development of early sedentary life in Mexico. A classic, in part, because threaded throughout the book are fictional pieces characterizing the clashes between traditional and modern social science approaches to Mesoamerican studies.

Harrison, Peter D. and Turner, B.L. II., eds. Pre-Hispanic Maya Agriculture. Albuquerque, NM: University of New Mexico Press, 1978. 403 p. ISBN 0-8263-0483-4. $20. LC 78-55703.   930•

Seventeen original articles that challenge the primacy of the traditional slash-and-burn view of ancient Maya subsistence. Provides a large amount of new information and interpretation on Maya agriculture and lowland environments. Although much of the data have been superseded, this volume illustrates the nature of conceptual changes involved in the subject.

Cultural Ecology

Roosevelt, Anna C. Parmana: prehistoric maize and manioc subsistence among the Amazon and Orinoco. (Studies in Archaeology). New York; London: Academic Press, 1980. 320 p. ISBN 0-12-595350-X. $35. LC 80-1678.   931

Challenges the prevailing manioc and hunter-gatherer-subsistence theme as it has been applied to cultural development in Amazonia. Argues that manioc is not suitable for growth in Amazonian floodplains and could not have been the subsistence base of the large populations that developed along them. Data from the Parmana area of the Orinoco Basin are used to test the maize population hypothesis.

Sanders, William T.; Parsons, Jeffrey R.; and Santley, Robert S. The Basin of Mexico: ecological processes in the evolution of civilization. (Studies in archaeology). New York; London: Academic Press, 1979. 561 p. ISBN 0-12-618450-X. Book and map $49.50. ISBN 0-12-618452-6. (map). $17.50. LC 79-11697.   932

Excellent synthesis of the cultural ecological history produced by the long-term Basin of Mexico Project. Traces the evolution of the Basin's civilizations primarily through an extensive site survey technique. Population numbers and location are traced through time and related to the rise of intensive, hydraulic systems of agriculture, to possible minor climatic cycles and cultural development. Important set of maps is included.

Turner, B.L. II. and Harrison, Peter D., eds. Pulltrouser swamp: ancient Maya habitat, agriculture, and settlement in northern Belize. (Texan Pan American series). Austin, TX: University of Texas Press, 1983. 294 p. ISBN 0-292-75067-6. $22.50. LC 83-5906.   933

Report of an intensive, interdisciplinary project addressing the natural and cultural history of the use of a wetland by the ancient Maya.

5. CULTURAL ECOLOGY*

Gregory W. Knapp

*In this section the editorial committee reduced the number of entries recommended by the contributor.

See also the entries in Part III, 4. Biogeography, and in Part IV, 3. Historical Cultural Geography (Old World), 4. New World Prehistory, and 6. Cultural Geography.

Bennett, Charles F. Jr. Man and Earth's ecosystems: an introduction to the geography of human modification of the Earth. New York; Chichester, U.K.: John Wiley and Sons, 1975. 331 p. ISBN 0-471-06638-9. $30.95. LC 75-22330.   934•

A unique introductory text which focuses on human impacts on the environment using a regional framework. Many maps and illustrations.

Bennett, John W. The Ecological transition: cultural anthropology and human adaption. Oxford, U.K.; New York: Pergamon Press, 1976. 378 p. ISBN 0-08-017867-7. $29. ISBN 0-08-017868-5. $16. (pbk). LC 74-30430.   935

The best single-author, one-volume survey of the theory and methods of cultural ecology.

Bennett, John W. Northern plainsmen: adaptive strategy and agrarian life. Chicago, IL; now Hawthorn NY: Aldine, 1969. 352 p. ISBN 0-88295-602-7. $23.95. ISBN 0-88295-603-5. $13.95. (pbk). LC 76-75043.   936•

An influential case study of the contrasting strategies and interactions of Indians, ranchers, farmers, and Hutterites in southwestern Saskatchewan. One of the best examples of the adaptive-dynamics approach to cultural ecology. Anthropologist Bennett has also provided one of the few models for the cultural ecological study of a modern society.

Boserup, Ester. The Conditions of agricultural growth: the economics of agrarian change under population pressure. Chicago, IL: Aldine, 1965. 124 p. ISBN 0-202-07003-4. $14.95. LC 65-19513. 937•

Succinct monograph by an economic historian outlining the theory that people adapt their agricultural technology and techniques in accordance with population rather than vice-versa. A key work in cultural ecology.

Carlstein, Tommy. Time resources, society, and ecology: on the capacity for human interaction in space and time. (Preindustrial society series v.1). London; Boston, MA: George Allen and Unwin, 1982. 444 p. ISBN 0-04-300082-7. $35. ISBN 0-04-300083-5. $15.95. (pbk). LC 81-19118. 938

An introduction to cultural ecology notable for its numerous ingenious figures depicting the use of space and time by traditional hunter-gatherers, pastoralists, and farmers. The study is more concerned with clarifying and recasting existing data in the language of time geography than with theory building.

Carr, Claudia. Pastoralism in crisis: the Dasanetch and their Ethiopian lands. (Research paper no. 180). Chicago, IL: University of Chicago, Department of Geography, 1977. 319 p. ISBN 0-89065-087-X. $10. (pbk). LC 77-1252. 939

Employs an innovative social-environmental systems approach to study a southwest Ethiopian pastoral society. Alternative courses of action for group survival and self-determination are explored.

Clarke, William. Place and people: an ecology of a New Guinean community. Berkeley, CA: University of California Press, 1971. 265 p. ISBN 0-520-01791-9. $39.50. LC 78-126764. 940

A study of the structure, functioning, and trends in the human ecosystem of the Ndwimba Basin of highland New Guinea, inhabited by the 154-member Bomagai-Angoiang ethnic community. The soundest cultural ecological study of New Guinea, a key region in the historical development of ideas in the subdiscipline.

Cohen, Mark Nathan. The Food crisis in prehistory: overpopulation and the origins of agriculture. New Haven, CT: Yale University Press, 1977. 341 p. ISBN 0-300-02016-3. $25. ISBN 0-300-02351-0. $10.95. (pbk). LC 76-41858. 941

Advances the thesis that the adoption of agriculture was only one of a long series of cultural adaptations to increasing population. A controversial work by an anthroplogist which has stimulated further research.

Cohen, Yehudi A., ed. Man in adaptation: the cultural present. 2nd ed. Chicago, IL; now Hawthorne, NY: Aldine, 1974. 602 p. (1st ed., 1968. 433 p.). ISBN 0-202-01109-7. $39.95. ISBN 0-202-01110-0. $16.95. (pbk). LC 68-6868. 942•

Valuable collection of anthropological and geographical contributions to the study of the cultural ecology of hunter-gatherers, pastoralists, and agriculturalists. Although dated, it is the latest general reader of this type available.

Cronon, William. Changes in the land: Indians, colonists, and the ecology of New England. (American Century Series). New York: Hill and Wang, 1983. 241 p. ISBN 0-8090-3405-0. $15.95. ISBN 0-8090-0158-6. $6.95. (pbk). LC 83-7899. 943•

A case study of the impacts of Indians and Europeans upon each other and on the New England environment. A historian's remarkable updating of the methodology of Sauer and the Berkeley School of Cultural Geography through mastery of the recent literatures of evolutionary ecology, ecological and economic anthropology, and economic history.

Gade, Daniel W. Plants, man, and the land in the Vilcanota Valley of Peru. The Hague, Netherlands: W. Junk, 1975. 240 p. ISBN 9-06-193207-6. Dfl 80. $42. LC 76-357725. 944

Study of the altitudinal variation of cultigens and cropping systems in a highland valley of southern Peru.

Cultural Ecology

Geertz, Clifford. Agricultural involution: the processes of ecological change in Indonesia. Berkeley, CA: University of California Press, 1963. 176 p. ISBN 0-520-00459-0. $6.95. (pbk). LC 63-20356. 945

A pioneering application of the ecosystem and succession concepts to the study of peasant agriculture in the tropics. Although much of the ecological analysis has been superseded by subsequent work, this anthropological case study remains exemplary in its ambition and scope.

Hames, Raymond B. and Vickers, William T. Adaptive responses of native Amazonians. (Studies in anthropology). New York; London: Academic Press, 1983. 516 p. ISBN 0-12-321250-2. $49. LC 82-18399. 946

The best starting point for those interested in the cultural ecology of peoples of the humid tropical lowlands.

Hanks, Lucien M. Rice and man: agricultural ecology in southeast Asia. (Worlds of man series). Chicago, IL: Aldine Atherton; later, Arlington Hts, IL: Harlan Davidson, 1972. 174 p. ISBN 0-88295-606-X. $16.95. ISBN 0-88295-607-8. $8.95. (pbk). LC 78-169512. 947 •

Traces the evolution of a Thai village from pioneer days to a densely populated community concentrating on the sequence of progressively more intensive rice-growing adaptations: shifting cultivation; broadcasting; and transplanting.

Jochim, Michael A. Strategies for survival: cultural behavior in an ecological context. New York; London: Academic Press, 1981. 233 p. ISBN 0-12-385460-1. $23. LC 81-7887. 948 •

Well-written introductory text by an anthropologist reflects the recent movement away from functionalist or 'ecosystemicist' approaches in cultural ecology. The text explores strategies for feeding, procurement, settlement, and 'maintenance' (territoriality, warfare, and exchange) within an evolutionary framework.

Kirkby, Anne V.T. The Use of land and water resources in the past and present, Valley of Oaxaca, Mexico. (Memoirs of the Museum of Anthropology no. 5). Ann Arbor, MI: University of Michigan, Museum of Anthropology, 1973. 174 p. LC 73-623353. 949

Studies peasant agriculture and physical geography of a valley in the southern highlands of Mexico. A model of population and settlement change is constructed and tested against archaeological data. The most creative and exemplary cultural ecological study of tropical highland America.

Klee, Gary A. World systems of traditional resource management. (Scripta series in geography). New York: Halsted Press of John Wiley and Sons, 1980. 290 p. ISBN 0-470-27008-X. $39.95. LC 80-17711. 950

Eleven essays by geographers discuss the physical settings and components of traditional agricultural, pastoral, fishing, forestry, and hunting technologies of major world regions. The first and only regional survey of paleotechnic resource management.

Knight, C. Gregory. Ecology and change: rural modernization in an African community. New York; London: Academic Press, 1974. 300 p. ISBN 0-12-785435-5. $46.50. LC 73-7446. 951

Study of the Nhiha of southwest Tanzania focuses on agricultural changes associated with both population growth and modernization based on the expansion of commercial coffee cultivation. Explores alternative models of changing spatial patterns of agricultural activities.

Mathewson, Kent. Irrigation horticulture in highland Guatemala: the Tablon system of Panajachel. Boulder, CO: Westview Press, 1984. 180 p. ISBN 0-86531-973-1. $20.50. LC 83-17043. 952

A microgeographic study of tablon raised-bed horticulture in a present day village of highland Guatemala, an example of extreme agricultural intensification or involution for small-scale commercial production which may have pre-Columbian antecedents.

Meggers, Betty Jane. Amazonia: man and culture in a counterfeit paradise. (Worlds of man series). Chicago, IL; later, Hawthorne, NY: Aldine Atherton, 1971. 171 p. ISBN 0-88295-608-6. $17.95. ISBN 0-88295-609-4. $7.95. (pbk). LC 74-141427.
953
Analyzes traditional indigenous economies in Amazonia as constituting two contrasting homeostatically adapted human ecosystems: the floodplains (varzea) and the interfluves.

Moran, Emilio F. Human adaptability: an introduction to ecological anthropology. Boulder, CO: Westview Press, 1982. 404 p. ISBN 0-86531-430-6. $13.95. LC 82-8605.
954 •
Introductory cultural ecology text. Discusses human biological and behavioral adaptations to problems presented by major biomes: arctic zones; high altitudes; arid lands; grasslands; and the humid tropics. A useful and appropriate text for a course in geographical cultural ecology.

Netting, Robert McC. Balancing on an Alp: ecological change and continuity in a Swiss Mountain community. Cambridge, U.K.; New York: Cambridge University Press, 1981. 278 p. ISBN 0-521-23743-2. $47.50. ISBN 0-521-28197-0. $17.95. (pbk). LC 81-358.
955
An anthropologist's historical-demographic analysis of the closed corporate peasant community of Torbel in the Swiss Alps. Suggests a balance between population and environment through homeostatic mechanisms including continuous outmigration.

Nietschmann, Bernard. Between land and water: the subsistence ecology of the Miskito Indians, eastern Nicaragua. New York: Seminar Press, 1973. 279 p. ISBN 0-12-785562-9. $42. LC 72-7703.
956 •
Excellent study of a Caribbean coastal community's changing utilization of rain forest, wetland, and ocean resources for subsistence and commercial exchange. Focuses in particular on the green turtle fishery. Exceptionally well-written.

Porter, Philip W. Food and development in the semi-arid zone of East Africa. (Foreign and comparative studies: African series no. 32). Syracuse, NY: Maxwell School of Citizenship and Public Affairs, Syracuse University, 1979. 107 p. ISBN 0-915984-54-7. $7.95. (pbk). LC 79-20312.
957
Focuses on the agroclimatology and drought adaptations of the semi-arid areas of Uganda, Kenya, and Tanzania. Gives particular attention to the relationships between the peoples of semi-arid areas and global political economy, national policy, and the peoples of adjacent wetter environments.

Rappaport, Roy A. Pigs for the ancestors: ritual in the ecology of a New Guinea people. 2nd ed. New Haven, CT: Yale University Press, 1984. 501 p. (1st ed., 1967. 311 p.) ISBN 0-300-03204-8. $7.95. (pbk). LC 83-51294.
958
Includes full text of the original 1967 edition plus fifteen new essays by the author. Essays serve to clarify the author's positions, rebut some critiques, and carry forward the discussion of other criticisms which the author now recognizes as valid. The original edition was a milestone study of cultural ecology which perhaps overestimated the homeostatic stability of New Guinea human ecosystems; the new edition thus simultaneously includes a historically important case study and a contribution to current debate over theory and methodology.

Rindos, David. The Origins of agriculture: an evolutionary perspective. Orlando, FL: Academic Press, 1984. 325 p. ISBN 0-12-589280-2. $32.50. LC 83-7165.
959
An application of Darwinian evolutionary theory (as opposed to traditional cultural evolutionary ideas) to the investigation of early domestication and reliance on agriculture. Argues that domestication has often been an unintentional result of coevolutionary human-plant relations, and that often unstable rather than stable agricultural systems have had the evolutionary advantage.

Sauer, Carl O. Selected essays 1963-1975. Berkeley, CA: Turtle Island Foundation, 1981. 391 p.

960 •

Contains seventeen essays, some written before 1963, on diverse topics: the American frontier; ecology of human origins; fire ecology; historical demography; and agricultural origins. Along with the author's earlier Land and Life *this constitutes the testament of a lifetime's work in geographical cultural ecology.*

Waddell, Eric. The Mound builders: agricultural practices, environment, and society in the central highlands of New Guinea. (American Ethnological Society monograph no. 53). Seattle, WA: University of Washington Press, 1972. 253 p. ISBN 0-295-95169. LC 70-159437.

961

Report of a massive and methodologically innovative project studying highland New Guinea ecology, adaptation, and settlement at a variety of scales.

Watts, Michael J. Silent violence: food, famine, and peasantry in northern Nigeria. Berkeley, CA: University of California Press, 1983. 687 p. ISBN 0-520-04323-5. $34.40. LC 82-11060.

962

A comprehensive historical study of food and famine in Nigeria grounded in the author's extensive field work and archival research, and reflecting an interest in Marxist political economy. One of the most intellectually wide-ranging studies yet to appear in geographical cultural ecology.

6. CULTURAL GEOGRAPHY

Marvin W. Mikesell

A. ATLAS

Al-Faruqi, Isma'il Ragi and Sopher, David E., eds. Historical atlas of the religions of the world. New York: Macmillan, 1974. 346 p. ISBN 0-02-336400-9. LC 73-16583

963 •

Maps and commentary by several authors on the origin and distribution of the world's major religions.

B. BOOKS

Ashworth, Georgina, ed. World minorities. Sunbury, U.K.: Quartemaine House, Ltd., 1977. 3 v. ISBN 0-905898-00-1. $30. (Available from State Mutual Book and Periodical Service, Ltd., New York). LC 78-310796.

964 •

A convenient collection of short articles on the status and aspiration of 109 minority groups defined mostly in reference to language or religion. Each article includes statistical information or estimates and suggestions for further reading.

Barth, Fredrik. Ethnic groups and boundaries: the social organization of culture differences. (Results of a symposium held at the University of Bergen, 1967). Boston: Little, Brown; London: Allen, Unwin, 1969. 153 p. ISBN 0-31608246-5. $8.95. (pbk). LC 78-446490.

965

A collection of essays on an issue of persistent importance in cultural geography.

Fischer, Eric. Minorities and minority problems. New York: Vantage Books, 1980. 475 p. ISBN 0-533-03610-0. LC 77-95437.

966 •

A valuable reference work which includes comment on the political and historical importance of minority groups and an encyclopedic account of the numerical strength and status of such groups in most of the world's countries.

Geertz, Clifford. The Interpretation of cultures: selected essays. New York: Basic Books, 1973. 470 p. ISBN 0-465-09719-7. $8.95. (pbk). LC 73-81196.

967

A collection of reprinted essays by an influential author on the meaning and application of the concept of culture.

Hobsbawn, Eric, and Ranger, Terence, eds. The Invention of tradition. Cambridge, U.K.; New York: Cambridge University Press, 1983. 320 p. ISBN 0-521-24645-8. $29.95. ISBN 0-521-26985-7. $9.95. (pbk). LC 82-14711.

968

A collection of seven scholarly and eminently readable essays that illustrate a common theme: traditions which appear or claim to be old are often quite recent in origin and may even be invented. Examples are offered from Scotland, Wales, England, continental Europe, India, and Africa.

Kaplan, David, and Manners, Robert A. Culture theory. Englewood Cliffs, NJ: Prentice-Hall, 1972. 209 p. ISBN 0-13-195511-X. $15.95. (pbk). LC 72-90.

969

A convenient account of the varying interpretation and use of the concept of culture by anthropologists. Chapter three on "Types of Culture Theory " offers especially valuable commentary for cultural geographers.

Mason, Philip. Patterns of dominance. London; New York: Oxford University Press, 1970. 377 p. ISBN 0-19-218186-6. LC 71-479726.

970

Perhaps the best single work on the cultural consequences of European colonization overseas.

Meinig, Donald W., ed. The Interpretation of ordinary landscapes: geographical essays. New York: Oxford University Press, 1979. 255 p. ISBN 0-19-502536-9. $9.95. (pbk). LC 78-23182.

971 •

A lively book consisting of nine essays by several well-known authorities on landscape interpretation. The chapters by Peirce Lewis "Axioms for Reading the Landscape" and D.W. Meinig "The Beholding Eye: Ten Versions of the Same Scene" are especially lucid and useful.

Richardson, Miles, ed. The Human mirror: material and spatial images of man. Baton Rouge, LA: Louisiana State University Press, 1974. 366 p. ISBN 0-8071-0074-9. $30. LC 73-90872.

972

A useful exploration by anthropologists and geographers of attitudes toward, and perceptions of, material culture. Also includes effective demonstration of the relationship between social structure and architecture and settlement patterns.

Riley, Carroll L.; Kelley, J. Charles; Pennington, Campbell W.; and Rands, Robert L. Man across the sea: problems of pre-Columbian contacts. Austin, TX: University of Texas Press, 1971. 552 p. ISBN 0-292-70117-9. $27.50. LC 70-149022.

973

A stimulating discussion or debate on evidence of contact between the Old World and the Americas before the time of Columbus. Chapter one "Diffusion Versus Independent Development: the Basis of Controversy" by Stephen C. Jett is a thoughtful and well-documented review of an important issue in cultural geography. The 21 articles include proponents both of contact and non-contact views. Excellent 71-page bibliography.

Sahlins, Marshall D. and Service, Elman R., eds. Evolution and culture. Ann Arbor, MI: University of Michigan Press, 1960. 131 p. ISBN 0-472-08775-4. $5. LC 60-7930.

974

A brief but effective examination of how the evolutionary impulses evident in all cultures can be discovered, examined, and placed within a theoretical structure.

Cultural Geography

Sauer, Carl O. Agricultural origins and dispersals: the domestication of animals and foodstuffs. (Bowman Memorial Lectures, series 2). 1st ed. New York: American Geographical Society, 1952. 110 p; 2nd ed. Cambridge, MA: MIT Press, 1969. 175 p; Reprinted with additions as Seeds, spades, hearths, and herds: the domestication of animals and foodstuffs. (Paperback series no. 111). Cambridge, MA: MIT Press, 1972. ISBN 0-262-69071-9. $3.95. (pbk). LC 69-10841.   975

A stimulating and controversial examination of the locational, cultural, and environmental consequences of the initial transition from food gathering and hunting to food production and animal husbandry. Includes attractive maps.

Schneider, Louis, and Bonjean, Charles M., eds. The Idea of culture in the social sciences. Cambridge, U.K.: Cambridge University Press, 1973. 149 p. LC 73-76175.   976

Essays on the use or possible use of the concept of culture in anthropology, sociology, political science, history, and geography. The chapter on geography "Place and Location: Notes on the Spatial Patterning of Culture" by David E. Sopher, includes a rich array of footnotes providing access to appropriate literature.

Schwind, Martin, ed. Religionsgeographie. (Wege der Forschung, 397). Darmstadt, W. Germany: Wissenschaftliche Buchgesellschschaft, 1975. 404 p. ISBN 3-534-06214-0. DM 73. LC 75-510594.   977•

Solid scholarly treatment of the geography of religion.

Sopher, David E. Geography of religions. (Foundation of cultural geography series). Englewood Cliffs, NJ: Prentice-Hall, 1967. 115 p. LC 67-13357.   978•

The only extended discussion in English of a key element of cultural geography. Chapters deal with geographic approaches to the study of religion, ecology, and religious institutions, the influence of religion on land tenure, land use and landscapes, and a number of other issues having to do with the religious organization of space and the distribution of religions.

Spencer, Joseph Earle, and Thomas, William L., Jr. Cultural geography: an evolutionary introduction to our humanized earth. New York: John Wiley and Sons, 1969. 591 p. ISBN 0-471-81550-0. LC 67-28950.   979•

A pioneer effort to survey the entire field of cultural geography from an evolutionary perspective, i.e., from prehistoric to modern times. Now useful more as a reference work than as a textbook. Numerous maps, charts, and photographs. The most original and useful part of the book (chapters 4-7) is devoted to "cultural divergence followed by cultural convergence."

Thomas, William L., Jr. Man's role in changing the face of the earth. Chicago, IL: University of Chicago Press, 1956. 1193 p. ISBN 0-226-79603-5. $45. LC 56-5865.
Vol. 1: ISBN 0-226-79604-3. 390 p. $10. (pbk).
Vol. 2: ISBN 0-226-79605-1. 391 p. $6.95. (pbk).   980•

A monumental work consisting of 52 essays and complementary discussion. Many of the essays are still very useful and suprisingly current after 30 years. Indispensable for students and teachers seeking understanding of the effect of human activities on major categories of natural environment, e.g., vegetation, wildlife, air, water, and non-renewable resources.

Wagner, Philip L. and Mikesell, Marvin W., editors and translators. Readings in cultural geography. Chicago, IL: University of Chicago Press, 1962. 589 p. ISBN 0-226-86931-8. $20. LC 62-9740.   981

A pioneer effort to survey the literature of cultural geography and define its character. The 34 essays, many translated from foreign languages, are grouped under four headings: orientation; cultural areas and distributions; cultural origins and dispersals; and landscape and ecology. The general introduction by the editors presents a retrospective view of the evolution of cultural geography as a special field of human geography.

7. SOCIAL GEOGRAPHY

David F. Ley

See also entries in Part IV, 6. Cultural Geography, 8. Behavioral Geography, and 20, E. Urban Social Geography, especially entries 1357-1363.

Berry, Brian J.L. The Open housing question: race and housing in Chicago, 1966-1976. Cambridge, MA: Ballinger, 1979. 517 p. ISBN 0-88410-429-X. $35. LC 79-14912.   982
An impressive study of Chicago's housing market emphasizing neighborhood racial change and barriers to racial integration.

Bourne, Larry S. The Geography of housing. (Scripta series in geography). New York: John Wiley and Sons, 1981. 288 p. ISBN 0-470-27058-6. $44.95. ISBN 0-470-27059-4. $19.95. (pbk). LC 80-19908.   983
Text. Reviews wide literature from a dominantly economic perspective.

Buttimer, Anne. Values in geography. (Commission on College Geography resource papers series no. 24). Washington, DC: Association of American Geographers, 1974. 58 p. ISBN 0-89291-071-2. $4. (pbk). LC 74-76634.   984
An important introduction and impetus to a 'values' or humanistic perspective in social geography.

Buttimer, Anne, and Seamon, David, eds. The Human experience of space and place. London: Croom Helm; New York: St. Martin's Press, 1980. 199 p. ISBN 0-312-39910-3. $29. LC 80-12173.   985
A set of essays with a social planning emphasis (alcoholism, the elderly, neighborhood planning, etc.).

Clarke, Colin; Ley, David; and Peach, Ceri, eds. Geography and ethnic pluralism. London; Boston, MA: Allen and Unwin, 1984. 294 p. ISBN 0-04-309107-5. $30. ISBN 0-04-309108-3. $14.95. (pbk). LC 84-6319.   986•
Eleven essays on geography of plural societies. Case studies from the Third World, the Americas, and Western Europe. Emphasis on problems of group integration in national society.

Claval, Paul. Principes de géographie sociale. Paris: Éditions M.-Th. Génin: Librairies Techniques, 1973. 351 p. FFr 123. LC 74-158040.   987
Broad treatment of social geography from a French point of view including discussion of the development and theory of the field.

Dennis, Richard, and Clout, Hugh. A Social geography of England and Wales. Oxford, U.K.; New York: Pergamon, 1980. 208 p. ISBN 0-08-021802-4. $36. ISBN 0-08-021801-6. $14.50. (pbk). LC 79-41690.   988
A good regional study emphasizing urban residential patterns.

Duncan, James S., eds. Housing and identity: cross cultural perspectives. London: Croom Helm, 1981; New York: Holmes and Meier, 1982. 250 p. ISBN 0-8419-0701-3. $25. LC 81-4837.   989
Eight substantial essays on relations between housing and identity: the meaning of home ownership; women's status and housing; housing and the elderly; etc. Well reviewed.

Geipel, Robert. Disaster and reconstruction: The Friuli, Italy, earthquake of 1976. Translated from German by Philip Wagner. London; Boston, MA: Allen and Unwin, 1982. 202 p. ISBN 0-04-904006-5. $40. ISBN 0-04-904007-3. $19.95. (pbk). LC 82-11625.   990
Excellent case study of the geography of the Friuli (Italy) earthquake and of the policies and practices of reconstruction. "A completely new book...rather than simply a translation of the [author's] previously published German and Italian research reports." (preface, p. xi)

Social Geography

Gould, Peter, and White, Rodney. Mental maps. Harmondsworth, U.K.; Markham, Ontario: Penguin, 1974. 203 p. ISBN 0-14-021688-X. £1.25. LC 75-4147.
Text. Much-quoted volume on spatial images and place references.
991 •

Herbert, David T. The Geography of urban crime. (Topics in applied geography series). London; New York: Longman, 1982. 120 p. ISBN 0-582-30046-0. $10.95. (pbk). LC 80-41890.
A clear exposition of a growing subfield in social geography. A very useful introductory text.
992 •

Herbert, David T. and Johnston, Ronald J., eds. Social areas in cities. London; New York: John Wiley, 1976. 2 v.
    Vol. 1: Spatial processes and form. ISBN 0-471-99417-0. $53.95.
    Vol. 2: Spatial perspectives on problems and policies. ISBN 0-471-37305-2. $61.95.
993

Herbert, David T., and Smith, David M., eds. Social problems and the city: geographical perspectives. London; New York: Oxford University Press, 1979. 271 p. ISBN 0-19-874079-4. $13.95. (pbk). LC 78-41144.
Thirteen essays, largely British, reviewing research in geography of health problems, ethnicity, housing, educational disadvantage, urban violence, etc. Useful reviews.
994

Holcomb, H. Briavel, and Beauregard, Robert A. Revitalizing cities. (Resource publications in geography series). Washington, DC: Association of American Geographers, 1981. 84 p. ISBN 0-89291-148-4. $5. (pbk). LC 81-69237.
A very comprehensive review of central and inner city revitalization in U.S. cities over the past 20 years. Best single source to date on this topic.
995 •

Institute of British Geographers, Women and Geography Study Group. Geography and gender: an introduction to feminist geography. London; Dover, NH: Hutchinson Education, 1984. 160 p. ISBN 0-09-156671-1. $4.95. (pbk). LC 84-12871.
Undergraduate text on feminist geography that argues that the implications of gender for the organization of space and society are as important as other social or economic factors. Outlines essential concepts and encourages the reader to pursue his or her own research interests by suggesting projects.
996

Jackson, Peter, and Smith, Susan J., eds. Social interaction and ethnic segregation. (Special publications of the Institute of British Geographers, series no. 12). London; New York: Academic Press, 1981. 235 p. ISBN 0-12-379080-8. $34.50. LC 81-67914.
British essays on ethnic spatial patterns and social relations.
997

Jackson, Peter, and Smith, Susan J. Exploring social geography. London; Boston, MA: Allen and Unwin, 1984. 239 p. ISBN 0-04-3011-69-1. $30. ISBN 0-04-301170-5. $9.95. (pbk). LC 83-25732.
Text. A review and discussion of perspectives and theoretical issues in social geography. Broad ranging and useful. An important book.
998

Jakle, John A.; Brunn, Stanley D.; and Roseman, Curtis C. Human spatial behavior: a social geography. North Scituate, MA: Duxbury Press, 1976. 315 p. ISBN 0-87872-098-7. $13.95. LC 75-11278.
Useful review of a narrow concept of social geography.
999 •

Jones, Emrys, and Eyles, John. An Introduction to social geography. London; New York: Oxford University Press, 1977. 273 p. ISBN 0-19-874062-X. $12.95. (pbk). LC 77-361794.
A useful Anglo-American text although dated in some respects.
1000

Knox, Paul. Social well-being: a spatial perspective. London; New York: Oxford University Press, 1975. 60 p. LC 76-354808.   1001•
Useful introductory text to the geography of the quality of life.

Knox, Paul. Urban social geography: an introduction. London; New York: Longman, 1981. 243 p. ISBN 0-582-30044-4. $16.95. (pbk). LC 80-40826.   1002
Useful text with British emphasis. Particularly good on equitable access to public facilities by different social groups.

Ley, David F. The Black inner city as frontier outpost: images and behavior of a Philadelphia neighborhood. (Monograph no. 7). Washington, DC: Association of American Geographers, 1974. 282 p. ISBN 0-89291-086-0. $6.95. (pbk). LC 74-82116.   1003
Case study of a Philadelphia ghetto neighborhood. Examines images of the inner city and reality of coping in a highly stressful environment.

Ley, David F. and Samuels, Marwyn S., eds. Humanistic geography: prospects and problems. Chicago, IL: Maaroufa Press, (now New York: Methuen) 1978. 337 p. ISBN 0-416-60101-4 no. 2862. $18.95. LC 78-52408.   1004
An influential set of essays identifying the humanistic perspective in social geography and elsewhere.

Lord, J. Dennis. Spatial perspectives on school desegregation and busing. (Resource papers for college geography no. 77-3). Washington, DC: Association of American Geographers, 1977. 36 p. ISBN 0-89291-124-7. $4. (pbk). LC 76-57034.   1005
Good discussion of one aspect of geography of education emphasizing trends in towns and counties of the American South.

Mazey, Mary Ellen, and Lee, David R. Her space, her place: A geography of women. (Resource publications in geography). Washington, DC: Association of American Geographers, 1983. 83 p. ISBN 0-89291-172-7. $5. (pbk). LC 83-22289.   1006•
An unabashedly feminist book, the first of its kind. The authors review existing materials and use traditional geographical themes (distribution, spatial analysis, diffusion, the cultural landscape, etc.) to suggest new directions in the study of the geography of women.

Mitchell, Bruce, and Draper, Dianne L. Relevance and ethics in geography. London; New York: Longman, 1982. 222 p. ISBN 0-582-30035-5. $28.50. LC 81-19386.   1007
Examines a number of methodological and ethical questions faced by social geographers.

Peach, Ceri, ed. Urban social segregation. London; New York: Longman, 1975. 444 p. ISBN 0-582-48088-4. $23. LC 74-80434.   1008
Essays in the human ecology tradition from geographers and sociologists.

Peach, Ceri; Robinson, Vaughan; and Smith, Susan., eds. Ethnic segregation in cities. Athens, GA: University of Georgia Press, 1981. 258 p. ISBN 0-8203-0599-5. $20. LC 81-14715.   1009
A collection of British and American papers on a topic of traditional emphasis to social geography by leaders in the field.

Porteous, J. Douglas. Environment and behavior: planning and everyday urban life. Reading, MA: Addison-Wesley, 1977. 446 p. ISBN 0-201-05867-7. $22.95. LC 76-1752.   1010•
Text. Good coverage of interdisciplinary behavior/environment literature, including design and policy issues of the built environment.

Robson, Brian Turnbull. Urban social areas. London; New York: Oxford University Press, 1975. 64 p. LC 76-352699.   1011
Succinct statement of relations between social structure and residential patterns. Very useful review and integration of literature.

Rowles, Graham D. Prisoners of space?: exploring the geographical experience of older people. Boulder, CO: Westview Press, 1980. 216 p. ISBN 0-89158-069-7. $30. $11.50. (pbk). LC 77-18655.
1012•
Detailed interpretation of geographical experience of elderly in a Massachussetts town. Sensitive and insightful.

Smith, Christopher J. Geography and mental health. (Resource paper no. 76-4). Washington, DC: Association of American Geographers, 1977. 51 p. ISBN 0-89291-119-0. $4. LC 76-29269.
1013
A good review commissioned as a college resource paper and useful for undergraduates.

Smith, David Marshall. Human geography: a welfare approach. London: Arnold; New York: St. Martin's Press, 1977. 402 p. ISBN 0-7131-5924-3. $27.50. LC 80-503096
1014
A thoughtful attempt to introduce concepts of social welfare to socio-spatial patterns. Discusses needs, wants, the geography of inequality, with empirical studies and suggestions for reform.

Watson, J. Wreford. (James). Social geography of the United States. London; New York: Longman, 1979. 290 p. ISBN 0-582-48197-X. $12.95. (pbk). LC 77-30747.
1015•
A personal but perceptive and highly readable account.

Western, John. Outcast Cape Town. Minneapolis, MN: University of Minnesota Press, 1981. 372 p. ISBN 0-8166-1025-8. $22.50. LC 81-14640.
1016•
An excellent study of social groups, political power, neighborhood, and spatial relations. Covers the Cape colored community and apartheid legislation in Cape Town.

White, Paul. The West European city: a social geography. London; New York: Longman, 1984. 269 p. ISBN 0-582-30047-9. $13.50. (pbk). LC 83-9382.
1017•
A useful book examining spatial structure of cities in Western Europe, with a social problems orientation to such issues as minority segregation, urban renewal, and suburbanization.

Wiseman, Robert F. Spatial aspects of aging. (Resource papers for college geography no. 78-4). Washington, DC: Association of American Geographers, 1978. 43 p. ISBN 0-89291-133-6. $4. (pbk). LC 78-59103.
1018•
Review of geography of the elderly.

8. BEHAVIORAL GEOGRAPHY

Roger M. Downs and Reginald G. Golledge*

*Reginald G. Golledge wishes to acknowledge the assistance of Helen Conclis and Nathan Gale.

See also the entries in Part III, 12. Natural Hazards, and in Part IV, 7. Social Geography, 9. Environmental Perception, and 23. Technological Hazards.

Boulding, Kenneth E. The Image: knowledge in life and society. Ann Arbor, MI: University of Michigan Press, 1956. 1961 (pbk). 175 p. ISBN 0-472-06047-3. $6.95. (pbk). LC 56-9720.
1019•
Simple and elegant statement of the role of the image in social and behavioral sciences. Wide-ranging, readable, and important to the evolution of the field. A remarkably prescient volume, it still has the power to inspire.

Carlstein, Tommy; Parkes, Don; and Thrift, Nigel, eds. Timing space and spacing time. London: Edward Arnold; New York: John Wiley, 1978. 3 v.
    Vol. 1: Making sense of time. ISBN 0-470-26511-6. $29.95.
    Vol. 2: Human activity and time geography. ISBN 0-470-26513-2. $49.95.
    Vol. 3: Time and regional dynamics. ISBN 0-470-26512-4. $29.95.   1020

Cox, Kevin R. and Golledge, Reginald G., eds. Behavioral problems in geography: a symposium. (Studies in geography, no. 17). Evanston, IL: Northwestern University. Department of Geography, 1969. 276 p. LC 72-27777.   1021
    A collection of early papers exploring the emerging field of behavioral geography.

Cox, Kevin R. and Golledge, Reginald G., eds. Behavioral problems in geography revisited. New York; London: Methuen, 1981. 290 p. ISBN 0-416-72430-2. $27.95. ISBN 0-416-72440-X. $13.50. (pbk). LC 82-176781.   1022
    Collection of twelve papers starting with four of the more widely referenced from the earlier Cox/Golledge volume, Behavioral Problems in Geography. *The four papers in section two represent research using theory, methods, and concepts stressed in the earlier volume. Section three includes critical evaluations of the contribution of behavioral research.*

Downs, Roger M. and Stea, David, eds. Image and environment: cognitive mapping and spatial behavior. Chicago, IL, later Hawthorne, NY: Aldine, 1973. 439 p. ISBN 0-202-10058-8. $34.95. LC 72-78215.   1023
    A collection of nineteen chapters, some reprinted and most original, which shaped the field of behavioral geography. Interdisciplinary in inspiration, it brought together theories and approaches from geography, psychology, sociology, and urban planning. Contains many of the "classic" contributions.

Downs, Roger M. and Stea, David. Maps in minds: reflections on cognitive mapping. New York: Harper and Row, 1977. 284 p. ISBN 0-06-041733-1. LC 76-57221.   1024
    Basic statement of the link between cognitive mapping and spatial behavior. Explicit linkage between geography and psychology, with illustrations and examples drawn from technical and popular sources. Designed for introductory-level audience.

Gold, John R. An Introduction to behavioural geography. Oxford, U.K.; New York: Oxford University Press, 1980. 290 p. ISBN 0-19-823233-0. $39.95. ISBN 0-19-823234-9. $12.95. (pbk). LC 79-41084.   1025 •
    Broad, lucid overview of the literature on spatial cognition, cognitive mapping, and spatial behavior as relating to landscape and natural hazard assessment, the designed environment, and locational decision making. An introductory text, sketchy with respect to analytic methods and key psychological theories. Excellent bibliography.

Golledge, Reginald G. and Rayner, John N., eds. Proximity and preference: Problems in the multidimentional analysis of large data sets. Minneapolis, MN: University of Minnesota Press, 1982. 310 p. ISBN 0-8166-1042-8. $35. LC 81-14634.   1026
    Collection of seventeen papers exploring the interface between geography and psychology. Focuses on multidimensional analysis methods largely derived from psychology and on the experimental design and measurement problems associated with large data sets in geography. Applications include preference functions and choice behavior.

Golledge, Reginald G. and Rushton, Gerard, eds. Spatial choice and spatial behavior: geographic essays on the analysis of preferences and perceptions. Columbus, OH: Ohio State University Press, 1976. 321 p. ISBN 0-8142-0241-1. $16.50. LC 76-889.   1027
    Collection of fifteen papers representative of analytic behavioral work in the late 1970's. Focuses on cognition, preference, individual decision making and choice, and the question of the interrelationship between perception, cognition, and spatial behavior. Introduces analytic techniques derived from experimental psychology but applied to real-world (rather than laboratory) environments.

Behavioral Geography

Hart, Roger. Children's experience of place. New York: Irvington (Distributed by Halsted Press), 1979. 518 p. ISBN 0-470-99190-9. $19.95. LC 77-21507.   1028
Massive study of children's behavior in their own spaces and places. Links geographical and psychological theory and methods to show the importance of the developmental approach in understanding spatial cognition and behavior. Excellent bibliography.

Lowenthal, David, ed. Environmental perception and behavior. (Research paper no. 109). Chicago, IL: University of Chicago, Department of Geography, 1967. 89 p. ISBN 0-89065-018-7. $10. (pbk). LC 66-29233.   1029
Seminal publication representing one of the first attempts by geographers and psychologists to interact academically on topics of mutual concern relating to the perception, interpretation, and use of large-scale environments. Both the humanistic and analytic research orientations are represented in the five articles included in the collection.

Lynch, Kevin. The Image of the city. Cambridge, MA: MIT Press, 1960. 194 p. ISBN 0-262-12004-6. $20. ISBN 0-262-62001-4. $5.95. (pbk). LC 60-7362.   1030•
Seminal work largely responsible for the surge of interest in geographic behavioral research. Introduced simple sketch mapping and analysis procedures which revealed the basic geometric elements (nodes, edges, landmarks, areas, and paths) in urban images. Highly readable and imaginatively illustrated.

Moore, Gary T. and Golledge, Reginald G., eds. Environmental knowing: theories, research, and methods. (Community Development Series, v. 23). Stroudsburg, PA: Dowden, Hutchinson, and Ross, 1976. 441 p. ISBN 0-87933-060-0. $23.95. (pbk). LC 76-4942.   1031
Interdisciplinary collection of 35 articles resulting from two meetings of the Environmental Design and Research Association, supplemented by substantive introductory and methodological sections by the editors. Part one concentrates on theories of environmental cognition and learning; part two presents different sociological, anthropological, literary and phenomenological perspectives; part three is mainly methodological.

Parkes, Don, and Thrift, Nigel. Times, spaces, and places: a chronogeographic perspective. Chichester, U.K.; New York; Toronto: John Wiley and Sons, 1980. 527 p. ISBN 0-471-27616-2. $88.95. LC 79-40523.   1032
Substantive work introducing the "chronogeographical" perspective, a generalization of time-geography as propounded by the Hägerstrand group. Applications focus on urban and social geography although the concepts and principles presented are much more general. Many illustrations with a useful methodological appendix.

Pocock, Douglas C. and Hudson, Ray. Images of the urban environment. London: Macmillan; New York: Columbia University Press; 1978. 181 p. ISBN 0-333-19198-6. $25. LC 78-312975.   1033
Introductory-level text typical of the British approach to behavioral geography. Emphasizes the affective, rather than cognitive and appraisive, dimensions of environmental images. Methodology is descriptive rather than analytic. Final chapters relate environmental images and planning policy, thus attempting the transition from individual to society.

Saarinen, Thomas F. Environmental planning: perception and behavior. Boston, MA: Houghton Mifflin, 1976. 262 p. ISBN 0-395-20618-9. $15.95. (pbk). LC 75-19533.   1034•
Significant early attempt at a comprehensive environmental planning text with pronounced behavioral bias. Focuses primarily on environmental perception at various scales but also addresses issues of environmental cognition and attitudes, especially towards natural hazards. Demonstrates the relevance of behavioral considerations for planning.

Sims, John H. and Baumann, Duane D. Human behavior and the environment: interactions between man and his physical world. Chicago, IL: Maaroufa Press, 1974. 354 p. LC 73-89450.   1035

Interdisciplinary collection of eighteen articles addressing a wide range of man-environment issues. Part one focuses on human manipulation of the environment; part two on the effects of the environment on man. Descriptive rather than analytic approach. Text particularly suitable for courses in environmental studies.

Walmsley, David J. and Lewis, George J. Human geography: behavioural approaches. London; New York: Longman, 1984. 195 p. ISBN 0-582-30091-6. $14.95. LC 83-13590.   1036•

Beginning-level text that is an alternative to Gold and Saarinen. Covers much of the basic material and makes good use of case studies.

9. ENVIRONMENTAL PERCEPTION

Thomas F. Saarinen

See also entries in Part IV, Section 8. Behavioral Geography.

A. SERIALS

Annual review of psychology. 1- (1950-). Annual. Annual Reviews, Inc., 4139 El Camino Way, Palo Alto, CA, 94306. ISSN 0066-4308.   1037

Contains periodical reviews of advances in environmental psychology.

Annual review of sociology. 1- (1975-). Annual. Annual Reviews, Inc., 4139 El Camino Way, Palo Alto, CA, 94306. ISSN 0360-0572.   1038

Contains reviews of advances in environmental sociology.

Environment and behavior. 1- (1969-). Bimonthly. Sage Publications, Inc., 275 South Beverly Drive, Beverly Hills, CA, 90212. ISSN 0013-9165.   1039

Earliest interdisciplinary journal entirely devoted to environment and behavior research.

Journal of architectural and planning research. 1- (1984-). Quarterly. Elsevier Scientific Publishers, 52 Vanderbilt Ave., New York, NY, 10017. ISSN 0738-0895.   1040

Most recent of the journals devoted to environment-behavior-design research with emphasis on architecture and planning.

Journal of environmental psychology. 1- (1982-). Quarterly. Academic Press Inc. (London) Ltd., 24-28 Oval Road, London, NW1 7DX, U.K. or 111 Fifth Ave., New York, NY, 10003. ISSN 0272-4944.   1041

Excellent new journal devoted to environment-behavior-design research. Strong international and interdisciplinary flavor.

B. OVERVIEW

Saarinen, Thomas F.; Seamon, David; and Sell, James L. Environmental perception and behavior: an inventory and prospect. (Research paper no. 209). Chicago, IL: University of Chicago, Department of Geography, 1984. 263 p. ISBN 0-89065-114-0. $10. (pbk). LC 84-2492.   1042

An AAG Environmental Perception Speciality Group effort to provide an inventory and prospect of the field, as well as the full range of philosophical perspectives being applied. Contains 18 chapters with contributions from 22 authors.

Environmental Perception

Fisher, Jeffrey D.; Bell, Paul A.; and Baum, Andrew. Environmental psychology. 2nd ed. New York: Holt, Rinehart, and Winston, 1984. 472 p. (1st ed., Bell, Paul A.; Fisher, Jeffery D.; and Loomis, Ross J. Environmental psychology, 1978. 457 p.) ISBN 0-03-059867-2. $29.95. LC 83-12911.
    Revised and updated text on environment and behavior studies with a heavy emphasis on the social psychological perspective.

1043

Gold, John R. (Robert). An Introduction to behavioural geography. Oxford, U.K.; New York: Oxford University Press, 1980. 290 p. ISBN 0-19-823233-0. $39.95. ISBN 0-19-823234-9. $12.95. (pbk). LC 79-41084.
    Text by British geographers which explains the underlying ideas, places development of the field in perspective, and reviews empirical studies.

1044 •

Holahan, Charles J. Environmental psychology. New York: Random House, 1982. 422 p. ISBN 0-394-32320-3. $26. LC 82-378.
    Text on environmental psychology which surveys related interdisciplinary areas and shows the role psychologists can play in resolving major environmental problems.

1045

Ittelson, William H.; Proshansky, Harold M.; Rivlin, Leanne G.; and Winkel, Gary H. An Introduction to environmental psychology. New York: Holt, Rinehart, and Winston, 1974. 406 p. LC 73-21953.
    First textbook which attempted to define and describe the field of environmental psychology. Described as multidisciplinary, concerned with social problems; focused on humans in everyday settings who are seen as playing an active role.

1046

Michelson, William. Man and his urban environment: a sociological approach. Reading, MA: Addison-Wesley, 1976. 273 p. ISBN 0-2014726-8. $13.95. LC 76-6012.
    Early integration of research linking social characteristics and activities of people with the home, neighborhood, and city. Written by a sociologist with an interdisciplinary perspective.

1047

Porteous, J. Douglas. Environment and behavior: planning and everyday urban life. Reading, MA: Addison-Wesley, 1977. ISBN 0-201-05867-7. $22.95. LC 76-1752.
    A broad review of the literature which focuses mainly at the urban scale.

1048 •

Proshansky, Harold M.; Ittelson, William H.; and Rivlin, Leanne G., eds. Environmental psychology: man and his physical setting. New York: Holt, Rinehart, and Winston, 1970. 690 p. ISBN 0-03-078970-2. LC 79-109238.
    First edited volume which sought to bring together papers of interdisciplinary research on environment-behavior-design. Contains many articles now considered classics.

1049

Saarinen, Thomas F. Environmental planning: perception and behavior. Prospect Heights, IL: Waveland Press, 1984. (Re-issue of Boston, MA: Houghton Mifflin, 1976.) 262 p. ISBN 0-88133-116-3. $15.95. LC 75-19533.
    First single-author textbook notable for its organization by scale from personal space and room geography through architectural space, small towns and neighborhoods, city, large conceptual regions, the nation, and the world.

1050 •

Michelson, William M., ed. Behavioral research methods in environmental design. (Community Development Series no. 8). Stroudsburg, PA: Dowden, Hutchinson, and Ross, (Distributed by Halsted Press), 1975. 307 p. ISBN 0-470-60145-0. $28.50. (pbk). LC 74-10937.

1051

Detailed discussion of a number of methodological approaches in terms of their relevance to research in environmental design with practical advice on their use and examples of applications.

Moore, Gary T. and Golledge, Reginald G., eds. Environmental knowing: theories, research, and methods. (Community Development Series, v. 23). Stroudsburg, PA: Dowden, Hutchinson, and Ross, 1976. 441 p. ISBN 0-87933-060-0. $23.95. (pbk). LC 76-4942.　　1052

A rigorous well-organized book containing thoughtful descriptions and assessments of a wide range of theories and methods utilized in behavioral geography research. More than forty separate articles, chapters, and commentaries are included by many of the most active researchers in the field.

Webb, Eugene J.; Campbell, Donald T.; Schwartz, Richard D.; and Sechrest, Lee. Unobtrusive measures: nonreactive research in the social sciences. Chicago, IL: Rand McNally, 1966. 225 p. LC 66-10806. 2nd ed. Nonreactive measures in the social sciences. Boston, MA: Houghton Mifflin, 1981. 394 p. ISBN 0-395-30767-8. $16.50. LC 80-50984.　　1053

An excellent source book of novel methods which stimulate creativity in research design. The authors direct attention to social science data not obtained by interview or questionaire including physical traces, archival material, and observation. They advocate multiple operationism to overcome weaknesses of any one method.

Whyte, Anne V.T. Guidelines for field studies in environmental perception. (MAB technical notes no. 5). Paris: UNESCO, 1977. 117 p. ISBN 9-231-01483-8. $9.25. (pbk). LC 78-323962.　　1054

A succinct discussion of a wide range of methods and techniques developed in many different disciplines which are considered suitable for measuring environmental perception, and for use cross-culturally.

E. EARLY EXAMPLES: CLASSICS

Boulding, Kenneth E. The Image: knowledge in life and society. Ann Arbor, MI: University of Michigan Press, 1956. 175 p. ISBN 0-472-06047-3. $6.95. (pbk). LC 56-9720.　　1055 •

Earliest volume to emphasize that behavior depends on the image, or subjective knowledge. Outlines vast array of topics on which research is needed.

Downs, Roger M. and Stea, David, eds. Image and environment: cognitive mapping and spatial behavior. Chicago, IL; now Hawthorne, NY: Aldine, 1973. 439 p. ISBN 0-202-10058-8. $34.95. LC 72-78215.　　1056 •

Excellent early volume on cognitive mapping which collected classic articles and important early works under one cover. Well-edited with extensive bibliography.

Glacken, Clarence J. Traces on the Rhodian shore: nature and culture in Western thought from ancient times to the end of the eighteenth century. Berkeley; Los Angeles: University of California Press, 1973. 763 p. ISBN 0-520-02367-6. $13.95. (pbk). LC 72-95298.　　1057

Classic volume tracing the history of Western thought with respect to three questions concerning the habitable earth and people's relationships to it: the idea of a designed earth; the idea of environmental influence; and the idea of man as a geographic agent.

Hall, Edward T. The Hidden dimension. Garden City, NJ: Doubleday, 1966. 201 p. ISBN 0-385-08476-5. $4.50. (pbk). LC 66-11173.　　1058 •

Classic work which introduced proxemics as the scientific study of space as a medium of interpersonal communication.

Environmental Perception

Klineberg, Otto. The Human dimension in international relations. New York: Holt, Rinehart, and Winston, 1964. 173 p. LC 64-11410.   1059•
An early readable account of the contribution psychology could make to international relations. While some research lines have been followed up, others remain unexamined.

Lowenthal, David, ed. Environmental perception and behavior. (Research paper no. 109). Chicago, IL: University of Chicago, Department of Geography, 1967. 88 p. ISBN 0-89065-018-7. $10. (pbk). LC 66-29233.   1060
Of interest as the product of the first session on environmental perception and behavior at a meeting of the Association of American Geographers.

Nash, Roderick. Wilderness and the American mind. 3rd ed. New Haven, CT: Yale University Press, 1982. 425 p. (1st ed., 1967. 300 p.) ISBN 0-300-02905-5. $30. ISBN 0-300-02910-1. $9.95. (pbk). LC 82-4874.   1061•
Updated edition of a well-written classic. Provides a history of American attitudes toward wilderness with exciting accounts of the key events and personalities from Old World roots to the present.

Sommer, Robert. Personal space: the behavioral basis of design. Englewood Cliffs, NJ: Prentice-Hall, 1969. 177 p. ISBN 0-13-657577-3. $3.95. (pbk). LC 69-11360.   1062•
Classic by a psychologist who advocated design based on users behavior. Discusses human spatial behavior and how it should be used as a guiding principle for various types of settings.

Wright, John Kirtland. Human nature in geography. Cambridge, MA: Harvard University Press, 1966. 361 p. LC 66-10128.   1063
A collection of fourteen papers on the impact of human emotions, motives, and behavior upon our geographical awareness by Wright who advocated such studies long before the emergence of the field of environmental perception.

F. REPRESENTATIVE WORKS

Altman, Irwin. The Environment and social behavior: privacy, personal space, territory, crowding. Monterey, CA: Brooks/Cole, 1975. 256 p. ISBN 0-8185-0168-5. $14.95. (pbk). LC 75-14724.   1064
Micro-scale treatment of human social behavior in relation to the environment. Integrates key concepts of privacy, personal space, territoriality, and crowding into a theoretical framework.

Appleyard, Donald. Liveable streets. Berkeley, CA: University of California Press, 1981. 364 p. ISBN 0-520-03689-1. $27.50. ISBN 0-520-04769-9. $14.95. (pbk). LC 78-54789.   1065
Excellent study of the quality of life in residential environments and the impacts of traffic and transportation. Explores what it is like to live on streets with different kinds of traffic and searches for ways to make more streets safe and liveable.

Brown, Lawrence A. Innovation diffusion: a new perspective. London; New York: Methuen, 1981. 345 p. ISBN 0-416-74270-X. $19.95. LC 80-49706.   1066
A review of the literature in a major area of behavioral applications. Brown considers various perspectives seen by social proponents such as the adoption perspective, the market and infrastructure perspective, the economic history perspective, and the development perspective.

Clarke, Colin; Ley, David; and Peach, Ceri., eds. Geography and ethnic pluralism. London; Boston, MA: Allen and Unwin, 1984. 294 p. ISBN 0-04-309107-5. $30. ISBN 0-04-309108-3. $14.95. (pbk). LC 84-6319.　　　　　1067

　　Chapters by some ten authors focus on separate, mainly Third World countries, examining pluralism at the national scale. The emphasis on sociology of place involves consideration of class, race, culture, ethnicity, caste, and gender, and includes discussions of such questions as construction of place, identity, and social worlds.

Craik, Kenneth H. and Zube, Ervin H., eds. Perceiving environmental quality: research and applications. (Environmental science research series, v. 9). New York: Plenum Press, 1976. 310 p. ISBN 0-306-36309-7. $35. LC 76-13513.　　　　　1068

　　Reports the findings of a series of workshops to examine the place of environmental perception in environmental quality indices. Chapters by individual authors deal with perceived environmental quality indices for scenic and recreation, residential and institutional, and air, water, and sonic environments.

Downs, Roger M. and Stea, David. Maps in minds: reflections on cognitive mapping. New York; Harper and Row, 1977. 284 p. ISBN 0-06-041733-1. LC 76-57221.　　　　　1069

　　A fascinating introductory text on cognitive mapping. The focus is on how humans process and organize spatial information and how the "cognitive maps" thus formed are used in everyday spatial behavior.

Jervis, Robert. Perception and misperception in international politics. Princeton, NJ: Princeton University Press, 1976. 445 p. ISBN 0-691-05656-0. $44. ISBN 0-691-10049-7. $16.50. (pbk). LC 76-3259.　　　　　1070

　　Explains how we develop various perceptions and misperceptions of other nations and how these images relate to decision-making in international politics.

Lowenthal, David, and Bowden, Martyn, eds. Geographies of the mind: essays in historical geosophy in honor of John Kirtland Wright. (American Geographical Society special publication no. 40). New York: Oxford University Press, 1976. 263 p. LC 75-10182.　　　　　1071

　　Essays in historical geosophy in honor of John Kirtland Wright which provide examples of the impact of environmental beliefs on human thought and action at the broad regional scale.

Meinig, Donald W., ed. The Interpretation of ordinary landscapes: geographical essays. New York: Oxford University Press, 1979. 255 p. ISBN 0-19-502536-9. $9.95. (pbk). LC 78-23182.　　　　　1072•

　　An excellent set of essays by leading scholars which examines landscapes as symbols; as expressions of cultural values, social behavior, and individual actions worked upon particular localities over a span of time.

Milbrath, Lester W. Environmentalists: vanguard for a new society. (SUNY series in environmental public policy). Albany, NY: State University of New York Press, 1984. 180 p. ISBN 0-87395-887-X. $29.50. ISBN 0-87395-888-8. $9.95. (pbk). LC 83-24250.　　　　　1073

　　Argues that a revolution in values is taking place in the world based on a new environmental paradigm. The environmentalists are in the vanguard of the movement towards a changed conception of the relationships of people to the earth.

Newman, Oscar. Defensible space: crime prevention through urban design. New York: Collier Books; Macmillan, 1973. 264 p. ISBN 0-02-000750-7. $8.95. (pbk). LC 74-157786.　　　　　1074•

　　Thought-provoking book on the relationship between the forms of our residential areas and crime. The author contends that crime can be inhibited by designing environments which encourage the inhabitant's latent territoriality and sense of community.

Environmental Perception

Norberg-Schultz, Christian. Genius loci: toward a phenomenology of architecture. New York: Rizzoli, 1980. 213 p. ISBN 0-8478-0287-6. $19.95. (pbk). LC 79-56612.　1075
　Book by a Norwegian architect which uses a phenomenological approach to investigate genius loci - the spirit of place. Contends that if we can understand genius loci, then the human world, especially the architecture, will be better in tune with the environment and therefore more liveable.

Parkes, Don, and Thrift, Nigel. Times, spaces, and places: a chronogeographic perspective. Chichester, U.K.; New York: John Wiley and Sons, 1980. 527 p. ISBN 0-471-27616-2. $88.95. LC 79-40523.　1076
　Focuses on diverse conceptions of time, objective and subjective, and how these relate to everyday use of urban spaces and districts.

Relph, Edward C. Place and placelessness. (Research in planning and design series no. 1). 2nd ed. London: Pion Limited, 1984, 166 p. (1st ed., 1976. 156 p.). ISBN 0-85086-111-X. £4. (pbk). LC 76-383276.　1077
　One of the first, and still the best statements about the role of "lived-space" and the "lived-world." Takes a phenomenological position and presents a wide-ranging critique of the modern landscape, pointing to the need for a sense of place and the erosion of distinctive places. Demanding reading but well worth the effort.

Saarinen, Thomas F. and Sell, James L. Warning and response to the Mount St. Helens eruption. (SUNY series in environmental public policy). Albany, NY: State University of New York Press, 1985. 240 p. ISBN 0-87395-915-9. $36.50. ISBN 0-87395-916-7. $12.95. (pbk). LC 84-46.　1078
　Good example of environmental perception approach in natural hazards. Traces the warning and response of individuals and major national agencies such as the USGS, USFS, and FEMA. Shows how major successes and failures of the warning system related to images.

Tuan, Yi-Fu. Space and place: the perspective of experience. Minneapolis, MN: University of Minnesota Press; London: Edward Arnold, 1977. 235 p. ISBN 0-8166-0808-3. $17.50. ISBN 0-8166-0884-9. $8.95. (pbk). LC 77-72910.　1079
　Illustrates the humanist approach. Explores the range of human experience with respect to space and place.

Western, John. Outcast Cape Town. Minneapolis, MN: University of Minnesota Press, 1981. 372 p. ISBN 0-8166-1025-8. $22.50. LC 81-14640.　1080 •
　Excellent book on the human tragedy associated with apartheid policies. Documents strong human attachment to place and the anguish experienced by those wrenched from their residential neighborhoods.

Whyte, William H. The Social life of small urban spaces. Washington, DC: Conservation Foundation, 1980. 125 p. ISBN 0-89164-057-6. $9.50. (pbk). LC 79-56569.　1081 •
　Clearly-written, well-illustrated volume which raises the question of why certain urban plazas are more used than others. Based on observations of major New York City plazas.

10. POPULATION GEOGRAPHY

Paul D. Simkins

A. BIBLIOGRAPHY

Population index. 1- (1935-). Quarterly. Office of Population Research, Princeton University and Population Association of America, Inc., 21 Prospect Ave., Princeton, NJ, 08544. ISSN 0032-4701.　　　　1082

Well-edited, comprehensive current bibliography covering all aspects of demography and related fields; the starting point for any major search of the literature. Annual index to authors and places, occasional research or review articles, selected population statistics, and professional news items are included.

B. SERIALS

Demography. 1- (1965-). Twice yearly, 1965-68. Quarterly since 1969. Population Association of America, P.O. Box 14182, Benjamin Franklin Station, Washington, DC, 20044. ISSN 0070-3370.　　　　1083

The principal American journal covering all aspects of population studies. Contains original reports of research and commentaries on articles.

Population bulletin. 1- (1945-). Quarterly. Population Reference Bureau, Inc., 2213 M Street, NW, Washington, DC, 20037. ISSN 0032-468X.　　　　1084

Each issue is devoted to a particular demographic topic written by authorities in population studies but intended for the general audience. Subscription includes annual World Population Data Sheet, *a valuable source of demographic data.*

Population and development review. 1- (1975-). Quarterly. Center for Policy Studies of the Population Council, One Dag Hammarskjold Plaza, New York, NY, 10017. ISSN 0098-7921.　　　　1085

Valuable articles on all aspects of population studies. Coverage tends to emphasize developing regions. Contains book reviews.

Population studies; a quarterly journal of demography. 1- (1947-). Quarterly. Population Investigation Committee, London School of Economics, Aldwych, London, WC2A 2AE, U.K. ISSN 0032-4728.　　　　1086

Consistently important substantive articles in the field of demography. Book reviews.

C. ATLAS

McEvedy, Colin, and Jones, Richard. Atlas of world population history. New York: Facts on File, 1978. 368 p. ISBN 0-87196-402-3. $19.95. LC 78-16954.　　　　1087

Presents estimates of population size of continents, major regions, and individual countries by various time intervals from 400 B.C. to 1975; projections to 2000.

D. GENERAL WORKS

Beaujeu Garnier, Jacqueline. Geography of population. Translated from French by S.H. Beaver. 2nd ed. London; New York: Longman, 1978. 400 p. (1st ed., 1967. 386 p.) ISBN 0-582-48570-3. $13.95. (pbk). LC 77-30726.　　　　1088

A synthesis of the author's regional study of population. Considers the controls and processes that affect population differentials.

Population Geography

Bennett, D. Gordon. World population problems: an introduction to population geography. Champaign, IL: Park Press, 1984. 250 p. ISBN 0-941226-04-2. $9.95. (pbk). LC 83-62691.
Introduces population geography by discussing world population problems in the past, present, and future.

1089

Boserup, Ester. Population and technological change: a study of long-term trends. Chicago, IL: University of Chicago Press, 1981. 255 p. ISBN 0-226-06673-8. $17.50. ISBN 0-226-06674-6. $7.95. (pbk). LC 80-21116.
An extension of theory set forth in the earlier and seminal work, Conditions of Agricultural Growth *(entries 937, 1310). Traces relationships between population growth and technological change from ancient society to, but not including, highly industrial societies.*

1090

Clarke, John I., ed. Geography and population: approaches and applications. Oxford, U.K.; New York: Pergamon Press, 1984. 245 p. ISBN 0-08-028781-6. $33. ISBN 0-08-028780-8. $11.50. (pbk). LC 83-22028.
A series of papers examining the field of population geography in general, and developments in population geography in nine individual countries including Japan, China, and the Islamic World.

1091 •

Clarke, John I. Population geography. 2nd ed. Oxford, U.K.; New York: Pergamon Press, 1972. 176 p. (1st ed., 1965. 164 p.). ISBN 0-08-016853-1. $21. ISBN 0-08-016854-X. $7.75. (pbk). LC 70-183339.
A general introduction to population geography. Differs from other standard texts in a more extended treatment of population characteristics.

1092 •

Demko, George J.; Rose, Harold M.; and Schnell, George A. Population geography: a reader. New York; London; Toronto: McGraw-Hill, 1970. 526 p. LC 78-91964.
Although somewhat dated, still a very useful assemblage of articles from scattered sources, with introductions, supplementary bibliographies by sections, and an overall selected bibliography of basic population works, periodicals, bibliographies, and data sources.

1093

Ehrlich, Paul R.; Ehrlich, Anne H.; and Holdren, John D. Ecoscience: population, resources, environment. San Francisco, CA: W.H. Freeman, 1977. 1051 p. ISBN 0-7167-0567-2. ISBN 0-7167-0029-8. $28.95. (pbk). LC 77-6824.
Concerns impact of population growth on resources and environmental quality. Final section on policies and possible solutions. Especially valuable for detailed examination of nutrient cycles, ecosystems, and resource availability.

1094

The Global 2000 report to the president. Barney, Gerald O., study director. Washington, DC: Government Printing Office, 1980-1981. 3 v. 800 p. New York: Pergamon. ISBN 0-08-025991-X. $83. LC 80-602859.
The three volumes consist of a summary report, technical report, and basic documentation. Geographers will find volume two, Entering the Twenty-First Century, *extremely helpful with respect to its assumptions and projections of future change in population, resource availability, and environmental pressures.*

1095

Griffin, Paul F., ed. Geography of population: a teachers guide. (National Council for Geographic Education Yearbook, 1970). Palo Alto, CA: Fearon, 1969. 370 p. LC 70-100992.
Contains 18 specially commissioned papers by specialists on population geography. Part I discusses the world's population; part II, giant population aggregates; part III, lands of high population density; part IV, lands of low population density; and part V, future prospects of world population. An appendix provides an annotated list of supplementary aids for the teacher.

1096 •

Hauser, Philip M., ed. World population and development: challenges and prospects. Syracuse, NY: Syracuse University Press, 1979. 683 p. ISBN 0-8156-2216-3. $18. ISBN 0-8156-2219-8. $9.95. (pbk). LC 79-15471.　　　　　1097
  A series of seventeen papers which examine the relationships between population and development. Papers on population redistribution, environmental and health consequences of population growth are especially helpful.

Hornby, William F. and Jones, Melvyn. An Introduction to population geography. (Studies in human geography no. 1). Cambridge, U.K.: New York: Cambridge University Press, 1980. 168 p. ISBN 0-521-21395-9. $13.95. (pbk). LC 78-74536.　　　1098 •
  Emphasizes population growth and distribution and population mobility. Each section is followed by case studies illustrating aspects of the two main themes. Most case studies are devoted to developed countries.

Jones, Huw R. A Population geography. London; San Francisco, CA; New York: Harper and Row, 1981. 330 p. ISBN 0-06-318188-6. $9.95. (pbk). LC 83-118273.　　1099
  Excellent introductory text in population geography with numerous maps and tables. Several illustrations concerned with Great Britain, but still suitable for American audiences, are included. Lengthy bibliography.

National Academy of Sciences. Rapid population growth: consequences and policy implications. Baltimore, MD: Johns Hopkins University Press, 1971. v.1. Summary and Recommendations, 116 p. ISBN 0-8018-1264-X. $5. (pbk). v.2. Research Papers, 700 p. ISBN 0-8018-1427-8. $8.50. (pbk). LC 75-148952.　　　1100
  Summary volume contains an excellent, succinct discussion of the consequences of rapid population growth. The second volume contains research papers from which the summary is prepared.

Newman, James L. and Matzke, Gordon E. Population: patterns, dynamics, and prospects. Englewood Cliffs, NJ: Prentice-Hall, 1984. 306 p. ISBN 0-13-687566-1. $27.95. LC 83-17629.　　　1101
  Basic text in population geography. In addition to the standard elements of demography, the book includes chapters on population policy and planning, and the relationships among population, resources, and politics.

Peters, Gary L. and Larkin, Robert P. Population geography: problems, concepts, and prospects. Dubuque, IA: Kendall/Hunt, 1983. 324 p. ISBN 0-8403-2925-3. $14.95. (pbk). LC 82-84517.　　　1102 •
  An introduction to population geography useful for undergraduate courses. Includes succinct treatment of basic elements together with a discussion of the interrelationships among population change, resource depletion, and food supply.

Ross, John A., ed. International encyclopedia of population. New York: Free Press, 1982. 2 v. 750 p. ISBN 0-02-927430-3. $125. LC 82-2326.　　　1103
  Defines and discusses demographic terms, measures, and theories. Geographers will find discussions of population characteristics of individual countries helpful.

Schnell, George A. and Monmonier, Mark S. The Study of population: elements, patterns, processes. Columbus, OH: Merrill, 1983. 371 p. ISBN 0-675-20046-6. $25.95. LC 82-62478.　　　1104 •
  A thorough coverage of major elements of population geography including population history, theories, and policy. Population distribution receives greater attention than is common in most population geography texts. An interesting feature is that world maps show countries scaled by population size rather than by area.

Population Geography

Shryock, Henry S. and Siegel, Jacob S. The Methods and materials of demography. New York; London: Academic Press, 1976. 577 p. ISBN 0-12-641150-6. $21. LC 76-18312.

1105

An excellent and comprehensive introduction to measures and methods used in population studies. The work is especially valuable for its step-by-step illustrations of methods using actual statistics. Although United States data are emphasized, numerous examples using other national data are presented.

United Nations. Department of Economic and Social Affairs. The Determinants and consequences of population trends: new summary of findings on interaction of demographic, economic, and social factors. (Population studies no. 50). New York: United Nations, 1973-1978. 2 v. 661; 155 p. (1st ed., United Nations. Department of Social Affairs. 1953. 404 p. [Population studies no. 17]) .LC 77-368556.

1106

A comprehensive review and summary of results of research in demographic, economic, and social interrelationships made after the first edition of 1953. Abundant references. Volume two consists of an extensive bibliography, and an author and subject index.

Weller, Robert H. and Bouvier, Leon F. Population: demography and policy. New York: St. Martin's Press, 1981. 365 p. ISBN 0-312-63114-6. $20.95. LC 80-52385.

1107

A general introduction to the field of demography emphasizing the basic elements of the discipline. The unusual features of the text include a chapter on contraceptive technology and two chapters on the causes and effects of age and sex composition.

Woods, Robert. Population analysis in geography. London; New York: Longman, 1979. 278 p. ISBN 0-582-48696-3. $15.95. (pbk). LC 78-41309.

1108

An introduction to demographic theory and methods of use to geographers. Concentrates on fertility, mortality, and population change.

Woods, Robert. Theoretical population geography. London; New York: Longman, 1982. 220 p. ISBN 0-582-30029-0. $11.95. (pbk). LC 81-2842.

1109

An examination of theories and models of population change, characteristics, and mobility with emphasis on the spatial perspective. Final chapter concerns the theories of the demographic transition and historical materialism.

The World Bank. World development report, 1984. New York: Oxford University Press, 1984. 286 p. ISBN 0-317-07231-5. $20. ISBN 0-19-520460-3. $8. (pbk).

1110

Part two of this annual report is devoted to population change and development which is concerned with population policy, consequences of rapid population growth, and the problems of slowing population growth. Excellent source of national demographic data.

E. SPECIAL TOPICS

Bulatao, Rodolfo A. and Lee, Ronald D., eds. Determinants of fertility in developing countries. New York; London: Academic Press, 1983. 2v. 830 p. LC 83-17135.
    Vol. 1: Supply and demand for children. $37.00. ISBN 0-12-140501-X;
    Vol. 2. Fertility regulation and institutional influences. $45.00. ISBN 0-12-140502-8.

1111

A series of papers on the factors that influence fertility levels and trends in lesser-developed countries. The combined volume gives a good survey of what is known concerning fertility determinants in developing areas.

Clarke, John I. Population geography and the developing countries. Oxford, U.K.; New York: Pergamon Press, 1971. 282 p. ISBN 0-08-016445-5. $21. ISBN 0-08-016446-3. $9.50. (pbk). LC 74-135413.

1112

An extended introduction treats characteristics and consequences of population growth in developing countries; the remaining chapters discuss the population geography of continents and major world regions.

Coale, Ansley J. Economic factors in population growth. (Proceedings of a conference held by the International Economic Association at Valescure, France, 1973). London: Macmillan; New York: Halstead Press, 1976. 600 p. ISBN 0-333-17955-2. LC 76-373733.   1113
  Nineteen papers which examine linkages between population and economic conditions and trends. Papers on fertility declines in Europe and the Third World and the role of population growth and movement on employment dominate the volume.

Easterlin, Richard A., ed. Population and economic change in developing countries. (Conference on population and economic change in lesser-developed countries, Philadelphia, 1976). Chicago, IL: University of Chicago Press, 1980. 581 p. ISBN 0-226-18026-3. $48. LC 79-12569.   1114
  Emphasizes causes and consequences of population change during the course of economic development. Includes chapters on fertility, mortality, migration, and population trends.

Eberstadt, Nick, ed. Fertility decline in the less developed countries. New York: Praeger, 1981. 370 p. ISBN 0-03-055271-0. $43.95. LC 80-23528.   1115
  Based on a symposium on fertility decline in the lesser developed countries sponsored by the American Association for the Advancement of Science, 1978. Individual papers concern fertility trends and their determinants.

Glass, D.V. and Revelle, Roger. Population and social change. London: Edward Arnold; New York: Crane, Russak, 1972. 520 p. ISBN 0-8448-0009-0. LC 72-80107.   1116
  A series of interesting papers on historical demography. Most concerned with preindustrial Europe, but includes studies of India and Japan.

Goldscheider, Calvin. Population, modernization, and social structure. Boston, MA: Little, Brown, and Co., 1971. 345 p. LC 74-165756.   1117
  Examines demographic processes from a sociological approach. Emphasizes the links among population processes, modernization, and social differentiation.

Grigg, David B. Population growth and agrarian change: a historical perspective. (Cambridge geographical series no. 13). Cambridge, U.K.; New York: Cambridge University Press, 1980. 340 p. ISBN 0-521-22760-7. $57. ISBN 0-521-29635-8. $19.95. (pbk). LC 79-4237.   1118
  Examines European population trends from medieval to modern times and their relationships with land availability and agricultural techniques. Final part points to similarities and differences of developing countries today. Lengthy bibliography.

Kosiński, Leszek A. and Prothero, R. Mansell, eds. People on the move: studies on internal migration. London; New York: Methuen, 1975. 393 p. ISBN 0-416-78410-0. $39.95. ISBN 0-416-83000-5. $18.95. (pbk). LC 75-312011.   1119
  Based on papers and discussions of meeting of I.G.U. Commission on Population Geography, University of Alberta, 1972. Of greatest value are papers on theoretical issues and empirical and comparative studies. The latter includes studies from both the developed and developing worlds.

Kosiński, Leszak, and Webb, John W., eds. Population at microscale. (New Zealand Geographical Society, Special publication no. 8). Christchurch, N.Z.: New Zealand Geographical Society, 1976. 193 p. NZ $11. (pbk). LC 77-363297. (IGU Commission on Population Geography, Department of Geography, University of Alberta, Edmonton, Alberta, Canada).   1120
  Based on a symposium organized by the IGU Commission on Population Geography at Massey University, Palmerston North, New Zealand, held in December, 1974. Sixteen papers by an international group of population geographers concerned with the study of population in very small areal units or of small groups and of the effect of scale on such studies.

Population Geography

Kritz, Mary M.; Keely, Charles B.; and Tomasi, Silvano M. Global trends in migrations:theory and research on international population movements. New York: Center for Migration Studies, 1981. 433 p. ISBN 0-913256-54-4. $14.95. ISBN 0-686-85840-9. $9.95. (pbk). LC 80-68399. 1121
> *A collection of sixteen essays on aspects of international migration including theoretical issues, current migration patterns, and migrant adaptations.*

Morrison, Peter A. Population movements: their forms and functions in urbanization and development. Liège, Belgium: Ordina Editions for International Union for the Scientific Study of Population, 1983. 351 p. 1122
> *Valuable papers on the role of migration in modernization theory, regional variability in rural-urban migration, planning, and policy issues.*

Simmons, Alan; Diaz-Briquets, Sergio; and Laquian, Aprodicio. Social change and internal migration: a review of research findings from Africa, Asia, and Latin America. Ottawa, Ontario: International Development Research Centre, 1977. 128 p. ISBN 0-88936-133-9. LC 79-308817. 1123
> *Arranged by continent, the study reviews literature in determinants of migration, migrant characteristics, consequences of migration, and policy implications. A commendable review of the literature on migration.*

Simon, Julian L. The Economics of population growth. Princeton, NJ: Princeton University Press, 1977. 555 p. ISBN 0-691-04212-8. $52. ISBN 0-691-10053-5. $19.50. (pbk). LC 75-15278. 1124
> *Divided into three parts: empirical analyses of the effects of population growth; effect of economic conditions on fertility; policies and methods for reducing fertility. Of interest because it stresses the benefits of population growth, a point of view counter to much of the literature.*

Tabah, Léon, ed. Population growth and economic development in the Third World. Liège, Belgium: Ordina Editions for International Union for Scientific Study of Population, 1979. 2 v. 816 p. ISBN 0-685-90713-9. Distributed in the U.S. by Unipub, New York. $50. set (pbk). LC 76-373401. 1125
> *A series of invited papers dealing with the interrelationships between population and economic development in developing countries.*

Trewartha, Glenn T. A Geography of population: world patterns. New York: John Wiley, 1969. 186 p. ISBN 0-471-88790-0. LC 75-88319. 1126 •
> *Topical in approach, this volume discusses the spatial distribution and arrangements of world population. Part I is devoted to population numbers in the past, in the modern period, and in current world distribution and density. Part II is devoted to biological and cultural characteristics.*

Trewartha, Glenn T. Less developed realm: a geography of its population. New York; London: John Wiley, 1972. 449 p. ISBN 0-471-88794-3. 0-471-88795-1. (pbk). LC 76-173680. 1127
> *Primarily a regional discussion of the spatial distribution features of population (dynamic and static) of the numbers and characteristics of population in Africa, tropical Latin America, and Asia (excluding Japan). Includes an introductory chapter with an analytical survey of the less developed realm in its entirety. Includes many maps and tables, bibliographies, and an index.*

Trewartha, Glenn T., ed. The More developed realm: a geography of its population. Oxford, U.K.; New York: Pergamon Press, 1978. 275 p. ISBN 0-08-020630-1. $16.25. (pbk). LC 76-39897. 1128
> *A companion volume to earlier studies by Trewartha on world patterns and the less developed realm. Consists of a series of papers on population distribution and characteristics in Europe, the Soviet Union, Anglo-America, Japan, and Oceania.*

White, Paul E. and Woods, Robert I., eds. The Geographical impact on migration. London; New York: Longman, 1980. 245 p. ISBN 0-582-48941-5. $33. LC 79-40463.   1129
Individual papers emphasize Western Europe although studies of migration in Bolivia and Uganda are included. Initial chapters discussing sources of data, migration typologies, and spatial patterns and impacts on migration are of special geographic interest.

Zelinsky, Wilbur; Kosiński, Leszek A.; and Prothero, R. Mansell, eds. Geography and a crowding world: a symposium on population pressures upon physical and social resources in the developing lands. New York: Oxford University Press, 1970. 601 p. LC 77-82999.   1130
The 33 topical and regional essays in this symposial volume represent the first systematic effort by geographers to deal with the conceptual and practical questions of population pressure in a comparative, international framework. Papers from a conference at Pennsylvania State University September 17-23, 1967, sponsored by the International Geographical Union, Commission on the Geography and Cartography of World Population.

11. POLITICAL GEOGRAPHY

Andrew M. Kirby

A. SERIAL

Political geography quarterly. 1- (1982-). Quarterly. Butterworth Scientific Ltd., Journals Division, P.O. Box 63, Westbury House, Bury Street, Guildford, Surrey, GU2 5BH, U.K. ISSN 0260-9827.   1131
A new periodical with articles on political geography.

B. BOOKS

Archer, J. Clark (John), and Taylor, Peter J. Section and party: a political geography of American Presidential elections from Andrew Jackson to Ronald Reagan. Chichester, U.K.; New York: Research Studies Press, Wiley, 1981. 271 p. ISBN 0-471-10014-5. $56. LC 81-182752.   1132
One of the few attempts to examine political-geographic concepts in a rigorous quantitative manner.

Bennett, D. Gordon, ed. Tension areas of the world: a problem oriented world regional geography. Champaign, IL: Park Press, 1982. 331 p. ISBN 0-941226-01-8. $19.95. LC 81-82632.   1133•
A text in terms of format and style, this is nonetheless an interesting volume with good essays. Hale's piece on the Middle East in particular is excellent.

Bennett, Robert J. The Geography of public finance: welfare under fiscal federalism and local government finance. London; New York: Methuen, 1980. 498 p. ISBN 0-416-73090-6. $19.95. ISBN 0-416-73100-7. $19.95. (pbk). LC 79-41792.   1134
A meticulous account of the intricacies of public finance with numerous empirical explorations which range widely through Europe and the U.S., but little on the underlying political issues.

Bergman, Edward F. Modern political geography. Dubuque, IA: W.C. Brown, 1975. 408 p. LC 74-82009.   1135•
A mix of traditional concerns and more recent approaches. Systems are employed, but the book also places emphasis on economic change underlaying political issues. As good as any other major text. Now a little dated.

Political Geography

Blowers, Andrew. The Limits to power: the politics of local planning policy. (Urban and Regional Planning Series, v. 21). Oxford, U.K.; New York: Pergamon, 1980. 216 p. ISBN 0-08-023016-4. $28. LC 79-40767.
An account of local politics written by a political geographer and politician. Material is British but has universal applications to situations of conflicts between jurisdictions.

1136

Brunn, Stanley D. Geography and politics in America. New York: Harper and Row, 1974. 443 p. LC 73-17670.
A well-written book which suffered from being way ahead of its time (in terms of an interest in the environment, for instance), and from attempting to cover too many issues, from law to future political organization.

1137•

Burnett, Alan D. and Taylor, Peter J., eds. Political studies from spatial perspectives: Anglo-American essays on political geography. Chichester, U.K.; New York: John Wiley and Sons, 1981. 519 p. ISBN 0-471-27909-9. $68.95. ISBN 0-471-27910-2. $23.95. (pbk). LC 80-41384.
An enormously influential collection of essays, many presented at the annual conference of the Institute of British Geographers in 1979. A uniformly high quality set of papers provides an overview of the possibilities of political geography in many different contexts, drawing examples and authors from many countries.

1138

Busteed, M. A. (Mervyn), ed. Developments in political geography. London; New York: Academic Press, 1983. 340 p. ISBN 0-12-148420-3. $42. LC 82-73230.
A rather disjointed collection of papers, some originally written in the late 1970's, plus an account of political geography by the editor.

1139

Clark, Gordon L. and Dear, Michael J. State apparatus: structures and language of legitimacy. Boston, MA: Allen and Unwin, 1984. 216 p. ISBN 0-04-320159-8. $29.95. ISBN 0-04-320160-1. $11.95. (pbk). LC 83-22357.
Virtually the first attempt by political geographers to begin to rewrite the theory of the state in a coherent manner. Certainly not the last word on the subject, but a clear indication of the importance of the topic.

1140

Cox, Kevin R. Conflict, power, and politics in the city: a geographic view. New York: McGraw Hill, 1973. 133 p. LC 72-6644.
A very influential book which focuses upon local politics an an adjunct to national political-geographic studies. Reflects the contemporary importance of the Vietnam War and the so called radical revolution in geography.

1141

Cox, Kevin R. Location and public problems: a political geography of the contemporary world. Chicago, IL: Maaroufa Press, now New York: Methuen; Oxford, U.K.: Blackwell, 1979. 352 p. ISBN 0-88425-015-6. $16.95. LC 78-71125.
Uses a trichotomy of scales (international, regional, and local) to examine problems of decision-making in a spatial context. Valuable efforts to range widely from country to country are undermined by the absence of a clear account of a theory which connects the examples.

1142

Cox, Kevin R., ed. Urbanization and conflict in market societies. Chicago, IL: Maaroufa Press, 1978. 255 p. ISBN 0-88425-007-5. LC 77-76158.
A collection of essays linked by a twin consideration of politics and urban issues. Papers by Harvey, Reynolds, and Walker are particularly important.

1143

Cox, Kevin R.; Reynolds, David R.; and Rokkan, Stein, eds. Locational approaches to power and conflict. Beverly Hills, CA: Sage; New York: Halsted Press, 1974. 345 p. ISBN 0-470-18122-2. Distributed by Krieger, Melbourne, FL. $19.50. LC 76-127983.
A very mixed bag. Good papers by Wolpert on locational conflict, and Cox and Reynolds on power and conflict. An attempt to mathematize political geography, which rarely works.

1144

Dear, Michael J. and Taylor, S. Martin. Not on our street: community attitudes to mental health care. London: Pion; New York: Methuen, 1982. 182 p. ISBN 0-85086-096-2. $19.95. LC 82-229194.　1145

A very detailed account of a research project examining community responses to an externality, in this case, mental health facilities. A fine example of mixing social and political theory with sophisticated research strategies. Generally very well received.

East, W. Gordon (William), and Prescott, J.R.V. Our fragmented world: an introduction to political geography. London: Macmillan, 1975. 276 p. ISBN 0-333-15109-7. LC 76-370778.　1146•

A text based upon the interrelationships between people and their territories. Useful for its emphasis upon land and the oceans.

Glassner, Martin Ira, and De Blij, Harm J. Systematic political geography. 3rd ed. New York: John Wiley and Sons, 1980. 537 p. (1st ed., 1967. 618 p.). ISBN 0-471-05228-0. $35.95. LC 79-26750.　1147•

Now in the third edition and approaching the third decade in print, this book has continued to evolve in line with the subject. In consequence it is perhaps overlong, and the most recent developments have yet to be assimilated, although undoubtedly they will be in the next edition. At present, it remains the most widely used text in the field.

Gottmann, Jean, ed. Centre and periphery: spatial variation in politics. (Sage Focus editions, v. 19). Beverly Hills, CA: Sage, 1980. 226 p. ISBN 0-8039-1344-3. $24. ISBN 0-8039-1345-1. $12. (pbk). LC 79-21564.　1148

A heavy emphasis upon the notion of core and periphery produces many different interpretations of the terms, and some contributions which seem antithetical. The volume as a whole is rather functionalist and tries unsuccessfully to stamp a geographical matrix on political evolution.

Gudgin, Graham, and Taylor, Peter J. Seats, votes, and the spatial organization of elections. London: Pion; New York: Methuen, 1979. 240 p. ISBN 0-85086-073-3. $25.50. LC 80-501951.　1149

A detailed quantitative monograph which explores the relations between spatial organization of elections, voting behavior, and electoral outcomes. Well received in political science, but too sophisticated for many geographer's tastes.

House, John W. Frontier on the Rio Grande: a political geography of development and social deprivation. Oxford, U.K.: Clarendon Press; New York: Oxford University Press, 1982. 281 p. ISBN 0-19-823237-3. $32.50. LC 82-3615.　1150•

One of several high-quality research works produced by John House towards the end of his life, this book comprises a modern interest in social issues with a more traditional occupation with frontiers and boundaries.

House, John W., ed. United States public policy: a geographical view. Oxford, U.K.: Clarendon Press; New York: Oxford University Press, 1983. 324 p. ISBN 0-19-874116-2. $32.50. ISBN 0-19-874117-0. $14.95. (pbk). LC 83-8071.　1151

House's last book, this edited collection deals with his two interests, namely applied and political geography. The result is a useful and up-to-date account of American urban and regional policy.

Johnston, Ronald J. Geography and the State: an essay in political geography. London: Macmillan; New York: St. Martin's Press, 1982. 283 p. ISBN 0-312-32172-4. $27.50. LC 82-10438.　1152

Johnston's polemic argues the case for the importance of developing a theory of the state within geography. The book has not pleased conservatives or Marxists, but has been applauded by those in the middle ground.

Political Geography

Johnston, Ronald J. Political, electoral, and spatial systems: an essay in political geography. Oxford, U.K.: Clarendon Press; New York: Oxford University Press, 1979. 221 p. ISBN 0-19-874071-9. £12.50. ISBN 0-19-874072-7. £5.50. (pbk). LC 78-40648.
An interesting book which examines political relations within the United Kingdom and the United States, drawing particularly upon electoral and spending data. Criticized for a lack of discussion of its assumptions, the book remains a source of ideas.

1153

Kirby, Andrew M. The Politics of location: an introduction. London; New York: Methuen, 1982. 199 p. ISBN 0-416-33900-X. $28. ISBN 0-416-33910-7. $13.95. (pbk). LC 82-8132.
Described by one political scientist as the best of a genre and one formerly eminent geographer as trivial, the book has also been criticized for daring to question the efficacy of geography.

1154

Morrill, Richard L., Knight, C. Gregory, ed. Political redistricting and geographical theory. (Resource publications in geography). Washington, DC: Association of American Geographers, 1981. 76 p. ISBN 0-89291-159-X. $5. (pbk). LC 81-69235.
Provides examples of redistricting practice in the U.S., but does not say a great deal about geographical theory.

1155 •

Muir, Richard. Modern political geography. London: Macmillan; New York: John Wiley and Sons, 1975. 262 p. ISBN 0-333-31128-0. £5.95. LC 75-332059.
Muir was almost a lone figure within British political geography in the 1970's. This book is a mix of styles and issues.

1156

Norris, Robert E. and Haring, Lloyd L. Political geography. Columbus, OH: C.E. Merrill, 1980. 316 p. ISBN 0-675-08223-4. $28.95. LC 79-91543.
An attempt to provide a text that integrates traditional and more systematic themes, the book has some valuable sections, particularly on federal spending and local organization.

1157 •

Pacione, Michael, ed. Progress in political geography. London; Dover, NH: Croom Helm, 1985. 288 p. ISBN 0-7099-2065-2. £22.50.
This volume contains Archer and Shelley's excellent historical review of political geography.

1158

Paddison, Ronan. The Fragmented state: the political geography of power. Oxford, U.K.: Basil Blackwell; New York: St. Martin's Press, 1983. 315 p. ISBN 0-312-30244-4. $25. LC 83-4490.
Very much a text, this book discusses the ways in which geographical space is fragmented, but leaves the reader awash amidst a number of competing explanations.

1159

Prescott, J.R.V. (John). Boundaries and frontiers. London: Croom Helm; Totowa, NJ: Rowman and Littlefield, 1978. 210 p. ISBN 0-8476-6086-9. $21.50. LC 78-322521.
Descriptive rather than contextual, but nonetheless a helpful reference to aspects of political geography.

1160

Prescott, J.R.V. (John). The Political geography of the oceans. Newton Abbott: David and Charles; New York: Halsted Press, 1975. 247 p. ISBN 0-470-69672-9. $24.95. LC 75-317876.
Recently described as admirable, Prescott's volume was clearly ten years ahead of its time in terms of its focus. Events have only recently caught up with his concerns.

1161

Short, John R. Introduction to political geography. London; Boston, MA: Routledge and Kegan Paul, 1982. 193 p. ISBN 0-7100-0964-X. $29.95. ISBN 0-7100-0965-8. $13.50. (pbk). LC 81-22712.
Generally well-received, this book is basically a political economy of international relations, with some national and local material added. Generally seen in the U.S. as anti-American in stance.

1162

Taylor, Peter J. and House, John W., eds. Political geography: recent advances and future directions. London: Croom Helm; Totowa, NJ: Barnes and Noble, 1984. 239 p. ISBN 0-389-20493-5. $27.50. LC 84-6493.

1163

A monument to John House's efforts to create a political geographic program within the International Geographic Union, this book of conference papers documents one such meeting. As with many such volumes, its very diversity is a weakness, although some good papers are included.

Taylor, Peter J. and Johnston, Ronald J. The Geography of elections. London: Croom Helm; New York: Holmes and Meier, 1979. 528 p. ISBN 0-8419-0495-2. $42. LC 79-4383.

1164

An excellent if overlong work on the geographical bases of electoral issues in many countries. Highly readable.

12. GENERAL ECONOMIC GEOGRAPHY

Peter O. Muller

A. BIBLIOGRAPHIES

Muller, Peter O. Locational analysis and economic geography: a comprehensive bibliography of recent literature on theory, techniques, and the spatial organization of agriculture, manufacturing, and transportation. Philadelphia, PA: Public Services Division, Samuel Paley Library, Temple University, 1972. 94 p. LC 78-109442. NUC 74-65837.

1165

A comprehensive listing of the literature of economic geography through early 1972, including books, articles, and monographs. Nearly 1500 entries (unannotated) are arranged into 15 categories that cover the nature of the field, textbooks and readers, location theory, analytical techniques, regional economic structure, growth and development, and principles and case studies in agricultural, manufacturing, and transportation geography.

Muller, Peter O., compiler. Locational analysis and economic geography: a comprehensive bibliography of recent literature on theory, techniques, and the spatial organization of agriculture, manufacturing, and transportation, supplement 1971-77. Philadelphia, PA: Public Services Division, Samuel Paley Library, Temple University, 1977. 57 p. LC 78-109442.

1166

An updating of the original bibliography published in 1972. Arranged within the identical 15 categories that covered the nature of the field, textbooks and readers, location theory, analytical techniques, regional economic structure, growth and development, and principles and case studies in agricultural, manufacturing, and transportation geography. Contains about 990 unannotated entries.

B. SERIALS

Economic geography. 1- (1925-). Quarterly. Clark University, 950 Main Street, Worcester, MA 01610. ISSN 0013-0095.

1167

A leading journal devoted to the field of economic geography. Formerly the mainstream journal in the field but since the mid 1970s has become more specialized on theory and technique including work in behaviorial and urban geography. Book reviews. Cumulative indexes: 1-25 (1925-1949), 26-41 (1950-1965).

Geo abstracts. Series C. Economic geography. 1- (1966-). 6 per annum. Geo Abstracts, Ltd., Regency House, 34 Duke Street, Norwich NR3 3AP, U.K. ISSN 0305-1919.

1168

Abstracts of literature in economic geography. Includes books, articles, and other noteworthy publications. Heavily oriented to the English-speaking countries of Europe, North America, and Australasia, it does attempt to include other sources as the availability of translators permits.

General Economic Geography

Regional studies (Regional Studies Association). 1- (1966-). Bimonthly. Pergamon Press, Headington Hill Hall, Oxford OX3 OBW, U.K., or Maxwell House, Fairview Park, Elmsford, NY 10523. ISSN 0034-3404.
    1169

Interdisciplinary and international journal devoted to modern approaches to regional analysis with many articles by geographers. Book reviews. Publications received.

TESG: Tijdschrift voor economische en sociale geografie. Netherlands journal of economic and social geography. 1- (1910-). Bimonthly. Secretary, Koninklijk Nederlands Aardrijkskundig Genootschap, Royal Dutch Geographical Society, Weteringschans 12, 1017 SG, Amsterdam, The Netherlands. ISSN 0040-747X.
    1170

A very useful international journal that publishes most of its articles in English. All approaches and methodologies are presented. This serial; has taken up some of the slack as the journal Economic Geography *has become extremely specialized in its focus on theory and techniques since the mid 1970's. Cumulative indexes: 1910-1934, 1935-1952.*

C. ATLASES

Ginsburg, Norton S. Atlas of economic development. Chicago, IL: University of Chicago Press, 1961. 119 p. LC Map 61-6.
    1171

A series of global maps accompanied by texts and tables portraying, analyzing, and ranking variables relating to economic development on a national basis. Although data are two decades old, the comparative insights and methodologies utilized are still an effective representation of world-scale patterns. A classic work that continues to offer a great deal to the economic geographer concerned with global modernization trends.

Oxford University Press. Oxford economic atlas of the world. 4th ed. London; New York: Oxford University Press, 1972. 239 p. (1st ed., 1954. 113, 152 p.). ISBN 0-19-894106-4. $39.95. ISBN 0-19-894107-2. $13.95. (pbk). LC 72-169337.
    1172•

The best atlas of its kind in the English language. A cartographic feast of world maps, most in color, that cover the environment, food, fiber, energy and mineral resources, industries, demography, transportation and communication, and health and sociopolitical patterns. A detailed gazetteer and more than 100 pages of comprehensive national economic statistics make this an outstanding reference on world economic geography.

D. GENERAL WORKS

Abler, Ronald F.; Adams, John S.; and Gould, Peter R. Spatial organization: the geographer's view of the world. Englewood Cliffs, NJ: Prentice-Hall, 1971. 587 p. LC 71-123081.
    1173

An especially well-written textbook deploying the scientific approach to human geography. Large sections of the book treat the scientific method, location analytic techniques, and the theoretical principles governing the distributions of economic activities. Chapters on perception and spatial diffusion are innovative and anticipate the widening of the field's behavioral frontiers after 1975.

Alexander, John W. and Gibson, Lay J. Economic geography. 2nd ed. Englewood Cliffs, NJ: Prentice-Hall, 1979. 480 p. (1st ed., 1963. 661 p.). ISBN 0-13-225151-5. $36.95. LC 78-15735.
    1174•

Classic textbook, the first to offer a locational-analytic approach to the field. Topical approach that admirably blends theory and empiricism, it is quite comprehensive though dated in its coverage. Excellent usage of techniques, maps, photographs, and tables. Second edition maintained North American focus. Now dated. Third edition, due in 1986, will expand to the global scene.

Berry, Brian J.L.; Conkling, Edgar C.; and Ray, D. Michael. The Geography of economic systems. Englewood Cliffs, NJ: Prentice-Hall, 1976. 529 p. ISBN 0-13-351296-7. $34.95. LC 75-15841.
A strongly conceptual presentation of the field emphasizing geographic and economic fundamentals, location theories, trading relationships, and regional economic structure. Also contains chapter-length case studies of the world's industrial-heartland and less-developed hinterland economies.

1175

Butler, Joseph H. Economic geography: spatial and environmental aspects of economic activity. New York: John Wiley and Sons, 1980. 402 p. ISBN 0-471-12681-0. $30.95. LC 80-14542.
An effective blending of theory and empiricism, supported by well-chosen case studies. Devotes more attention than usual to the important subjects of contemporary urbanization and environmental externalities.

1176•

Chisholm, Michael. Geography and economics. 2nd ed. London: Bell, 1970. 219 p. (1st ed., 1966. 230 p.). ISBN 0-7135-1820-0. £7.95. LC 73-550033.
One of the first attempts by a geographer to apply the principles of economics in a systematic way to the explanation of the location of human activities. Stresses the role of government on economic location patterns as well as the effects of various private-sector pricing policies.

1177

Cole, John P. The Development gap: a spatial analysis of world poverty and inequality. Chichester, U.K.; New York: John Wiley and Sons, 1980. 454 p. ISBN 0-471-27796-7. $56.95. LC 80-40284.
A well-organized global overview of inequalities in regional economic development. The world is divided into 100 cells, each representing about one percent of humanity, and each unit is analyzed in terms of its constituent resources and relative standing in the overall global framework. Well supported with maps, charts, and tables.

1178•

Conkling, Edgar C. and Yeates, Maurice H. Man's economic environment. New York: McGraw-Hill, 1976. 308 p. ISBN 0-07-012408-6. LC 75-22469.
A concise introduction to the theoretical foundations of the space-economy, activity-location decision making, transportation, trade, and development. A particularly effective blending of principles, concepts, and empirical examples demonstrating the utility of economic geography for solving practical problems.

1179

de Souza, Anthony R. and Foust, J. Brady. World-space economy. Columbus, OH: C.E. Merrill, 1979. 615 p. ISBN 0-675-08292-7. $26.95. LC 78-65767.
A thorough review of the field from a global perspective. An innovative feature is the addition of non-traditional radical critiques. A particularly generous coverage of underdevelopment also distinguishes this book.

1180

Dohrs, Fred E. and Sommers, Lawrence M., eds. Economic geography: selected readings. New York: Crowell, 1970. 470 p. LC 70-77337.
A collection of 28 articles that represent both the traditional and theoretical approaches to the field. Coverage includes introduction, primary, secondary, and tertiary economic activities (agriculture, manufacturing, and services), and economic regions.

1181

Eliot Hurst, Michael E. A Geography of economic behavior: an introduction. London: Prentice-Hall; North Scituate, MA: Duxbury Press, 1974. 427 p. ISBN 0-13-351692-X. LC 75-316572.
The first comprehensive attempt to link economic geography to the behavioralist approach of contemporary human geography. Although somewhat dated, it is still a valuable guide to this literature. An especially well-organized treatment of locational decision-making.

1182

General Economic Geography

Lloyd, Peter E. and Dicken, Peter. Location in space: a theoretical approach to economic geography. 2nd ed. London; New York: Harper and Row, 1977. 474 p. (1st ed., 1972. 292 p.). ISBN 0-06-044048-1. $29.95. LC 77-23256.   1183
   A comprehensive and well-documented review of the entire field that makes especially effective use of economic spatial models in its presentations. Contains a good balance of economics as well as economic geography, and goes further than most standard textbooks.

McCarty, Harold H. and Lindberg, James B. A Preface to economic geography. Englewood Cliffs, NJ: Prentice-Hall, 1966. 261 p. LC 66-13726.   1184 •
   A minor classic that sets forth, with unusual lucidity, an extremely well-organized outline of this often complex subdiscipline. Reviews the various modern approaches to the subject, including hypothesis formulation and testing. Provides newcomers with a useful framework for exploring the field.

Mc Nee, Robert B. A Primer on economic geography. New York: Random House, 1971. 208 p. ISBN 0-394-31019-5. LC 74-124663.   1185 •
   A compact but lively treatment of the field from an urban perspective. The metropolis is viewed as the key link in the spatial production system, and the analysis carried to the macro-scale to review the space-economy of developed countries.

O'Sullivan, Patrick E. Geographical economics. New York: Halsted Press, John Wiley and Sons, 1981. 199 p. ISBN 0-470-27122-1. $31.95. LC 80-39881.   1186
   A theoretical work that explores the linkages between the space-economy and the social welfare of individuals, aimed at deriving policies of growth that benefit all social groups. In the context of an integrated welfare economics/geography approach, locational processes are reviewed employing a wide variety of analytical techniques, and the effects of political decisions are weighed.

Smith, Robert H.T.; Taaffe, Edward J.; and King, Leslie J., eds. Readings in economic geography: the location of economic activity. Chicago, IL: Rand McNally, 1968. 406 p. LC 68-10126.   1187
   Still the only work of its particular genre that has appeared to date. A very valuable collection of studies, mostly from the late 1950's and 1960's, that lucidly demonstrate the utility of the locational-analysis approach to the subject. Although limited to works of logical positivism, this is still a useful anthology that captures some of the best work of the first generation of theoretical geographers.

Thoman, Richard S. and Corbin, Peter B. The Geography of economic activity. 3rd ed. New York: McGraw-Hill, 1974. 420 p. (1st ed., 1962. 602 p.). ISBN 0-07-064207-9. $37.50. LC 74-5756.   1188
   A well-balanced treatment of the subject, stressing theory and models, but extensively supported by an empirical treatment of the subject matter (which is now rather dated). Innovative feature is the treatment of many specific agricultural systems and industries.

Wheeler, James O. and Muller, Peter O. Economic geography. New York: John Wiley and Sons, 1981. 395 p. ISBN 0-471-93760-6. $28.95. LC 80-21536.   1189 •
   A standard textbook that in addition to coverage of agriculture, manufacturing, and service activities, also contains chapters on energy, contemporary American urbanization, and applied economic geography. Transportation is particularly emphasized throughout. The orientation is heavily toward the current North American scene.

Wilbanks, Thomas J. Location and well-being: an introduction to economic geography. San Francisco, CA: Harper and Row, 1980. 462 p. ISBN 0-06-167404-4. $26.95. LC 79-26646.   1190 •

An unusually broad introduction to the subject draws upon the most innovative work in human geography of the past two decades. A most thoughtful and interesting approach. The topics that are stressed and integrated are locational well-being, economic environments, locationsl advantages and disadvantages, government and public policy, and spatial planning.

E. SPECIAL TOPICS

Barlowe, Raleigh. Land resource economics: the economics of real estate. 3rd ed. Englewood Cliffs, NJ: Prentice-Hall, 1978. 653 p. (1st ed., 1958. 585 p.). ISBN 0-13-522532-9. $29.95. LC 78-4146. 1191

A comprehensive survey of the economics of land resources, a field closely related to many aspects of economic geography. Provides extensive background to understanding patterns of land-use, including competition among various activities, institutional forces, property and ownership rights, taxation, planning, and public policy-making.

Foust, J. Brady, and de Souza, Anthony R. The Economic landscape: a theoretical introduction. Columbus, OH: C.E. Merrill, 1978. 344 p. ISBN 0-675-08432-6. LC 78-201. 1192

A concise overview of economic-spatial theory and the basic models of regional economic structuring. Includes many useful examples and demonstrations of the spatial organization of various economic activities and the locational decision-making process.

Hoover, Edgar M. An Introduction to regional economics. 2nd ed. New York: Alfred Knopf, 1975. 395 p. (1st ed., 1971. 395 p.). ISBN 0-394-31899-4. $22. LC 74-26840. 1193

A lucidly written introduction to the location of economic activities; a classic work by a scholar whose previous research helped provide the foundations for contemporary economic geography. A tour-de-force of theoretical and empirical analysis, it covers locational decision-making, transport patterns, land-use structures at various scales, agglomeration, urbanization, regionalization, spatial change, and policy-making.

Isard, Walter. Introduction to regional science. Englewood Cliffs, NJ: Prentice-Hall, 1975. 506 p. ISBN 0-13-493841-0. $29.95. LC 74-22031. 1194

A comprehensive overview of this interdisciplinary social science, so closely associated with modern economic geography, by the scholar whose name is synonymous with the evolution of this field since the 1950's. Theories, techniques, and case studies pertaining to the structure and performance of the space-economy over time.

Jumper, Sidney R.; Bell, Thomas L.; and Ralston, Bruce A. Economic growth and disparities: a world view. Englewood Cliffs, NJ: Prentice-Hall, 1980. 472 p. ISBN 0-13-225680-0. $33.95. LC 79-23728. 1195•

An approach to the subject that is both global in scope and organized around the study of differences in levels of development. A strong conceptual foundation is supported by ample statistical documentation. Themes stressed include growth of the space-economy, international disparities in wealth and economic-spatial patterns, development strategies, and the global variations of economic activities.

Lösch, August. The Economics of location. Translated from the 2nd revised German edition by William H. Woglom and Wolfgang P. Stolper. New Haven, CT: Yale University Press, 1954. 520 p. LC 52-9268. (Die räumliche Ordnung der Wirtschaft. Jena: G. Fischer, 1944. 380 p.). 1196

The classical theoretical statement on the location of economic activities and the normative spatial organization of the economic landscape. Shows how the various components of the space-economy (agriculture, industry, and services), are interrelated, and how these activities are arranged into idealized economic regions.

163

Agricultural Geography

Toyne, Peter. Organisation, location, and behaviour: decision-making in economic geography. London: Macmillan; New York; Toronto: Halsted Press, John Wiley and Sons, 1974. 285 p. ISBN 0-470-88100-3. LC 73-22708.   1197
  A brief, highly theoretical treatment of the field which is heavily steeped in economic principles. Limited empirical coverage is mainly devoted to Western European examples.

Webber, Michael J. Impact of uncertainty on location. Canberra, Australia: Australian National University Press; Cambridge, MA; London: MIT Press; 1972. 310 p. ISBN 0-7081-0815-6. ISBN 0-262-23054-2. $27.50. LC 78-159392. LC 75-181862.   1198
  An attempt to generalize location theory to include uncertainty in the decision-making process of economic activities. Includes sophisticated theoretical and methodological treatment in which a model is developed that combines individual aspects of uncertain decision-making with spatial diffusion processes, suggesting that such integration can lead to a better general dynamic model of location. A classic in contemporary locational analysis.

Yeates, Maurice H. An Introduction to quantitative analysis in human geography. New York; Montreal: McGraw-Hill, 1974. 300 p. ISBN 0-07-072251-X. LC 75-4143.   1199
  A lucid non-mathematical introduction to what are still the basic techniques of statistical analysis in economic geography.

13. AGRICULTURAL GEOGRAPHY

Owen J. Furuseth

A. BIBLIOGRAPHIES

Bibliography of agriculture. 1- (1942-). Monthly. U.S. Department of Agriculture Library, Technical Information Systems, Washington, DC; Oryx Press, 2214 N. Central at Encanto, Suite 103, Phoenix, AZ 85004. ISSN 0006-1530.   1200
  Provides indexing to approximately 123,000 records per year from about 6,000 journals, as well as books and government publications. A comprehensive and invaluable resource.

Biological and agricultural index. 50- (1964-). Monthly except August. Continues Agricultural Index, v. 1-49 (1916-1964). H.W. Wilson Co., 950 University Ave., Bronx, NY, 10452. ISSN 0006-3177.   1201
  A cumulative subject index to more than 100 periodicals covering a wide range of agricultural topics, including physical and social sciences.

Blanchard, J. Richard, and Farrell, Lois. Guide to sources for agricultural and biological research. Berkeley, CA: University of California Press, 1981. 735 p. ISBN 0-520-03226-8. $47.50. LC 81-3018.   1202
  Includes more than 5000 bibliographic citations to important reference sources in agriculturally related research, including agricultural geography. Most citations are for works published between 1959 and 1979.

Rechcigl, Miloslav, Jr. World food problems: a selective bibliography of reviews. Cleveland, OH: CRC Press, 1975. 271 p. ISBN 0-87819-066-X. LC 74-30748.   1203
  A bibliographic review of more than 5000 publications dealing with world food problems.

Agricultural Geography

B. ATLASES

Bertin, Jacques; Hémardinquer, Jean-Jacques; Keul, Michael; and Randles, W.G.L. Atlas of food crops. Translated from French. (École Pratiques des Hautes Études: section 6). Paris: Mouton; Hawthorne, NY: Walter de Gruyter, 1971. 41 p. ISBN 0-90-2791-798-1. $51. LC 75-149077. (Atlas des cultures vivrières).　　　1204

Eighteen maps accompanied by a brief narrative depicting the origin and geographical diffusion of major world food crops. An excellent bibliography. Contains more than 1800 references.

International Association of Agricultural Economists. World atlas of agriculture. Novara, Italy: Istituto Geografico de Agostini, 1969. 62 plates. LC 77-653907.　　　1205

An impressive cartographic presentation of global agricultural land use and topographic relief maps. Somewhat dated, but the only cartographic compilation of global agricultural data. Accompanied by monographs on major world regions: 4 v. 2508 p.

　　Vol. 1: *Europe, U.S.S.R., Asia Minor. 1972.*
　　Vol. 2: *South and East Asia, Oceania. 1973.*
　　Vol. 3: *The Americas, 1970.*
　　Vol. 4: *Africa, 1976.*

Johnson, Hugh. A World atlas of wine: a complete guide to the wines and spirits of the world. rev. ed. New York: Simon and Schuster, 1981. 288 p. (1st ed., 1971) ISBN 0-686-78792-7. $59. ISBN 0-67122417-4. $8.95. (pbk). LC 78-324831.　　　1206

Detailed maps and commentary on the vintages of all the more notable grape-growing regions of the world, with a brief treatment of distilled liquors. A handsome volume.

C. SERIES

Geography of world agriculture. 1- (1972-). Irregular series. Editor: György Enyedi. Akadémiai Kiadó, Box 24, H-1363, Budapest, Hungary.
　　Vol.　1: **Poland.** Jerzy Kostrowicki, 1972.
　　Vol.　2: **Southeast of the United States.** James R. Anderson, 1973.
　　Vol.　3: **Land supply and specialization.** Norbert Csáki, 1974.
　　Vol.　4: **California.** Howard F. Gregor, 1974.
　　Vol.　5: **Denmark.** Aage H. Kampp, 1975.
　　Vol.　6: **Finland.** Uuno Varjo, 1977.
　　Vol.　7: **Ireland.** Desmond A. Gillmor, 1977.
　　Vol.　8: **Eastern Siberia.** Vladimir P. Shotskii, 1979.
　　Vol.　9: **Australia.** Peter Scott, 1981.
　　Vol. 10: **Canada.** Michael J. Troughton, 1982.　　　1207

D. BOOKS

Andreae, Bernd. Farming development and space, a world agricultural geography. Translated from German by Howard F. Gregor. Berlin; New York: de Gruyter, 1981. 345 p. ISBN 0-3-11-007632-2. $39. LC 81-4755.　　　1208

A unique text combining a systematic theory of farming development with detailed discussion of world agricultural problems. Reflecting contemporary German economic geography, a climatic zone framework is used to delimit world agricultural space. Cultural, political, and economic aspects of individual agricultural geographies are developed. A detailed bibliography and generous use of illustrations enhance this text.

Agricultural Geography

Bayliss-Smith, Timothy P. The Ecology of agricultural systems. 2nd ed. (Cambridge topics in geography no. 2). Cambridge, U.K.; New York: Cambridge University Press, 1982. 112 p. ISBN 0-521-23125-6. $13.95. ISBN 0-521-29829-6. $7.95. (pbk). LC 82-1132.   1209
   A detailed discussion and analysis of agricultural systems in the British tradition of ecological analysis. Examples of pre-industrial, semi-industrial, and full industrial agricultural systems are provided, followed by an ecological analysis of farming practices. Examples are international and the reader is given detailed insight into non-Western agricultural systems.

Chisholm, Michael. Rural settlement and land use: an essay in location. 3rd ed. London: Hutchinson; Atlantic Highlands, NJ: Humanities Press, 1979. 189 p. (1st ed. 1966. 207 p.) ISBN 0-09-139770-7. $17. LC 79-317579.   1210
   Agricultural location and rural settlement theory are detailed. Numerous global examples are employed to illustrate theory. An important text for students of agricultural geography.

Cox, George W. and Atkins, Michael D. Agricultural ecology: an analysis of world food production systems. San Francisco, CA: W.H. Freeman, 1979. 721 p. ISBN 0-7167-1046-3. LC 78-25745.   1211 •
   A well-written introductory text on agricultural ecology. An interdisciplinary work incorporating both physical and social factors affecting agricultural development. A valuable complement to economically oriented agricultural geographies.

de Blij, Harm J. Wine: a geographic appreciation. Totowa, NJ: Rowman and Allanheld, 1983. 239 p. ISBN 0-86598-091-8. $18.95. LC 82-20648.   1212
   A comprehensive geographic treatment of viticulture, including matters botanical, historical, physical, political, and economic. Contains glowing accounts of the regional geographies of the significant vintages and the resulting cultural landscapes.

de Blij, Harm J. Wine regions of the Southern Hemisphere. Totowa, NJ: Rowman and Allanheld, 1985. 255 p. ISBN 0-8476-7390-1. $32.95. LC 84-17949.   1213
   A geographical look at the Southern Hemisphere describing the enormous diversity of the wine growing regions in Chile, Argentina, Brazil, South Africa, Zimbabwe, Australia, and New Zealand. Physical geography and microclimates, cultural landscapes and vinicultural regions, emerging regional identities, and representative wines are discussed. Each topic is presented in an understandable language with practical information to add to the knowledge of wine history and selection. Shows that wines embody a fascinating summary of each region's civilization and the people creating them.

Ebeling, Walter. The Fruited plain, the story of American agriculture. Berkeley, CA: University of California Press, 1979. 433 p. ISBN 0-520-03751-0. $32.50. LC 78-62837.   1214 •
   The history of American agricultural development from neolithic beginnings to the present. Geographical regions form the organizational structure. Emphasis on the socioeconomic and environmental impacts of agriculture as it has evolved in the U.S.

Eckholm, Erik P. Losing ground: environmental stress and world food prospects. New York: W.W. Norton, 1976. 223 p. ISBN 0-393-06410-7. $7.95. (pbk). Also Oxford, U.K.; Elmsford, NY: Pergamon, 1978. 223 p. ISBN 0-08-021496-7. $32. ISBN 0-08-021495-9. $11.25. (pbk). LC 77-30755.   1215 •
   A detailed presentation of the negative ecological trends undermining global food-production systems. Specific discussion of the impacts of deforestation, siltation, erosion, desertification, and salinization on long term agricultural output. Extensive bibliography.

Enyedi, György, and Volgyes, Ivan. The Effect of modern agriculture on rural development. Oxford, U.K.; Elmsford, NY: Pergamon, 1982. 330 p. ISBN 0-08-027179-0. $36. LC 80-25232.   1216

Includes 23 papers originally presented at the fourth international meetings of the Commission on Rural Development, International Geographical Union. The focus is on the role of agriculture in the process of rural transformation. While the quality of individual contributions varies, overall this book is quite useful.

Furuseth, Owen J. and Pierce, John T. Agricultural land in an urban society. (Resource publications in geography). Washington, DC: Association of American Geographers, 1982. 89 p. ISBN 0-89192-149-2. $5. (pbk). LC 82-18424.

1217•

An analysis of the agricultural land use and policies to protect farmland from urbanization in the United States and Canada. Land use and research theory are introduced and supplemented by empirical data and case studies. About 200 references.

Gregor, Howard. Geography of agriculture: themes in research. Englewood Cliffs, NJ: Prentice-Hall, 1970. 181 p. ISBN 0-13-351320-3. LC 73-127855.

1218•

One of the best overall introductory texts on agricultural geography. A comprehensive presentation of the research approach and issues that comprise agricultural geography. Most of the examples and literature are focused on North America. Extensive bibliography.

Grigg, David B. The Agricultural systems of the world: an evolutionary approach. (Cambridge geographical series no. 5). London; New York: Cambridge University Press, 1974. 358 p. ISBN 0-521-09843-2. $21.95. (pbk). LC 73-82451.

1219

Present characteristics and distribution of major global agricultural systems are presented. Part one is focused on the history, diffusion, and evolution of modern agriculture, while part two is an in-depth analysis of individual systems. An extensive international bibliography of more than 1200 references is provided.

Grigg, David B. The Dynamics of agricultural change, the historical experience. London: Hutchinson Educational, 1982; New York: St. Martin's Press, 1983. 260 p. ISBN 0-312-22316-1. $27.50. LC 82-24034.

1220

A systematic historical geography of agriculture, with particular emphasis on the impact of technology and economic trends on agricultural evolution. Although systematic, the text is almost exclusively oriented to the European and American experience. Heavy emphasis on models of agrarian change and economic geography mark this effort. A stimulating interpretation of historical agricultural development.

Manshard, Walther. Tropical agriculture: a geographical introduction and appraisal. Translated from German by D.A.M. Naylon. London; New York: Longman, 1974. 226 p. ISBN 0-582-48187-2. $17.95. LC 73-88902. (Agrargeographie der Tropen. Mannheim, W. Germany: Bibliographisches Institut, 1968. 307 p.).

1221

A comprehensive synopsis of tropical agriculture, with emphasis on the social, economic, and environmental factors affecting agricultural geography. Heavy use of specific regional examples rather than quantitative models. Includes 55 figures and photographs. Excellent bibliography.

Morgan, William B. Agriculture in the Third World, a spatial analysis. London: Bell and Hyman; Boulder, CO: Westview, 1978. 290 p. ISBN 0-89158-820-5. ISBN 0-71351078-1. £8.75. LC 78-24064.

1222

Agricultural development and the special problems of agricultural geography are the concerns of this excellent text. The relevance of agricultural theory to the reality of agricultural practice is a particular concern. More than 300 international references.

Morgan, William B. and Munton, R.J.C. Agricultural geography. London: Methuen, 1971; New York: St. Martin's Press, 1972. 175 p. ISBN 0-312-01470-8. $19.95. LC 79-884547.

1223

A basic systematic text on agricultural geography. A strong emphasis on the farm as an economic unit, resulting in a strong economic geography orientation. Reliance on European and North American data for illustrative purposes.

Newbury, Paul A.R. A Geography of agriculture. Estover, Plymouth, U.K.: MacDonald and Evans; Philadelphia, PA: International Ideas, Inc., 1980. 326 p. ISBN 0-7121-0733-9. $23.50. (pbk). NUC 80-360980.　　　　　　　　　　　　　　　1224

Written for British students preparing for university and college exams. A comprehensive work which covers both theoretical and systematic aspects of agricultural geography. Detailed international case studies of agricultural systems are provided.

Organisation for Economic Co-operation and Development. Agriculture in the planning and management of peri-urban areas. Paris: Organisation for Economic Co-operation and Development; Washington, DC: OECD Publications and Information Center, 1979. 2 v. 795 p. ISBN 9-264-11893-4. $5. LC 79-313532.　　　　　　1225

Volume one provides an economic, behavioral, and political overview of the problems facing agriculturalists in peri-urban areas. Policy options are addressed. Volume two presents 20 case studies from Western Europe, Australia, New Zealand, Japan, Canada, and the U.S. Clearly, one of the most comprehensive treatments of this topic.

Organisation for Economic Co-operation and Development. Land use policies and agriculture. Paris: Organisation for Economic Co-operation and Development; Washington, DC: OECD Publications and Information Center, 1976. 84 p. ISBN 9-264-11558-7. LC 77-352522.　　　　　　　　　　　　　　　　　　　　　　1226

This short book effectively summarizes the issues and trends in agricultural land alienation in the OECD nations. Various policy approaches for dealing with this issue are discussed.

Symons, Leslie. Agricultural geography. rev. ed. London: G. Bell; Boulder, CO: Westview Press, 1979. 285 p. (1st ed., 1967. 272 p.). ISBN 0-89158-499-4. $37. LC 78-26680.　　1227 •

An updated edition of the seminal introductory text in agricultural geography. A broad systematic treatment, supplemented by case studies. Concepts and methods used in agricultural realization and classification are provided. Detailed bibliography.

Tarrant, John R. Agricultural geography: problems in modern geography. Newton Abbot, U.K.: David and Charles; New York: Halsted Press, John Wiley and Sons, 1980. 279 p. (1st ed., 1974. 279 p.). ISBN 0-7153-6286-0. $31.50. LC 74-158669.　　　1228

A methodological approach to agricultural geography structured around model building, hypothesis testing, and the development of analytical techniques for agricultural geography. Issues in contemporary agriculture, including marketing and land alienation are discussed, and analytical methods dealing with these questions are introduced. A unique quantitative approach to an agricultural and geography text.

14. INDUSTRIAL GEOGRAPHY

John Rees

A. BIBLIOGRAPHIES

Miller, E. Willard, and Miller, Ruby M. Industrial location and planning. Monticello, IL: Vance Bibliographies, 1984. 4 v.

Vol. 1: **P-1373. Theory, models, and factors in localization.** 49 p. 541 references. ISBN 0-88066-833-4. $7.50.

Vol. 2: **P-1374. Localization, growth, and organization.** 54 p. 590 references. ISBN 0-88066-834-2. $8.25.

Vol. 3: **P-1375. Regions and countries.** 97 p. 1090 references. ISBN 0-88066-835-0. $15.

Vol. 4: **P-1376. Industries.** 34 p. 371 references. ISBN 0-88066-836-9. $5.25.　　　　　　　　　　　　　　　　　　　　　　　　　　　1229

A set of four bibliographies on various aspects of industrial location and planning. Comprehensive coverage of articles, books, government reports, and other publications from 1977 to 1983. Worldwide in scope.

Wheeler, James O. Industrial location: a bibliography, 1966-1972. (Exchange bibliography, 436). Monticello, IL: Council of Planning Librarians, 1973. 68 p. $6.50. LC 75-307926.
  1230
  About 750 entries.

B. BOOKS

Bluestone, Barry, and Harrison, Bennett. The Deindustrialization of America: plant closings, community abandonment, and the dismantling of basic industry. New York: Basic Books, 1982. 323 p. ISBN 0-465-01590-5. $19.95. LC 82-70844.
  1231•
  One of the first books to argue that the American economy is deindustrializing, with large implications for different regions. Analyzes reasons why management closes plants in boomtowns as well as in declining areas, the impact on workers and their families, and the national future. Looks at prospects for reindustrialization in an industrial policy context. A controversial book.

Collins, Lyndhurst, and Walker, David F., eds. Locational dynamics of manufacturing activity. London; New York: John Wiley and Sons, 1975. 402 p. ISBN 0-471-16582-4. LC 73-21939.
  1232
  Collection of original papers by international scholars on the locational dynamics of manufacturing--theoretical, methodological, and empirical issues. Chapters deal with innovation, government policy, behavioral approaches, corporate strategy, systems approaches, and forecasting.

Gillespie, A. E., ed. Technological change and regional development. (London papers in regional science no. 12). London: Pion Ltd.; New York: Methuen, 1983. 171 p. ISBN 0-85086-107-1. $14. (pbk). LC 84-129658.
  1233
  One of the first collections of papers on technological changes in industry and their regional development implications. Papers examine research and development funding, the diffusion of industrial innovations, the employment implications of technological change, and the changing role of government.

Hamilton, F.E. Ian, ed. Industrial change: international experience and public policy. London; New York: Longman, 1978. 183 p. ISBN 0-582-48593-2. $11.95. (pbk). LC 77-11075.
  1234
  A collection of essays on industrial change by members of the Commission on Industrial Systems of the International Geographical Union. Papers examine energy and environmental issues, the role of multinational enterprises, and industrial change and regional policy in the U.K., West Germany, and the U.S. case studies include Hungary, Poland, Italy, Africa, and India.

Hamilton, F.E. Ian, and Linge, Godfrey J.R., eds. Spatial analysis, industry, and the industrial environment: progress in research and applications. Chichester, U.K.; New York: John Wiley and Sons, 1979-1984. 3 v. ISBN 0-471-90431-7. $128.95 (for complete set).
  Vol. 1: **Industrial systems.** 1979. 304 p. ISBN 0-471-99738-2. $69.96. LC 78-10298.
  Vol. 2: **International industrial systems.** 1981. 652 p. ISBN 0-471-27918-8. $78.95.
  Vol. 3: **Regional economics and industrial systems.** 1984. 680 p. ISBN 0-471-10271-7. $96.95.
  1235
  Series concerned with critical evaluation and extension of industrial location theory, methods of analysis, production systems, and case studies of particular industries, organizations, and regions. Contributions by international specialists.

Industrial Geography

Hamilton, F.E. Ian. Spatial perspectives on industrial organization and decision-making. London; New York: John Wiley and Sons, 1974. 533 p. ISBN 0-471-34715-9. £30.35. LC 73-14379.　　　　1236

A volume of original contributions by international researchers with focus on the spatial behavior of industrial organizations; responds to the inadequacies of classical industrial location theory. Essays focus on developing new hypotheses, corporate growth, location decision-making, ownership and locational change, and ways that firms adapt to different economic and political environments. Best collection on the behavioral approach to industrial location studies.

Hewings, Geoffrey J.D. Regional industrial analysis and development. London: Methuen; New York: St. Martin's Press, 1977. 180 p. ISBN 0-312-66910-0. $19.95. LC 77-76803.　　　　1237

One of the better texts on the various approaches to regional industrial analysis. Includes economic base and trade flow models, inter-industry input-output models, and the theoretical and policy context for analysis.

Keeble, David. Industrial location and planning in the United Kingdom. London; New York: Methuen, 1976. 317 p. ISBN 0-416-80060-2. $17.95. (pbk). LC 77-369081.　　　　1238

One of the best country-based studies of industrial location. Discusses industrial decentralization in a center-periphery context, taking into account the impact of industrial structure, the components of industrial change, and the role of government policy. Presents four industry case studies. Discusses implications for regional development and planning. Useful for academics and practitioners.

Massey, Doreen B. and Meegan, Richard. The Anatomy of job loss: the how, why, and where of employment decline. London; New York: Methuen, 1982. 258 p. ISBN 0-416-32350-2. $26. ISBN 0-416-32360-X. $12.95. (pbk). LC 82-2259.　　　　1239

Examines the causes and geographical differences in industrial job losses in the British context. Shows that different kinds of production changes may result in different geographical patterns of decline. Industrial case studies focus on nationalization, technical change, and intensification of the work force. A provocative neo-Marxist interpretation.

McDermott, Philip J. and Taylor, Michael. Industrial organisation and location. (Cambridge Geographical Series no. 16). Cambridge, U.K.; New York: Cambridge University Press, 1982. 226 p. ISBN 0-521-24671-7. $47.50. LC 81-21586.　　　　1240

One of the few texts in industrial geography that explores the links between industrial organizations and the environments in which they operate. Ideas about linkages and information flows are merged with organization theorists' concern with contingency models of organization structure. Ideas tested in context of British electronics industry.

Miller, E. Willard (Eugene). Manufacturing: a study of industrial location. University Park, PA; London: Pennsylvania State University Press, 1977. 286 p. ISBN 0-271-01224-2. LC 76-1849.　　　　1241

Divided into three parts: industrial location theory; factors of industrial localization; and locational analysis of selected industries.

Moriarty, Barry M. Industrial location and community development. Chapel Hill, NC: University of North Carolina Press, 1980. 381 p. ISBN 0-8078-1400-8. $25. ISBN 0-8078-4064-5. $9.95. (pbk). LC 79-16029.　　　　1242

Explains the concepts, materials, and methods designed to make industrial and community development programs more effective. Shows how to bring key interest groups together--industrialists looking for sites, economic development agents, and community planners. Useful for practitioners and academics.

Oakey, Raymond P. High technology small firms: innovation and regional development in Britain and the United States. New York: St. Martin's Press; London: F. Pinter, 1984. 179 p. ISBN 0-312-37239-6. $25. LC 83-40705.　　　　1243

One of the first studies to look at geographical differences in innovation potential, with focus on high technology small firms in Britain and California. Examines agglomeration effects, linkages, research and development funding, labor, and financial issues.

Rees, John; Hewings, Geoffrey J.D.; and Stafford, Howard A., eds. Industrial location and regional systems: spatial organization in the economic sector. Brooklyn, NY: J.F. Bergin, 1981. 260 p. ISBN 0-89789-008-6. $27.95. LC 80-28399.   1244

Collection of original papers mostly by American industrial geographers focusing on theoretical issues in interpreting industrial systems (structural theory, industry growth, linkages, flexibility) and empirical analyses of industrial change (including Canadian acquisitions, rural branch plants, the influence of government policy in the U.K. and Canada, and research and development in the U.S.).

Sant, Morgan E.C. Industrial movement and regional development: the British case. (Urban and Regional Planning no. 11). Oxford; New York: Pergamon, 1975. 253 p. ISBN 0-08-017965-7. $28. LC 75-7830.   1245

Examines the way industrial mobility operates in Britain and how it responds to regional policy. Study integrates temporal and spatial distributions of industrial analysis in a time series framework. Looks at industrial movement as a component of a larger national economic system.

Schmenner, Roger W. Making business location decisions. Englewood Cliffs, NJ: Prentice-Hall, 1982. 268 p. ISBN 0-13-545863-3. $30.95. LC 81-22657.   1246

A thorough guide for executives charged with making business location decisions. Also useful for industrial development agencies. Discusses how America's largest companies make locational choices, also the changing geography of manufacturing in the U.S. and the nature of plant closings. Of practical use to both academics and practitioners.

Smith, David M. Industrial location: an economic geographical analysis. 2nd ed. New York: John Wiley and Sons, 1981. 492 p. ISBN 0-471-06078-X. $36.95. LC 80-19231.   1247

One of the best texts dealing with the development of industrial location theory in a least-cost framework. Places the plant location problem in context, and develops a variable cost model of industry location. Includes operational models and other empirical applications. Deals with the impact of industry and industrial development strategies but not how decision makers actually behave.

Stafford, Howard A. Principles of industrial facility location. Atlanta, GA: Conway Publications, 1980. 275 p. ISBN 0-910436-08-8. $25. (pbk). LC 80-26737.   1248 •

Succinct survey of the 'art' of industrial location decision-making. Considers location factors, area attributes, and government influences. Covers comparative cost and capital budgeting techniques as they influence location decision mechanisms. Useful for both practitioners and scholars.

Townroe, Peter M. Industrial movement: experience in the U.S. and U.K. Farnborough, U.K.: Saxon House; Brookfield, VT: Gower, 1979. 250 p. ISBN 0-566-00279-5. $41. LC 79-318486.   1249

A comprehensive synthesis of components of industrial change in two major industrialized societies: the U.K. and the U.S. Focus is on intra-regional and inter-regional industrial movement, relating these processes to theories of industrial location and location decision-making procedures, while examining past decision experiences of plants. Useful for practitioners and scholars.

Watts, Hugh D. The Branch plant economy: a study of external control. London; New York: Longman, 1981. 104 p. ISBN 0-582-30028-2. £5.50 (pbk). LC 80-42201.   1250

The most comprehensive study of the impact of external ownership of industry on regional development, with empirical focus on the British brewing industry. Reviews the impact of ownership patterns on employment, linkage patterns, and technology potential.

15. TRANSPORTATION

James O. Wheeler

A. BIBLIOGRAPHIES

Adams, Russell B. and Geyer, Mark C. Transportation: a geographical bibliography. Minneapolis, MN: University of Minnesota, Department of Geography, 1970. 154 p.
1251
Available from the University of Minnesota, Department of Geography on a limited basis, this bibliography contains more than 1800 citations in the area of transportation geography.

Olsson, Gunnar. Distance and human interaction: a review and bibliography. (Bibliography Series no. 2). Philadelphia, PA: Regional Science Research Institute, 1965. 112 p. LC 68-6340.
1252
A critical review of the role of distance in location theories and migration and diffusion models, with special emphasis on gravity and potential models. Nearly 400 bibliographic entries.

Siddall, William R. Transportation geography: a bibliography. 3rd ed. Manhattan, KS: Kansas State University Library, 1969. 94 p. (1st ed., 1964. 64 p.). ISBN 0-686-20817-X. $2.50. (pbk). LC 73-633686.
1253
About 1700 articles or books in English, primarily from the period 1950-1969, mainly organized according to type of transportation: ocean shipping, seaports, inland waterways, railroads, highways, pipelines, and air transportation but including also general and regional studies in transportation geography.

Wheeler, James O. Spatial studies in transportation: introduction and annotated bibliography. (Exchange bibliography no. 324-325). Monticello, IL: Council of Planning Librarians, 1972. 161 p. NUC 74-4822.
1254
Contains 1000 annotated citations classified into 21 categories.

B. GENERAL WORKS

Daniels, P.W. and Warnes, A.M. Movement in cities: spatial perspectives on urban transport and travel. London; New York: Methuen, 1980. 395 p. ISBN 0-416-35620-6. $19.95. (pbk). LC 79-40867.
1255
A comprehensive coverage of urban transportation based on an extensive survey of the literature. Explains courses of urban travel, travel change and techniques to predict future travel, and policy features of urban transport problems.

Eliot Hurst, Michael E., ed. Transportation geography: comments and readings. New York: McGraw-Hill, 1974. 528 p. ISBN 0-07-019190-5. LC 73-8630.
1256
A general collection of readings in transportation geography with a critique of the field.

Haggett, Peter, and Chorley, Richard J. Network analysis in geography. London: Edward Arnold; New York: St. Martin's Press, 1969. 348 p. ISBN 0-7131-5359-4. LC 77-480965.
1257
A comprehensive treatment of network models in geography, including topologic and geometric structure, the relationships between network structure and flows, and the growth and transformations of networks. An analytic and conceptual approach.

Lowe, John C. and Moryadas, S. The Geography of movement. Prospect Heights, IL: Waveland, 1984. 333 p. Reprint of 1975 ed., Boston, MA: Houghton Mifflin. ISBN 0-88133-100-7. $24.95. LC 74-15401.
1258
A modern treatment of transportation geography organized by concepts and providing an extensive coverage of the literature.

Taaffe, Edward J. and Gauthier, Howard L., Jr. Geography of transportation. Englewood Cliffs, NJ: Prentice-Hall, 1973. 226 p. ISBN 0-13-351387-4. $17.95. (pbk). LC 72-8995.　　　　　　　　　　　　　　　　　　　　　　　　　　　1259

　　A widely used treatment of classical topics and quantitative and conceptual models in transportation geography emphasizing the gravity model, graph-theoretic measures, and allocation models.

White, Henry P. and Senior, M.L. Transport geography. London; New York: Longman, 1983. 224 p. ISBN 0-582-30025-8. $13.95. (pbk). LC 82-15332.　　　　1260

　　Consists of three parts: the basic historical, technological, physical, economic, political and social, and morphological factors; locational studies; and selected quantitative approaches. Extensive bibliography with emphasis on British sources.

C. SPECIAL TOPICS

Altschiller, Donald. Transportation in America. (The Reference Shelf, vol. 54, no. 3). New York: H.W. Wilson, 1982. 204 p. ISBN 0-8242-0667-3. $7. LC 82-11190.　　　1261•

　　A collection of essays originally published in a variety of outlets focusing on problems and prospects for American transportation.

Appleton, James H. A Morphological approach to the geography of transport. (Occasional papers in geography no. 3). Hull, U.K.: University of Hull, 1965. 44 p. LC 67-118684.　　　　　　　　　　　　　　　　　　　　　　　　　　　1262

　　An essay calling for greater emphasis on morphological studies in transport geography.

Becht, J. Edwin. A Geography of transportation and business logistics. Dubuque, IA: Wm C. Brown, 1970. 118 p. ISBN 0-697-05152-8. LC 70-118884.　　　　　1263

　　Description of transportation modes, largely with a U.S. context, with some focus on historical trends and on economic (business) aspects.

Berry, Brian J.L. Essays on commodity flows and the spatial structure of the Indian economy. (Research paper no. 111). Chicago, IL: University of Chicago, Department of Geography, 1966. 334 p. LC 66-29563.　　　　　　　　　　　　　1264

　　An application of factor analysis and input-output analysis to extensive commodity flow data for India. Forty tables and 101 figures.

Blonk, W.A.G., ed. Transport and regional development: an international handbook. Farnborough, U.K.: Saxon House; Brookfield, VT: Gower, 1979. 338 p. ISBN 0-556-00285-X. $47.50. LC 80-471441.　　　　　　　　　　　　　　1265

　　A collection of writings on theoretical aspects of the relationships between transport and regional policy and on these relationships in twelve countries.

Burns, Lawrence D. Transportation, temporal, and spatial components of accessibility. Lexington, MA: Lexington Books, 1979. 152 p. ISBN 0-669-02916-5. LC 79-1725.　　1266

　　An analysis built upon Hägerstrand's time-space model focusing conceptually upon human accessibility and presenting different analytical strategies for time-space tradeoffs.

Button, Kenneth John, and Gillingwater, David, eds. Transport, location, and spatial policy. Aldershot, Hants, U.K.; Brookfield, VT: Gower, 1983. 260 p. ISBN 0-566-00527-1. $35.50. LC 82-243738.　　　　　　　　　　　　　　1267

　　A collection of original essays on applied transport economics, planning, and geography examining the ways in which decisions and structures in the transport sectors have impacts in other sectors, especially cities and regions as well as spatial policy in general.

Transportation

Chisholm, Michael, and O'Sullivan, Patrick. Freight flows and spatial aspects of the British economy. (Cambridge Geographical Studies no. 4). London, U.K.; New York: Cambridge University Press, 1973. 141 p. ISBN 0-521-08672-8. LC 72-83592.   1268
   One of the few comprehensive case studies of national inter-regional freight flows (rail and road), using gravity and linear programming models to test relationships between distance and transport costs.

Dickey, John W. and others. Metropolitan transportation planning. 2nd ed. Washington, DC: Hemisphere Publishing Corporation; New York: McGraw-Hill, 1983. 607 p. (1st ed., Washington DC: Scripta Book Co. 1975. 562 p.) ISBN 0-89116-271-2. ISBN 0-07-016816-4. $39.95. LC 82-23319.   1269
   A survey of urban transportation planning. Interdisciplinary in approach.

Hay, Alan M. Transport for the space economy: a geographical study. London: Macmillan; Seattle, WA: University of Washington Press, 1973. 192 p. ISBN 0-295-95306-3. $25. LC 73-181082.   1270
   A conceptual treatment within the framework of geography and economics, organized around the elements of transport demand, vehicle capacity, flows, and networks.

Hoyle, Brian S., ed. Transport and development. London: Macmillan; New York: Barnes and Noble, 1973. 230 p. ISBN 0-06-493030-0. LC 74-194426.   1271
   A collection of major articles previously published largely in the late 1960's and early 1970's dealing with transport development in various parts of Africa, Latin America, and Australia.

O'Sullivan, Patrick E. Transport policy: geographic and economic planning aspects. London: Batsford Academic and Educational; Totowa, NJ: Barnes and Noble, 1980. 313 p. ISBN 0-389-20143-X. $26.50. LC 81-459646.   1272
   A broad treatment of transportation policy linking conceptual and applied approaches.

O'Sullivan, Patrick E.; Holtzclaw, Gary D.; and Barber, Gerald. Transport network planning. London: Croom Helm, 1979. 187 p. ISBN 0-85664-742-X. £14.95. LC 79-322642.   1273
   A study of transport investment and development focusing on planning models.

Pederson, Eldor O. Transportation in cities. New York: Pergamon Press, 1980. v. 3. 87 p. ISBN 0-08-024666-4. $13.75. LC 79-22726.   1274
   A study of the relationships between transportation and land uses in urban areas emphasizing historical and future changes and contrasting transportation in wealthy and Third World nations.

Potts, Renfrey B. and Oliver, Robert M. Flows in transportation networks. (Mathematics in science and engineering, v. 90). New York: Academic Press, 1972. 192 p. ISBN 0-12-563650-4. $55. LC 70-182673.   1275
   Mathematical treatment of transportation network theory, traffic assignment, and trip distribution models.

Pushkarev, Boris S. and Zupan, Jeffrey M. Public transportation and land use policy. Bloomington, IN: Indiana University Press, 1977. 242 p. ISBN 0-253-34682-7. $27.50. LC 76-29299.   1276
   Emphasizes the supply and demand for transit service as related to urban densities.

Stutz, Frederick P. Social aspects of interaction and transportation. (Resource paper no. 76-2). Washington, DC: Association of American Geographers, 1976. 74 p. ISBN 0-89291-117-4. $4. (pbk). LC 76-19932.   1277
   An examination of social travel linkages in cities, transportation and deprived people, impacts of transportation on the social and physical environment, and the role of transport planning policies on environmental and social equity.

Taaffe, Edward J.; Garner, Barry J.; and Yeates, Maurice H. The Peripheral journey to work; a geographic consideration. Evanston, IL: Northwestern University Press for the Transportation Center at Northwestern University, 1963. 125 p. LC 63-13481.   1278

A study of commuting to a west suburban area of Chicago, utilizing a gravity model approach.

Ullman, Edward L. American commodity flow: a geographic interpretation of rail and water traffic based on principles of spatial interchange. Seattle, WA: University of Washington Press, 1957. 215 p. LC 67-9184.   1279

An important geographic treatment of American rail and water transportation, and the problems of spatial interaction, with particular emphasis on flow maps of the origin and destination of commodities among the 48 states.

Wheeler, James O., ed. Proceedings of the International Geographical Union Working Group on transport geography. (January 1982 and September 1983 proceedings). Athens, GA: University of Georgia, Department of Geography, 1983. 205 p.   1280

A collection of thirteen papers dealing with topics in urban transportation.

Wheeler, James O. The Urban circulation noose. North Scituate, MA: Duxbury Press, 1974. 137 p. ISBN 0-87872-056-1. LC 73-89887.   1281

A treatment of mobility in urban areas with an emphasis on the social problems and impacts.

16. MARKETING GEOGRAPHY

See also Part V, Section 1. Applied Geography, especially entries 1474-1476, 1478-1481, and 1485.

Berry, Brian J.L. Geography of market centers and retail distribution. Englewood Cliffs, NJ: Prentice-Hall, 1967. 146 p. ISBN 0-13-351304-1. $17.95. (pbk). LC 67-13355.   1282

Davies, Ross L. Marketing geography with special reference to retailing. London; New York: Methuen, 1976. 300 p. ISBN 0-416-70700-9. $14.95. (pbk). LC 77-364245.   1283

Scott, Peter. Geography and retailing. London: Hutchinson; Chicago: Aldine, 1970. 192 p. LC 79-477565.   1284

Vance, James E., Jr. The Merchant's world: a geography of wholesaling. Englewood Cliffs, NJ: Prentice-Hall, 1970. 167 p. LC 73-132041.   1285

17. COMMUNICATION

Donald G. Janelle and Ronald F. Abler

Hiltz, Starr Roxanne, and Turoff, Murray. The Network nation: human communication via computer. Reading, MA: Addison-Wesley, 1978. 528 p. ISBN 0-201-03140-X. $38.50. ISBN 0-201-03141-8. $26.50. (pbk). LC 78-12365.
    An indepth review of computerized conferencing. Focused primarily on the exchange of scientific research information, examples of management and service to handicapped populations are also considered. An evaluation of existing systems, guidelines on setting up an information exchange, and a discussion of regular policies are provided.

1286

Martin, James. Telematic society: a challenge for tomorrow. 2nd ed. Englewood Cliffs, NJ: Prentice-Hall, 1981. 244 p. (1st ed., The wired society, 1978. 300 p.). ISBN 0-13-902460-3. $25. LC 81-5219.
    A non-technical description of how new telecommunications and computer technologies are being used to alter the organization of industry and banking, the delivery of medical and educational services, and the nature of home life and travel behavior.

1287

Masuda, Yoneji. The Information society: as post-industrial society. Translated from Japanese by Bernard Halliwell. New Brunswick, NJ: Transaction Books; Tokyo, Japan: Institute for Information Society, 1980. 171 p. ISBN 0-930242-15-7. $12.50. (pbk). LC 81-176767.
    A somewhat visionary set of forecasts and extrapolations of the effects of computers and telecommunications on many aspects of social, economic, and political life.

1288

Meier, Richard L. A Communications theory of urban growth. Cambridge, MA: MIT Press for MIT and Harvard University, Joint Center for Urban Studies, 1962. 184 p. LC 62-20480 rev.
    A seminal introduction to the role of information exchange and the role of communications in the function and development of cities. Relationships between communications, land use, and transportation are considered.

1289

Nilles, J.M.; Carlson, F.R., Jr.; Gray, P.; and Hanneman, G.J. The Telecommunications-transportation tradeoff. New York: John Wiley and Sons, 1976. 196 p. ISBN 0-471-01507-5. $44.50. LC 76-18107.
    Investigates the uses of private telecommunications networks for banking and insurance firms in the Los Angeles area to assess corporate decentralization policies for branch facilities. The impacts on the firm's organizational needs, the employee job satisfaction, commuting costs, and energy use are considered. Surveys of users of Interactive Instructional Television are analyzed to determine the desirability of such systems.

1290

Pool, Ithiel de Sola, ed. The Social impact of the telephone. Cambridge, MA: MIT Press, 1977. 502 p. ISBN 0-262-16066-8. ISBN 0-262-66048-2. $9.95. (pbk). LC 77-4110.
    An outstanding collection of twenty-one original papers on the role of the telephone in shaping society. Considers the historical development of the telephone, its social uses, its impact on the daily lives of people, and its facilitation of diverse structural changes within cities.

1291

Saunders, Robert J.; Warford, Jeremy J.; and Wellenius, Bjorn. Telecommunications and economic development. Baltimore, MD: Johns Hopkins University Press, 1983. 395 p. ISBN 0-8018-2828-7. $32.50. ISBN 0-8018-2829-5. $14.95. (pbk). LC 82-49065.
    A thorough exploration of the subject with due attention to the spatial factor and much cross-national analysis.

1292

Toffler, Alvin. The Third wave. New York: William Morrow, 1980. 544 p. ISBN 0-688-03597-3. $14.95; New York: Bantam Books, 1981. ISBN 0-553-24698-4. $4.95. (pbk). LC 79-26690.   1293 •

A provocative synthesis of the role of new technologies and new value systems in shaping the way society organizes its production, consumption, and politics. The interrelated roles of the media, new modes of communications, and the micro-computer receive particular attention.

18. INTERNATIONAL TRADE AND LOCATION

James E. McConnell

Behrman, Jack N. Industrial policies: international restructuring and transnationals. Lexington, MA: Lexington Books, 1984. 254 p. ISBN 0-669-08275-9. $23.50. LC 83-49533   1294

A compelling argument for the need to restructure industrial production internationally to achieve industrial integration and enhance global efficiency. Emphasis is given to how national policies can be designed to use transnational corporations effectively to bring about an integrated global system of industrial production.

Britton, John N.H. and Gilmour, James M. The Weakest link: a technological perspective on Canadian industrial underdevelopment. Ottawa, Canada: Science Council of Canada, 1978. 216 p. ISBN 0-660-10100-9. LC 79-321656.   1295

A perceptive accounting of how foreign investment can directly affect a country's international posture on commodity trade, technological development, employment structure, and industrial location and growth. Special emphasis upon the Canadian manufacturing sector is included.

Caves, Richard E. Multinational enterprise and economic analysis. Cambridge, U.K.; New York: Cambridge University Press, 1982. 346 p. ISBN 0-521-24990-2. $34.50. ISBN 0-521-27115-0. $9.95. (pbk). LC 82-4543.   1296

An excellent review and assessment of industry analysis and firm-level research in economics and business on the international behavior of multinational enterprises. The book contains a 30-page bibliography with more than 500 references.

Dunning, John H., ed. Economic analysis and the multinational enterprise. London; Boston: Allen and Unwin, 1981. 405 p. ISBN 0-04-330246-7. $37.50. LC 75-300865.   1297

Original writings by noted international economists that bridge the gap between the behavioral theory of the firm and the empirical study of multinational enterprises. Emphasis is given to the adequacy of economic analysis in accounting for the international location of production, international trade, and the industrial organization of transnational corporations.

Dunning, John H., ed. International production and the multinational enterprise. London; Boston: Allen and Unwin, 1981. 439 p. ISBN 0-04-330319-6. $40. ISBN 0-04-330320-X. $18.50. (pbk). LC 81-8101.   1298

Contains fifteen papers by Dunning and others which were originally published between 1973 and 1981. Primary focus is upon Dunning's "eclectic" theory of international production, which emphasizes firm-specific, country-specific, and internalization advantages in explaining international patterns of corporate behavior.

International Trade

Ghertman, Michel, and Leontiades, James, eds. European research in international business. Amsterdam, The Netherlands: North-Holland; New York: Elsevier North-Holland, 1978. 368 p. ISBN 0-444-85089-9. £32.75. LC 77-16087.　　　　1299

Seventeen position papers by noted scholars on the progress of European research on international business and the multinational corporation are contained in this revealing volume. Emphasis is upon the internationalization process of the firm, technology transfer, political risk assessment, and international marketing and production strategy.

Hymer, Stephen Herbert. The International operations of national firms: a study of direct foreign investment. Cambridge, MA: MIT Press, 1976. 253 p. ISBN 0-262-08085-0. $20. LC 75-33365.　　　　1300

First to demonstrate that the location of foreign direct investment could not be coupled with portfolio investments in a unified theory of international capital movements. Instead, direct investment activity belongs to the theory of industrial organization and to the analysis of market imperfections.

Robock, Stefan H. and Simmonds, Kenneth. International business and multinational enterprise. 3rd ed. Homewood, IL: Richard D. Irwin, 1983. 777 p. (1st ed., 1973. 652 p.). ISBN 0-256-02514-2. $29.95. LC 82-83424.　　　　1301

A premier international text on the movement of merchandise, human resources, technology, and ownership across national boundaries. Major emphasis is upon developing a comprehensive "geobusiness" theory of the international locational adjustment of industrial organizations.

Rugman, Alan M. Inside the multinationals: the economics of internal markets. New York: Columbia University Press, 1981. 179 p. ISBN 0-231-05384-3. $28.50. LC 81-7691.　　　　1302

Builds upon the market imperfections theory of foreign direct investment, and argues for a general theory of internalization to account for the international location of production, foreign trade, and technology transfers by multinational industrial enterprises.

Ryans, Cynthia C. International business reference sources: developing a corporate library. Lexington, MA: Lexington Press, 1983. 195 p. ISBN 0-669-06612-5. $22. LC 82-49323.　　　　1303 •

Compiled by an international librarian and bibliographer, this volume contains essential sources of information for the United States international trade community. Includes government publications, journals, annuals, business directories and handbooks, and an appendix of names and addresses of organizations where country and product data may be obtained.

Schollhammer, Hans. Locational strategies of multinational firms. (CIB study no. 1). Los Angeles: Pepperdine University, Center for International Business, 1974. 50 p. NUC 75-138851.　　　　1304

Comprehensive survey of the international locational strategies of 140 multinational corporations. International location decisions are described as a function of three sets of locational factors: country constraints; product characteristics; and company attributes.

United Nations Centre on Transnational Corporations. Transnational corporations in world development: third survey. New York: Unipub, 1983. 385 p. ISBN 0-317-00916-8. $38. (pbk). LC 83-213889.　　　　1305 •

Presents a comprehensive description and assessment of the role of transnational corporations in the global economy. Contains extensive tables and text on the growth and diversification of multinational corporations, foreign investments, interactions with host-country governments, and trade and technology transfers.

19. DEVELOPMENT

Philip W. Porter

A. ATLAS

Kidron, Michael, and Segal, Ronald. State of the world atlas. 2nd ed. New York: Simon and Schuster; London: Heinemann, 1984. (1st ed., 1981) 176 p. ISBN 0-671-42438-6. $16.95. ISBN 0-671-42439-4. $9.95. (pbk). LC 84-675087.　　　　　　　　1306 •
A collection of 66 colorful, provocative maps showing development topics in ways not commonly shown, for example: tax havens; the arms trade; sexual discrimination in employment; environmental pollution; foreign currency; liabilities of banks; and the cost of credit. Valuable interpretative notes are provided.

B. BOOKS

Abu-Lughod, Janet, and Hay, Richard, Jr. Third World urbanization. London; New York: Methuen, (originally Chicago: Maaroufa Press), 1977. 395 p. ISBN 0-416-60141-3. $9.95. (pbk). LC 76-53367.　　　　　　　　1307
A set of readings about urbanization and development ranging from Ibn-Khaldun to Lenin to HABITAT: the UN Conference on Human Settlements. Presentation is historical in parts one and two, and covers new theories and policies in parts three and five. Part four on urban problems treats regional imbalances, migration, jobs and class stratification, and housing.

Amin, Samir. Unequal development: an essay on the social formations of peripheral capitalism. Translated from French by Brian Pearce. New York: Monthly Review Press; Hassocks, U.K.: Harvester Press, 1976. 440 p. ISBN 0-85345-433-7. $8.50. (pbk). LC 77-357662. (Le développement inégal. Paris: Les Éditions de minuit, 1973. 365 p.).　　1308
A highly readable critique of development economics, mounted from the periphery (rather than the center), an area Amin sees as a frequent locus of origins and innovations. He discusses central and peripheral capitalist modes of production. Using these analyses he explores the "development of underdevelopment" and its consequences for social groups in Third World countries. He chides geographers for being "content to juxtapose facts, while the basic question of their discipline--how natural conditions act upon social formations--remains almost unanswered."

Berry, Brian J.L. The Human consequences of urbanization: divergent paths in the urban experience of the twentieth century. New York; London: Macmillan, 1973. 205 p. ISBN 0-333-11347-0. LC 74-150016.　　　　　　　　1309
A general overview which challenges "conventional wisdom" and argues that urbanization is several different processes arising out of differences in time and culture. He contrasts urban processes and human consequences as evidenced in North America (run by competitive bargaining and interest-group politics), the Third World (governed by unconstrained growth despite authoritarian forms of government), and post World War II Europe, including Eastern Europe and the USSR (guided by institutions which regulate and direct urban change in the public interest).

Boserup, Ester. The Conditions of agricultural growth: the economics of agrarian change under population pressure. London: Allen and Unwin; Hawthorne, NY: Aldine, 1965. 124 p. ISBN 0-202-07003-4. $14.95. LC 65-19513.　　　　　　　　1310
A brilliant, provocative synthesis of ideas on agricultural intensification, population pressure, and social and economic change. Boserup's argument inverts that of Malthus. She develops a typology of agricultural systems arranged along a continuum of frequency of use. For each system there are appropriate tools and methods. She posits "expectations" of social and political development given agricultural change. A strength of the book is its specificity in linking social institutions with mobilization of labor.

179

Development

Boserup, Ester. Woman's role in economic development. London: Allen and Unwin; New York: St. Martin's Press, 1974. 283 p. ISBN 0-312-88655-1. $10.95. LC 73-475138.   1311
  Geographers have not written book-length analyses of this important aspect of development. Boserup's ground-breaking study has inspired a decade of research on women in development. It explores in rural and urban areas, in developed countries and underdeveloped countries, the consequences in women's lives of economic development, which changes the nature of the division of labor. Development projects, particularly those based on mechanization, have often deprived women of their productive roles in society, thereby reducing their power and their participation.

Brookfield, Harold C. Interdependent development. London: Methuen; Pittsburgh, PA: University of Pittsburgh Press, 1975. 234 p. ISBN 0-8229-1118-3. $11.95. ISBN 0-416-78070-9. $11.95. (pbk). LC 75-333162.   1312
  Centers around "ideas of development" and presumes a considerable background on development issues and global inequality. Brookfield relies heavily on a reinterpretation of Schumpeter, and gives special attention to dualism, "two-sector economies," and center-periphery relations. According to Brookfield, development theory and practice have been found wanting and a new paradigm is emerging. He wants to incorporate anti-positivist views into development thinking.

Buchanan, Keith M. The Transformation of the Chinese earth: aspects of the evaluation of the Chinese earth from earliest times to Mao Tse-tung. London: Bell and Sons; New York: Praeger, 1970. 336 p. ISBN 0-7135-1549-X. £9.50. LC 79-506022.   1313 •
  A perceptive account of fundamental transformations in China of the relations of people to environment and of people to one another following the revolution which brought Mao Zedong to power in 1949. In exploring the diversity of China, Buchanan features the government's immense mobilization of labor for development, done on a self-reliant basis, largely independent of foreign aid programs. The book has 84 of Buchanan's distinctive readable maps, and 63 well-chosen photographs.

Castro, Josue de. The Geography of hunger. Boston: Little Brown, 1952. 337 p. Reprinted with an introduction by Jean-Pierre Berlan, as The Geopolitics of Hunger. New York: Monthly Review Press, 1977. 524 p. ISBN 0-85345-357-8. $18.50. ISBN 0-85345-456-6. $7.50. (pbk). LC 52-5012.   1314 •
  A classic, non-technical study by a geographer of development and inequality. Castro deals with environmental, Malthusian, and social causes of hunger, the nature and expression of dietary food insufficiencies, case studies and other work on food shortages, hunger, and their underlying causes. He is anti-Malthusian, and insists that hunger is not a natural necessity but a "man-made plague." A final section deals with technological and political change.

Chisholm, Michael. Modern world development: a geographical perspective. London: Hutchinson; Totowa, NJ: Barnes and Noble, 1982. 216 p. ISBN 0-389-20320-3. $28.50. ISBN 0-389-20321-1. $14.50. (pbk). LC 82-11404.   1315
  An extended essay on wealth and poverty of nations, using the core-periphery model for analysis of development patterns. Chisholm is optimistic, noting that the core of developed nations has expanded. He argues that for most Third World countries, agriculture must play a key role in development. He places great emphasis on natural resources and culture, and accords little explanatory importance to exploitation of the periphery by the core.

Cusack, David F., ed. Agroclimate information for development: reviving the green revolution. Boulder, CO: Westview Press, 1983. 397 p. ISBN 0-86531-429-2. $22.50. LC 82-51104.   1316
  These symposium proceedings on Agroclimate Models and Information Systems, Caracas, September, 1981, contain articles on climatic information and world agriculture, world food problems, agroclimate models, case studies, and a section on social science perspectives on applications. A leading theme is the use of inexpensive instruments and microcomputers in agricultural planning and management in less developed countries.

de Souza, Anthony R. and Porter, Philip W. The Underdevelopment and modernization of the Third World. (Resource paper no. 28). Washington, DC: Association of American Geographers, Commission on College Geography, 1974. 94 p. LC 74-20053. 1317

A college-level critique of geographical and other treatments of development and modernization. There are three parts: attributes and theories of underdevelopment, with extensive review of development literature in geography, political science, economics, sociology, rural sociology, and psychology; the nature and causes of development, including external relations, internal relations (colonial organization of space and labor), and urban and national development in the post-independence period; and competing paradigms in the geography of modernization: spatial, ecological, and Marxist.

Frank, Andre Gunder. Capitalism and underdevelopment in Latin America: historical studies of Chile and Brazil. Rev. ed. New York: Monthly Review Press; Harmonsworth, U.K.: Penguin Books, 1971. 368 p. (1st ed., 1967. 298 p.) ISBN 0-85345-093-5. $6.95. (pbk). LC 73-174622. 1318

Frank coined the phrase "the development of underdevelopment" and popularized among readers of English the idea of underdevelopment as a process arising out of relations between places, some of which develop at the expense of other places. An early and influential study, the book draws its examples mainly from Brazil and Chile.

Furtado, Celso. Economic development of Latin America: historical background and contemporary problems. Translated by Suzette Macedo. (Latin America Studies no. 8). 2nd ed. Cambridge, U.K.; New York: Cambridge University Press, 1976. 317 p. (1st ed., 1970. 271 p.) ISBN 0-521-21197-2. $47.50. ISBN 0-521-29070-8. $12.95. (pbk). LC 76-14914. (Formação econômica da América Latina. Rio de Janeiro, 1969. 365 p. 2nd ed. 1970. 365 p.). 1319

First published in 1969 in Portuguese (and in a Spanish translation), this is probably the most influential study of Latin American development. It deals with the conquest and formation of states, the international division of labor, agrarian systems, and the industrialization process. Furtado provides an economic history of the development experiences of individual countries, each unique, with special reference to the institutional nature of obstacles to development.

Gran, Guy. Development by people: citizen construction of a just world. New York: Praeger, 1983. 480 p. ISBN 0-03-063294-3. $33.95. ISBN 0-03-063296-X. $11.95. (pbk). LC 82-22442. 1320

Gran's position is that development must be participatory and can only be accomplished by the democratization of power. Three "exclusionary forces" (state/corporate, cultural, and bureaucratic), found both in capitalist West and socialist East, stand in the way. There are instructive case studies from Zaire (AID, IMF), Indonesia, and Thailand (World Bank), and Peru. Gran's plan for an alternative people-centered development, using ideas of Maslow, Galtung, Illich, and Freire, is scale sensitive, and, he argues, in the long run productive and practical. Includes a useful bibliography of about 2000 items.

Grossman, Lawrence S. Peasants, subsistence ecology, and development in the highlands of Papua New Guinea. Princeton, NJ: Princeton University Press, 1984. 302 p. ISBN 0-691-09406-3. $32.50. LC 84-42581. 1321

A study of conflicts between subsistence and commodity production (coffee and cattle) in a highland village (Kapanara) from a holistic perspective which combines cultural ecology and the complex links of political economy. Grossman examines in historical depth the interplay of subsistence activity and cash-economy "bisnis", the role of government, and growing differentiation in Kapanaran society.

Johnson, Edgar A.J. The Organization of space in developing countries. Cambridge, MA: Harvard University Press, 1970. 452 p. ISBN 0-674-64338-0. $27.50. LC 74-122216. 1322

By an economist, this book draws heavily on the work of geographers in seeking to use spatial design for market systems and functional economic areas in the development of less-developed countries. For Johnston, a key to economic development is the relations of town and country. An early, thought-provoking treatment of the geography of development.

Development

Knight, C. Gregory. Ecology and change: rural modernization in an African community. New York: Academic Press, 1974. 300 p. ISBN 0-12-785435-5. $46.50. LC 73-7446.   1323

A study of rural modernization in Mbozi, Tanzania, which combines ethnogeographical analysis of Nyiha farming systems with a study of the many sources of change at work, transforming the economy. Knight presents four models of change, including a spatial one, and considers the implications of change on spatial organization, economy, ecology, and culture.

Mabogunje, Akinlawon L. The Development process: a spatial perspective. London: Hutchinson University Library; New York: Holmes and Meier, 1981. 383 p. ISBN 0-8419-0659-9. $32.50. ISBN 0-8419-0660-2. $18.50. (pbk). LC 80-19939.   1324

A book by a leading scholar that explores development at three levels: rural development; urban development; and national integration. Draws examples from developed as well as less developed countries, and emphasizes spatial, economic, and social processual links between urban and rural areas. "The thrust of this volume is thus on how to use spatial forms, structures, and organizations to concentrate the energies of people in underdeveloped countries to engage in their own development."

McGee, Terence G. The Urbanization process in the Third World: explorations in search of a theory. London: Bell, 1971. 179 p. ISBN 0-7135-1623-2. £6.50. LC 70-871905.   1325

A pioneering work on urbanization processes in Third World countries, with emphasis on Southeast Asia. McGee discusses migration and the continuing relations migrants have with rural areas--a reevaluation of the "folk-urban continuum." He writes on comparatively neglected sectors of the urban economy: the bazaar system, street hawkers, and the important large informal tertiary or service sector.

Popkin, Samuel. The Rational peasant: the political economy of rural society in Vietnam. Berkeley, CA: University of California Press, 1979. 306 p. ISBN 0-520-03561-5. $25.50. ISBN 0-520-03954-8. $7.95. (pbk). LC 77-83105.   1326

A lengthy rebuttal to Scott's The Moral Economy of the Peasant, that argues that the romantic ideal of village society, so often encountered in connection with writings on the moral economy, must be abandoned. Popkin posits a political economy in opposition to a moral economy, and explores the rationality of peasants in their approach to markets, village life, and political relations.

Porter, Philip W. Food and development in the semi-arid zone of East Africa. (Foreign and Comparative Studies Program, African series no. 32). Syracuse, NY: Maxwell School of Citizenship and Public Affairs, Syracuse University, 1979. 107 p. ISBN 0-915984-54-7. $7.50. (pbk). LC 79-20312.   1327

A case study that examines both environment and political economy, and shows why the seven million people living in semi-arid areas of East Africa face severe problems of inequality, food insecurity, and underdevelopment. The history of policies and performance of the Kenyan and Tanzanian economies is traced for the colonial and post-colonial periods.

Santos, Milton. The Shared space: the two circuits of the urban economy in underdeveloped countries. Translated from French and adapted by Chris Gerry. London; New York: Methuen, 1979. 266 p. ISBN 0-416-79660-5. $13.95. ISBN 0-416-79670-2. $13.95. (pbk). LC 76-452757. (L'éspace partagé: les deux circuits de l'économie urbaine des pays sous-développés. Paris: M-Th. Genin. 1975. 405 p.).   1328

Santos seeks to develop a "theoretical framework which can be used for the analysis of the spatial implications of urbanization" with special reference to the Third World. Like McGee, he pays particular attention to the articulation of two circuits of urban economies in underdeveloped countries: the lower circuit--the indigenous domestic urban economy, and the upper circuit, which functions within a national and international framework. The study makes important contributions to our understanding of urban development in less developed countries.

Scott, James C. The Moral economy of the peasant: rebellion and subsistence in Southeast Asia. New Haven, CT: Yale University Press, 1976. 246 p. ISBN 0-300-01862-2. ISBN 0-300-02190-9. $7.95. (pbk). LC 75-43334.

1329

A non-Marxian analysis of the adaptability of the peasant under changing political and economic regimes. The author presents two case studies, one on Vietnam and one on Burma. He places "the subsistence ethic at the center of the analysis of peasant politics" as a means of understanding peasant rebellions and politics during the colonial era. The colonial economy disrupted peasant attempts to preserve the "norm of reciprocity" and the "right to subsistence," two key ideas.

Slater, David. Underdevelopment and spatial inequity: approaches to the problems of regional planning in the Third World. (Progress in Planning; v.4, part 2). Oxford, U.K.; Elmsford, NY: Pergamon Press, 1975. p. 97-167. ISBN 0-08-018769-2. LC 76-382761.

1330

Slater's analysis merges a Marxist critique with a spatial analysis of development and underdevelopment processes. It emphasizes the spatial and regional inequalities in underdeveloped countries generated through colonialism or dependency, illustrated respectively by case studies of Tanzania and Peru.

Smith, David M. Where the grass is greener: living in an unequal world. London: Croom Helm; Harmondsworth U.K.: Penguin, 1979; Baltimore, MD: Johns Hopkins University Press, 1982. 386 p. ISBN 0-8018-2883-X. $8.95. (pbk). LC 82-47979.

1331•

An inventive, imaginative, non-technical analysis of inequality treated both in the developed and underdeveloped world at several scales: among nations; within nations; and in cities and neighborhoods. Smith asks: "who gets what where?" The answers, based on case studies, emphasize inequality and different life chances at the individual level. Especially suitable for secondary school libraries.

Szentes, Tamas. The Political economy of underdevelopment. Translated by I. Vegas; translation revised by A. Gardiner. 4th rev. and enl. ed. Budapest, Hungary: Akadémiai Kiadó, 1983. 425 p. (1st ed., 1971. 328 p.) ISBN 9-630-52891-6. £28.75. LC 83-218505.

1332

The analysis is divided into two parts: a critique of theories of underdevelopment as limited by demographic and environmental features, as an historic "vicious cycle," as culturally prevented, as stages of growth (à la Rostow), as externally generated; and a historical examination of both internal and external causes of underdevelopment, seen as a process rather than an initial state. Although the writing is sometimes wooden, the book presents a substantive and probing analysis of development theory as it stood at the beginning of the 1970's.

Vogeler, Ingolf, and de Souza, Anthony R., eds. Dialectics of Third World development. Montclair, NJ: Allanheld, Osmun, and Co., 1980. 349 p. ISBN 0-916672-33-6. $24.50. ISBN 0-916672-35-2. $9.50. (pbk). LC 79-53704.

1333

A book of readings carefully chosen to present three arguments about Third World development: conservative, liberal, and radical. The readings cover climate and resources, culture and tradition, colonialism, population, tourism, and imperialism. The editors provide commentary, introductions, and "awareness exercises."

Watts, Michael. Silent violence: food, famine, and peasantry in northern Nigeria. Berkeley, CA: University of California Press, 1983. 687 p. ISBN 0-520-04323-5. $34.50. LC 82-13384.

1334

A richly detailed analysis of "the changing form and character of food crises in capitalist relations developed historically" in Sokoto Caliphate, Hausaland, northern Nigeria. Watts describes economic and social relations of production and reproduction prior to the colonial period and then traces how colonial policy enabled merchant capital to acquire ascendancy. This book is important as a case study which attempts to merge environmental and political economic approaches to the study of development.

20. URBAN GEOGRAPHY

John S. Adams*

*David Ley, Risa Palm, and Truman Hartshorn assisted in the establishment of priorities for the titles to be included.

A. SERIALS

Environment and planning. A. International journal of urban and regional research. 1- (1969-). Monthly. Pion, 207 Brondesbury Park, London, NW2 5JN, U.K. ISSN 0308-518X.
Articles on urban and regional research, many by geographers, utilizing models and modern quantitative techniques. Expensive. Book reviews.
1335

Geo abstracts. Series F. Regional and community planning. 1- (1972-). 6 per annum. Geo Abstracts, Ltd., Regency House, 34 Duke Street, Norwich, NR3 3AP, U.K. ISSN 0305-1943.
Abstracts of books and articles. See also Series C, Economic geography, and Series D, Social and historical geography, both of which also include abstracts on phases of urban geography.
1336

Urban geography. 1- (1980-). Quarterly. V.H. Winston and Sons, 7961 Eastern Avenue, Silver Spring, Maryland, 20910. ISSN 0272-3638.
Research articles and book reviews in the field of urban geography.
1337

B. TEXTBOOKS

Berry, Brian J.L. and Horton, Frank E. Geographic perspectives on urban systems with integrated readings. Englewood Cliffs, NJ: Prentice-Hall, 1970. 564 p. LC 72-86517.
Offers an excellent (although somewhat dated) summary of urban-economic geographic research and a particularly strong introduction to locational analysis techniques, but offers little discussion of other topics (perception, environmental problems, etc.).
1338

Bourne, Larry S., ed. Internal structure of the city: readings on urban form, growth, and policy. 2nd ed. New York; Toronto; London: Oxford University Press, 1982. 629 p. (1st ed., 1971. 528 p.) ISBN 0-19-503032-X. $15.95. (pbk). LC 81-18931.
Collection of papers covering all the major theoretical frameworks of urban geography. The book includes a number of influential authors and essays, and provides an excellent reference in contemporary urban geography.
1339

Brunn, Stanley D. and Wheeler, James O., eds. The American metropolitan system: present and future. New York: Halsted Press, John Wiley and Sons, 1980. 216 p. ISBN 0-470-27018-7. LC 80-36824.
Thirteen chapters address economic, social, and political problems of U.S. urban systems: the impacts of slow growth; inner-city revitalization; mortgage-lending practices; geopolitical fragmentation; corporate control; communication technology; and regional and metropolitan impacts of energy supplies and demands.
1340 •

Christian, Charles M. and Harper, Robert A., eds. Modern metropolitan systems. Columbus, OH: C.E. Merrill, 1982. 495 p. ISBN 0-06-759-892-0. $25.95. LC 81-84853.
Nineteen contributors cover a broad range of urban processes and interactions in the metropolitan systems of the United States.
1341 •

Hartshorn, Truman A. Interpreting the city: an urban geography. New York: John Wiley and Sons, 1980. 498 p. ISBN 0-471-05637-5. $30.95. LC 79-19544.   1342•
    A textbook in urban geography. Emphasizes urban problem-solving. A particularly complete treatment of land use in the Western city. Numerous maps and diagrams.

Herbert, David T. and Thomas, Colin J. Urban geography: a first approach. Chichester, U.K.; New York; Toronto: John Wiley and Sons, 1982. 508 p. ISBN 0-471-10137-0. $49.95. ISBN 0-471-10138-9. $23.95. (pbk). LC 81-16041.   1343
    A survey of the field of urban geography. Organized as an introductory text, but offers considerably more detail on some topics (particularly urban social geography) than on others.

Jones, Emrys. Towns and cities. London: Oxford University Press, 1966. 152 p. Reprint: Westport, CT: Greenwood Press, 1981. 152 p. ISBN 0-313-22724-1. $19.50. LC 80-24687.   1344
    Introduces the reader to urban geography covering such topics as the process of urbanization, preindustrial cities, city classification, and the city in its region. Important chapter on the Western city.

Muller, Peter O. Contemporary suburban America. Englewood Cliffs, NJ: Prentice-Hall, 1981. 218 p. ISBN 0-13-170647-0. $16.95. (pbk). LC 80-25653.   1345•
    Good general text dealing with the problems, changes, and developments that have taken place in the burgeoning suburban areas of American cities.

Palm, Risa. The Geography of American cities. New York; Oxford, U.K.: Oxford University Press, 1981. 365 p. ISBN 0-19-502785-X. $21.95. LC 80-14038.   1346•
    A complete and highly readable introduction to the urban geography of the North American city.

Yeates, Maurice H. and Garner, Barry J. The North American city. 3rd ed. San Francisco, CA; New York; London: Harper and Row, 1980. 536 p. (1st ed., 1971. 536 p.) ISBN 0-06-047334-7. $31.95. LC 80-10703.   1347•
    An urban geography textbook in three parts, dealing with: the city system; the internal structure of urban areas; and the urban dilemma.

C. CITY DEVELOPMENT THROUGHOUT HISTORY

Berry, Brian J.L., ed. Urbanization and counterurbanization. (Urban Affairs Annual Reviews, vol. 11). Beverly Hills, CA: Sage Publications, 1976. 334 p. ISBN 0-8039-0499-1. $28. ISBN 0-8039-0682-X. $14. (pbk). LC 76-15864.   1348
    Leading social scientists examine economic, social, and political impacts of urbanization in developed and developing countries and regions. Good mix of studies on theories and policies related to contrasting paths of urban development.

Mumford, Lewis. The City in history: its origins, its transformations, and its prospects. New York: Harcourt, Brace and World, 1961. 657 p. ISBN 0-15-618035-9. $9.95. (pbk). LC 61-7689.   1349•
    Classic work on the forms and functions of Western cities. Addresses basic questions such as: What is a city? How did it come into existence? What processes does it further? and, What purposes does it fulfill?

Sjoberg, Gideon. The Preindustrial city, past and present. Glencoe, IL: Free Press, 1960. 353 p. ISBN 0-02-928980-7. $9.95. (pbk). LC 60-10903.   1350•
    Analyzes the urban structure of pre-industrial societies. Introduces the inverse concentric theory of urban growth.

Urban Geography

Vance, James E., Jr. This Scene of man: the role and structure of the city in the geography of Western civilization. New York: Harper and Row, 1977. 437 p. ISBN 0-06-167407-9. LC 77-852.

1351 •

Detailed treatment of urban morphogenesis--focused on the role and purpose of cities in Western societies and the industrial processes societies have used to create and transform the physical structure of their cities.

Castells, Manuel. The City and the grassroots: a cross-cultural theory of urban social movements. (California series in urban development, vol. 2). Berkeley, CA: University of California Press; London: Edward Arnold, 1983. 450 p. ISBN 0-520-04756-7. $29.95. LC 84-672639.

1352

Castells, Manuel. The Urban question: a Marxist approach. Cambridge, MA: MIT Press, 1977. 502 p. ISBN 0-262-03063-2. ISBN 0-262-53035-X. $12.50. (pbk). LC 77-75345.

1353

Develops tools of research while criticizing traditional categories used by social sciences and mass media to diagnose urban problems. Examines urban sociological literature of several countries in relation to different themes.

Cox, Kevin R., ed. Urbanization and conflict in market societies. Chicago, IL: Maaroufa Press, 1978. 255 p. ISBN 0-88425-007-5. LC 77-76158.

1354

Collection of ten essays presenting alternative perspectives on urban problems. Five essays assume a Marxist orientation. The others present a cross-section of other views. The book concentrates on specific problems in the provision of urban housing.

Johnston, Ronald J. The American urban system: a geographical perspective. London: Longman; New York: St. Martin's Press, 1982. 348 p. ISBN 0-312-03124-6. $19.95. ISBN 0-312-03125-4. $11.95. (pbk). LC 81-51842.

1355

Knowledge gained from urban empirical studies is placed into a political-economic format emphasizing the effects on cities of changes in the American capitalist system.

Saunders, Peter R. Social theory and the urban question. London: Hutchinson; New York: Holmes and Meier, 1981. 310 p. ISBN 0-8419-0622-X. $32.50. ISBN 0-8419-0623-8. $14.50. (pbk). LC 80-21654.

1356

Agnew, John; Mercer, John; and Sopher, David, eds. The City in cultural context. Boston, MA: Allen and Unwin, 1984. 299 p. ISBN 0-04-301176-4. $30. ISBN 0-04-301177-2. $14.95. (pbk). LC 84-2851.

1357 •

The exceptional richness of this collection derives not only from the eleven chapters that make up the bulk of the volume, but also from well-reasoned opening and closing statements by the authors. The cultural and historical perspective evident in the volume is a welcome complement to the economic and contemporary orientation of most general works devoted to urban geography.

Harvey, David. Social justice in the city. Baltimore, MD: Johns Hopkins University Press; London: Edward Arnold, 1973. 336 p. ISBN 0-8018-1524-X. $29.50. ISBN 0-8018-1688-2. $8.95. (pbk). LC 73-7183.

1358

Detailed discussion of the spatial, economic, political, and historical aspects of cities from a Marxist perspective.

Herbert, David T. Urban geography: a social perspective. London: Newton Abbott; Vancouver: David and Charles, 1972; New York: Praeger, 1973. 320 p. ISBN 0-7153-5771-9. ISBN 0-7153-7518-0. (pbk). LC 72-83562.

1359 •

An introduction to urban social geography emphasizing residential, social, and movement patterns within the city. Discussion concentrates on the North American city.

Johnston, Ronald J. Urban residential patterns: an introductory review. London: Bell; New York: Praeger, 1971. 382 p. ISBN 0-7135-1675-5. LC 71-172079.   1360
Survey and interpretation of research in urban social geography. Concentrates on residential patterns in the western city.

LaGory, Mark, and Pipkin, John. Urban social space. Belmont, CA: Wadsworth, 1981. 356 p. ISBN 0-534-00864-X. LC 80-29675.   1361
An attempt to integrate research findings from a variety of perspectives (urban sociology, human ecology, and urban and behavioral geography) into one theoretical framework. Questions of how space is arranged and used in cities, and how images of spatial structure interact with spatial decision-making to create urban forms.

Lake, Robert W. The New suburbanites: race and housing in the suburbs. New Brunswick, NJ: Rutgers University, Center for Urban Policy Research, 1981. 303 p. ISBN 0-88285-072-5. LC 80-21659.   1362
Discusses the combination of race and housing in the suburban environment contrasting one against the other and comparing differences that are evident from one community to another.

Ley, David F. A Social geography of the city. New York: Harper and Row, 1983. 449 p. ISBN 0-06-384875-9. $19.50. (pbk). LC 82-11986.   1363 •
A textbook in urban social geography integrating behavioral, humanistic, and radical perspectives on the city. Excellent treatments of the sense of place in the city, and the functioning of informal social groups and complex organizations.

F. ENVIRONMENTAL RELATIONS IN THE CITY

Berry, Brian J.L. and Horton, Frank E. Urban environmental management: planning for pollution control: an original text with integrated readings. Englewood Cliffs, NJ: Prentice-Hall, 1974. 425 p. ISBN 0-13-939611-X. $33.95. LC 74-1315.   1364
Text with integrated readings: environmental beliefs and perceptions; environmental effects of urbanization; air resources and pollution; water pollution and management; municipal water supplies; solid wastes; noise pollution; pollution and health; and environmental law.

Detwyler, Thomas R.; Marcus, Melvin G.; and contributors. Urbanization and environment: the physical geography of the city. Belmont, CA: Duxbury Press, 1972. 287 p. ISBN 0-87872-034-0. LC 72-75110.   1365 •
An evaluation of how physical features and processes influence the growth and function of cities, and of how people have changed the natural environment by urbanization. The book addresses the interaction between people and urban climate, geology, and vegetation. A good basic introduction.

G. PERCEPTION AND BEHAVIOR

Chapin, F. Stuart, Jr. (Francis). Human activity patterns in the city: things people do in time and in space. New York: Wiley Interscience, 1974. 272 p. ISBN 0-471-14563-7. $76.50. LC 74-5364.   1366 •
Detailed examination of human activity patterns in a time-space perspective. Discusses why these patterns are important. Applications of the time-space approach to urban planning.

Urban Geography

Eliot Hurst, Michael E. I came to the city: essays and comments on the urban scene. Boston, MA: Houghton Mifflin, 1975. ISBN 0-395-04402-2. 401 p. LC 74-9338.   1367•
A text that helps the student see the city as an entity, a social and cultural place, an economic and political place. Explores the urban landscape from a human viewpoint. Uses essays, photographs, and poetry to focus on the experience of the past and present city, and prospects for the future. Original, imaginative, readable book.

Lynch, Kevin. The Image of the city. Cambridge, MA: MIT Press, 1960. 194 p. ISBN 0-262-12004-6. $20. ISBN 0-262-62001-4. $5.95. (pbk). LC 60-7362.   1368•
Classic work on the interpretation of urban landscapes. Lynch outlines the perceptual processes that contribute to the formation of personal images of the city. Introduces research techniques (such as the "mental map") for the assessment of perceptual images.

H. SPECIAL TOPICS

Adams, John S., ed. Urban policy making and metropolitan dynamics: a comparative geographical analysis. (Association of American Geographers. Comparative metropolitan analysis project). Cambridge, MA: Ballinger, 1976. 576 p. ISBN 0-88410-426-5. LC 76-25165.   1369
Analyzes progress made in the U.S urban policy in the 1960's in the nation's major metropolitan areas. Each chapter focuses on one issue--land speculation, abandoned housing, housing and urban renewal, environmental quality, public and parochial school attendance, crime, health, open space, transportation, the elderly, governance, and gerrymandering.

Berry, Brian J.L. Comparative urbanization: divergent paths in the twentieth century. 2nd ed. London: Macmillan; New York: St. Martin's Press, 1981. 235 p. (1st ed., 1973. 205 p.). ISBN 0-312-15475-5. $15.95. ISBN 0-312-15476-3. $10.95. (pbk). LC 82-61562.   1370
Examination of the different processes and consequences of urbanization throughout the world. Concentrates on the role of planning in Western and non-Western cities, and outlines four fundamental approaches to planning in a global context.

Berry, Brian J.L. The Geography of market centers and retail distribution. Englewood Cliffs, NJ; London: Prentice-Hall, 1967. 147 p. ISBN 0-13-351304-1. $17.95. (pbk). LC 67-13355.   1371
The classic introduction and summary presentation of central place theory, its applications in the U.S. and elsewhere in the world.

Bourne, Larry S. The Geography of housing. New York: Halsted Press, John Wiley and Sons, 1981. 288 p. ISBN 0-470-27058-6. $44.95. ISBN 0-470-27059-4. $19.95. (pbk). LC 80-19908.   1372
Comprehensive survey of housing patterns and processes in the Western city. Focuses on the production and distribution of housing, the development of housing patterns in cities, and policy implications of changes in the housing stock.

Murphy, Raymond E. The Central business district. Chicago, IL: Aldine-Atherton, 1972. 193 p. ISBN 0-202-10032-4. LC 76-159598.   1373•
A synthesis of geographic research into the characteristics and problems of the central business district. Now quite dated, it still offers a good description of the functions of the modern "downtown" area.

Perin, Constance. Everything in its place: social order and land use in America. Princeton, NJ: Princeton University Press, 1977. 291 p. ISBN 0-691-09372-5. $28.50. ISBN 0-691-02819-2. $9.95. (pbk). LC 77-72133.   1374•
Describes symbolic meanings of renters and homeowners in American society. A fresh and stimulating look at housing policy questions in the U.S., and the ways that the physical arrangement of land expresses American ideals, assumptions, and beliefs.

Rose, Harold M. The Black ghetto: a spatial behavioral perspective. New York: McGraw-Hill, 1971. 147 p. ISBN 0-07-053613-9. LC 70-179713.   1375
A "territorial view" of the black ghetto as a social area composed of people of the same race and culture. Specific chapters treat population growth and residential development, the economy of the ghetto, and the role of institutions (education, health care, and safety) in the ghetto.

I. THE CITY SYSTEM IN NORTH AMERICA

Abler, Ronald F.; Adams, John S.; and Lee, Ki-Suk, eds. A Comparative atlas of America's great cities: twenty metropolitan regions. (Association of American Geographers. Comparative metropolitan analysis project). Minneapolis, MN: University of Minnesota Press, 1976. 503 p. ISBN 0-8166-0753-2. $95. LC 76-14268.   1376•
Presents comparative data in maps and text on the twenty most populous metropolitan regions of the U.S., including details about population, housing, transportation, land use, social characteristics, health, unemployment, poverty, and other issues in major cities and their suburbs.

Adams, John S., ed. Contemporary metropolitan America: twenty geographical vignettes. (Association of American Geographers. Comparative metropolitan analysis project). Cambridge, MA: Ballinger, 1976. 4 v. ISBN 0-88410-425-7. LC 76-56167.
    Vol. 1: **Cities of the nation's historic metropolitan core.** 354 p.
        Boston, New York, Philadelphia, and Central Connecticut.
    Vol. 2: **Nineteenth century ports.** 314 p.
        Baltimore, New Orleans, and San Francisco-Oakland.
    Vol. 3: **Nineteenth century inland centers and ports.** 507 p.
        Pittsburgh, St. Louis, Cleveland, Chicago, Detroit,
        Minnneapolis-St. Paul, and Seattle.
    Vol. 4: **Twentieth century cities.** 350 p.
        Dallas-Fort Worth, Miami, Houston, Atlanta,
        Los Angeles-Long Beach, and Washington, DC.   1377•
Presents a detailed geographical portrait of each of twenty major American cities.

Dunn, Edgar S., Jr. The Development of the U.S. urban system. Baltimore, MD: Johns Hopkins University Press, Resources for the Future. V. 1.: 1980. 198 p. and microfiche. ISBN 0-8018-2196-7; V. 2.: 1983. 266 p. and microfiche. ISBN 0-8018-2638-1. $115. LC 79-2180.   1378
Volume one offers an explanation of the urban system based on the interdependence of the parts with themselves and the whole, and indicates the complex links between different urban subsystems. The book includes many detailed graphics, and a microfiche data set. Volume two is concerned particularly with industry sector shifts, with an explanation of change patterns in industrial sectors, their linkages with transportation and resources, and their expressions in individual industry groups.

Gottmann, Jean. Megalopolis: the urbanized northeastern seaboard of the United States. New York: Twentieth Century Fund, 1961; Cambridge, MA: MIT Press, 1964. 810 p. LC 61-17298.   1379•
Classic work on the development of "a new order in the organization of inhabited space," the almost continuous chain of cities along the northeastern seaboard from Boston to Washington. Aspects of the growth of this megalopolis, from the radical changes in land use and economy to the dynamics of interaction within the urbanized strip are documented.

Nader, George A. Cities of Canada. Toronto, Ontario: Macmillan of Canada, 1975. 2 v. ISBN 0-77051-029-9. LC 75-8440.   1380•
    Vol. 1: **Theoretical, history, and planning perspectives.** 388 p.
    Vol. 2: **Profiles of fifteen metropolitan centres.** ISBN 0-7705-1239-9.
        Can $11.95. LC 74-10295.
A comprehensive survey of the historical evolution of Canadian cities, the structure of individual cities, and possible solutions to current urban problems. Valuable as a general reference source for Canadian cities.

Urban Geography

Pred, Allan R. Urban growth and city systems in the United States, 1840-1860. Cambridge, MA: Harvard University Press, 1980. 282 p. ISBN 0-674-93091-6. $28. LC 80-12098.
  Examines the growth of U.S. cities from 1840 to 1860, and notes that the largest cities (concentrated in the Northeast) absorbed the overwhelming majority of urban growth. The author attributes this to the development of extensive lines of communication between large cities, fostering interdependence and the spread of innovation. 1381

Reps, John W. Cities of the American West: a history of frontier urban planning. Princeton, NJ: Princeton University Press, 1979. 827 p. ISBN 0-691-04648-4. $110. LC 78-51187.
  Important study of the development of urban places in the American West. "A major focus of this study is the physical patterns on which the towns and cities of the trans-Mississippi West grew." 1382 •

Reps, John W. Tidewater towns: city planning in colonial Virginia and Maryland. Williamsburg, VA: Colonial Williamsburg Foundation, 1972. 345 p. ISBN 0-910412-87-1. LC 77-154342.
  A significant and interesting study of colonial town planning in the Tidewater region of Virginia and Maryland around Chesapeake Bay and along its river estuaries. 1383

Reps, John W. Views and viewmakers of urban America: lithographs of towns and cities in the United States and Canada, notes on the artists and publishers, and a union catalog of their work, 1825-1925. Columbia, MO: University of Missouri Press, 1984. 570 p. ISBN 0-8262-0416-3. $89.50. LC 83-6495.
  Valuable union catalog of 4480 published lithographic views of cities and towns in the United States and Canada. In addition, the volume includes an introductory essay entitled The Making and Selling of Urban Views, *biographies of 51 viewmakers, and alphhabetical indexes to the text and the catalog.* 1384

Sawers, Larry, and Tabb, William, eds. Sunbelt/snowbelt: urban development and regional restructuring. New York: Oxford University Press, 1984. 431 p. ISBN 0-19-503264-0. $19.95. ISBN 0-19-503265-9. $10.95. (pbk). LC 82-14145.
  Identifies and addresses growth from New England to the sunbelt. Looks at urban policy in the U.S. and economic conditions affecting urban development. Bibliography. 1385

United States. Library of Congress. Geography and Map Division, Reference and Bibliography Section. Fire insurance maps in the Library of Congress: plans of North American cities and towns produced by the Sanborn Map Company. Washington, DC: Library of Congress, 1981. 773 p. ISBN 0-8444-0337-7. $32. LC 80-607938.
  Indispensable guide to the fire insurance maps on file in the Library of Congress published by the Sanborn Map Company. The collection described consists of a uniform series of large-scale maps dating from 1867 to the present and depicting the commercial, industrial, and residential sections of some twelve thousand cities and towns in the United States, Canada, and Mexico. Accompanying the checklist is an introductory essay by Walter W. Ristow. 1386

J. EUROPE

Burtenshaw, David; Bateman, M.; and Ashworth, Gregory J. The City in West Europe. New York; Chichester, U.K.: John Wiley and Sons, 1981. 340 p. ISBN 0-471-27929-3. $47.95. LC 80-41589.
  Examination of the special aspects of the West European city, emphasizing the effects of historical growth on modern urban planning problems. 1387 •

Dickinson, Robert E. The West European city: a geographical interpretation. 2nd ed. London: Routledge; New York: Humanities, 1964. 582 p. (1st ed., 1951. 580 p.) LC 51-8811.
  Explores the nature of the Western European city with emphasis on urban history. Includes information on most major cities in Western Europe. 1388 •

Hall, Peter G. London 2000. 2nd ed. London: Faber and Faber; New York: Praeger, 1969. 287 p. (1st ed., 1963. 220 p.). ISBN 0-571-09705-7. LC 75-78629.
1389 •
   The nature of the London region and projected problems by the year 2000. The central problem--shall London grow?--is discussed in terms of the vastly different consequences of whether or not the London region expands in areal size, function, and population.

Hall, Peter G. and Hay, Dennis. Growth centres in the European urban system. Berkeley, CA: University of California Press; London: Heinemann Educational Books, 1980. 278 p. ISBN 0-520-04198-4. $34. LC 81-135069.
1390 •
   Analysis of evolving urban systems regions in northern, central, and southern Europe. Detailed examinations of population and economic shifts on a regional and country-by-country basis.

Thompson, Ian B. The Paris basin: problem regions of Europe. 2nd ed. London; New York: Oxford University Press, 1981. 48 p. (1st ed., 1973. 48 p.) ISBN 0-19-913278-X. £2.75. ISBN 0-19-913069-8. £2.75. (pbk). LC 84-672044.
1391
   Short analysis of Paris and the Paris basin. Emphasis on planning problems of Greater Paris and strategies for solutions including new towns and the future role of the Lower Seine Valley.

Van Den Berg, Leo; Drewett, R.; Klassen, L.H.; Rossi, A.; and Vijverberg, C.H.T. Urban Europe: a study of growth and decline. Vol. 1. Oxford, U.K.; New York: Pergamon Press, 1982. 162 p. ISBN 0-08-023156-X. $44. LC 81-81233.
1392
   Part one of the book outlines three sequential stages of urban development: urbanization, suburbanization, and desuburbanization (characterized by metropolitan decline). Part two offers an empirical analysis to support this model (based on European data). Part three evaluates European urban policies, and speculates on the prospects for a future "reurbanization" stage.

K. CITIES OF THE WORLD

Brunn, Stanley D. Urbanization in developing countries: an international bibliography. (Latin American Studies Center Research Report no. 8). East Lansing, MI: Michigan State University, Latin American Studies Center and Center for World Affairs, 1971. 693 p. LC 79-172535.
1393
   An unannotated list of over 7,000 publications on all facets of urbanization in Latin America, Africa, and Asia.

Brunn, Stanley D. and Williams, Jack F. Cities of the world: world regional urban development. New York; London: Harper and Row, 1983. 506 p. ISBN 0-06-381225-8. $20.95. (pbk). LC 82-11689.
1394 •
   A study in comparative world urban development. Roughly equal space is given to twelve major regions (North America, Western Europe, Eastern Europe, the Soviet Union, Oceania, Latin America, Subsaharan Africa, North Africa, the Middle East, South Asia, Southeast Asia, and East Asia). The book is an excellent reference source in world urbanization, and includes an extensive bibliography.

Davis, Kingsley. World urbanization 1950-1970. (Population monograph series nos. 4 and 9). Berkeley, CA: University of California, Institute of International Studies, v. 1., 1969; v. 2., 1972. Reprint: Westport, CT: Greenwood Press, 1976. 670 p. ISBN 0-8371-8828-8. $62.50. LC 76-4570.
1395
   Volume one contains basic population data for 1950, 1960, and 1970, while volume two provides an analysis of the relationships and trends.

Hall, Peter G. The World cities. 2nd ed. New York: McGraw-Hill; London: Weidenfeld and Nicolson, 1977. 271 p. (1st ed., 1966. 256 p.) ISBN 0-297-77310-0. ISBN 0-297-77311-9. (pbk). LC 78-303702.
1396 •
   A brief treatment of seven major world urban centers: London; Paris; Randstad (The Netherlands); Rhine-Ruhr; Moscow; New York; and Tokyo.

Urban Geography

Drakakis-Smith, David. Urbanization, housing, and the development process. New York: St. Martin's Press, 1980. 234 p. ISBN 0-312-83519-1. $36. LC 80-52163.　1397

Dwyer, Denis J., ed. The City in the Third World. New York: Barnes and Noble, 1974. 253 p. LC 74-181583.　1398 •
Essays discussing the urbanization process in developing countries, and general forms of urbanization. Specific problems caused by rapid growth of cities are looked at in detail.

Ghosh, Pradip K., ed. Urban development in the Third World. (International Development Resource Books no. 2). Westport, CT: Greenwood Press, 1984. 546 p. ISBN 0-313-24138-4. $45. LC 83-22859.　1399

Hance, William A. Population, migration, and urbanization in Africa. New York: Columbia University Press, 1970. 450 p. LC 75-116378.　1400 •
Excellent readings on the general character of African urban systems, their problems, and possible solutions. Includes profiles of selected African cities.

Ma, Laurence J.C. Cities and planning in the People's Republic of China: an annotated bibliography. Washington, DC: U.S. Dept. of Housing and Urban Development, Office of Policy Development and Research (now part of the Dept. of Health and Human Services), 1980. 62 p. LC 81-601007.　1401
Useful compendium of 194 publications about cities, past and present, in China. The title is somewhat misleading.

McGee, Terence G. The Southeast Asian city: a social geography of the primate cities in Southeast Asia. London: Bell; New York: Praeger, 1967. 204 p. LC 67-99423.　1402 •
A study of the development of dominant "primate" cities in the nations of Southeast Asia. Includes discussion of the growth of primate cities, their economic, demographic, and residential patterns, and prospects for the future. The classic work on the subject.

McGee, Terence G. The Urbanization process in the Third World: explorations in search of a theory. London: G. Bell and Sons, 1971. 179 p. ISBN 0-7135-1623-2. £6.50. LC 70-871905.　1403
Explains how urbanization in the Third World differs from what occurred in the West and why theoretical models based on the Western urbanization experience cannot be applied to the Third World. Offers a convincing explanation of why urbanization in developing countries does not induce social change.

Neutze, Graeme Max. Urban development in Australia: a descriptive analysis. Sydney, Australia; London: Allen and Unwin, 1977. 258 p. ISBN 0-86861-000-3. £13. ISBN 0-86861-008-9. Aus $14.95. £7.50. (pbk). LC 77-371990.　1404 •
The best general treatment of the structural development of Australian cities.

Noble, Allen G. and Dutt, Ashok K., eds. Indian urbanization and planning: vehicles of modernization. (University of Akron. Geography Department research series). New Delhi, India: Tata-McGraw-Hill, 1977. 366 p. Rs. 120. NUC 80-446024. (Distributed in the United States by Akron University Bookstore, Akron, OH).　1405
Over twenty original articles on various aspects of Indian city structure and urban economic and environmental values.

Rose, Arthur J. Patterns of cities. Melbourne, Australia: Thomas Nelson, 1967. 237 p. LC 68-85414.　1406 •
Well-illustrated overview of urbanism and the spatial location of cities, as well as their internal structures and hierarchies. Primary focus on Australia.

Stöhr, Walter. Regional development: experiences and prospects in Latin America. (Regional Planning Series no. 3). Paris; Hawthorne, NY: Mouton, 1975. 186 p. ISBN 90-2797-661-9. $20. LC 75-319315.

1407

Systematic analysis of regional planning in Latin America. Outlines the spatial pattern of development, and planning strategies adopted to reduce regional disparities.

21. RECREATIONAL GEOGRAPHY

Lisle S. Mitchell

A. RECREATION

Chubb, Michael, and Chubb, Holly R. One third of our time? An introduction to recreation behavior and resources. New York; Chichester, U.K.: John Wiley and Sons, 1981. 742 p. ISBN 0-471-15637-X. $34.50. LC 80-25131.

1408 •

Near-encyclopedic review of behavioral and resource aspects of recreation. Emphasis is on facts and figures, agencies and users. Excellent introduction to the field. Strong geographical content, as well as recognition of other social science contributions.

Clawson, Marion, and Knetsch, Jack L. Economics of outdoor recreation. Baltimore, MD: Johns Hopkins University Press for Resources for the Future, 1966. 328 p. ISBN 0-8018-1302-6. $5.45. (pbk). LC 66-16040.

1409

A dated but thorough coverage of recreational economic principles. Spatial distribution of recreation phenomena and consumption patterns are discussed. Resource problems and implications for policy formulation and research endeavors are presented.

Coppock, John Terence, and Duffield, B.S. Recreation in the countryside: a spatial analysis. London: Macmillan; New York: St. Martin's Press. 1975. 262 p. ISBN 0-312-66605-5. $26. LC 75-9115.

1410

Jubenville, Alan. Outdoor recreation planning. Philadelphia, PA: W.B. Saunders, 1976. 399 p. ISBN 0-7216-5228-X. LC 75-5051.

1411 •

Presents a framework for outdoor recreational planning with an emphasis on the theories, principles, and application of planning. Guidelines for the planning of various sites and facilities are discussed. An excellent collection of planning reports are included.

Ng, David, and Smith, Stephen L.J., eds. Perspectives on the nature of leisure research. Waterloo, Ontario, Canada: University of Waterloo Bookstore, 1982. 263 p. ISBN 0-88898-039-6. Can $7.50. (pbk).

1412

Review essays on leisure research in geography, economics, psychology, communications, mathematics, history, and social work. An excellent chapter on the philosophy of leisure research and a well-structured concluding chapter.

Outdoor Recreation Resources Review Commission. Outdoor recreation for America: a report to the President and to the Congress. Washington, DC: U.S. Government Printing Office, 1962. 245 p. LC 62-60017.

1413

A dated but highly useful summary volume of the largest, most comprehensive recreation research effort undertaken in the U.S. Twenty-seven supporting volumes to the summary volume, dealing with specific topics (e.g., provision, participation, multiple use, open space), and particular regions (e.g., metropolitan regions, Alaska, and the Northeast), are included.

Recreational Geography

Smith, Stephen L.J. Recreation geography. New York; London: Longman, 1983. 220 p. ISBN 0-582-30050-9. $11.95. (pbk). LC 81-20749.   1414
A systematic examination of the methods used by geographers in their study of recreation. Methods include descriptive, explanatory, and predictive techniques. Policy related methods are also reviewed with examples from North America and Europe.

B. TOURISM

Gunn, Clare A. Vacationscape: designing tourist regions. Austin, TX: University of Texas, Bureau of Business Research, 1972. 238 p. LC 72-241.   1415 •
Provides valuable insights into the complicated process of planning vacation landscapes. Emphasis is placed upon a regional design system for tourist regions. A human ecological approach is advocated balancing human, environmental, social, political, and economic considerations. Well-illustrated but limited bibliography and index.

Mathieson, Alister, and Wall, Geoffrey. Tourism: economic, physical, and social impacts. London; New York: Longman, 1982. 208 p. ISBN 0-582-30061-4. $11.95. (pbk). LC 81-17135.   1416
Excellent introductory overview of the effects of tourism on regional and national economics, physical environment, and social structures. Presentation is balanced between negative and positive effects. Emphasis is on substantive findings and concepts rather than methods.

Matley, Ian M. The Geography of international tourism. (Resource papers for college geography no. 76-1). Washington, DC: Association of American Geographers, 1976. 40 p. ISBN 0-89291-110-7. $4. (pbk). LC 76-18390.   1417
Considers geographic aspects of international tourism: tourism and the regional economy; physical and cultural factors influencing the location of tourism; and the geography of tourism as an applied field.

Pearce, Douglas G. Tourist development. London; New York: Longman, 1981. 112 p. ISBN 0-582-30053-3. $9.95. (pbk). LC 81-3717.   1418
Focuses on the structures and processes of tourist development and planning, especially in terms of the contributions of geographers to tourist planning. Strongest examples are from New Zealand and France.

Robinson, Harry. A Geography of tourism. London: Macdonald and Evans. 1976. 476 p. ISBN 0-7121-0721-5. LC 76-366730.   1419

Turner, Louis, and Ash, John. The Golden hordes: international tourism and the pleasure periphery. London: Constable; New York: St. Martin's Press, 1976. 319 p. ISBN 0-312-33740-X. LC 76-25498.   1420
A comprehensive coverage of tourism impacts including the Soviet bloc and Western societies, as well as the Third World. An extensive range of sources are utilized and a number of imaginative ideas and real world examples are cited. Factual mistakes and an inconsistent Marxist bias flaws the manuscript.

C. NATIONAL PARKS

Foresta, Ronald A. America's national parks and their keepers. Baltimore, MD: Johns Hopkins University Press for Resources for the Future, 1984. 382 p. ISBN 0-915707-02-0. $45. ISBN 0-915707-03-9. $11.95. (pbk). LC 83-43262.   1421 •
An excellent geographic analysis of the changing role of U.S. national parks and the National Park Service. The focus is on the effect of changing perceptions of civilization and nature, and social changes. Includes a lengthy bibliography.

Nelson, J. Gordon; Needham, Roger D.; Nelson, Shirley H.; and Scace, Robert C. Canada's national parks: today and tomorrow, Conference Two, ten years later. Waterloo, Ontario, Canada: University of Waterloo, 1979. 2 v. 525 and 313 p.   1422 •
Conference proceedings that include papers, critiques, and discussions of such issues as international perspectives and examinations, indigenous and other uses of parks, people and technology in parks, and the future of national parks. The focus is on changes and developments in park policy, planning, and management since 1968.

D. WILDERNESS

Hendee, John C.; Stankey, George H.; and Lucas, Robert C. Wilderness management. (Miscellaneous publication no. 1365). Washington, DC: U.S. Department of Agriculture, Forest Service, 1978. 381 p. LC 77-600039.   1423
The only comprehensive work on wilderness. Includes the historical roots of the wilderness concept, the origin of wilderness legislation, the existing wilderness system, its use, and possible approaches to planning and management. The distribution of wilderness and an analysis of typical distributions of recreational use are included. Extensive bibliographies accompany each chapter.

Ittner, Ruth; Patter, Dale R.; Agree, James K.; and Aschell, Susie, eds. Recreational impact on wildlands. (Conference proceedings, Recreational Impact on Wildlands, Oct. 27-29, 1978, Seattle, Washington). Portland, OR: U.S. Forest Service, Pacific Northwest Region and National Park Service, Pacific Northwest Region, 1979. 341 p. GPO 13.2:W65/10. OCLC 6190210.   1424
Contains 57 papers concerned with the perception of impact, research findings, experiences in preventing and remedying impacts, approaches to future management, research, and the role of the wildland user. Emphasis is on the physical and biological changes resulting from recreation in wildlands.

E. SPORT

Rooney, John F. Jr. A Geography of American sport: from Cabin Creek to Anaheim. Reading, MA: Addison-Wesley, 1974. 306 p. ISBN 0-201-06491-X. $15.95. LC 73-16555.   1425 •
The first major geographic text on American sport. Emphasis in on major university and professional American sports: baseball, football, and basketball. Includes a good analysis of regional patterns and spatial interactions of individual athletes, teams, and leagues.

22. MEDICAL GEOGRAPHY

Gerald F. Pyle

A. ATLASES

Howe, George M. National atlas of disease mortality in the United Kingdom. Rev. and enl. ed. London; New York: Nelson, 1970. 197 p. (1st ed., 1963. 111 p.) ISBN 0-17-152008-4. LC 78-451480.   1426
Illustrates geographic patterns of disease mortality in the United Kingdom. The maps are clear and include cancer (four types), circulatory system (two types), respiratory system (three types), ulcers, diabetes, infant mortality, accidents, and suicide. Appendix shows death rates in selected areas in the United Kingdom.

Mason, Thomas J., et al. Atlas of cancer mortality for U.S. counties: 1950-1969. (DHEW publication no. (NIH) 75-80). Washington, DC: National Cancer Institute, Epidemiology Branch, 1975. 103 p. LC 78-302362.   1427
The 66 computer-generated color maps and appended table in this innovative atlas do more than depict spatial patterns of mortality from major forms of cancer for white males and females. They suggest intriguing research problems for the medical geographer.

Medical Geography

May, Jacques M., ed. Atlas of the distribution of diseases. New York: American Geographical Society, 1950-1955. 17 plates.   1428
Maps of world distribution of a disease and often its vectors: poliomyelitis, cholera, malaria, helminthiases, dengue and yellow fever, plague, leprosy, human starvation and deficiency diseases, rickettsial diseases, arthropod-borne viral infections, leishmaniasis, and spirochetal diseases (yaws, etc.). Each plate also includes a comprehensive list of sources.

Rodenwaldt, Ernst, ed. Welt-Seuchen Atlas. (World atlas of epidemic diseases). Hamburg, W. Germany: Falk-Verlag, 1952-1956. 3 v. LC Map-53-185.   1429
A comprehensive atlas of the historical and contemporary disease patterns throughout the world with special emphasis on Europe. Painstaking detail is evident in this excellent atlas.

B. GENERAL WORKS

Ackerknecht, Erwin H. History and geography of the most important diseases. New York: Hafner, 1965. 210 p. LC 20093.   1430
A very readable survey of the history, geographical distribution, and current etiology of a number of communicable, chronic, and nutritional diseases.

Eyles, John, and Woods, Kevin J. The Social geography of medicine and health. New York: St. Martin's Press; London: Croom Helm, 1983. 272 p. ISBN 0-312-73292-9. $32.50. LC 83-2921.   1431
Consists of an examination of some social scientific approaches to medical geography. The authors present some of the geopolitical issues involved in health care delivery in different parts of the world.

Henschen, Folke. The History and geography of diseases. Translated from Swedish by Joan Tate. London: Longman, 1966; New York: Delacorte, 1967. 344 p. LC 67-13149. (Sjukdomarnas historia och geografi. Stockholm: Bonniers, 1963. 287 p.).   1432
Causes, origins, and historical development of infectious and non-infectious diseases; their geographical distribution and frequency within population groups distinguished by race, religion, occupation, and social and hygienic standards. Many good illustrations. Of interest to a wide readership.

Hunter, John M., ed. The Geography of health and disease: papers of the first Carolina Geographical Symposium (1974). (Studies in geography no. 6). Chapel Hill, NC: University of North Carolina, Dept. of Geography, 1974. 193 p. LC 74-620073.   1433 •
A series of studies on the status of medical geography during the 1970's. Includes the geography of infectious and chronic diseases, psychosocial stress, and health care delivery.

Learmonth, Andrew T.A., ed. The Geography of health: selected international congress papers. Oxford, U.K.; New York: Pergamon Press, 1981. 262 p. ISBN 0-08-027434-X. £14.50.   1434
Contains seven groups of papers that constituted the bulk of the medical geography contributions from the 24th International Geographical Congress in Tokyo in 1980. A wide range of international case studies is included.

Learmonth, Andrew T.A. Patterns of disease and hunger: a study in medical geography. London; North Pomfret, VT: David and Charles, 1978. 256 p. ISBN 0-7153-7538-5. $27. LC 77-91741.   1435 •
This work is an excellent introduction to some of the central themes in medical geography. Included are discussions of infectious as well as chronic diseases and a chapter on the geography of hunger. Various case studies are also incorporated into this volume.

May, Jacques M. Ecology of human disease. New York: MD Publications, 1959.
327 p. LC 58-13432.   1436
*Discusses relationships between disease and geographic factors. Ecology of chol-
era, brucellosis, poliomyelitis, tuberculosis, leprosy, bacillary dysentery, salmonello-
sis, amebiasis, yaws, nematode infections, scarlet fever, measles, and trachoma.
Some maps are included.*

May, Jacques M., ed. Studies in disease ecology. (Studies in medical geography, vol.
2). New York: Hafner, 1961, 613 p. ISBN 0-02-848980-2. $27.95. LC 61-13002.   1437
*Ecological studies by fourteen contributors of the following diseases: smallpox,
dengue, filariasis, onchocerciasis, tularemia, relapsing fever, hydatidosis, malaria,
trypanosomiasis, schistosomiasis, leishmaniasis (kala-azar), scrub typhus, and
plague. Includes maps, diagrams, and tables.*

McGlashan, Neil D., ed. Medical geography: techniques and field studies. London:
Methuen, 1972. 336 p. ISBN 0-416-80480-2. £3.25. (pbk). LC 73-151622.   1438
*A collection of readings, some original and some reprinted from journal articles, in-
dicative of British approaches to medical geography during the 1960's. The chapters
consist of a wide range of subjects from disease atlases to diffusion processes.*

McGlashan, Neil D. and Blunden, John R., eds. Geographical aspects of health.
London; Orlando, FL: Academic Press, 1983. 392 p. ISBN 0-12-483780-8. $45.95.   1439
*This volume consists of a three-part work dedicated to the 30-year career of Andrew
Learmonth. The first, and most interesting part, consists of contributions from au-
thorities in various parts of the world. The remainder of the book should be read once
by beginning students.*

McNeill, William H. Plagues and peoples. Garden City, NY: Doubleday (Anchor
Press), 1976. 368 p. ISBN 0-385-11256-4. ISBN 0-385-12122-9. $5.50. (pbk). LC 76-2798.   1440
*Suggestive interpretation of the role of infectious diseases in human history, partic-
ularly in the conquest by Europeans of other peoples.*

Meade, Melinda S., ed. Conceptual and methodological issues in medical geography.
(Studies in geography no. 15). Chapel Hill, NC: University of North Carolina, Dept. of
Geography, 1980. 301 p. LC 80-620034.   1441
*Fourteen contributions on a wide range of approaches to medical geography indic-
ative of trends during the 1980's. Topics include river blindness, nutrition, ethnomedi-
cine, modeling chronic diseases, disease diffusion, and mental health.*

Phillips, David R. Contemporary issues in the geography of health care. Norwich,
U.K.: GEO Books; New York: State Mutual Book and Periodical Service, 1981. 191 p.
ISBN 0-86094-062-4. $25.   1442
*An interesting review of approaches to the geography of health care. British and
American health care systems are compared.*

Pyle, Gerald F. Applied medical geography. New York: Halsted Press, John Wiley
and Sons, 1979. 282 p. ISBN 0-470-26643-0. LC 78-27856.   1443
*Consists of a cross-sectional approach to medical geography as accomplished in
North America during the late 1960's and 1970's. Included are discussions of international disease classification systems, disease mapping, disease ecology in North
America, disease diffusion mechanisms, variable methods of association, improved
health care delivery, and geocoding applications.*

Pyle, Gerald F., ed. New directions in medical geography. Oxford, U.K.; New York:
Pergamon Press, 1980. 93 p. ISBN 0-08-025817-4. $25.   1444
*Includes medical geography papers from the 75th anniversary meeting of the Asso-
ciation of American Geographers held in Philadelphia in April, 1979. Contributions
include studies of sleeping sickness, morbidity surveys, cardiovascular mortality, and
attitudes toward the mentally ill.*

Medical Geography

Shannon, Gary W. and Dever, G.E. Alan. Health care delivery: spatial perspectives. New York: McGraw-Hill, 1974. 141 p. ISBN 0-07-056411-6. LC 73-11322.   1445
Serves as an introduction to the geography of health care. The health care delivery system within the United States is examined and some international comparisons are made.

Stamp, Sir Laurence Dudley. The Geography of life and death. Ithaca, NY: Cornell University Press, 1964. 160 p. LC 65-17544.   1446
An excellent, but dated, introduction to medical geography. The book includes short discussions of a number of major world diseases and makes good use of maps showing disease distribution and malnutrition patterns.

C. SPECIAL TOPICS

Cliff, Andrew D.; Haggett, Peter; Ord, J.K.; and Versey, G.R. Spatial diffusion: an historical geography of epidemics in an island community. (Cambridge geographical series no. 14). Cambridge, U.K.; New York: Cambridge University Press, 1981. 238 p. ISBN 0-521-22840-9. $47.50. LC 80-41329.   1447
Includes a review of spatial diffusion theories and how they help explain the geographical spread of infectious diseases. Long-term statistics on the incidence of measles in Iceland are used in empirical testing. An excellent synthesis of leading mathematical statements on disease diffusion is included.

Earickson, Robert. The Spatial behavior of hospital patients: a behavioral approach to spatial interaction in metropolitan Chicago. (Research paper no. 124). Chicago, IL: University of Chicago, Department of Geography, 1970. 138 p. ISBN 0-89065-031-4. $10. (pbk). LC 79-104877.   1448
An interesting comparison of patient travel to hospitals within a large American metropolitan area. Models of patient travel are tested and compared.

Gesler, Wilbert M. Health care in developing countries. (Resource publications in geography). Washington, DC: Association of American Geographers, 1984. 88 p. ISBN 0-89291-182-4. $5. LC 84-16890.   1449 •
A synthesis of various approaches to health care in developing nations in several parts of the world. Explains differences in both health problems and treatment philosophies in developing and developed countries and includes methods for evaluating treatment systems.

Prothero, R. Mansell (Ralph). Migrants and malaria in Africa. London: Longman, 1965. 142 p.; Pittsburgh, PA: University of Pittsburgh Press, 1965. 148 p. LC 68-12979. LC 66-85339.   1450
Study of population migration patterns in Africa and how they relate to the distribution and control of malaria. Well supported by tables and maps.

Pyle, Gerald F. Heart disease, cancer, and stroke in Chicago: a geographical analysis with facilities, plans for 1980. (Research paper no. 134). Chicago, IL: University of Chicago, Department of Geography, 1971. 292 p. ISBN 0-89065-041-1. $10. (pbk). LC 77-167941.   1451
Utilizes regression techniques in modeling; the future use of specialized treatment facilities. Suggestions are made for methods of locating health care facilities.

Stock, Robert F. Cholera in Africa: diffusion of the disease 1970-1975, with particular emphasis on West Africa. (African environment, special report no. 3). London: International African Institute, 1976. 127 p. ISBN 0-85302-050-7. £5.95. (pbk). LC 77-361351.   1452
Explains the diffusion of cholera in Africa during the early 1970's. Radial, linear, and multiple epicenter diffusion patterns are identified.

23. TECHNOLOGICAL HAZARDS

Roger E. Kasperson

See also Part III, Section 10, F. Nuclear Energy, entries 776-779.

Burton, Ian; Fowle, C.D.; and McCullough, R.S., eds. Living with risk: environmental risk management in Canada. (Environmental monograph series no. 3). Toronto, Canada: University of Toronto, Institute for Environmental Studies, 1982. 247 p.
A collection of 18 papers growing out of a series of study group meetings held at the Institute for Environmental Studies during the winter of 1980-81. Seven of the papers treat such thematic issues as the nature of risk, risk perception, decision-making for risk, and risk-taking. Eleven papers present case studies of such technological hazards as smoking; saccharin; 2,4,5-T; PCB's; and the transportation of gasoline.

1453

Crandall, Robert W. and Lave, Lester B., eds. The Scientific basis of health and safety regulations. Washington, DC: Brookings Institution, 1981. 309 p. ISBN 0-8157-1600-1. $28.95. ISBN 0-8157-1599-4. $11.95. (pbk). LC 81-10224.
With side-by-side chapters presenting the perspectives of the health scientist, the economist, and the regulator, this book examines five technological hazards--automobile crashes (and passive restraints), cotton dust, saccharin, waterborne carcinogens, and sulfur dioxide. The justification and adequacy of regulatory decisions are explored in each perspective.

1454

Douglas, Mary T. and Wildavsky, Aaron. Risk and culture: an essay on the selection of technological and environmental dangers. Berkeley, CA: University of California Press, 1982. 221 p. ISBN 0-520-04491-6. $14.95. LC 81-16318.
A controversial essay which argues that the American preoccupation with environmental risks can be understood only through cultural interpretation. Sectorial interests, in this view, have led to the American preoccupation with environmental threats, rather than the objective reality posed by such dangers. While this volume aims to understand the social forces which speak on behalf of environmental protection in America, the supporting arguments and documentation themselves suffer from selective bias.

1455

Hohenemser, Christoph, and Kasperson, Jeanne X., eds. Risk in technological society. (AAAS selected symposium 65). Boulder, CO: Westview Press for the American Association for the Advancement of Science, 1982. 339 p. ISBN 0-86531-316-4. $31.50. LC 81-14745.
This volume, developed out of a 1980 symposium, treats systematic issues in the structure of technological risk and public response to risks as well as the major failures represented by Three Mile Island and Love Canal. Particularly useful are the lengthy editorial introductions to the sections and the extensive bibliographic materials.

1456

International Atomic Energy Commission. Management of wastes from uranium mining and milling. Vienna, Austria: International Atomic Energy Agency, 1982. 735 p. ISBN 9-200-20282-9. LC 83-132235.
Overview of the worldwide status of management of radioactive tailings from the mining and milling of uranium ore. Background on the 'front-end' of the nuclear cycle. Papers submitted during a symposium held in Albuquerque, New Mexico in 1982.

1457

Kasperson, Roger E., ed. Equity issues in radioactive waste management. Cambridge, MA: Oelgeschlager, Gunn, and Hain, 1983. 381 p. ISBN 0-89946-055-0. $30. LC 81-18702.
An analysis of the equity issues posed by three clusters of risk and societal impacts: equity among places (the locus problem), equity among generations (the legacy problem), and equity between workers and publics (the labor/laity problem). The various chapters address both empirical and moral dimensions of these problems, concluding with overall recommendations for planning a socially acceptable management system.

1458

Technological Hazards

Kates, Robert W., ed. Managing technological hazard: research needs and opportunities. (Program on Technology, Environment, and Man, monograph no. 25). Boulder, CO: University of Colorado, Institute for Behavioral Science, 1977. 169 p. LC 77-17756.   1459
  An early effort to define the state of knowledge and needed research, funded by the National Science Foundation. The three workshops addressed hazard identification, risk estimation and evaluation, and risk communication and decision-making. Many of the needs identified have emerged as major areas of risk research in the subsequent eight years.

Kates, Robert W.; Hohenemser, Christoph; and Kasperson, Jeanne X., eds. Perilous progress: technology as hazard. Boulder, CO: Westview, 1984. 460 p. ISBN 0-8133-7025-6. $32.50. (pbk). LC 84-25623.   1460
  The culmination of a many-year trend of geographic research into hazards dealing first with environmental, then with technological hazards. Many papers deal with nuclear power.

Lagadec, Patrick. Major technological risk: an assessment of industrial disaster. Translated by H. Ostwald from French. Oxford, U.K.; New York: Pergamon Press, 1982. 516 p. ISBN 0-08-028913-4. $66. LC 82-3678. (Le risque technologique majeur, Paris: Pergamon Press France, 1981. 700 p.).   1461 •
  A broad-based, international analysis of industrial disasters which surveys experience over the past three centuries and examines in detail the cases of Flixborough, Seneso, Amoco Cadiz, Three Mile Island, and the Mississauga derailment. Subsequent chapters treat the potentials and limitations of various management options, the roles of regulators, operators, and citizens, and the handling of crises. An indispensable work on technological disasters.

National Research Council, Board on Radioactive Waste Management, Panel on Social and Economic Aspects of Radioactive Waste Management. Social and economic aspects of radioactive waste disposal: considerations for institutional management. Washington, DC: National Academy Press, 1984. 175 p. ISBN 0-309-03444-2. $14.50. (pbk). LC 84-60101.   1462
  With three geographers (Roger E. Kasperson, John E. Seley, and Julian Wolpert) on the panel, geographical aspects of radioactive waste disposal receive detailed attention. Of particular interest are chapters treating the waste management facility network and the site-related impacts of a waste repository. A series of maps in appendix A show the results of a simulation of the waste disposal system performed by Oak Ridge National Laboratory.

National Research Council, Committee on the Institutional Means for Assessment of Risks to Public Health. Risk assessment in the federal government: managing the process. Washington, DC: National Academy Press, 1983. 191 p. ISBN 0-309-03349-7. $11.75. (pbk). LC 83-80381.   1463
  An influential report which has stimulated new approaches to risk questions, particularly in the U.S. Environmental Protection Agency under Ruckelshaus. Emphasis is given to the inherent uncertainty in risk assessment, the need to choose among analytic methods, and the importance of making assumptions explicit. The report concludes that risk assessment and risk management should be sharply contrasted and that uniform risk assessment guidelines should be developed.

Openshaw, Stan; Steadman, Philip; and Greene, Owen. Doomsday: Britain after nuclear attack. Oxford, U.K.; New York: Basil Blackwell, 1983. 296 p. ISBN 0-631-13393-3. $24.95. ISBN 0-631-13394-1. $7.95. (pbk). LC 83-201237.   1464
  A largely geographic appraisal of man's potential impact on the earth. An important new avenue of geographic research which has developed out of earlier research into such nuclear topics as risk acceptability and evacuation behavior.

Pasqualetti, Martin J. and Pijawka, K. David, eds. Nuclear power: assessing and managing hazardous technology. Boulder, CO: Westview, 1984. 423 p. ISBN 0-86531-811-5. $31. (pbk). LC 83-21797.   1465 •

Brings together geographical contributions to the hazards and social impacts of, policy problems in, and public response to nuclear power. Of interest is not only the range of work presented, but the detailed analysis of several specialized problems, paticularly emergency response planning, socio-economic impact analysis, and waste disposal.

U.S. Congress, Office of Technology Assessment. Acid rain and transported air pollutants: implications for public policy. (OTA-0-204). Washington, DC: U.S. Government Printing Office, 1984. 323 p. $9.50. LC 84-601073.   1466

A searching analysis of the scientific evidence and uncertainties of this volatile risk problem as well as the major associated policy options. Of particular interest to geographers will be the discussion of the regional distribution of risk (chapter 5) and its entrance into policy (chapter 6) and legislative deliberations (chapter 7).

Zeigler, Donald J.; Johnson, James H., Jr.; and Brunn, Stanley D. Technological hazards. (Resource publications in geography (1983) no. 2). Washington, DC: Association of American Geographers, 1983. 103 p. ISBN 0-89291-173-5. $5. (pbk). LC 83-22356.   1467 •

An introductory essay on technological hazards which presents the variety of issues and concepts which have emerged in geographic research. Special attention is given to evacuation, an issue on which the authors have worked, and a chapter on "risk mosaics" which treats spatial characteristics of technological risks. A helpful bibliography is included.

PART V. APPLIED GEOGRAPHY

1. GENERAL APPLIED GEOGRAPHY

John W. Frazier

A. SERIALS

Applied geography: an international journal. 1- (1981-). Quarterly. Butterworth Scientific, Ltd., Westbury House, Bury Street, P.O. Box 63, Guildford GU2 5BH, U.K. ISSN 0143-6228. 1468
  Papers on a broad spectrum of topics in applied geography and the management of environmental resources.

Applied geography conferences. (edited by John W. Frazier and others). 1- (1977-). Annual. State University of New York at Binghamton, Department of Geography, Binghamton, NY 13901. ISSN 0192-8996. 1469
  Collection of edited papers from the first three annual conferences held at Binghamton, New York. Papers cover major topics of geography and include academic and private perspectives.

B. ARTICLES

Berry, Brian J.L. "Creating future geographies." Annals of the Association of American Geographers, vol. 70, no. 4, 1980, p. 449-458. 1470
  An important work that offers a future focus of geography. Provides applied geography with a key role. Gives an interpretation of the practitioner's world and suggests that fundamental geographic research should contribute to practice.

Epstein, Bart J. "Geography and the business of retail site evaluation and selection." Economic geography, vol. 47, no. 2, 1971, p. 192-199. 1471
  A comprehensive treatment of factors considered in retail site planning: economic, physical, and psychological. Excellent check list.

Frazier, John W. "Pragmatism: geography in the real world." in **Harvey, M.E. and Holly, B.P., eds.** Themes in geographic thought. London: Croom Helm, 1981. p. 61-72. 1472
  Describes applied geography pursuits within the framework of pragmatism. Uses the case study approach.

Green, H.L.; Applebaum, William; and Dupree, H. "When are store locations good? The case of the National Tea Company in Detroit." Professional geographer, vol. 30, no. 2, 1978, p. 162-167. 1473
  An important application of a theory of locational advantage. Demonstrates that entrepreneurial shortcomings negate a good location.

C. BOOKS

Applebaum, William. Guide to store location research with emphasis on supermarkets. Reading, MA: Addison-Wesley (sponsored by Super Market Institute, Inc.), 1968, 269 p. LC 68-55321. 1474
  Deals with conceptual, technical, and applied aspects of store location research. It provides a detailed description of methodologies and applications.

Applebaum, William. Hardware retailing strategy cases. Indianapolis, IN: Russell R. Mueller Retail Hardware Research Foundation, 1971. 103 p. NUC 72-10676. 1475
  Produced for hardware store decision makers, this volume provides the case study method. Case studies on relocation and acquisition decisions.

General Applied Geography

Applebaum, William. Shopping center strategy, a case study of the planning, location, and development of the Del Monte Center, Monterey, California. New York: International Council of Shopping Centers, 1970. 202 p. LC 75-139281.　　　　1476

A very unusual and worthwhile account of retail location and development. The book has three parts: a theoretical teaching case that presents development problems and state of the art solutions; a follow-up research that evaluates the actual development and appraises the validity of prior market analysis; and a reflection by the experts of parts one and two.

Burton, Ian, and Kates, Robert W., eds. Readings in resource management and conservation. Chicago, IL: University of Chicago Press, 1965. 609 p. ISBN 0-226-08237-7. $15. LC 65-14427.　　　　1477 •

An edited volume of readings that emphasizes the application of science for the better use of water and land.

Cohen, Saul B., ed. Store location research for the food industry. (National-American Wholesale Grocers Association special report). New York: National-American Wholesale Grocers Association (NAWGA), 1961. 131 p.　　　　1478

Research techniques appropriate for surveying market areas and for site evaluation of the food industry are systematically reviewed. These are a useful guideline for the beginner.

Davies, Ross L. and Rogers, David S., eds. Store location and store assessment research. New York; Chichester, U.K.: John Wiley and Sons, 1984. 375 p. ISBN 0-471-90381-7. $34.95. LC 83-21614.　　　　1479

Focuses on basic business geography and offers an applied perspective. The problem solving approach is attractive. Topics include retailing trends and patterns, impact of the developmental process, store location strategies, and store assessment and forecasting techniques.

Dawson, John A., ed. Retail geography. London: Croom Helm; New York: John Wiley and Sons, 1980. 248 p. ISBN 0-470-27014-4. $37.95. LC 81-453726.　　　　1480

A review of the literature that comprises retail geography. Much of the volume deals with conventional topics such as consumer behavior, retail organization, and locational analysis. It also contains a significant look at policies including those of retail price, business structure, and location.

Dawson, John A. Shopping centre development. (Topics in applied geography). London; New York: Longman, 1983. 124 p. ISBN 0-582-30068-1. $11.95. (pbk). LC 82-12674.　　　　1481

Provides an overview of existing shopping center types. While there is no applied research in this volume it is worthwhile to future practitioners because it not only covers locational and developmental processes, but identifies policy issues related to this industry's growth.

Dawson, John A. and Doornkamp, John C., eds. Evaluating the human environment: essays in applied geography. London: Edward Arnold, 1973. 288 p. ISBN 0-7131-5677-5. £6.95. LC 73-180991.　　　　1482

Covers a wide range of topics such as agriculture, water, urban land evaluation, marketing, air pollution, and health and recreation. It is an attempt to demonstrate the utility and relevance of geography to societal problems. Chapters stress human-environmental linkages and the value of the geographical perspective in identifying and solving problems.

Frazier, John W., ed. Applied geography: selected perspectives. Englewood Cliffs, NJ: Prentice-Hall, 1982. 333 p. ISBN 0-13-040451-9. $29.95. LC 81-10542.　　　　1483

An edited volume with thirteen chapters divided into four parts: historical and contemporary frameworks for applied geography defined as an applied science; a demonstration of applied geography in city and regional planning; an illustration of environmental applications; and dealing with problems and prospects for those wishing to become applied geographers. The case studies are good and plentiful. Bibliographic references are extensive.

Lea, David A.M., ed. Geographical research: application and relevance. Vancouver, B.C.: Pacific Science Congress, 13th, 1975. 85 p. ISBN 0-85834-109-3. LC 77-359839.   1484
A collection of nine essays rooted in a special session of the 13th Pacific Science Conference (1975). *They result from disciplinary introspection and raise critical questions regarding geography's role in solving problems in the developing world.*

Nelson, Richard L. The Selection of retail locations. New York: F.W. Dodge Corporation, 1958. 422 p. LC 58-10539.   1485
A description of methods utilized in the assessment of specific sites for various businesses. This also includes an analysis for planned shopping centers.

Sant, Morgan E.C. Applied geography: practice, problems, and prospects. London; New York: Longman, 1982. 152 p. ISBN 0-582-30040-1. $30. ISBN 0-582-30041-X. $11.95. (pbk). LC 80-41371.   1486
A personalized view of what contemporary applied geography is and should be. Strengths include the description of the frameworks of public policy and the factors that influence decision making. Sant illustrates well the importance of four public policy elements that have geographical dimensions: individual and group desires; resources; information; and institutional components.

Stamp, Sir Laurence Dudley. Applied geography. (Pelican Books A449). Harmondsworth, Middlesex, U.K.: Penguin Books, 1960. 207 p. LC 62-2158.   1487
Views the purpose of applied geography as applying geographical methods toward the understanding and interpretation of the real world. Presents the land classification of Britain and concludes that cartographic analysis can help solve problems while assisting in the progress of the human race.

White, Gilbert F. Choice of adjustment to floods. (Research paper no. 93). Chicago, IL: University of Chicago, Department of Geography, 1964. 150 p. LC 64-25664.   1488
Floodplain managers of property choose among eight possible adjustments to floods. This examination of decision maker's choices reveals factors affecting resource policy/management decisions.

White, Gilbert F. Human adjustment to floods: a geographical approach to the flood problem of the United States. (Research paper no. 29, reprint of 1942 dissertation). Chicago, IL: University of Chicago, Department of Geography, 1953. 225 p.   1489
Major emphasis on the nature and extent of flooding in the United States, the variation in human adjustments to flooding, and the public policy implications. Coverage is comprehensive.

Winters, Harold A. and Winters, Marjorie K., eds. Applications of geographic research. Viewpoint from Michigan State University. East Lansing, MI: Michigan State University, Department of Geography, 1977. 171 p. LC 77-78459.   1490
A collection of very broadly based viewpoints that suggests that geographic concepts and methods are widely applicable. While no applied research appears in this volume it does result in quite a catalogue of problems to which applied geographers can attempt to contribute solutions.

2. PLANNING

Kenneth E. Corey*

*Acknowledgement is made for suggestions from Howell S. Baum, Robert Beauregard, Brian J.L. Berry, Jean Gottmann, H. Briavel Holcomb, Donald A. Krueckeberg, Melvin R. Levin, Alvin R. McNeal, Frank Popper, and Mary Vance.

A. SERIALS

American Planning Association. Journal. (formerly American Institute of Planners. Journal). 1- (1925-). Quarterly. American Planning Association, 1313 E. 60th Street, Chicago, IL 60637. ISSN 0194-4363.　　　　　　　　　　　　　　　1491
　　The essential scholarly periodical in American planning. In addition to research articles, it includes reviews of recent planning books and tables of contents from selected periodicals in urban studies and planning.

Association of Collegiate Schools of Planning. Bulletin. 1- (1963-). Quarterly. Association of Collegiate Schools of Planning (ACSP), School of Urban Planning, Michigan State University, East Lansing, MI 48824. ISSN 0004-5675.　　　　　　　　1492
　　The periodical that tracks the developments of North American planning education. Includes short research articles.

Ekistics: reviews on the problems and science of human settlements. 1- (1955-). Bimonthly. Athens Center of Ekistics 24, Strat. Syndesmou Street, Box 471, Athens 136, Greece. ISSN 0013-2942.　　　　　　　　　　　　　　　　　　　1493
　　A journal that features small articles and digest articles on topics of international planning issues. Many issues are organized around a theme.

Environment and planning. A. International journal of urban and regional research. 1- (1969-). Monthly. Pion, 207 Brondesbury Park, London NW2 5JN, U.K. ISSN 0308-518X.　　　　　　　　　　　　　　　　　　　　　　　　　1494
　　Articles on urban and regional research, many by geographers, utilizing models and modern quantitative techniques.

Futures: the journal of forecasting and planning. 1- (1968-). Bimonthly. Butterworth Scientific, Ltd., Journals Division, P.O. Box 63, Westbury House, Bury Street, Guildford GU2 5BH, U.K. ISSN 0016-3287.　　　　　　　　　　　　　　1495
　　Treats future research from a scientific and systems analytic perspective. Leading futurist thinkers regularly contribute articles to this serial.

Planning. 1- (1972-). Monthly. American Planning Association, 1313 E. 60th Street, Chicago, IL 60637. ISSN 0001-2610.　　　　　　　　　　　　　　　　　1496•
　　A news and brief-article magazine for the North American professional and citizen planner. A brief book review and discussion of current comprehensive planning issues characterize the magazine.

Planner. 1- (1973-). Monthly. The Royal Town Planning Institute, 26 Portland Place, London W1 N 4BE, U.K. ISSN 0309-1384.　　　　　　　　　　　　　　1497
　　A periodical written for the British planning professional. Short planning practice articles are supplemented by news items on planning and the Royal Town Planning Institute *and its branches, and by book reviews.*

Third world planning review. 1- (1979-). Quarterly. Liverpool University Press, Box 147, Liverpool L69 3BX, U.K. ISSN 0142-7849.　　　　　　　　　　　　1498
　　Focuses on planning in developing countries. Contains research articles and book reviews of international planning literature on issues of development and underdevelopment at local and national scales.

Town planning review. 1- (1910-). Quarterly. Liverpool University Press, Box 147, Liverpool L69 3BX, U.K. ISSN 0041-0020. 1499
One of Britain's leading planning research journals. Consists principally of research articles and book reviews of current British and international planning literature.

Vance bibliographies. Vance, Mary, ed. 1- (1978-). Irregular and frequent. Vance Bibliographies, P.O. Box 229, Monticello, IL 61856. 1500
An essential reference series for planning researchers. Various authors produce many individual bibliographies on a wide range of planning topics.

B. BOOKS

Argyris, Chris, and Schon, Donald A. Theory in practice: increasing professional effectiveness. (Higher education series). San Francisco, CA: Jossey-Bass, 1974. 224 p. ISBN 0-87589-230-2. $16.95. LC 74-3606. 1501
Introduces an action framework and method for analyzing the planning theories which planners actually follow in practice. The framework distinguishes between espoused theories and theories in use.

Baum, Howell S. Planners and public expectations. Cambridge, MA: Schenkman, 1983. 307 p. ISBN 0-87073-634-5. $15.95. ISBN 0-87073-635-3. $9.95. (pbk). LC 82-16961. 1502
Examines planner's cognitive maps and their tacit models of the planning process. The research finds that, even though nearly all planners work in complex organizations, most planners think little about organizational issues or decision making when they work.

Benveniste, Guy. The Politics of expertise. 2nd ed. San Francisco, CA: Boyd and Fraser, 1977. 232 p. (1st ed., Berkeley, CA: Glendessary Press. 1972. 232 p.) ISBN 0-87835-067-5. $18.75. ISBN 0-87835-060-8. $9.95. (pbk). LC 77-9200. 1503
Treats the role of the expert in influencing public and private policy. Explains the limits of technocracy and cautions against over reliance on rationality. Value judgements and political commitments can contribute to defining the role of the planning expert in rapidly changing technological societies.

Benevolo, Leonardo. The History of the city. Translated from Italian by Geofrey Culverwell. London: Scolar Press; Cambridge, MA: MIT Press, 1980. 1011 p. ISBN 0-262-02146-3. $125. LC 79-90966. (Storia della città. 2nd ed. Roma: Laterza, 1976. 1008 p.) 1504•
This monumental work provides a basic history of the man-made environment found in cities from prehistory to today. It includes a large number of illustrations accompanied by brief text. European city history is stressed.

Bennis, Warren G.; Benne, Kenneth D.; Chin, Robert; and Corey, Kenneth E., eds. The Planning of change. 3rd ed. New York: Holt, Rinehart, and Winston, 1976. 517 p. (1st ed., 1961. 781 p.) ISBN 0-03-089518-9. $28.95. LC 75-41359. 1505
Since the publication of the first edition of this title it has been the standard in the planned change tradition of American applied behavioral science. The third edition has a planned community change theme.

Berry, Brian J.L. Comparative urbanization: divergent paths in the twentieth century. rev. and enl. 2nd ed. New York: St. Martin's Press; London: Macmillan, 1981. 235 p. ISBN 0-312-15475-5. $15.95. ISBN 0-312-15476-3. $10.95. (pbk). LC 82-61562. 1506
A revised and updated edition of The Human Consequences of Urbanization, *(1973. 205 p.). The point of this book is that urbanization has varied from culture to culture. Urbanization is examined in market economies, welfare states, and developing countries.*

Berry, Brian J.L. The Open housing question: race and housing in Chicago, 1966-1976. Cambridge, MA: Ballinger, 1979. 517 p. ISBN 0-88410-429-X. $35. LC 79-14912.   1507

Draws together diverse reports and research findings on a metropolitan Chicago area initiative by The Leadership Council for Metropolitan Open Communities *to test the effectiveness of a market system fair housing service. Over a six-year period, beginning in 1968, Berry evaluates whether this was a useful social experiment.*

Berry, Brian J.L. and Silverman, Lester P., eds. Population redistribution and public policy. Washington, DC: National Academy Press, 1980. 351 p. ISBN 0-309-02926-0. $14.50. LC 79-25533.   1508

A collection of papers the purpose of which is to examine the effects of U.S. population redistribution on public services, institutions, and policy, and the effect and role of public policy changes on population redistribution that have taken place since 1970.

Boyer, M. Christine. Dreaming the rational city: the myth of American city planning. Cambridge, MA: MIT Press, 1983. 331 p. ISBN 0-262-02186-2. $27.50. LC 83-5402.   1509

A view of city planning and its mentality from the late nineteenth century to the Second World War. Explores the quest for order in the American city. Questions whether planning is a field and whether planners have developed an identity as a profession. Provides an answer to the question, "What impact has planning had on American cities?"

Caro, Robert A. The Power broker: Robert Moses and the fall of New York. New York: Vintage Books, 1975. 1246 p. ISBN 0-394-72024-5. $14.95. LC 75-9557.   1510

The Pulitzer Prize winning book on the role that power and personality can play in shaping a city. It provides the reader with extraordinary insights into the backrooms of the political machine.

Carr, James H. and Duensing, Edward, eds. Land use issues of the 1980's. New Brunswick, NJ: Rutgers University, Center for Urban Policy Research, 1983. 325 p. ISBN 0-88285-085-7. $12.95. LC 83-1888.   1511

This collection of readings explores the factors that impact on development choices throughout the 1980's. The readings are organized under the main headings of: factors affecting land use demand; institutional controls on the supply of land; modifying land use regulation; and future land use considerations. The book includes a useful bibliography on the subject.

Catanese, Anthony J. and Snyder, James C., eds. Introduction to urban planning. New York: McGraw-Hill, 1979. 354 p. ISBN 0-07-010228-7. $31.50. LC 78-13275.   1512

A collection of readings by thirteen contributors compiled to provide an introductory university level survey of the physical elements of the built environment. The treatment is historical and cultural, theoretical and methodological, and concerned with functional planning practices.

Chapin, F. Stuart, Jr. (Francis), and Kaiser, Edward J. Urban land use planning. 3rd ed. Urbana, IL: University of Illinois Press, 1979. 565 p. (1st ed., New York: Harper, 1957, 397 p.) ISBN 0-252-00580-5. $22.50. ISBN 0-252-00791-3. (wkbk). $6.95. LC 64-18666.   1513

A standard in U.S. urban land use planning. Core planning is treated conceptually, by the basic studies required to develop land use plans, and by approaches to plan formulation and evaluation.

Clavel, Pierre; Forester, John; and Goldsmith, William W., eds. Urban and regional planning in an age of austerity. (Policy studies in urban affairs). Oxford, U.K.; New York: Pergamon Press, 1980. 391 p. ISBN 0-08-025539-6. $46. ISBN 0-08-025540-X. $8. (pbk). LC 79-21416.   1514

A collection of papers from a conference on radical planning theory and practice for progressive planners. Offers political alternatives for U.S. planners. Includes several essays on Marxist and socialist interpretations of U.S. planning issues.

Faludi, Andreas, ed. Planning theory. Oxford, U.K.; New York: Pergamon Press, 1973. 399 p. ISBN 0-08-017741-7. $30. ISBN 0-08-017756-5. $12.75. (pbk). LC 73-11236.   1515
Contains a selection of some of the most influential articles in planning thought through the 1960's. The introduction outlining planning theory is a useful opening to this collection.

Friedmann, John, and Weaver, Clyde. Territory and function: the evolution of regional planning. London: Edward Arnold; Berkeley, CA: University of California Press, 1979. 234 p. ISBN 0-520-03928-9. $36. ISBN 0-520-04105-4. $7.95. (pbk). LC 80-453989.   1516
Regional planning, little practiced in the United States, receives treatment in this book as doctrine rather than practice. Basic needs and self help are stressed. A paradigm shift in regional planning is discussed.

Gottmann, Jean. The Coming of the transactional city. (Monograph series no. 2; occasional papers in geography no. 5). College Park, MD: University of Maryland, Institute for Urban Studies, 1983. 106 p. ISBN 0-913749-001. $12. (pbk). LC 83-50449.   1517
Examines the impact of the rise of employment in services and information occupations in the contemporary metropolis. An extremely valuable aid to practicing planners. Complete bibliography of Jean Gottmann's publications is included.

Hall, Peter G. Urban and regional planning. New ed. Harmondsworth, U.K.: Penguin Books, 1980. 312 p. (1st ed., 1974. 312 p.) ISBN 0-14-080399-8. £5.95. LC 75-323656.   1518
A British view of the evolution of modern planning. The book includes a chapter each on planning in Western Europe and the United States. Spatial planning is the central theme of the book.

Hamblin, Roger, ed. Guide to graduate education in urban and regional planning. 3rd ed. Atlanta, GA: Association of Collegiate Schools of Planning, 1978. 401 p. ISBN 0-918286-15-8. $5. LC 78-71242.   1519•
Ideal for the prospective graduate planning student. Eighty-three graduate programs in the U.S., Canada, and one from England are described. An essential information resource for those interested in American planning education.

Hanson, Royce, ed. Rethinking urban policy: urban development in an advanced economy. Washington, DC: National Academy Press, 1983. 215 p. ISBN 0-309-03426-4. $16.50. LC 83-19422.   1520
A compilation by an eminent committee of scholars that addresses the interaction of urban policy and the new urban system being created by the growing service economy. Key themes include: economic strategy, investing private and public capital, investing in the urban labor force, stabilizing metropolitan economies, and fostering local institutions to manage the transition.

Holcomb, H. Briavel, and Beauregard, Robert A. Revitalizing cities. (Resource publications in geography series). Washington, DC: Association of American Geographers, 1982. 84 p. ISBN 0-89291-148-4. $5. (pbk). LC 81-69237.   1521•
A treatment of the North American city as a place where decline can be and has been arrested. Includes a geographic perspective, a historical perspective, the causes of revitalization, and an incorporation of a social justice perspective.

Jacobs, Allan B. Making city planning work. Chicago, IL: American Society of Planning Officials, 1978. 323 p. ISBN 0-918286-12-3. $15.95. (pbk). LC 78-72577.   1522•
Documents what it is like to be planning director of San Francisco. It also includes typical case studies of the issues faced by a big American city planning department. The personal insights by Jacobs provide the reader with a unique view inside the world of public politics and professional planning.

Planning

Krueckeberg, Donald A., ed. The American planner: biographies and recollections. New York: Methuen, 1983. 433 p. ISBN 0-416-33360-5. $29.95. LC 82-6461.   1523 •
A history of American urban planning through the eyes of selected planners: John Nolen, Edward Bassett, Alfred Bettmann, Charles Dyer Norton, Benton MacKaye, Henry Wright, Rexford Guy Tugwell, Harland Bartholomew, Coleman Woodbury, Charles Abrams, and others.

Krueckeberg, Donald A., ed. Introduction to planning history in the United States. New Brunswick, NJ: Rutgers University, Center for Urban Policy Research, 1983. 302 p. ISBN 0-88285-083-0. LC 82-14572.   1524 •
For the beginning student in urban planning. Treats American planning history from 1840 to the present. A bibliography of U.S. planning history is included.

Krueckeberg, Donald A. and Silvers, Arthur L. Urban planning analysis: methods and models. New York: John Wiley and Sons, 1974. 486 p. ISBN 0-471-50858-6. $37.50. LC 74-7087.   1525
An introduction to contemporary planning methods and planning models. The principal themes are planning process, information, decision models, and an extensive range of quantitative methods for the planner. It includes a useful set of appendices for the beginning student of quantitative methods.

Levin, Melvin R. Community and regional planning: issues in public policy. 3rd ed. New York: Praeger, 1977. 278 p. (1st ed., 1969. 305 p.) ISBN 0-275-23690-0. LC 76-12862.   1526
A series of domestic planning issues from the mid-1970's is treated in a hard-hitting well-written manner. Principal themes include the position of planners and the planning profession, the role of government in planning, the gap between elites and the populace, and the need for foresight, intelligence, and innovation in planning.

Perloff, Harvey S. Planning the post-industrial city. Chicago, IL: Planners Press, 1980. 327 p. ISBN 0-918286-21-2. $25.95. LC 80-67753.   1527
One of the last major statements by the late dean of American planning education. Addresses the need for the practice of planning to change so as to reflect the transformation associated with the shift to a service economy. Represents a linking of rational comprehensive planning with the needs to develop a new paradigm for planning. U.S. case studies of recent planning experiments enrich this volume.

Popper, Frank J. The Politics of land use reform. Madison, WI: University of Wisconsin Press, 1981. 321 p. ISBN 0-299-08530-9. $25. ISBN 0-299-08534-1. $8.50. (pbk). LC 80-23255.   1528
Examines the U.S. land-use reform movement. Assesses the effects of state land use programs as cases of centralized regulation. Written from the perspective of an environmentalist, it treats land-use reform as a case study of the Nader approach to regulation.

Porteous, J. Douglas. Environment and behavior: planning and everyday urban life. Reading, MA: Addison-Wesley, 1977. 446 p. ISBN 0-201-05867-7. $22.95. LC 76-1752.   1529 •
A comprehensive overview of the multidisciplinary field of urban man-environment relations. The principal treatments of the book include: behavior, environment, and planning. Written to be understood by all readers.

Scott, Mel. American city planning since 1890. (California studies in urbanization and environmental design). Berkeley, CA: University of California Press, 1969. 745 p. ISBN 0-520-02051-0. $14.95. (pbk). LC 70-84533.   1530
A comprehensive treatment of American city planning history from the late nineteenth century into the 1960's. A standard in the field.

Slater, David C. Management of local planning. (Municipal management series). Washington, DC: International City Management Association, 1984. 288 p. ISBN 0-87326-031-7. $34. ISBN 0-87326-032-5. $28. (pbk). LC 84-6577.　　　　1531

Offers the reader ways to incorporate planning into the management of U.S. local governments. Material is based on interviews with experienced planners and managers from throughout the U.S. In addition to planners, managers and students, members of boards, and commissions and councils will benefit from reading this volume.

So, Frank S.; Stollman, Israel; Beal, Frank; and Arnold, David S., eds. The Practice of local government planning. (Municipal management series). Washington, DC: International City Management Association, 1979. 676 p. ISBN 0-87326-020-1. $37. LC 79-21380.　　　　1532

The latest in the unofficial series of "green books" published over the years on planning in local government. A "must" reference for the planner's library. Chapters cover the principal functions and procedures of contemporary U.S. local-scaled planning practice.

Van Lierop, Wal F.J. and Nijkamp, Peter, eds. Locational developments and urban planning. (NATO Advanced Study Institute series). Rockville, MD: Sijthoff and Noordhoff, 1981, 531 p. ISBN 90-286-2651-4.　　　　1533

Gives comparative international information on residential mobility shifts and current and future urban policy issues including new theories and methods for analyzing changes in urban locational patterns.

Weaver, Clifford L. and Babcock, Richard F. City zoning: the once and future frontier. Chicago, IL: Planners Press, 1980. 328 p. ISBN 0-918286-17-4. $18.95. LC 79-90347.　　　　1534•

A readable treatment of the current and possible system of land use regulation in the U.S. The focus is on city land use and the style is pragmatic, prescriptive, and fresh.

C. SERVICE

Planning advisory service. American Planning Association, 1313 E. 60th Street, Chicago, IL 60637.　　　　1535

Planning Advisory Service (PAS) subscribers receive eight to ten technical planning reports each year. In addition, PAS staff will assist in the development of zoning and subdivision codes, give advice, provide peer exchange, and document and mail the Planning Advisory Service Memo *monthy. A valuable service for planning practitioners and librarians.*

3. GEOGRAPHY IN EDUCATION

Joseph P. Stoltman

A. BIBLIOGRAPHIES

Ball, John M. Bibliography for geographic education. 2nd ed. Athens GA: University of Georgia, Geography Curriculum Project, 1969. 109 p. (1st ed., 1968. 92 p.). LC 77-631753.　　　　1536

Annotated listings particularly for curriculum and for teaching methods. Emphasis is on American materials.

Lukehurst, Clare Therese, and Graves, Norman J. Geography in education. A bibliography of British sources 1870-1970. Sheffield, U.K.: Geographical Association., 1972. 86 p. ISBN 0-900395-42-7. LC 72-197205.　　　　1537

More than 1400 unannotated entries arranged by broad subjects: geography; nature and evolution; the curriculum, content, and objectives of geography; mental development in relation to geography teaching; teaching methods (by far the longest section); geography teaching overseas; geography and international understanding; and testing in geography.

Geography in Education

Sperling, Walter. Geographiedidaktische Quellenkunde. Internationale Basisbibliographie und Einführung in die wissenschaftlichen Hilfsmittel (Ende des 17. Jahrhunderts bis 1978). (Beiheft 4 zum BIB-Report). Duisburg, West Germany: Verlag für Pädagogische Dokumentation, 1978. 897 p. (ISSN 0342-0582). DM 102. LC 78-40933.   1538
More than 4275 numbered annotated entries in 93 classes, which are in alphabetical order, including many devoted to the teaching of geography, pedagogy, teaching aids, and bibliographies. Extensive textual introduction to the field of geography teaching and large section of sample pages from basic sources. Covers the entire period of modern geography from the end of the seventeenth century to 1978. Emphasis is on German periodicals.

Sperling, Walter. Geographieunterricht und Landschaftslehre. Sachstandsbericht und bibliographisches Handbuch 1968 bis 1979-80. (Beihefte zum BIB-Report). Duisburg, West Germany: Verlag für Pädagogische Dokumentation, 1981-1984. 5 Vols. (ISSN 0342-0582).   1539
Covers the German language literature on geography teaching 1968 to about 1980. Includes both systematic fields of geography and regional geography as well as many aspects of geography instruction. Headings for each section are in German, English, French, and Russian. Vols. 1-2 cover the teaching of geography and study of landscapes and regions in general. Vols. 3-4 are devoted to the literature on professional preparation and professional work materials for teaching geography. Vol. 5 is an index.

B. SERIALS

Association of American Geographers. Resource papers. 1-28 (1968-1974); **Resource papers for college geography,** 1975-1978: Four each year; **Resource publications in geography,** 1981- . Four each year but not numbered. Association of American Geographers, 1710 16th Street N.W., Washington, DC 20009. ISSN 0066-9369.   1540 •
Brief analyses on the entire range of concepts and theoretical developments in the discipline. Valuable for instructional background and reading materials for beginning and advanced students. An important instructional resource.

Geographical education. 1- (1969-). Annual. Australian Geography Teachers Association, Geographical Society of New South Wales, Gloucester Street, Sydney, 2000, Australia. ISSN 0085-0969.   1541
Presents research and development articles in geographical education. Most of the entries are from Australia but occasionally New Zealand, Western Europe, and North America are represented. Items are of high scholarship.

Journal of geography. 1- (1902-). 7 per annum. National Council for Geographic Education, Western Illinois University, Macomb, IL 61455. ISSN 0022-1341.   1542 •
Particularly for teachers of elementary, secondary, and college geography. Short articles on geography and teaching. Numerous notes and reviews. Cumulative indexes: 1-69 (1902-1970), 71-75 (1972-1976).

Note: Also of value are the leading journals for geographical education in Britain, France, and Germany: Geography (41), Information géographique, 1- (1936-), and Geographische Rundschau, 1- (1949-).

C. BOOKS

Bale, John, ed. The Third World: issues and approaches. Sheffield, U.K.: The Geographical Association, 79 p. 1983. ISBN 0-900395-84-2. £3.45. (pbk).   1543
Focuses principally on the way in which issues germane to the developing nations of the world may be identified and integrated into the curriculum. Articles focusing on such issues are followed by syllabuses, teaching ideas, and resources pertinent to geographical education.

Bartlett, Leo, and Cox, Bernard. Learning to teach geography: practical workshops in geographical education. Brisbane, Australia: John Wiley and Sons, 1982. 292 p. ISBN 0-471-33384-0.　　　　1544

Provides the pre-service and in-service teacher with an opportunity to address important issues in geographical education. Issue related tasks are to be completed either by the individual or as a team member. Issues presented include content selection, curriculum decisions, and classroom approaches. Practical examples for the geography classroom are paramount.

Biddle, Donald S. and Deer, Christine E., eds. Readings in geographical education: selections from Australian and New Zealand sources. Sydney, Australia: Whitcombe and Tombs for the Australian Geographical Teachers Association, 1973. 2 v. 462 p. ISBN 0-7233-5281-X. LC 71-383416.　　　　1545

An excellent and representative selection of papers which first appeared in serials and focus upon geographical education. Arrangement is within the following headings: purpose; curriculum and development; systems; teaching strategies in school; teaching strategies in the field; and testing and evaluation. A key item in geographical education.

Chorley, Richard J. and Haggett, Peter., eds. Frontiers in geographical teaching. 2nd ed. London: Methuen, 1970. 385 p. (1st ed., 1965. 378 p.) LC 76-503388.　　　　1546

A classic set of papers prepared for pre-service and in-service teachers during an extension course at Cambridge University. Reflects creative and projective views of geography which continue to prevail in the 1980's. In three parts. The first two review concepts and techniques in research geography whereas the third addresses geographic education at the secondary and tertiary levels in the United Kingdom.

Cole, Ann, and Axeman, Lois. Children are children are children: an activity approach to exploring Brazil, France, Iran, Japan, Nigeria, and the U.S.S.R. Boston, MA: Little, Brown, 1978. 212 p. ISBN 0-316-15114-9. $11.95. ISBN 0-316-15113-0. $7.95. (pbk).　　　　1547•

Describes the life and customs of Brazil, France, Iran, Nigeria, Japan, and the Soviet Union with suggested related activities and projects.

Fien, John; Gerber, Rodney; and Wilson, Peter, eds. The Geography teacher's guide to the classroom. Melbourne, Australia: The Macmillan Company of Australia Pty Ltd., 1984. 316 p. ISBN 0-333-35660-8.　　　　1548

Provides an excellent perspective upon theory and practice in geographical education. Directed principally toward pre-service geography teachers studying in colleges and universities. Provides numerous examples of the way in which content preparation, educational theory, and classroom instruction in the school interact with one another. A valuable source of information and ideas for geographical educators.

Gerber, Rodney, and Biddle, Donald S., eds. Geographical issues. Brisbane, Australia: Australian Geography Teachers Associations, 1980. 175 p. ISBN 0-909867-28-3.　　　　1549

Instructional resource materials in the form of well-written background papers on physical geography. The edited papers focus on three aspects of physical geography in a teaching context: perceptions of the earth's surface; changes in the earth's surface; and studying physical geography through a systems approach.

Graves, Norman J., ed. Computer assisted learning in geographical education. London, U.K.: University of London. Institute of Education, 1984. 146 p. ISBN 0-85473-188-1.　　　　1550

The papers in this volume result from a Conference on Computer Assisted Learning sponsored by the Commission on Geographical Education, International Geographical Union. A review of the state-of-the-art development of computer assisted learning software in geography using microcomputers. Excellent references to geography software are contained in the bibliography.

Geography in Education

Graves, Norman J., ed. Geography in education. 3rd ed. London: Heinemann Educational Books Ltd., 1984. 246 p. (1st ed., 1975. 232 p.) ISBN 0-435-35315-2. £5.95. (pbk). LC 81-101808.

1551

An important work analyzing the role of geography in education as it has been affected by the numerous changes in the discipline, as well as those changes which have altered the field of education during the past two decades. A synthesis of the ways in which geography fits into and benefits the present school curriculum is a key element of the book.

Graves, Norman J., ed. New UNESCO source book for geography teaching. (Commission on Geographical Education of the International Geographical Union). Harlow, Essex: Longman Group Limited; Paris: UNESCO Press, 1982. 394 p. ISBN 0-582-36122-2. $17.95. LC 83-169878.

1552

The international emergence of school geography reflecting the theories and principles concerning the spatial aspects of human behavior is a major goal underlying the development of this book. New concepts reflecting changes in the discipline are presented. The contributing authors have selected significant ideas and woven them into a highly useful, yet comprehensive format for geographers teaching at all levels of education.

Graves, Norman J., ed. Research and research methods in geographical education. London: University of London Institute of Education, 1984. 221 p. ISBN 0-85473-187-3.

1553

Papers of the Commission on Geographical Education, International Geographical Union, to report the progress made by researchers in ten different nations.

Haubrich, Hartwig, ed. International focus on geographical education. (Studien zur internationalen Schulbuchforschung, v. 34). Braunschweig, West Germany: Georg-Eckert-Institut für internationale Schulbuchforschung, 1982. 292 p. ISBN 3-88304-234-X.

1554

Descriptive papers prepared by geographical educators from thirty different nations reporting upon the development and state of geography instruction. Provides the best single comparative account of geography teaching internationally, but lacks a concise synthesis by the editor.

Horst, Oscar H. and Stoltman, Joseph P. New themes in instruction for Latin American geography. (Special publications of the conference of Latin Americanist Geographers). Muncie, IN: Conference of Latin Americanist Geographers, 1982. 116 p. OCLC 88-00645.

1555

Focuses on contemporary issues in the teaching of geography. A collection of papers from a conference held in 1981 at which representatives from several organizations, including the International Geographical Union and the Latin American Studies Association, exchanged ideas regarding instructional materials design and classroom implementation in Latin American geography courses.

Huckle, John, ed. Geographical education: reflection and action. Oxford, U.K.: Oxford University Press, 1983. 156 p. ISBN 0-19-913281-X. £4.50. (pbk).

1556

Commissioned essays written for secondary school teachers. Focuses on recent developments in geography. The central thesis is that geography curriculum should be restructured to emphasize a more constructive contribution to human development and social justice.

Joint Committee on Geographic Education. Guidelines for geographic education: elementary and secondary schools. Washington, DC: Association of American Geographers; Macomb, IL: National Council for Geographic Education, 1984. 26 p. ISBN 0-89291-185-9. $3. (pbk). LC 84-21706.

1557 •

Presents the case for improved geographic education in the U.S. Defines five fundamental themes for geographic education and suggests how these can be implemented in each grade level in the schools through learning outcomes. Demonstrates the importance of geographic learning and shows the relationship of geography to other subjects and to the preparation of careers.

London. University. Institute of Education. Handbook for geography teachers, prepared by the Standing Committee in Geography, M. Long, gen. ed. 6th ed. London: Methuen, 1974. 724 p. (1st ed., 1932 ed. by D.M. Forsaith; 2nd ed., 1955 ed. by G. Cons). ISBN 0-423-88830-7. LC 75-312586.   1558

Valuable and detailed collaborative work with sections devoted to the teaching of geography and syllabuses of work; outdoor geography; indoor geography; visual aids; geographical societies; book list for the primary stage; book list for the secondary stage; book list for teachers and sixth forms; geography in literature, exploration, and travel; official publications; and addresses of publishers. Although written primarily for use in Great Britain, it has much material of value to teachers of geography in other countries.

Manson, Gary A. and Ridd, Merrill K., eds. New perspectives on geographical education: putting theory into practice. Dubuque, IA: Kendall-Hunt, 1977. 214 p. ISBN 0-8403-1782-4. $8.95. (pbk). LC 77-82416.   1559

Addresses questions on what to teach and how to teach geography at the precollegiate level. The contributions by the different authors do not provide simple answers to these complex questions, but rather offer an array of ideas related to content and method for geographic education.

Marsden, William E., ed. Evaluating the geography curriculum. Edinburgh, U.K.; New York: Oliver and Boyd, 1976. 312 p. ISBN 0-05-002900-2. LC 76-103.   1560

Probably the most substantive treatment of educational evaluation as it applies to the geography curriculum. Treatment begins by addressing theory and practice, continues through aims and objectives, places geography as a field of learning in the school curriculum, provides a thorough discussion of assessment forms in geography, and ends with a synthesis of geography as a school subject.

Marsden, William E., ed. Teacher education models in geography: an international comparison. Kalamazoo, MI: Western Michigan University, 1984. 155 p. ISBN 0-912244-17-8.   1561

Presents twelve different models for geography teacher training representing different nations. The degree of similarity among many of the models is striking. Differences between the models result primarily from pre-teacher training studies in geography which several models incorporate.

Natoli, Salvatore J. et al., eds. Experiences in inquiry: HSGP and SRSS. Boston, MA: Allyn and Bacon, 1974. 267 p. LC 73-87822.   1562

Presents exemplary learning activities selected from the High School Geography Project (HSGP) and Sociological Resources for the Social Studies (SRSS), both of which were national secondary school curriculum projects during the 1960's. The activities represent those judged to be especially appropriate for teaching concepts from geography and sociology.

Slater, Frances. Learning through geography. London: Heinemann Educational Books, 1982. 147 p. ISBN 0-435-35715-8. £5.95. (pbk).   1563

Focuses on issues such as waste disposal, environmental impact, landscape evaluation, and skill development in geography. Examples of both scientific and humanistic approaches to geography are included. An important resource for in-service and pre-service teachers of geography.

Stoltman, Joseph P., ed. International research in geographic education: spatial studies development in children and teacher classroom style in geography. Kalamazoo, MI: Western Michigan University, Department of Geography, 1976. 248 p. LC 76-372270.   1564

Two sets of empirical research reports are contained in this volume. The first set addresses the spatial stages development of elementary school-age children using a Piagetian model. The second set reports on the various teacher classroom styles used in teaching geography at the secondary level using a directive-nondirective pedagogical scale. Reports from researchers in several different nations are included.

Walford, Rex, ed. Signposts for geography teachers. (Charney Manor Conference papers, 1980). Harlow, Essex: Longman Group Limited, 1981. 222 p. ISBN 0-582-35334-3. £8.75. ISBN 0-582-35335-1. £4.95. (pbk). LC 82-177446.　　　　　　　　　1565

Emphasis is on the future for geography teaching based upon an analysis of the recent past. Does not provide a survey of current teaching practices, but instead indicates new avenues which are developing for classroom geography. The three parts of the book describe the new directions to which the signposts of geography teaching are pointing.

Winston, Barbara J. Map and globe skills: K-8 teaching guide. (Topics in Geography no. 7). Macomb, IL: National Council for Geographic Education, 1984. 47 p.　　　1566

Addresses the importance of sequence in designing a map. In addition, the globe skills component of grades K-8 curriculum is discussed. Specific skills and sequences, along with suggested instructional guides are included.

4. MILITARY GEOGRAPHY

Harold A. Winters

A. BIBLIOGRAPHY

Peltier, Louis C. Bibliography of military geography. Washington, DC: Association of American Geographers, 1962. 76 p. NUC 65-55488.　　　　　　　　　　　1567

Represents the most complete published bibliography of pre-1960 books and articles on military grography written in English. Also contains some references in foreign languages, mainly German and French.

B. ATLASES

Esposito, Vincent J. The West Point atlas of American wars. New York: Praeger, 1957. 2 v. LC 59-7452.　　　　　　　　　　　　　　　　　　　　　　　1568•

Contains 412 maps of military operations from the American Revolutionary War through the Korean War accompanied by a concise and useful descriptive text.

Kidron, Michael, and Smith, Dan. War atlas. New York: Simon and Schuster, 1983. ISBN 0-671-472-49-6. $19.95. LC 83-573.　　　　　　　　　　　　　　1569•

A small but current thought-provoking atlas of the present world situation regarding the military and related fields.

Nebenzahl, Kenneth, and Higginbotham, Don. Atlas of the American Revolution. Chicago, IL: Rand McNally, 1974. 218 p. ISBN 0-528-83465-7. LC 74-6976.　　　1570•

An annotated atlas of the American Revolution. Maps are reproductions of contemporary maps, and are tactical rather than strategic.

U.S. War Department. The Official military atlas of the Civil War. New York: Arno Press, 1978. 175 colored plates. ISBN 0-405-11198-3. LC 78-16801.　　　　　1571•

Reprint of the 1891-1895 edition published under the title Atlas to Accompany the Official Records of the Union and Confederate Armies. *The most detailed atlas yet published on the Civil War. Consists of reproductions of maps compiled by both Union and Confederate soldiers.*

C. BOOKS

Clausewitz, Carl von. On war. Translated from German and edited by Michael Howard and Peter Paret. Princeton, NJ: Princeton University Press, 1976. 717 p. ISBN 0-691-05657-9. $31. LC 75-30190. (Vom Kriege. Berlin: F. Dümmler, 1832-34. 3 v.). 1572
Famous treatise on military theory, strategy, and tactics. Also addresses key geographical concepts in war.

Cline, Ray S. World power trends and U.S. foreign policy for the 1980's. Boulder, CO: Westview, 1980. 228 p. ISBN 0-89158-917-1. $23. ISBN 0-89158-790-X. $11.50. (pbk). LC 79-26790. 1573
A futuristic analysis of geopolitics and military problems with special reference to the United States. Contains many geographical dimensions.

Cohen, Saul B. Geography and politics in a world divided. 2nd ed. New York: Oxford University Press, 1973. 334 p. (1st ed., New York: Random House, 1963. 347 p.) ISBN 0-19-501695-5. $10.95. (pbk). LC 73-77923. 1574
Geopolitical view of the world that is relevant to both geography and the military.

Dorpolen, Andreas. The World of General Haushofer: geopolitics in action. New York: Farrar and Rinehart, 1942. 337 p. Reprinted: Port Washington, NY: Associated Faculty Press, 1966. 337 p. ISBN 0-8046-0112-7. $25. LC 66-21393. 1575
Analysis of the controversial and subjective application of geopolitics.

Dupuy, T. N. (Trevor Nevitt). Numbers, predictions, and war: using history to evaluate combat factors and predict the outcome of battles. Indianapolis, IN: Bobbs-Merrill, 1979. 244 p. ISBN 0-672-52131-8. LC 77-5243. 1576
Standard work on the use of the quantitative results of recent conflicts to construct computer models of the patterns of possible future conflicts.

Earle, Edward Mead, ed. Makers of modern strategy: military thought from Machiavelli to Hitler. (Princeton paperbacks, 260). 2nd ed. Princeton, NJ: Princeton University Press, 1973. 553 p. (1st ed., 1943. 553 p.) ISBN 0-691-06907-7. $55. ISBN 0-691-01853-7. $12.50. (pbk). NUC 75-105362. 1577
Summarizes and analyzes the treatises of major military theorists through World War II. Geographically pertinent. Reprint of the 1943 edition.

Garver, J.B., Jr. and Galloway, G.E., Jr. Readings in military geography. West Point, NY: U.S. Military Academy, 1984. 444 p. 1578
Excellent up-to-date collection of thirty-eight articles relating to various aspects of military geography.

Goetzmann, William H. Army exploration in the American West, 1803-1863. Lincoln, NE: University of Nebraska, 1979. 489 p. ISBN 0-8032-2013-7. $29.50. ISBN 0-8032-7003-8. $7.95. (pbk). LC 79-14764. 1579
Historical review of exploration in the American West that gives insight into the importance of exploration, mapping, and knowledge of topography to the Army and the nation.

Gray, Colin S. The Geopolitics of the nuclear era: heartland, rimlands, and the technological revolution. (Strategy paper no. 30). New York: Crane, Russak, and Co., 1977. 67 p. ISBN 0-8448-1258-7. $4.95. (pbk). LC 77-83666. 1580
Geopolitical approach to Soviet power and the heartland as compared to the United States rimland maritime alliance. Short but timely. Thought provoking.

Jeffries, William W., ed. Geography and national power. 4th ed. Annapolis, MD: U.S. Naval Academy, Department of English, History, and Government, 1967. 184 p. (1st ed., 1953. 100 p.) LC 68-24. 1581
Outlines the impact of geography and geopolitics on national policies and power.

Military Geography

Johnson, Douglas W. Battlefields of the World War. (American Geographical Society. Research series no. 3). New York: American Geographical Society; London: Oxford University Press, 1921. 648 p.   1582
  Dated but excellent analysis of terrain factors and the development of battle in western and southern Europe during WWI.

Jomini, Henri. Jomini's Art of war. Translated from French by G.H. Mendell and W.P. Craighill. Philadelphia, PA: J.B. Lippincott, 1862. 410 p. (Reprint: Westport, CT: Greenwood, 1971. 410 p.) ISBN 0-8371-5014-0. $35. LC 68-54793. (Précis de l'art de la guerre. Paris, France: Anselin, 1838. 2 v.).   1583
  Interesting contrast to Clausewitz that puts heavy emphasis on geography and military operations.

Lucas, James Sidney. War on the Eastern Front: 1941-1945: the German soldier in Russia. London: Jane's, 1979; New York: Bonanza Books, 1982. 214 p. ISBN 0-354-01255-X. ISBN 0-517-38285-7. LC 82-4228.   1584 •
  A well-written and graphic portrayal of Operation Barbarossa, the German invasion of the USSR in World War II. In addition to reviewing strategy and duties, Lucas addresses geographical as well as political considerations in planning. Recounts the effects of terrain and climate, mud and dust, and ice and snow on soldiers and machinery from diaries, interviews, etc.

Marshall, S. L. (Samuel). Battles in the monsoon: campaigning in the central highlands, Vietnam, summer 1966. New York: William Morrow, 1967. 408 p. (Reprint: Vietnam War series no. 5. New York: William Morrow and Co., 1984. 408 p.) ISBN 0-89839-075-3. $18.95. LC 67-15157.   1585 •
  Review of campaigns in the central highlands of Vietnam that has geographical implications.

Nettesheim, Daniel Dick. Topographical intelligence and the American Civil War. Fort Leavenworth, KS: U.S. Army Command and General Staff College, 1978. 101 p.   1586
  Master of Military Art and Science thesis dealing with Civil War history, maps, and military topography.

O'Sullivan, Patrick, and Miller, Jesse W., Jr. The Geography of warfare. London: Croom Helm; New York: St. Martin's Press, 1983. 172 p. ISBN 0-312-32276-3. $19.95. LC 82-42771.   1587
  A study of the impact of geography on strategy, operations, tactics, logistics, and intelligence. Historical examples and contemporary studies are used to develop general principles on the impact of terrain, political and economic geography, and technological change on the patterns of war.

Peltier, Louis C. and Pearcy, G. Etzel (George). Military geography. Princeton, NJ: Van Nostrand, 1966. 176 p. LC 66-4413.   1588
  Describes the utility of geography in combat and presents the geography of political-military relations. Somewhat dated.

Progress Publishers, ed. Marxism-Leninism on war and army: a Soviet view. (Soviet military thought, no. 2). Translated from Russian by Donald Danemanis. Washington, DC: U.S. Government Printing Office, 1974. 335 p. ISBN 0-8464-0613-6. $17.95. LC 75-600605.   1589
  Soviet view of war and the use of crimes and military power involving some geographical dimensions.

Tzu, Sun. The Art of war. Translated from Chinese by Samuel B. Griffith. Edited by James Clavell. New York: Delacorte Press, 1983. 82 p. ISBN 0-440-00243-5. $8.95. LC 82-19939.   1590
  Sun Tzu's 6th century B.C. treatise addressing military theory including the application of geography. This first organized and written theory on war is still studied at military schools around the world.

PART VI. REGIONAL GEOGRAPHY

1. THE ANCIENT AND MEDIEVAL WORLD

Richard W. Stephenson

Avery, Catherine B., ed. The New century handbook of classical geography. New York: Appleton-Century-Crofts, 1972. 362 p. ISBN 0-390-66930-X. LC 78-189006.
   Gazetteer of classical place names. Most entries derived from The new century classical handbook.
1591

Bunbury, Edward H. A History of ancient geography among the Greeks and Romans, from the earliest ages till the fall of the Roman Empire. 2nd. ed. London: J. Murray, 1883. 2 v. Reprinted: Atlantic Heights, NJ: Humanities, 1979. 2 v. (1st ed., 1879). ISBN 90-7026-511-7. $185.25. LC 60-749.
   Detailed account of geography among the Greeks and Romans. Basic work on the classical period of geography. Includes useful map reconstructions of the known world as conceived by classical scholars.
1592

Cary, Max. The Geographic background of Greek and Roman history. Oxford, U.K.: Clarendon Press, 1949. 331 p. Reprinted: Westport, CT: Greenwood Press, 1981. 331 p. ISBN 0-313-23187-7. $37.50. LC 81-7170.
   Valuable study of the geography of the ancient world.
1593

Dicks, D.R., ed. The Geographical fragments of Hipparchus. (University of London classical studies, 1). London: Athlone Press, 1960. 214 p. LC 60-2886.
   Traces the "Life and works" of Hipparchus, "Greek geography up to the time of Hipparchus," "Hipparchus as geographer," "The arrangement of the geographical fragments," and "The value of the stade." This is followed by the Greek text of Hipparchus and the English translation.
1594

Dion, Roger. Aspects politiques de la géographie antique. Paris: Les Belles Lettres, 1977. 305 p. French book code 5102348. FFr 145. LC 78-388304
   An unusual and interesting reinterpretation of classic Greek sources of geographical information, including The Iliad, The Odyssey, *and* The travels of Pytheas.
1595

Fabre, Paul. Les Grecs et la connaissance de l'Occident. Le mythe Occidental. Paris: Université, 1977. 2 v.: 362; 280 p. (Available from Services de Reproduction des Thèses. Université de Lille III, Lille, France).
   Detailed survey of Greek geographical knowledge of the "West," both literary and historical. (originally presented as author's thesis, Université de Paris I, 20 June, 1977).
1596

Grant, Michael. Ancient history atlas. New York: Macmillan, 1972. 87 p. Reprinted: Long Island City, NY: S.J. Dorst, 1981. 87 maps. ISBN 0-915262-73-8. $10. (pbk). LC 73-654430.
   Clear, concise atlas of the classical world consisting of 87 black and white maps.
1597•

Hammond, Nicholas G.L. Atlas of the Greek and Roman world in antiquity. Park Ridge, NJ: Noyes Press, 1981. 56 p. ISBN 0-8155-5060-X. $48. LC-81-675203.
   A well-executed atlas containing maps ranging from the Neolithic age to the sixth century A.D. More than 10,000 sites are depicted on the maps and listed in the gazetteer at the end of the volume. Each map was compiled by a scholar with specialized knowledge of the area or subject.
1598•

Ancient and Medieval World

Heidel, William A. The Frame of ancient Greek maps: with a discussion of the discovery of the sphericity of the Earth. (Research series no. 20). New York: American Geographical Society, 1937. 141 p. Reprinted: New York: Arno Press, 1976. 141 p. (History of ideas in ancient Greece: an Arno Press collection). ISBN 0-405-07312-7. $9.00. LC 75-13271.   1599

A valuable, well-footnoted study on early Greek cartography.

Kimble, George H. T. Geography in the middle ages. London: Methuen, 1938. 272 p. Reprinted: New York: Russell and Russell, 1968. 272 p. LC 68-10930.   1600

A scholarly survey; invaluable for its listing of medieval texts, as well as Arab works.

Matthew, Donald. Atlas of medieval Europe. (Culture atlas series). New York: Facts on File, 1983. 240 p. ISBN 0-87196-133-4. $35. LC 82-675303.   1601 •

Valuable and attractive atlas of the Middle Ages containing 53 colored maps, many colored photographs, a chronological table, and text that is designed "to set the visual assets of the atlas in context."

Russell, Josiah C. Medieval regions and their cities. Newton Abbot, U.K.: David and Charles, William Dawson; Bloomington, IN: Indiana University Press, 1972. ISBN 0-7153-5278-4. £6. ISBN 0-253-33735. LC 72-178144; LC 70-172025.   1602

An interesting "study of late medieval regionalism, an examination of the spatial distribution of cities, a discussion of other regional characteristics, and an explanation of the possible reasons for the development of regions at that time."

Stillwell, Richard, ed.; MacDonald, William L., associate ed.; and McAllister, Marian H. The Princeton encyclopedia of classical sites. Princeton, NJ: Princeton University Press, 1976. 1019 p. ISBN 0-691-03542-3. $175. LC 75-30210.   1603

Important gazetteer of place names from the classical period. Names keyed to 24 page map section at end of volume.

Warmington, Eric H. Greek geography. London; Toronto: J.M. Dent and Sons; New York: E.P. Dutton, 1934. 269 p. Reprinted: New York: AMS Press, 1973. 269 p. (The library of Greek thought, vol. 7). ISBN 0-404-07805-2. $21.50. LC 70-177849.   1604

Fine source book for the geographical concepts of early Greek writers, translated into English. Arranged into four parts: I. Cosmology and the older ideas of climatology, geology, and physical geography; dawn of scientific geography; II. Climatology; physical and political geography; III. Exploration and growth of knowledge; descriptive or topographic geography; IV. Mathematical geography with cartography.

Wright, John K. The Geographical lore of the time of the Crusades: a study in the history of medieval science and tradition in Western Europe. (Research series no. 15). New York: American Geographical Society, 1925. 563 p. Reprinted: New York: Dover; Gordon Press, 1965. 563 p. ISBN 0-8490-0217-6. $59.95. LC 65-12262.   1605

European science and beliefs in the period 1100-1250. Principally historical, but relates geography to classical and medieval thought, religion, and science. An excellent demonstration of how a first-rate historian of geography handles basic documents. The Dover edition includes new introduction by Clarence J. Glacken.

2. NORTH AMERICA OR ANGLO-AMERICA

A. ATLASES   RICHARD W. STEPHENSON AND WILBUR ZELINSKY

National Geographic Society. Atlas of North America: space-age portrait of a continent. Washington, DC: National Geographic Society, 1985. 256 p. ISBN 0-87044-606-1. $29.95.ISBN 0-87044-607-X. $39.95. (deluxe).   1606•
Portrays North America (Canada, United States, Mexico, Central America, and West Indies) on a regional basis through map coverage, textual description, and more than 100 remote sensing images, almost all in color. Most remote sensing scenes are accompanied by instructive locator maps. In addition to political and physical maps, thematic maps illustrate continental tectonics, climate, energy and mineral resources, population and land use, environmental stress points, and transportation networks. Metropolitan maps and texts portray 82 major cities, with four pages of maps and text describing major parks. Includes thirteen regional travel guide sections listing more than 1000 attractions and a special guide to viewing the earth from space.

Oxford University Press, Cartographic Department. Oxford regional economic atlas: the United States and Canada. Advisory editors: John D. Chapman and John C. Sherman. 2nd ed. London; New York: Oxford University Press, 1975. 128 p. (1st ed., 1967. 128 p.). ISBN 0-19-894308-3. $14.95. (pbk). LC 76-362820.   1607•
Includes plans of larger cities, regional relief maps, and many aspects of physical geography, demography, agriculture, forestry, fishing, fuels and energy, mining and manufacturing, and transport. Gazetteer index locates about 10,000 places by latitude and longitude.

Rooney, John F., Jr.; Zelinsky, Wilbur; and Louder, Dean R., eds. This remarkable continent: an atlas of United States and Canadian society and culture. College Station, TX: Texas A and M Press, 1982. 316 p. ISBN 0-89096-111-5. $45. LC 80-6113.   1608•
This remarkable publication includes 387 maps grouped under 13 headings: (1) General cultural and popular regions; (2) Settlement patterns; (3) Division of the land; (4) Structures; (5) Social organizations and behavior; (6) Language and place names; (7) Ethnicity; (8) Religion; (9) Politics; (10) Foodways; (11) Music and dance; (12) Sports and games; and (13) Place perception. The text accompanying the maps-13 autonomous and generally effective essays-provides orientation and commentary.

B. GENERAL WORKS   WILBUR ZELINSKY

Watson, J. Wreford (James). North America: its countries and regions. 2nd rev. ed. London: Longmans; New York: F.A. Praeger, 1967. 881 p. (1st ed., 1963. 854 p.). LC 67-21758.   1609
Topical and regional presentation by a perceptive British geographer. Devotes more space to social and cultural items than is customary in most North American texts. Brief chapter on Mexico included.

White, C. Langdon (Charles); Foscue, Edwin J.; and McKnight, Tom L. Regional geography of Anglo-America. 6th ed. Englewood Cliffs, NJ: Prentice-Hall, 1985. 640 p. (1st ed., 1943. 898 p.). ISBN 0-13-770892-0. $35.95. LC-78-23338.   1610•
Richly detailed, durable standard text with a regional approach.

C. PHYSICAL GEOGRAPHY   HAROLD A. WINTERS

Bryson, Reid A. and Hare, F. Kenneth, eds. Climates of North America. (World survey of climatology, Vol. 11., H.E. Landsburg, gen. ed.). New York: American Elsevier Publishing Co., 1974. 420 p. ISBN 0-444-41062-7. $112.75. LC 78-477739.   1611
Authoritative text, extensive tables, and many maps. Bibliography with 355 references.

North America

Braun, Emma Lucy. Deciduous forests of eastern North America. Philadelphia, PA: Blakiston, 1950. 596 p. Reprinted: New York: Hafner (Macmillan), 1967. 596 p. ISBN 0-0-02-841910-3. $31.95. LC 64-20220.
    Detailed and classic study of eastern North America's vegetation with special emphasis on the central Appalachians.
1612

Daubenmire, Rexford F. Plant geography with special reference to North America. New York: Academic Press, 1978. 338 p. ISBN 0-12-204150-X. $45. LC 77-75570.
    Floristic and ecologic aspects of plants and their distribution in North America. An excellent reference.
1613

Hunt, Charles B. Natural Regions of the United States and Canada. San Francisco, CA: W.H. Freeman, 1974. 725 p. ISBN 0-7167-0255-X. $29.95. LC 75-3760.
    Regional approach to physical environmental systems.
1614•

King, Philip B. The Evolution of North America. Rev. ed. Princeton, NJ: Princeton University Press, 1977. 197 p. (1st ed., 1959. 189 p.). ISBN 0-691-08195-6. $44. LC 77-71987.
    A very well-written summary of geologic evolution of North America.
1615

Shelford, Victor E. The Ecology of North America. Urbana, IL: University of Illinois Press, 1963. Reprinted: 1978. 610 p. LC 63-7255.
    A detailed and well-referenced intermediate to advanced textbook stressing the ecology and biogeography of North America. About 1000 references.
1616

Vale, Thomas R. Plants and people: vegetation change in North America. (Resource publications in geography). Washington, DC: Association of American Geographers, 1982. 88 p. ISBN 0-89291-151-4. $5. LC 82-8865.
    A timely and well-documented essay focusing on various forms of inadvertant or purposeful alteration of North American vegetation (including major sections on grazing, logging, and fire suppression) placed in a series of alternating perspectives representing several ecological "schools of thought" on patterns of vegetation recovery. An up-to-date bibliography of 481 references is particularly useful.
1617•

Vankat, John L. The Natural vegetation of North America: an introduction. New York: John Wiley and Sons, 1979. 261 p. ISBN 0-471-01770-1. $16.95. LC 78-31264.
    A text about North American vegetation patterns that is easy to understand for introductory students yet contains sufficient depth and detail for advanced students. Six chapters dealing with basic concepts and seven chapters describing major vegetation formations are presented. The environment, structure, distribution, adaptations, and human interactions of each are examined. Many maps, photographs, and other figures; sources at chapter ends.
1618

D. HUMAN GEOGRAPHY   WILBUR ZELINSKY

Coan, Otis W. and Lillard, Richard G. America in fiction; an annotated list of novels that interpret aspects of life in the United States, Canada, and Mexico. 5th ed. Palo Alto, CA: Pacific Press, 1967. 232 p. (1st ed., Stanford, CA: Stanford University Press, 1941. 180 p.). LC 66-28118.
    Useful guide to American life as revealed by novelists. Arranged by area, period, and subject.
1619•

Driver, Harold E. Indians of North America. 2nd rev. ed. Chicago, IL: University of Chicago Press, 1969. 667 p. (1st ed., 1961. 667 p.). ISBN 0-226-16466-7. $25. LC 79-76207.
    A comprehensive comparative description and analysis of Native American culture from the Arctic to Panama. An authoritative treatment that takes up in turn each of a score or more of major topics.
1620•

Garreau, Joel. The Nine nations of North America. Boston, MA: Houghton Mifflin, 1981. 427 p. ISBN 0-395-29124. $14.95. LC 80-28556. 1621•
A successful, highly readable and semi-popular account by a journalist of "the way North America really works"--at least in regional terms. Garreau maps, describes, and interprets nine semi-autonomous "nations" within the United States, Canada, Mexico, and the Caribbean that cohere socially, culturally, and economically with little regard for state, provincial, or other political boundaries: in essence a functional-vernacular geography.

Sealock, Richard B.; Sealock, Margaret M.; and Powell, Margaret S. Bibliography of place name literature, United States and Canada. 3rd. ed. Chicago, IL: American Library Association, 1982. 435 p. (1st ed., Sealock, Richard B. and Seely, Pauline A., 1948. 331 p.). ISBN 0-8389-0360-6. $30. LC 81-22878. 1622•
The 4830 entries are arranged by state and province in this definitive bibliography. Author and subject indexes.

Sturtevant, William C., gen. ed. Handbook of North American Indians. Washington, DC: Smithsonian Institution, 1978- . 20 v. (6 v. as of 1985). LC 77-17162.
    Vol. 5: **Arctic.** Damas, David, ed. 1985. 862 p. ISBN 0-87474-185-8. $29.
    Vol. 6: **Subarctic.** Helm, June, ed. 1981. 837 p. ISBN 0-87474-186-6. $25.
    Vol. 8: **California.** Heizer, Robert F., ed. 1978. 800 p. ISBN 0-87474-188-2. $25.
    Vol. 9: **Southwest.** Ortiz, Alfonso, ed. 1980. 701 p. ISBN 0-87474-189-0. $23.
    Vol. 10: **Southwest.** Ortiz, Alfonso, ed. 1983. 868 p. ISBN 0-87474-190-4. $25.
    Vol. 15: **Northeast.** Trigger, Bruce G., ed. 1979. 924 p. ISBN 0-87474-195-5. $27. 1623
This monumental project is providing us with the most thorough and authoritative account one could ask for concerning the aboriginal populations of the United States, Canada, and northern Mexico before and after European conquest. Volumes 1 to 4 and 16 to 19 deal with general and topical items; Volumes 5 to 15 cover individual regions; and Volume 20 will index the others. Each volume contains numerous chapters by specialists.

Yeates, Maurice, and Garner, Barry. The North American city. 3rd ed. New York: Harper and Row, 1980. 557 p. (1st ed., 1971. 563 p.). ISBN 0-06-047334-7. $31.95. LC 80-10703. 1624
Probably the most substantial text on the subject in the economic-positivistic mode. Such topics as systems of cities, their external relations, and internal structure and operation receive thorough treatment, but there is scant attention to social and cultural issues.

Yeates, Maurice. North American Urban patterns. London: Edward Arnold, 1980. 169 p. $30.95. LC 80-17708. 1625
Definitions and descriptions of the major urban regions of North America, the economic changes that have occurred since 1920, explanations for the changes, and some public policy issues that require discussion.

E. HISTORICAL GEOGRAPHY   ROBERT D. MITCHELL

Gibson, James R., ed. European settlement and development in North America: essays on geographic change in honour and memory of Andrew Hill Clark. Toronto; Buffalo, NY: University of Toronto Press, 1978. 231 p. ISBN 0-8020-3357-1. $8.50. (pbk). LC 78-8335. 1626
Contains a series of essays on the settling of North America in honor of America's leading historical geographer. Topics include Russia and France in North America, early Pennsylvania, American cultural regions, southern rice production, the fur trade in early Canada, ethnic settlements in the great plains, Victorian cities, and appreciations of Andrew Clark.

United States

Modelski, Andrew M. Railroad maps of North America: the first hundred years. Washington, DC: Library of Congress, 1984. 186 p. ISBN 0-8444-0396-2. $28. LC 82-675134.

1627 •

This excellent atlas contains reproductions of 92 maps selected from the 5000 railroad maps preserved in the collections of the Library of Congress. Each map includes descriptive text and bibliographical notes and an introductory essay traces the history of railroad mapping in North America.

Sauer, Carl O. Northern Mists. Berkeley, CA: University of California Press, 1968. 204 p. Reprinted: Berkeley, CA: Turtle Island Foundation, 1973. 216 p. ISBN 0-913666-00-9. $3.50. (pbk). LC 68-16757.

1628

Describes European exploration of the north Atlantic between the sixth and fifteenth centuries, before Columbus's voyages. Focus on Irish monks in the exploration of Iceland, Greenland, and possibly Newfoundland and New England; the Vikings; British sailors; and Portuguese fishermen and navigators.

Sauer, Carl O. Sixteenth-century North America: the land and the people as seen by the Europeans. Berkeley; Los Angeles, CA: University of California Press, 1971. 319 p. ISBN 0-520-01854-0. $31. LC 75-138635.

1629

Reconstructs the pre-European geography of North America based on explorers' accounts of the physical environment and Indian cultures. Focus on the Atlantic coast, Florida, Spanish involvement in the interior and Far West, and the English settlements at the end of the century.

Sauer, Carl O. Seventeenth-century North America. Berkeley, CA: Turtle Island Foundation, 1980. 295 p. ISBN 0-91-3666-23-8. $19.95. LC 77-82309

1630

This work, published posthumously, focuses on French and Spanish activities in the seventeenth century. The emphasis is on the relationship between the Europeans on the one hand and the flora, fauna, and peoples of North America on the other.

3. THE UNITED STATES

Note: For geographical serials see the index and especially entries 29, 39, 49.

A. BIBLIOGRAPHIES   WILBUR ZELINSKY

Freidel, Frank B., ed. Harvard guide to American history. Rev. ed. Cambridge, MA: Belknap Press of Harvard University Press, 1974. 2 v. 1290 p. (1st ed. edited by Handlin, Oscar and others, 1954. 689 p.). ISBN 0-674-37560-2. $60. LC 72-81272.

1631

Many thousands of citations of books and articles (without annotations) of special value to students of American history--and geography--arranged by area, topic, and period. A basic bibliographic tool.

Harris, Chauncy D. Bibliography of geography. Part 2. Regional. Volume 1. The United States of America. (Reasearch paper no. 206). Chicago, IL: University of Chicago, Department of Geography, 1984. 178 p. ISBN 0-89065-086-1. $10. (pbk). LC 76-1910.

1632 •

A list of 974 publications dealing solely or substantially with the geography of the United States. Many entries are accompanied by descriptive annotations, and there is a strong emphasis throughout on bibliographic materials. In this uniquely valuable guide, every significant aspect of American geography is covered. Entries arranged by topic and region.

Post, Joyce A. and Post, Jeremiah B. Travel in the United States: a guide to information sources. (Geography and travel information guide series, vol. 3). Detroit, MI: Gale Research, 1981. 578 p. ISBN 0-8103-1423-1. $55. LC 81-4375.

1633 •

The most useful and comprehensive volume of its kind; some 2000 annotated references for the traveler, arranged by area and category, covering books, atlases, magazines, maps, publishers, tourist organizations, and state parks and forests.

U.S. Library of Congress, General References and Bibliography Division. A Guide to the study of the United States of America; representative books reflecting the development of American life and thought. Washington, DC: Government Printing Office, 1960. 1193 p. LC 60-60009.

1634

A selection of 6487 titles, liberally annotated, that afford an excellent introduction to every form of inquiry concerning the United States. Of special value are chapters 6 and 12, dealing with the geography and local history.

U.S. Library of Congress, General References and Bibliography Division. Basler, Roy P., ed. A Guide to the study of the United States of America: representative books reflecting the development of American life and thought. Supplement, 1956-1965. Washington, DC: Library of Congress, 1970. 526 p. ISBN 0-8444-0164-1. LC 60-60009.

1635

This supplement provides 2943 additional titles published over the span of one decade. The arrangement is by subject.

Vinge, Clarence L., and Vinge, Ada G. United States Government publications for research and teaching in geography and related social and natural sciences. Totowa, NJ: Littlefield, Adams & Co., 1967. 360 p. LC 67-8346.

1636

Identifies out of more than half a million publications issued by the Federal Government of the United States in the period 1945-1966 about 3500 considered to be of special value to geographers. Arranged by departments and bureaus.

B. ATLASES   RICHARD W. STEPHENSON AND WILBUR ZELINSKY

Abler, Ronald F. and Adams, John S., eds. A Comparative atlas of America's great cities: twenty metropolitan regions. (AAG, Comparative metropolitan analysis project). Minneapolis, MN: University of Minnesota Press, 1976. 503 p. ISBN 0-8166-0753-2. $95. LC 76-14268.

1637•

Uniformly designed maps for 20 major metropolitan areas depicting, as of 1970, the details of relief, land use, population characteristics, socio-economic attributes, transportation and communication, employment and poverty, segregation, and urban renewal and redevelopment. A major tool for studying the anatomy and physiology of large American cities.

Arbingast, Stanley A., et al. Atlas of Texas. 5th ed. Austin, TX: University of Texas at Austin, Bureau of Business Research, 1976. 179 p. (1st ed., Texas resources and industries: selected maps of distribution, 1955. 42 maps). ISBN 0-87755-261-4. $29.95. LC 76-24780.

1638•

Revision of popular state atlas. "This revised edition...includes many areas of information not covered in previous editions, notably a completely new section of cultural and historical maps."

Arkansas Department of Planning. Atlas of Arkansas. Little Rock, AR: Department of Planning, 1973. 99 p. LC 74-621314.

1639•

"This document was designed as a general thematic atlas of the State of Arkansas. It was perceived and developed to serve the needs of legislators, state officials, educators, students, interested citizens, and members of the business community... Each map is supported by a brief statement and/or table."

Caldwell, Harry H. Idaho economic atlas. Moscow, ID: Idaho Bureau of Mines and Geology, 1970. 82 p. LC 70-654385.

1640•

"Though this volume is entitled an economic atlas, it also deals with the physical and cultural environment of the state." Atlas is designed primarily for use in schools.

United States

Cappon, Lester J.; Petchenik, Barbara Bartz; and Long, John H., eds. Atlas of
early American history: the revolutionary era, 1760-1790. Princeton, NJ: Princeton University Press, 1976. 157 p. ISBN 0-691-04634-4. $200. LC 75-2982.
1641•
    *A major contribution to the cartographic depiction of geographic phenomena in
early America. The atlas contains seventy-four pages of maps showing aspects of Indian groupings, late colonial population, settlement, roads, economic and cultural
activities, as well as military activities during the Revolutionary War, and post-war
conditions.*

Clay, James W.; Orr, Douglas M. Jr.; and Stuart, Alfred W. North Carolina atlas:
portrait of a changing southern state. Chapel Hill, NC: University of North Carolina
Press, 1975. 331 p. ISBN 0-8078-1244-7. $8.95. LC 75-6984.
1642•
    *Thematic atlas divided into six parts: introduction; human settlement and profile;
physical resources and environmental quality; the economy; services and amenities;
and retrospect and prospect.*

Collins, Charles W. An Atlas of Wisconsin. 2nd ed. Madison, WI: American Printing
and Publishing, 1972. 187 p. (1st ed., 1968. 191p.). LC 73-159553.
1643•
    *Revised edition of a thematic atlas first published in 1968. Maps are arranged in six
sections: population; physical geography; climate; agriculture; industry; and conservation-tourism.*

Cross, Ralph D., ed.; Wales, Robert W.; and Traylor, Charles T. Atlas of Mississippi. Jackson, MS: University Press of Mississippi, 1974. 187 p. ISBN 0-87805-061-2.
LC 74-78569.
1644•
    *Thematic atlas divided into eight sections: physical features; history; population
and social institutions; natural resources and resource utilization; transportation and
communication; manufacturing; services; and Mississippi: future perspective.*

Donley, Michael W.; Allan, Stuart; Caro, Patricia; and Patton, Clyde P. Atlas of
California. Portland, OR: Professional Book Center, Inc., 1979. 191 p. ISBN 0-943226-02-3. $29.95. LC 79-84439.
1645•
    *Important thematic atlas, attractively executed, displaying "the distributions and
concentrations of the physical, human, and economic elements which make up the
state."*

Durrenburger, Robert W. and Johnson, Robert B. California: patterns on the land.
5th ed. Palo Alto, CA: Mayfield Publishing Co., 1976. 134 p. (1st ed., Los Angeles:
Brewster Publishing Co., Atlas and Maps Division, 1957. 59 p.). ISBN 0-87484-385-5.
LC 78-318665.
1646•
    "With the fifth edition of California: patterns on the land, *The California Council
for Geographic Education assumes the responsibility for the continuing revision of
this atlas."*

Economic atlas of Nebraska. Lawson, Merlin P., project director, and Lonsdale,
Richard E. Lincoln, NE: University of Nebraska Press, 1977. 165 p. ISBN 0-8032-0911-8. $18.95. LC 76-30887.
1647•
    *"The organization of the atlas into eleven parts generally reflects the volume's purpose. Within the eleven parts approximately 60 different topics are discussed, accompanied by 117 maps and graphs and 51 tables." This is one of a series of three atlases
published in 1977 by the University of Nebraska Press. The other two are the* Climatic
atlas of Nebraska *and* Agricultural atlas of Nebraska.

Facts on File, Inc. State maps on file. New York: Facts on File, Inc., 1984. 7 v. ISBN 0-87196-894. $250 for the set. LC 84-675108.
    Vol. 1: **New England.** ISBN 0-8160-0117-0. $55.
    Vol. 2: **Mid-Atlantic.** ISBN 0-8160-0118-9. $55.
    Vol. 3: **Southeast.** ISBN 0-8160-0119-7. $55.
    Vol. 4: **Midwest.** ISBN 0-8160-0120-0. $55.
    Vol. 5: **Mountain & Prairie.** ISBN 0-8160-0121-9. $55.
    Vol. 6: **Southwest.** ISBN 0-8160-0122-7. $55.
    Vol. 7: **West.** ISBN 0-8160-123-5. $55.
    1648 •
    This work consists of simple black-and-white maps of each state depicting such topics as counties and county seats, congressional districts, major rivers and waterways, climate, industry, population density, etc.

Goodman, James Marion. The Navajo atlas. Norman, OK: University of Oklahoma Press, 1982. 109 p. ISBN 0-8061-1621-8. $22.50. LC 81-40287.
    1649 •
    Graphic representation of Navajo resources, demography, history, and political organization. Each map is accompanied by informative background text.

Goodman, Lowell R. and Eidem, R.J. The Atlas of North Dakota. Fargo, ND: North Dakota Studies, 1976. 111 p. LC 77-352606.
    1650 •
    Thematic atlas divided into six sections: introduction; physical; settlement and population; social and cultural; economic; and administrative.

Hartman, Charles W. and Johnson, Philip R. Environmental atlas of Alaska. 2nd ed. Seattle, WA: University of Washington Press, 1978. 95 p. ISBN 0-295-96186-4. $22.50. LC 72-626165.
    1651 •
    "This atlas gives an overall picture of many aspects of physical Alaska...Much of the material in this atlas was obtained directly from published sources."

Highsmith, Richard M. Jr. and Kimerling, A. Jon, eds. Atlas of the Pacific Northwest. 6th ed. Corvallis, OR: Oregon State University Press, 1979. 135 p. (1st ed., 1953. 118 p.). ISBN 0-87071-408-2. $20. ISBN 0-87071-409-0. $10. (pbk). LC 80-100716.
    1652 •
    Popular regional atlas. "This edition is essentially all new. It is a reflection of improved cartographic technology as well as the dynamics of regional change."

Hogan, Edward P.; Opheim, Lee A.; and Zieske, Scott H. Atlas of South Dakota. Dubuque, IA: Kendall-Hunt Publishing Co., 1970. 137 p. ISBN 0-8403-0210-X. LC 71-654017.
    1653 •
    "This atlas was developed as a work atlas for those individuals concerned with the physical, cultural, and economic aspects of South Dakota. It is generally designed to serve the needs of students, educators, state officials, interested citizens, visitors to the states, and members of the business community."

Kahrl, William L., ed. The California water atlas. Sacramento, CA: The Governors Office of Planning and Research, (distributed by William Kaufmann), 1979. 118 p. ISBN 0-913232-68-8. LC 78-620062.
    1654 •
    Beautifully designed atlas employing the latest in cartographic techniques to delineate in a single volume "how water works in the State of California."

Karan, P.P. and Mather, Cotton, eds. Atlas of Kentucky. Lexington, KY: University Press of Kentucky, 1977. 182 p. ISBN 0-8131-1348-2. LC 76-24337. (available on microfilm from Books on Demand, Ann Arbor, MI. ISBN 0-8357-9782-1. $50.30).
    1655 •
    Atlas includes 142 colored maps of Kentucky and "provides a concise and up-to-date view of the Commonwealth's physical, human, and economic patterns. As a reference document it is a graphic representation of the various phases of Kentucky's geography and development."

Kingsbury, Robert C. An Atlas of Indiana. (Occasional publication no. 5). Bloomington, IN: Department of Geography, Indiana University, 1970. 94 p. LC 74-653653.
    1656
    Black-and-white maps arranged in six sections: the state; the physical landscape; population characteristics; cities, trade, and industries; farming and agricultural products; and transportation and recreation.

United States

Lineback, Neal G. and Traylor, Charles T., eds. Atlas of Alabama. University, AL: University of Alabama Press, 1973. 134 p. ISBN 0-8173-9000-6. $9.75. LC 72-11148.   1657 •
Atlas was designed to "objectively describe the physical, cultural, and economic characteristics of the State in a manner that would involve the most effective use of maps, graphs and text."

Loy, William G. and Stuart, Allan. Atlas of Oregon. Eugene, OR: University of Oregon, 1976. 215 p. ISBN 0-87114-076-4. LC 76-13964.   1658 •
The finest atlas of an American state. Attractively designed maps and text are arranged in four parts: human geography; economic systems; natural environment; and facts, figures, and place names. The latter section includes a listing of 1600 place names, all shown on the U.S. Geological Survey's 1:500,000 scale map reproduced on pages 184-199.

Martis, Kenneth C. and Rowles, Ruth Anderson, eds. The Historical atlas of U.S. Congressional Districts. New York: Free Press; London: Collier Macmillan, 1982. 302 p. ISBN 0-02-920150-0. $150. LC 82-70583.   1659 •
The first definitive congressional district atlas of the United States. Part one traces the history and development of congressional districts and geographical representation in the House of Representatives; part two contains congressional district maps and membership lists; and part three is the legal history of congressional redistricting laws for each state.

Meeks, Harold A. The Geographical regions of Vermont: a study in maps. (Geography publications at Dartmouth no. 10). Hanover, NH: Dartmouth College, Department of Geography, 1975. 182 p. LC 75-24869.   1660 •
"Assembled in this work are nearly all relevant systematic (dealing with topics) maps made since 1930 which relate to the physical and cultural patterns of Vermont. Many other maps are newly drawn based on the most recent statistical information, largely of 1970 vintage."

Miller, Eugene Willard. Socioeconomic patterns of Pennsylvania: an atlas. Harrisburg, PA: Commonwealth of Pennsylvania, 1975. 228 p. (for sale by Bureau of Management Services, State Book Store). LC 75-622145.   1661 •
Ninety-nine black-and-white maps designed to "provide a measure of some of the social and economic differences and similarities that exist between counties and regions in Pennsylvania."

Pfanstiehl, Margaret Rockwell, ed. The Tactile and large print atlas of the State of Maryland. Silver Spring, MD: Washington Ear, Inc., 1985. 3 parts.
      Part 1. Braille Atlas, 41 p.
      Part 2. Commentary in Print, 47 p.
      Part 3. Commentary on four audio cassettes.
Atlas in braille, raised line, large print plus commentary on four cassettes and in regular print. $23. Atlas in large print plus commentary on four cassettes and in regular print. $18. LC 84-675224.   1662
An innovative new atlas of Maryland combines braille, raised lines, large print and voice-indexed cassettes to present geographic, historical, and demographic information for easy understanding by low visioned and blind readers. Includes 21 maps (seven state and 14 regional), a unique touch coordinate system, an index with 230 entries, and an instructional voice commentary for each map with additional historic, geographical, and demographic information. Also includes maps showing the major political boundaries, cities, towns, waterways, railroads, highways and roads, parks, cultural and educational institutions, and historical information from early resident Indian tribes through colonial settlements to the present. The commentary is also available in print for sighted family members, teachers, and friends. A large-print-only version of the maps is available for low vision and fully sighted readers. The audio commentary is voice-indexed on four cassettes for quick reference.

Rafferty, Milton D.; Gerlach, Russel L.; and Hrebec, Dennis J. Atlas of Missouri. Springfield, MO: Aux-Arc Research Associates, 1970. 88 p. LC 78-654233.   1663 •
Simple black-and-white maps depicting the physical, economic, political, and social character of Missouri.

Rand McNally & Co. Rand McNally commercial atlas and marketing guide. Chicago, IL: Rand McNally, 1876- . Annual. $135. (available on a lease basis).   1664 •
Large, double-page state maps showing counties, townships, and 110,000 cities, towns, villages, and crossroads settlements. Alphabetical index for each state, with populations of places. The most useful and comprehensive gazetteer for U.S. settlements. Data on business centers and counties.

Reader's Digest Association. These United States: our nation's geography, history, and people. Pleasantville, NY: Reader's Digest Association, 1968. 236 p. LC MAP 68-02.   1665 •
The best and most extensive of the unofficial national atlases of the United States. Wide national and regional coverage of physical, historical, social, and economic topics in maps, photographs and text. Imaginative graphic design.

Rizza, Paul F. and Hughes, James C. Pennslyvania atlas: a thematic atlas of the Keystone State. 2nd rev. ed. Grove City, PA: Ptolemy Press, 1982. 127 p. (1st ed., Slippery Rock, PA: Department of Geography, Slippery Rock State College, 1975. 108 p.). ISBN 0-933550-02-2. $16. LC 83-684876.   1666 •
Thematic atlas containing black-and-white maps arranged in seven sections: physical patterns; population; political and social patterns; economic patterns; agriculture; transportation and communications; and tourism and recreation.

Smith, Allen R. Connecticut: a thematic atlas. [n.p.]. 1974. 90 p. LC 75-311340.   1667 •
Contains more than 80 black-and-white maps prepared "to give more people a better understanding and appreciation of the State of Connecticut."

Sommers, Lawrence M., ed. Atlas of Michigan. East Lansing, MI: Michigan State University Press, 1977. 242 p. ISBN 0-87013-205-9. $19.95. LC 77-74637.   1668 •
"This full color Atlas of Michigan portrays the major characteristics of the state's natural, man-made, and human resources, largely in graphic form. A combination of maps, graphs, photographs, text, and sketches reveals the complexity of this Great Lakes state."

Taylor, Robert L.; Edie, Milton J.; and Gritzner, Charles F. Montana in maps. Bozeman, MT: Big Sky Books, Montana State University, 1974. 76 p. LC 74-27527.   1669 •
"The title and portions of the material in this atlas are taken from Montana in maps by Nicholas Helburn, Milton J. Edie, and Gordon W. Lightfoot (Bozeman, MT: The Reasearch and Endowment Foundation at Montana State College, 1962. 84 p.).

Thompson, Derek, and Murphey, Charles E., eds. and Wiedel, Joseph W. Atlas of Maryland. Commemorative edition. College Park, MD: University of Maryland, 1977. 116 p. ISBN 0-918512-02-6. $4.50. (pbk). LC 76-57793.   1670 •
"This atlas sets out in graphic form many aspects of contemporary Maryland; its people, their quality of life, and the environment in which they live."

U.S. Bureau of the Census. 1978 census of agriculture. Vol. 5. Special report. Part 1. Graphic summary. Washington, DC: Government Printing Office, 1982. 177 p. LC 79-600215.   1671
The 300 black-and-white and colored maps in this splendid atlas depict every significant aspect of contemporary American agriculture using the county as the areal unit.

United States

U.S. Geological Survey. The National atlas of the United States of America. Washington, DC: U.S. Geological Survey, 1970. 417 p. LC 79-654043.
At long last a national atlas of which all Americans can be proud. In this lavishly produced, encyclopedic reference volume, the coverage is excellent for physical and economic topics, but rather less thorough for historical, social, and cultural items. Superb gazetteer-index with 41,000 entries.
1672 •

University of Hawaii at Manoa. Department of Geography. Atlas of Hawaii. 2nd. ed. Honolulu, HI: University of Hawaii Press, 1983. 238 p. (1st ed., 1973. 322 p.). ISBN 0-8248-0837-1. $29.95. (pbk). LC 82-675462.
Attractive atlas edited by R. Warwick Armstrong and containing maps drawn under the direction of James A. Bier. Atlas is divided into the following sections: reference maps; the natural environment; the cultural environment; the social environment; and appendices.
1673 •

Wahlquist, Wayne L. Jr., ed.; Greer, Deon; Gurgel, Klaus D.; and Christy, Howard A. Atlas of Utah. Provo, UT: Brigham Young University Press for Weber State College, 1981. 300 p. ISBN 0-842-1831-1. $49.95. LC 81-675069.
Colored thematic maps and informative accompanying text comprehensively covering the land, people, social institutions, government, economy, and recreation. Maps and text by specialists in each topic. Colored and black-and-white photographs. Tables. Extensive bibliography.
1674 •

Williams, Jerry L. and McAllister, Paul E., eds. New Mexico in maps. Albuquerque, NM: University of New Mexico Press, 1979. 177 p. ISBN 0-8263-0601-2. $14.95. LC 81-675344.
Thematic atlas containing black-and-white maps, divided into the following categories: natural environment; historical landscapes; population characteristics; economic characteristics; and recreation and government.
1675 •

Wood, Roland, and Fernald, Edward A. The New Florida atlas: patterns of the Sunshine State. St. Petersburg, FL: Trend House, 1974. 119 p. ISBN 0-88251-028-2. LC 72-96324.
Atlas is divided into the following topics: population; housing; income; education; government and politics; physical characteristics; water; climate; transportation and communication; tourism and recreation; economic activity; employment; health; history; topographic mapping; and aerial photography.
1676 •

Wright, Marion I. and Sullivan, Robert J. The Rhode Island atlas. Providence, RI: Rhode Island Publications Society, 1982. 239 p. ISBN 0-917012-56-6. $22.50. (pbk). LC 80-52911.
"This atlas attempts to provide a new perspective on both the physical and the cultural attributes of Rhode Island." It includes 124 maps and 30 graphs arranged in six units: introduction; the physical realm; places and names; the people; the economy; and transportation.
1677 •

C. STATISTICS   WILBUR ZELINSKY

U.S. Bureau of the Census. Statistical abstract of the United States. Washington, DC: Government Printing Office, 1879- . Annual. ISSN 0081-4741. LC 04-018089.
The one publication that is absolutely indispensable for anyone concerned with the U.S. In addition to the 1500 tables furnishing data on virtually every conceivable subjects, this volume is a guide to sources.
1678 •

U.S. Bureau of the Census. County and city data book, 1983. 10th ed. Washington DC: Government Printing Office, 1983. 996 p. (1st ed., 1949. Preceded by County data book, 1947. 431 p. and Cities supplement, 1944. 47 p.).ISSN 0081-4741. $24. LC 53-4576. 1679•
A veritable cornucopia of geographical information. The three enormous tables for states, counties, and cities present statistics on 216, 216, and 170 items, respectively.

U.S. Bureau of the Census. State and metropolitan area data book, 1982: a statistical abstract supplement. Washington, DC: Government Printing Office, 1982. 611 p. (1st ed., 1979). ISSN 0276-6566. $15. LC 80-600018. 1680•
Valuable companion to County and city data book, 1983 *providing detailed data on 320 items for 318 standard metropolitan areas and for each county within these areas, 73 items for 429 central cities, and 2018 items for the United States as a whole, for four major regions, nine smaller regions, 50 states, and the District of Columbia.*

U.S. Bureau of the Census. Historical statistics of the United States: colonial times to 1970. Bicentennial Edition. 3rd ed. Washington, DC: Government Printing Office, 1975. 2 v. (1st ed., 1949. 363 p.).Stock no. 003-024-00120-9. $35. LC 75-38832. House document number: 93-78. 1681•
More than 12,500 time series, mainly annual, on social and economic development of the United States from 1610 to 1970.

U.S. Bureau of Mines. Minerals yearbook. Washington, DC: Government Printing Office, 1932/33- . Annual. 3 v. ISSN 0076-8952. LC 33-026551. 1682
    Vol. 1: **Metals and minerals.**
    Vol. 2: **Area reports domestic.**
    Vol. 3: **Area reports international.**
Detailed annual review of the mineral industry of the United States and the world. Recent annual yearbooks consist of three volumes, one devoted to individual metallic and non-metallic mineral commodities, one to area reports for the United States, and one to area reports international with chapters on individual countries. The Centennial Edition 1981 *(1982-1983) is especially valuable.*

D. PHYSICAL GEOGRAPHY   HAROLD A. WINTERS

Curran, H. Allen (Harold); Justus, Philip J.; Young, Drew M.; and Garver, John B. Atlas of landforms. 3rd. ed. New York: Wiley, 1984. 165 p. (1st ed., U.S. Military Academy, West Point. Department of Earth, Space, and Graphic Sciences. Atlas of Landforms by James L. Scovel, and others, 1965. 164 p.). ISBN 0-471-87434-5. $35.95. LC Map 83-675974. 1683•
An atlas of maps, landform images, diagrams, and descriptions of many landform types, most within the United States.

Fenneman, Nevin M. Physiography of western United States. New York: McGraw-Hill, 1931. 534 p. LC 31-4608. 1684
This volume, and its companion work on the eastern United States, is still a useful text and reference.

Fenneman, Nevin M. Physiography of eastern United States. New York: McGraw-Hill, 1938. 714 p. LC 38-9303. 1685
See preceding entry.

Foth, Henry D. and Schaffer, John W. Soil geography and land use. New York: Wiley, 1980. 484 p. ISBN 0-471-01710-8. $38.95. LC 79-27731. 1686
A good basic book on the classification and distribution of soils with unusually heavy emphasis on the United States.

United States

Harris, David V. The geologic story of the national parks and monuments. 3rd ed. Fort Collins, CO: Colorado State University Foundation Press, 1980. 322 p. ISBN 0-471-09764-0. $26.50. LC 81-112429.   1687•
After a 30-page introduction giving a basic review, the national parks and monuments are described in 21 chapters. A very useful reference book.

Küchler, A. William (August). Potential natural vegetation of the conterminous United States. 2nd ed. (Special publication no. 36). New York: American Geographical Society, 1975. (1st ed., 1964). LC 79-692016.   1688
Map at the scale of 1:3,168,000 and a 50-page manual of exploration and description.

Pirkle, E.C. (Earl), and Yoho, W.H. Natural landscape of the United States. 3rd ed. Dubuque, IA: Kendall-Hunt, 1982. 399 p. ISBN 0-8403-2591-6. $19.95. (pbk).   1689•
A brief introductory and elementary description of landform regions of the United States.

Thornbury, William D. Regional geomorphology of the United States. New York: Wiley, 1965. 609 p. LC 65-12698.   1690
General description and genetic analysis of landform regions of the United States.

U.S. Soil Survey Staff. Soil taxonomy. (Agricultural handbook no. 436). Washington, DC: U.S. Department of Agriculture, Soil Conservation Service, 1975. 754 p. Stock no. 001-000-02597-0. $17.95.   1691
A detailed and highly useful description of the basic system of soil classification with references to soils of North America and the United States.

Wright, Herbert E. Jr. and Porter, S.C. Late Quaternary environments of the United States. Minneapolis, MN: University of Minnesota Press. 1983. 2 v.
Vol. 1: 480 p. ISBN 0-8166-1169-6. $45. LC 83-5804.
Vol. 2: 384 p. ISBN 0-8166-1171-8. $45. LC 83-5804.   1692
More that 30 studies describe various aspects of the late-glacial and post-glacial conditions in the United States. Topics include glaciation, pro-glacial environments, coastal and marine morphology, Pleistocene biota and climate, paleoecology and archaeology.

E. HUMAN GEOGRAPHY   WILBUR ZELINSKY

American Heritage cookbook and illustrated history of American eating and drinking. New York: Simon and Schuster, 1964. 629 p. LC 64-21278.   1693
Although a trifle folksy and regrettably undocumented, this is the most comprehensive effort to date to paint a broad picture of the history and regionalism of American foodways. Lavishly illustrated.

Barone, Michael, and Ujifusa, Grant. The Almanac of American politics, 1985. Washington, DC: National Journal, 1983. 1402 p. ISBN 0-89234-030-4. $35. ISBN 0-89234-031-2. $22.50. (pbk).   1694
The current national and state-by-state political scene but most of all, fine thumbnail sketches of social, economic, and political conditions in each of the 435 congressional districts--all in lively prose.

Belasco, Warren James. Americans on the road: from auto camp to motel, 1910-1945. Cambridge, MA.: MIT Press, 1979. 212 p. ISBN 0-262-02123-4. $25. LC 79-15304.   1695
A splendid account of the formative stages of mass automobility in the U.S. Using as his central theme the tension between liberty and order, Belasco has produced the definitive statement on the impact of early motor touring on the roadside landscape and its relationship to American social history.

Brunn, Stanley D. Geography and politics in America. New York: Harper and Row, 1974. 443 p. ISBN 0-06-041018-3. LC 73-17670.

1696

A wide-ranging and perceptive text that treats, inter alia, electoral, legislative, and administrative behavior, international relations, territoriality, boundary questions, and political cultures and preceptions.

Carney, George O., ed. The Sounds of people and places: readings in the geography of music. Washington, DC.: University Press of America, 1978. 336 p. ISBN 0-8191-0394-2. $14. (pbk). LC 78-304155.

1697

In this first book-length treatment of the geography of music, Carney has assembled the best of a sparse literature. The fourteen essays deal almost exclusively with the American scene and emphasize country, gospel, and popular idioms. Useful lists of references and resources.

Clay, Grady. Close-up: how to read the American city. New York: Praeger; London, Pall Mall Press, 1973. 192 p. Reprinted: Chicago, IL:University of Chicago Press, 1980. (Phoenix edition). 192 p. ISBN 0-226-10945-3. $9.95. (pbk). LC 79-26307.

1698

A highly original, literally eye-opening "Baedeker to the commonplace," useful effort by "an urban journalist and professional observer" to come to grips visually with the recent structure and changes of American cities. The chapter titles include: fixes, epitome districts, fronts, strips, beats, stacks, sinks, and turf. Imaginatively illustrated.

Davis, George A. and Donaldson, O. Fred. Blacks in the United States: a geographic perspective. Boston, MA: Houghton Mifflin, 1975. 270 p. ISBN 0-395-14066-8. LC 74-14362.

1699

The materials on Black history, migration, economics, housing, education, and the arts add up to a readable and stimulating text in this first book-length effort to explore the social and historical geography of America's largest minority group.

Elazar, Daniel J. American federalism: the view from the states. 3rd ed. New York: Harper and Row, 1984. 336 p. (1st ed., 1966. 228 p.). ISBN 0-06-041884-2. $12.95. (pbk). LC 83-26392.

1700

In addition to covering the theory and practice of federalism in state-local and federal-state relationships, this influential publication delves into the geography of political cultures by mapping and discussing three major modes of behavior: the individualistic, moralistic, and traditionalistic.

Gastil, Raymond D. Cultural regions of the United States. Seattle, WA: University of Washington Press, 1976. 382 p. ISBN 0-295-95426. $21. LC 75-8933.

1701

A sociologist has produced the first book-length treatment of the subject using both the topical and regional approach. Encyclopedically rich in facts and references but thin in terms of new ideas.

Gaustad, Edwin, S. Historical atlas of religion in America. Rev. ed. New York: Harper and Row, 1976. 189 p. (1st ed., 1962. 179 p.). ISBN 0-06-063089-2. LC 76-25947.

1702

An expert analysis in text, map, and graph of the historical geography of the principal American denominations and some general aspects of American religion.

Glassie, Henry. Pattern in the material folk culture of the Eastern United States. (American Folklore Society Series). Philadelphia, PA: University of Pennsylvania Press, 1971. 316 p. ISBN 0-8122-7569-1. $20. LC 75-160630.

1703

An illustrated essay on the regionalization and significance of structures and handicrafts based on the author's field observations of houses, furniture, tools, decorative items, and other artifacts. A bibliographic treasure trove.

233

United States

Golant, Stephen M., ed. Location and environment of elderly population. (A Halsted Press book). New York; Toronto; London: John Wiley, 1979. 214 p. ISBN 0-470-26788-7. $17.95. LC 79-13621.

Papers by 18 contributors on the residential location and migration patterns of the U.S. elderly population, their environmental context, and planning methodologies and strategies.

1704

Harries, Keith D. The Geography of crime and justice. New York: McGraw-Hill, 1974. 125 p. ISBN 0-07-026749-9. LC 73-11265.

An analysis of place-to-place variations in the occurrence of crime and in the quality of law enforcement and the judicial process with a very strong emphasis on the American scene. The spatial scale of treatment ranges from the international through the interstate and intermetropolitan levels to the intraurban.

1705

Harries, Keith D. and Brunn, Stanley D. The Geography of laws and justice: spatial perspectives on the criminal justice system. New York; London: Praeger, 1978. 175 p. ISBN 0-03-022331. LC 77-25460.

A pioneering attempt to measure and describe the enormous spatial variations in the United States in the operation of the judicial system: regional differences in statutes, jury selection, sentence disparity, discretion for prosecutors, and possible court reform and reorganization.

1706 •

Hart, John Fraser. The Look of the land. Englewood Cliffs, NJ: Prentice-Hall, 1975. 210 p. ISBN 0-13-540534-3. ISBN 0-13-540526-2. (pbk). LC 74-20995.

A somewhat eclectic but informative set of twelve essays on the rural geography of the United States and Great Britain, with special emphasis on the former. Among the topics covered are land division, farm tenure, farm management, rural structures and villages, mining, forestry, and recreation.

1707

Hawley, Amos H. and Mazie, Sara Mills, eds. Nonmetropolitan America in transition. (Institute for Research in Social Science Series). Chapel Hill, NC: University of North Carolina Press, 1981. 833 p. ISBN 0-8078-1490-3. $27.50. LC 81-3511.

This collection of 21 original essays by 29 authors, mostly economists and sociologists, is the most comprehensive and useful recent book on nonmetropolitan America. The coverage is broad. The basic topics are population deconcentration, the changing structure of economic opportunity, differential access to opportunity, the amenities, and environmental impact and planning as related to growth.

1708

Jackson, John Brinckerhoff. The Necessity for ruins and other topics. Amherst, MA: University of Massachusetts Press, 1980. 136 p. ISBN 0-87023-291-6. $10. LC 79-23212.

These nine superbly written provocative essays by the founder of modern landscape analysis and criticism range widely and deeply, including such topics as European gardens and theater scenery, but the dominant concern is the American landscape. Indispensable for anyone curious about the evolution and meaning of the man-made scene.

1709

Jackson, John Brinckerhoff. Discovering the vernacular landscape. New Haven, CT: Yale University Press, 1984. 192 p. ISBN 0-300-03138-6. $16.50. LC 83-21925.

Can there ever be too much of the prose and thoughts of J.B. Jackson? Here we have fourteen of the most recent of his reflections on the evolution and implications of ordinary American landscapes. A dominant theme is the tension between the 'Landscape One' of land-based locals and the 'Landscape Two' imposed by central authority. "Wisdom" is not too strong a descriptor.

1710

Kurath, Hans. A Word geography of the Eastern United States. (Michigan University studies in American English, no. 1). Ann Arbor, MI: University of Michigan Press, 1966. 88 p. ISBN 0-472-08532-8. $9.95. LC 49-50233.

This first major project of the Linguistic atlas of the United States *analyzes through maps and text the spatial distribution of selected elements in the American vocabulary and offers a set of linguistic regions for the Atlantic seaboard.*

1711

Lewis, Peirce F.; Lowenthal, David; and Tuan, Yi-Fu. Visual blight in America. (Commission on college geography, resource paper no. 23). Washington, DC: Association of American Geographers, 1973. 48 p. ISBN 0-89291-070-4. $4. (pbk). LC 73-88850.   1712
The three essays by Lewis, Lowenthal, and Tuan and the commentaries by Donald Meinig and J.B. Jackson offer rich, thoughtful insights into an important, but contentious, set of problems in environmental ethics and appraisal that geographers and citizens can ill afford to evade.

McAlester, Virginia, and McAlester, Lee. A Field guide to American houses. New York: Alfred A. Knopf, 1984. 525 p. ISBN 0-394-51032-1. $30. ISBN 0-394-73969-8. $19.95. (pbk). LC 82-48740.   1713
Arguably the best in its genre, a guide that identifies and places in their historic, regional, and architectural contexts the houses built for American families (rich, poor, and in-between), in city and countryside, from the 17th Century to the present. Several maps appear amidst the profusion of illustrations. Useful bibliography.

Morrill, Richard L. and Wohlenberg, Ernest H. The Geography of poverty in the United States. New York: McGraw-Hill, 1971. 148 p. ISBN 0-07-043130-2. $6.95. (pbk). LC 72-178931.   1714
First book-length attempt at describing and explaining the geographic patterns of poverty within the U.S. and an evaluation of existing and proposed anti-poverty measures from a geographic viewpoint.

Muller, Peter O. Contemporary suburban America. Englewood Cliffs, NJ: Prentice-Hall, 1981. 218 p. ISBN 0-13-170647-0. $16.95. LC 80-25653.   1715
The first extended geographical treatment of an increasingly important sector of the American scene: the suburban zone surrounding our cities.

Murphy, Raymond E. The American city: an urban geography. 2nd ed. New York: McGraw-Hill, 1974. 556 p. (1st ed., 1966. 464 p.). ISBN 0-07-044063-8. LC 73-17318.   1716
A comprehensive text that deals with all major aspects of the macrogeography of the American urban system and the internal morphology and functions of contemporary cities.

Platt, Rutherford H. and Macinko, George, eds. Beyond the urban fringe: land use issues of nonmetropolitan America. Minneapolis, MN: University of Minnesota Press, 1983. 416 p. ISBN 0-8166-1099-1. $39.50. LC 83-3518.   1717
This collection of 27 essays on a wide variety of topics is the best book currently available dealing with the subject within an interdisciplinary framework. The central theme is obviously land-use problems, but it is one that involves all manner of political, social, ecological, and technological matters.

Raitz, Karl B. and Hart, John Fraser. Cultural geography on topographic maps. New York: John Wiley and Sons, 1975. 139 p. ISBN 0-471-70595-0. LC 75-12524.   1718•
Accompanying the full-color reproductions of sections of 30 quite varied U.S. and Canadian topographic maps are capsule accounts of the areas in question and a set of exercises for the student. This is a useful cartographic approach to the study and interpretation of North American regions. Index of cultural features and selected bibliography.

Rooney, John F., Jr. A Geography of American sport: from Cabin Creek to Anaheim. Reading, MA: Addison Wesley, 1974. 306 p. ISBN 0-201-06491-X. $15.95. LC 73-16555.   1719
An initial examination of the geographical aspects of American sports, and an interesting, stimulating beginning it is. The origins and diffusion, spatial organization, source areas of atheletes, recruiting patterns, and women's sports are the main topics. Baseball, football, basketball, golf, and college wrestling are the major sports considered.

United States

Roseman, Curtis C. Changing migration patterns within the United States. (Resource papers for college geography, no. 77-2). Washington, DC: Association of American Geographers, 1977. 34 p. ISBN 0-89291-123-9. $4. (pbk). LC 76-57033.   1720

    A concise exposition of migration theory and an account of recent trends in population redistribution. An item of special concern is the 'turnaround', the reversal in the 1970s of the traditional net movement from nonmetropolitan to metropolitan areas.

Smith, David M. The Geography of well-being in the United States: an introduction to territorial social indicators. New York: McGraw-Hill, 1973. 144 p. ISBN 0-07-058550-4. ISBN 0-07-058551-2. (pbk). LC 72-6605.   1721

    A pioneering attempt to apply the methods of the social indicators movement to the geographical analysis of the well-being of Americans at various areal scales from the state level to the intraurban. Useful for its methodological discussions and empirical findings and also as the necessary prelude to policy formulation.

Spengler, Joseph J. Population and America's future. San Francisco, CA: W.H. Freeman, 1975. 260 p. ISBN 0-7167-0744-6. $9.95. (pbk). LC 75-14031.   1722

    A thoughtful and readable examination of current and prospective population questions by an eminent economic demographer. Two of the eight chapters deal with the spatial distribution and "megalopolitanization" of Americans and the interplay between population and the natural environment.

Stewart, George R. U.S. 40: Cross section of the United States of America. Boston, MA: Houghton Mifflin, 1953. 311 p. Reprinted: Westport, CT: Greenwood Press, 1973. 311 p. ISBN 0-8371-6655-1. $37.50. LC 72-11338.   1723

    A classic exercise in reading and interpreting the mostly man-made aspects of the American landscape as the author progresses from the Atlantic coast to the Pacific coast. Not recommended for travelers with tunnel vision. The photographs are integral portions of this opus, and must be perused with care.

Stewart, George R. American ways of life. Garden City, NY: Doubleday, 1954. 310 p. Reprinted: New York: Russell and Russell, 1971. 310 p. ISBN 0-8462-1503-9. $14. LC 77-102546.   1724

    A perceptive, highly readable set of essays on some major aspects of American culture relying mostly on the historical approach. By a writer who was novelist, historian, linguist, and all-round student of Americana.

Stewart, George R. Names on the land: a historical account of place-naming in the United States. 4th ed. San Francisco, CA: Lexikos, 1982. 560 p. (1st ed., New York: Random House, 1945. 418 p.). ISBN 0-938530-02-X. $10. LC 82-6578.   1725

    The best general chronicle of the naming of the major features of the American land by its European settlers. A highly readable account.

Stewart, George R. American place-names: a concise and selective dictionary for the continental United States of America. New York: Oxford University Press, 1970. 550 p. ISBN 0-19-500121-4. $25. LC 72-83018.   1726

    The culminating work of a lifelong study of American toponymy. The 12,000 entries, concerned mainly with generic terms and the more widespread words and names used for the specific portion of place-names, provide information on origins, namers, and general significance of the items.

Stilgoe, John R. Common landscape of America, 1580 to 1845. New Haven, CT: Yale University Press, 1982. 512 p. ISBN 0-300-02699-4. $40. LC 81-16367.   1727

    A highly detailed, well-written, and quite perceptive description and interpretation of the evolution of virtually every phase of the early vernacular landscapes of America. Stilgoe also takes careful note of the regional dimension. Although this is the first monographic treatment of the subject, it is close to being definitive. Rich bibliography.

Stilgoe, John R. Metropolitan corridor: railroads and the American scene. New Haven, CT: Yale University Press, 1983. 416 p. ISBN 0-300-03042-8. $29.95. LC 83-3585.   1728
Using a wide variety of material, Stilgoe has produced the definitive treatment of the impact of the railroad system on landscape, society, and culture during the half-century (1880-1930) when it was the dominant agent in transforming and modernizing American life. Abundant references and remarkable illustrations.

Taeuber, Conrad F. and Taeuber, Irene B. The Changing population of the United States. (U.S. Bureau of the Census, Census monograph series). New York: John Wiley, 1958. 357 p. Reprinted: New York: Russell and Russell, 1975. 357 p. ISBN 0-8462-1747-3. $20. LC 73-84763.   1729
An authoritative account, organized chiefly along historical lines, of the major characteristics of the American population from 1790 to 1950. Some maps and tables.

Tunnard, Christopher, and Reed, Henry H. American skyline: the growth and form of our cities and towns. New York: New American Library of World Literature, 1956. 224 p. LC 56-3993.   1730
The only well-rounded discussion of the evolution of the morphology of American urban places--and a good one. Many valuable sidelights on the general historical geography of the nation.

Vale, Thomas R. and Vale, Geraldine R. U.S. 40 today: thirty years of landscape change in America. Madison, WI: University of Wisconsin Press, 1983. 208 p. ISBN 0-299-09480-4. $27.50. LC 83-50081.   1731
Follows the route and matches selected photographs of U.S. 40 thirty years after George R. Stewart perceptively reported on the dynamics of landscape change. See entry 1723.

Watson, J. Wreford (James). Social geography of the United States. London; New York: Longman, 1979. 290 p. ISBN 0-582-48197-X. $12.95. (pbk). LC 77-30744.   1732
An unconventional, innovative approach to such contentious topics as the treatment of Native Americans; the Black condition in America; poverty and crime; the problems of the young and the aged; the women's movement; clashes among cultures, religions, and classes; and the counterculture. Watson makes liberal use of major novelists and poets in this disturbing, but thought-provoking, essay in social geography.

Watson, J. Wreford (James), and O'Riordan, Timothy, eds. The American environment: perceptions and policies. London; New York: John Wiley, 1976. 340 p. ISBN 0-471-92221-8. LC 74-32224.   1733
Perception is indeed the operational term in this volume. The nineteen essays concentrate on the ways in which North American landscapes and resources have been imagined, perceived, and experienced and how these relationships have affected the development of the environment. A useful review of resource management policy and philosophy.

W.P.A. Writers' Program. The American guide series. 1935-1942.   1734
No standard entry is feasible for this indispensable collection of scores of guidebooks for states, regions, and cities published during the latter part of the Great Depression. Although much of the information is obsolete, the essays on major aspects of the history, geography, society, and culture of the area and spot data on points of interest still have not been superseded by any other series. A number of volumes have been re-issued by commercial publishers. For a complete list see: WRITERS' PROGRAM. Catalogue, W.P.A. Writers' Program Publications, the American Guide Series, the American Life Series (Washington: Government Printing Office, 1942) 54 p. LC-42-37616.

United States

Zelinsky, Wilbur. The Cultural geography of the United States. Englewood Cliffs, NJ: Prentice-Hall, 1973. 176 p. ISBN 0-13-195499-4. $17.95. LC 72-4503.　　1735

　　The four chapters in this pioneering conspectus of the subject deal with the origins and evolution of the American population and their cultural system, the peculiar identity of that system, the geographical processes operating to shape it, and the regionalization of American culture. Useful annotated bibliography.

F. ECONOMIC GEOGRAPHY　　　　　　　　　　　　　　　WILBUR ZELINSKY

Clawson, Marion. America's land and its uses. (Resources for the Future series). Baltimore, MD: Johns Hopkins Press, 1972. 166 p. ISBN 0-8018-1330-1. $5. LC 70-167985.　　1736

　　A succinct and authoritative account concerning the history of American land use and current issues involving urban, recreational, crop, grazing, forest, and other lands.

Cuff, David J. and Young, William J. The United States energy atlas. New York: Free Press, 1980. 416 p. ISBN 0-02-691250-3. $85. LC 80-24942.　　1737

　　By means of maps, graphs, and ample text this first publication of its kind presents the current and prospective situation for each of the significant renewable and nonrenewable energy sources from a geographic viewpoint.

Dunn, Edgar S. Jr. The Development of the U.S. urban system. (Resources for the Future series). Baltimore, MD: Johns Hopkins University Press, 1980-1983. 2 v.

　　Vol. 1: Concepts, structures, regional shifts. 1980. 626 p. ISBN 0-8018-2196-7.
　　Vol. 2: Industrial shifts, implications. 1983. 386 p. ISBN 0-8018-2638-1. $115.
　　LC 79-2180.　　1738

　　A comprehensive, quantitative, historical economic geography of the urban system in the U.S. with emphasis on the growth, development, and shifts within the system between 1940 and 1970. A treasure trove of data for the serious student of regional and urban analysis.

Jackson, Richard H. Land use in America. (Scripta geography series). New York: Halsted Press, 1982. 226 p. ISBN 0-470-27363-1. $22.95. (pbk). LC 80-20184.　　1739

　　A penetrating account of land-use problems, with special attention to environmental protection, the suburbs, and prime agricultural lands, and how they are being managed or mismanaged by Federal, state, and local agencies.

Marschner, Francis J. Land use and its patterns in the United States. (U.S. Department of Agriculture handbook no. 296). Washington, DC: Government Printing Office, 1971. 277 p. (Updated version of handbook no. 135). LC 72-606046.　　1740

　　The first 100 pages of this classic monograph treat land use patterns and land survey from the perspective of their historical development. The remainder of the volume is a choice selection of air photos showing the patterns and the general appearance of various portions of rural America.

Miller, E. Willard. Manufacturing: a study in industrial location. University Park, PA: Pennsylvania State University Press, 1977. 286 p. ISBN 0-271-01224. LC 76-1849.　　1741

　　Although this volume begins with an exposition of classical location and equilibrium theories, in the main it consists of a locational analysis of eight major American manufacturing industries and recent changes therein. It is the most recent and comprehensive treatment of the geography of this segment of the economy.

Perloff, Harvey S. et al. Regions, resources, and economic growth. Lincoln, NE: University of Nebraska Press, 1965. (First issued in 1960 for Resources for the Future). 716 p. LC 60-12311.　　1742

　　A major study that examines regional differences in economic growth in the United States and the impact of changing locations of industry on regional economics as of the 1950's. Includes an evaluation of various methodologies.

Vogeler, Ingolf. The Myth of the family farm: agribusiness' dominance of United States agriculture. Boulder, CO: Westview Press, 1981. 352 p. ISBN 0-89158-910-4. $32. LC 80-21091.

1743

An able fact-laden, neopopulist tract in defense of the family farm and basic ecological and social good sense and against agribusiness. Much useful information on the structure and operation of the rural economy.

G. HISTORICAL GEOGRAPHY   ROBERT D. MITCHELL

Allen, John L. Passage through the garden: Lewis and Clark and the image of the American Northwest. Urbana, IL: University of Illinois Press, 1975. 360 p. ISBN 0-252-00397-7. $32.95. LC 74-14512.

1744

The principal geographical study of Lewis and Clark's expedition to the American Northwest between 1804 and 1806. Focuses on the relationship between exploration and landscape imagery with emphasis on images of the Northwest as a passageway and as a garden.

Blouet, Brian W. and Lawson, Merlin P., eds. Images of the plains: the role of human nature in settlement. Lincoln, NE: University of Nebraska Press, 1975. 214 p. ISBN 0-8032-0839-1. $17.95. LC 74-76130.

1745

A series of papers on the Great Plains arranged around themes of exploration, resource evaluation, governmental appraisal, climatic hazards, desert and garden, and adaptations to reality.

Blouet, Brian W. and Luebke, Frederick C., eds. The Great Plains: environment and culture. Lincoln, NE: University of Nebraska Press, 1979. 246 p. ISBN 0-8032-1155-4. $18.95. LC 79-1152.

1746

A series of multi-disciplinary papers on the cultural heritage of the Plains. The principal focus is on the changing interpretations of the region between environmental perspectives and cultural perspectives.

Bowen, William A. The Willamette Valley: migration and settlement on the Oregon frontier. Seattle, WA: University of Washington Press, 1978. 134 p. ISBN 0-295-95590-2. $22.50. LC 77-15183.

1747

A study of the Oregon frontier during the 1840's using a series of forty-five carefully reconstructed maps. The text focuses on population distribution, sources of immigrants, and agricultural characteristics.

Conzen, Michael P. Frontier farming in an urban shadow: the influence of Madison's proximity on the agricultural development of Blooming Grove, Wisconsin. Madison, WI: State Historical Society of Wisconsin, 1971. 235 p. LC 73-158576.

1748

This book focuses on the transformation of a Wisconsin frontier town into a commercial farming area under the impact of an urban center. Topics include land ownership, social and demographic characteristics, agricultural production and growth, and changing market connections.

De Vorsey, Louis Jr. The Indian boundary in the southern colonies, 1763-1775. Chapel Hill, NC: University of North Carolina Press, 1966. 267 p. LC 66-25356.

1749

The evolution of the southern Indian boundary utilizing extensive documentary and cartographic sources for Virginia, the Carolinas, Georgia, and east and west Florida. The principal focus is on the conception of the boundary and its impact on the Cherokee, the Creek, and the Choctaw.

Earle, Carville V. The Evolution of a tidewater settlement system: All Hallow's Parish, Maryland, 1650-1783. (Research paper no. 170). Chicago IL: University of Chicago, Department of Geography, 1975. 239 p. ISBN 0-89065-077-2. $10. (pbk). LC 75-11554.

1750

A detailed examination of a southern colonial settlement system with particular attention to the geographical impact of tobacco production. Topics include changing population, land ownership, plantation system, slavery, dispersed settlement, and decentralized trade.

Francaviglia, Richard V. The Mormon landscape: existence, creation, and perception of a unique image in the American West. (Studies in social history no. 2). New York: AMS Press, 1978. ISBN 0-404-16020-4. $32.50. LC 77-83791.　　　　1751
Examines the significance of the Mormons in creating a distinctive landscape in the American west, associated especially with rural-village settlement and irrigated agriculture.

Grim, Ronald E. Historical geography of the United States: a guide to information sources. (Geography and travel information guide series, vol. 5). Detroit, MI: Gale Research Co., 1982. 291 p. ISBN 0-8103-1471-1. $55. LC 82-15674.　　　　1752
Valuable, up-to-date bibliography pertaining to historical geography organized into three units: part 1, cartographic sources; part 2, archival and other historical sources; and part 3, selected literature in historical geography. Work concludes with an addendum and author, title, and subject indexes.

Hewes, Leslie. The Suitcase farming frontier: a study in the historical geography of the central Great Plains. Lincoln, NE: University of Nebraska Press, 1973. 281 p. ISBN 0-8032-0825-1. $21.50. LC 72-85031.　　　　1753
Examines the role of non-resident farmers in transforming western Kansas and eastern Colorado from livestock rearing and general farming to wheat specialization between 1920 and 1950.

Hilliard, Sam B. Hog meat and hoecake: food supply in the old South, 1840-1860. Carbondale, IL: Southern Illinois University Press, 1972. ISBN 0-8093-0512-7. $17.95. LC 75-156778.　　　　1754
This study examines the significance of food supply in the Southern agricultural system. It focuses on food self-sufficiency, regional food production, food habits, food consumption, and trade.

Jackson, John B. American space: the centennial years, 1865-1876. New York: W.W. Norton and Co., 1972. 254 p. ISBN 0-393-09382-4. $4.95. (pbk). LC 72-5266.　　　　1755
A series of essays examining the major landscape changes in the decade following the Civil War. Seven regions are examined in detail: Northwest, Midwest, New England, South, Plains, California, and New York.

Jakle, John A. Images of the Ohio Valley: a historical geography of travel, 1740 to 1860. New York: Oxford University Press, 1977. 217 p. ISBN 0-19-502241-6. $9.95. (pbk). LC 77-9570.　　　　1756
This book offers a reconstruction of various aspects of the Ohio Valley using about four hundred travel accounts written between 1740 and 1860. Topics include travelers' biases, modes of transportation, images of Indian life, military affairs, and physical, rural, and urban landscapes.

Johnson, Hildegard Binder. Order upon the land: the U.S. rectangular land survey and the upper Mississippi country. New York: Oxford University Press, 1976. 268 p. ISBN 0-19-501913-X. $9.95. LC 75-32362.　　　　1757
Interprets the effects of the rectangular survey system on southeast Minnesota and adjacent areas during the nineteenth and twentieth centuries. Emphasizes the presurvey settlement pattern, the effects of the survey, more recent modifications, and the contrasts between the hilly terrain and the geometric symmetry of the survey.

Jordan, Terry G. German seed in Texas soil: immigrant farmers in nineteenth-century Texas. Austin, TX: University of Texas Press, 1966. 237 p. ISBN 0-292-72707-0. $7.95. (pbk). LC 66-15703.　　　　1758
Examines German settlement in south-central Texas between 1850 and 1880. Compares German agricultural practices in different environments in Texas with those of southern American white and black farmers and evaluates the extent of German assimilation.

Jordan, Terry G. Trails to Texas: southern roots of western cattle ranching. Lincoln, NE: University of Nebraska Press, 1981. 220 p. ISBN 0-8032-2554-7. $17.95. LC 80-14169. 1759

The author argues that Anglo herding practices, diffused from South Carolina, amalgamated with Hispanic herding traditions in eastern Texas to form the hybrid herding characteristics of open-range ranching on the Great Plains during the nineteenth century.

Lemon, James T. The Best poor man's country: a geographical study of early southeastern Pennsylvania. 2nd ed. New York: W.W. Norton, 1976. 295 p. (1st ed., 1972. 295 p.). ISBN 0-8018-1189-9. $25. ISBN 0-393-00804-5. $7.95. (pbk). LC 77-165352. 1760

Prize-winning study of the settlement and transformation of southeastern Pennsylvania between 1680 and 1800. Examines the physical background, characteristics of early settlers, population patterns, land tenure, rural and urban settlement, and regional variations. Argues that a liberal individualistic philosophy was the key to the region's development.

McManis, Douglas R. Colonial New England: a historical geography. New York, Oxford University Press, 1975. 159 p. LC 74-21824. 1761

The first general treatment of New England by a historical geographer. A survey of colonial New England that emphasizes pre-European settlement contact, Puritan settlement and population, and economic activities including farming, fishing, commerce, forestry, shipbuilding, manufacturing, and transportation.

Meinig, Donald W. The Great Columbia Plain: a historical geography, 1805-1910. Seattle, WA: University of Washington Press, 1983. 598 p. ISBN 0-295-96044-2. $14.95. (pbk). LC 68-11044. 1762

A detailed, chronological reconstruction of eastern Washington and adjacent Oregon during the nineteenth century. Emphasizes physical and Indian landscapes, exploration, fur trade and mission period, military campaigns and territorial organization, settlement and economic activity (mining, ranching, wheat), railroads, and towns.

Meinig, Donald W. Imperial Texas: an interpretive essay in cultural geography. Austin, TX: University of Texas Press, 1969. 145 p. ISBN 0-292-73807-2. $6.95. (pbk). LC 69-18807. 1763

A historical interpretation of Texas' cultural geography. Author discusses population origins, settlement characteristics, trade and circulation patterns during the Spanish-Mexican period, the Republic and statehood years, the post-Civil War era, and the twentieth century.

Meinig, Donald W. Southwest: three peoples in geographical change, 1600-1970. New York: Oxford University Press, 1971. 151 p. ISBN 0-19-501289-5. $9.95. (pbk). LC 71-125508. 1764

Focuses on a historical interpretation of the cultural and social geography of Indians, Hispanos, and Anglos, principally in New Mexico and Arizona for the periods 1820-1900 and 1900-1970. Emphasizes demographic, political, and economic patterns of interaction within a regional context.

Merrens, Harry Roy. Colonial North Carolina in the eighteenth century: a study in historical geography. Chapel Hill, NC: University of North Carolina Press, 1964. 293 p. LC 64-13555. 1765

A study of the changing geography of North Carolina before the American Revolution, especially between 1750 and 1775. Examines administrative and political organization, environmental perception, population, economic activities, urban development, and decentralized trade.

United States

Mitchell, Robert D. Commercialism and frontier: perspectives on the early Shenandoah Valley. Charlottesville, VA: University Press of Virginia, 1977. 251 p. ISBN 0-8139-0661-X. $14.95. (pbk). LC 76-26610.
    1766
Examines the Shenandoah Valley of Virginia during the eighteenth century from the perspective of the relationships between economic development, social stratification, and settlement structure. Emphasizes environmental factors, population and settlement, land acquisition and speculation, social structure, pioneer economy, commercial agriculture, and related transport, urban, and trading networks.

Olson, Sherry H. Baltimore: the building of an American city. Baltimore, MD: Johns Hopkins Press, 1980. 432 p. ISBN 0-8018-2224-6. $28.50. LC 79-21950.
    1767
Examines the physical growth of Baltimore from the early eighteenth century until the present. Growth and expansion are associated with long-term business and building cycles and their effects on morphology and internal structure of the city. Profusely illustrated.

Olson, Sherry H. The Depletion myth: a history of railroad use of timber. Cambridge, MA: Harvard University Press, 1971. 228 p. ISBN 0-674-19820-4. $16.50. LC 70-148940.
    1768
A case study, based principally on the Burlington Northern Railroad, of the use of timber for railroad construction. Principal emphasis is on the late nineteenth and early twentieth centuries when timber demand reached its peak, a level well below that required to deplete America's forests.

Pred, Allan R. Urban growth and the circulation of information: the United States system of cities, 1790-1840. (Studies in urban history series). Cambridge, MA: Harvard University Press, 1973. 348 p. ISBN 0-674-93090-8. $22.50. LC 73-76384.
    1769
Investigates the relationship between information circulation and the inter-dependent growth of large cities between 1790 and 1840. Focus on newspapers, postal service, domestic trade, and inter-urban travel in this pre-telegraphic period.

Pred, Allan R. Urban growth and city-systems in the United States, 1840-1860. (Studies in urban history series). Cambridge MA: Harvard University Press, 1980. 297 p. ISBN 0-674-93091-6. $28. LC 80-12098.
    1770
Focuses on urban growth and systems development in the key 1840-1860 pre-Civil War period. Argues that this was a time of rapid urban growth and of commercial to industrial transition. Provides case studies of Boston, Philadelphia, Buffalo, Cincinnati, New Orleans, and Charleston.

Reps, John W. The Making of urban America: a history of city planning in the United States. Princeton, NJ: Princeton University Press, 1965. 574 p. ISBN 0-691-04525-9. $90. LC 63-23414.
    1771
Traces the traditions that influenced American town planning from its European background to the twentieth century. Excellent overview of American urban planning. Includes more than 300 reproductions and town plans.

Rosenkrantz, Barbara G. and Koelsch, William A., eds. American habitat: a historical perspective. New York: Free Press, 1973. 372 p. LC 72-90281.
    1772
A collection of previously published materials on American perceptions of environment. Materials are organized under habitats perceived, ordered, and managed; experienced, occupied, and shaped; viewed and recorded; and chosen, valued, and affirmed.

Wacker, Peter O. The Musconetcong Valley of New Jersey: a historical geography. New Brunswick, NJ: Rutgers University Press, 1968. 207 p. ISBN 0-8135-0575-5. $17.50. LC 68-18694.
    1773
A study of the changing cultural landscape of a watershed region of northern New Jersey during the eighteenth century. Emphasis on the physical geography, aboriginal occupance, population, land ownership, agriculture, pioneer house and barn types, charcoal iron industry, marketing, and transportation.

Wacker, Peter O. Land and people: a cultural geography of preindustrial New Jersey: origins and settlement patterns. New Brunswick, NJ: Rutgers University Press, 1975. 499 p. ISBN 0-8135-0742-1. $42.50. LC 75-12650. 1774

A cultural and historical study of New Jersey prior to 1820 with emphasis on ethnic and cultural diversity, altered Indian landscapes, European and African groups, land subdivision, and settlement patterns.

Ward, David. Cities and immigrants: a geography of change in nineteenth-century America. New York: Oxford University Press, 1971. 164 p. ISBN 0-19-501284-4. $9.95. (pbk). LC 74-124612. 1775

An examination of the historical geography of urban growth between 1820 and 1920 with emphasis on external relations and internal differentiation of cities, immigrant residential patterns, and intra-urban transportation networks.

Ward, David, ed. Geographic perspectives on America's past: readings on the historical geography of the United States. New York: Oxford University Press, 1979. 364 p. ISBN 0-19-502353-6. LC 77-28302. 1776

A collection of previously published articles that provides an anthology of readings on the historical geography of the United States. The articles are grouped under three headings: the land and its people; the regional mosaic; and urbanization.

Wishart, David J. The Fur trade of the American West, 1807-1840: a geographical synthesis. Lincoln, NE: University of Nebraska Press, 1979. 237 p. ISBN 0-8032-4705-2. $17.95. LC 78-62915. 1777

A geographical examination of the western fur trade during the early nineteenth century. Emphasis on the geographical context and spatial organization of the fur trade. Emergence of two production systems, the upper Missouri fur trade and the Rocky Mountain trapping system, both focused in St. Louis.

H. NORTHEAST   GEORGE K. LEWIS

Alexander, Lewis M. The Northeastern United States. (New searchlight series). 2nd ed. New York: Van Nostrand, 1976. 142 p. (1st ed., 1967. 122 p.). ISBN 0-442-29749-1. LC 75-41660. 1778

A general overview of a region rarely treated as a single geographic entity. Captures the principal trends in the area although the data are changing. Physical descriptions and general economic and social characteristics are sound. Useful bibliography.

Carey, George W. A Vignette of the New York-New Jersey metropolitan region. (Comparative metropolitan analysis project). Cambridge, MA: Ballinger, 1976. 75 p. ISBN 0-88-410-436-2. $8.95. LC 76-4796. 1779

Still up-to-date in overall description although data are now old. Covers physical setting, economy, transportation, problems and planning attempts.

Conzen, Michael, and Lewis, George K. Boston: a geographical portrait. (Comparative metropolitan analysis project). Cambridge, MA: Ballinger, 1976. 87 p. ISBN 0-88410-432-X. $8.95. LC 76-4791. 1780

A general geographical description of the Boston region, including much of eastern Massachusetts. Contemporary developments in population, employment, transportation, and regional political structure are discussed in detail.

Eisenmenger, Robert W. The Dynamics of growth in New England's economy, 1870-1964. (New England research series no. 2). Middletown, CT: Wesleyan University Press, 1967. 201 p. ISBN 0-8195-8013-9. $16. LC 66-23926. 1781

An exhaustive review of a century of change in New England concentrating on significant land use changes and the industrial employment sector. Each relevant factor is analyzed separately with summary sections on contemporary characteristics of the region. Excellent bibliography. (This reference, unfortunately, misses most of the 'high-tech' period, but lays a strong historical foundation for its comprehension today.)

U.S. Middle West

Estall, Robert C. New England: a study in industrial adjustment. New York: Praeger, 1966. 296 p. LC 66-12483.
    A detailed review of the growth of manufacturing in New England with special emphasis on the industries dominant before the second world war. Emerging structural changes between 1945 and 1960 are discussed against the growth of 'high tech'. Foresees current economic development but statistical background is out of date.

1782

Gottmann, Jean. Megalopolis: the urbanized Northeastern seaboard of the United States. New York: The Twentieth Century Fund, 1961. 810 p. LC 61-17298.
    The classic overview of the urban region that stretches from Washington DC to Boston. Supporting data are now old, but the general material is surprisingly current. Physical, historical, social, political, and economic aspects are discussed individually in detail and then reviewed in their regional or megalopolitan context.

1783

Meyer, David R. From farm to factory to urban pastoralism: urban change in central Connecticut. (Comparative metropolitan analysis project). Cambridge, MA: Ballinger, 1976. 57 p. ISBN 0-88410-441-9. $8.95. LC 76-3740.
    An overview of the Hartford region as it undergoes extraordinary land use and population changes.

1784

Thompson, John H., ed. Geography of New York State. 2nd ed. Syracuse, NY: Syracuse University Press, 1977. 543 p. (1st ed., 1966. 534 p.). ISBN 0-8156-2182-5. $25. LC 77-4337.
    A comprehensive description of New York State by a team of writers. Data on population, employment, and industrial production.

1785

Whitehill, Walter Muir. Boston: a topographical history. 2nd ed. Cambridge, MA: Harvard University Press, 1968. 299 p. (1st ed., 1959. 244 p.). ISBN 0-674-07950-7. $18.50. ISBN 0-674-07951-5. $8.95. (pbk). LC 69-13769.
    The classic description of the present topography of the inner part of Boston city and the processes by which today's geographical landscape was shaped. Profusely illustrated with photographs and maps.

1786

I. MIDDLE WEST AND GREAT PLAINS   JESSE H. WHEELER, JR.

Akin, Wallace E. The North Central United States. (Searchlight books no. 37). Princeton, NJ: Van Nostrand Company, 1968. 160 p. LC 68-27134.
    Dated in many ways, but still a useful short overview of traditional conceptions. Topical organization: physical environment; settlement history; agriculture and agricultural regions; minerals; forestry; commodity processing and distribution; the heartland and the American scene. Clear maps; short bibliography.

1787

Borchert, John R. and Yaeger, Donald P., eds. Atlas of Minnesota resources and settlement. rev. ed. Minneapolis, MN: University of Minnesota, 1969. 262 p. (1st ed., 1968. 243 p.). LC 75-653995.
    Contains a large number of black and white maps (some with red overprint), interspersed with short text. Concentrates on physical, economic, and urban topics. Most maps are single-distribution maps of the state, but others are included. Aging, but still a valuable repository of basic data.

1788

Cutler, Irving. Chicago: metropolis of the mid-continent. 3rd ed. Dubuque, IA: Kendall-Hunt (under the auspices of the Geographic Society of Chicago.), 1982. 319 p. (1st ed., Chicago, IL: Geographic Society of Chicago, 1973. 128 p.). ISBN 0-8403-2645-9. $11.95. LC 75-40940.
    Historically oriented survey by a geographer. Topics include: the physical setting; general evolution of Chicago; ethnic groups and their settlement patterns; the economy; transportation; expansion of the metropolitan area; and planning for the future. Maps, photos, bibliography; miscellaneous data in tables and appendices.

1789

Dillon, Lowell I. and Lyon, Edward E., eds. Indiana: crossroads of America. (Regional geography series). Dubuque, IA: Kendall-Hunt, 1978. 160 p. ISBN 0-8403-1893-6. $11.95. LC 78-59281.

1790

Topically organized geographical introduction: geology; drainage; streams and lakes; weather and climate; vegetation; soils; transportation; agriculture; minerals; manufacturing; population; outdoor recreation. Graphics; bibliography.

Flader, Susan L., ed. The Great Lakes forest: an environmental and social history. Minneapolis, MN: University of Minnesota Press (in association with the Forest History Society), 1983. 336 p. ISBN 0-8166-1089-4. $29.50. LC 82-24714.

1791

A collection of 19 papers from an interdisciplinary Forest History Society symposium in 1979, edited by a well-known environmental historian. Topics include: the forest the settlers saw; forest succession; human impact on ecosystems; wildlife; the Indian experience; the logging era; changing land use and policies; response of forest industries to environmental change; social adjustments; prospects; perceptions and values.

Garland, John H., ed. The North American Midwest: a regional geography. New York: John Wiley and Sons; London: Chapman and Hall, 1955. 252 p. LC 55-9845.

1792

A multi-authored survey, now outdated in many ways but not replaced by any newer counterpart. Still quite valuable for general concepts and physical and historical detail, but must be used very selectively. Maps; selected bibliography.

Gerlach, Russel L. Immigrants in the Ozarks: a study in ethnic geography. Columbia, MO: University of Missouri Press, 1976. 232 p. ISBN 0-8262-0201-2. $18. LC 75-4328.

1793

A clearly written and well-researched study presenting introductory material on Ozark settlement history followed by discussions of particular ethnic groups emphasizing Germans, but with sections on Swedes, French, Swiss, Poles, Italians, and a chapter on Amish and Mennonites. Strongly rural focus; covers ethnic migrations, distributions, settlement patterns, landscapes, land use, social characteristics, and acculturation. Well-designed maps; photos; tables; very full bibliography.

Malin, James C. The Grassland of North America: prolegomena to its history, with addenda and postscript. Gloucester, MA: Peter Smith, 1967. 490 p. (1st ed., Lawrence, KS: Author, 1947. 398 p.). ISBN 0-8446-1296-0. $12. LC 67-4595.

1794

A wide-ranging, evocative book, considered a classic, by an ecologically oriented historian. Essentially an interpretive review of literature and thought on the grassland and kindred subjects citing scholars from many disciplines. Immense bibliography including several pages of citations to works by geographers.

Marshall, Howard Wight. Folk architecture in Little Dixie: a regional culture in Missouri. Columbia, MO: University of Missouri Press, 1981. 160 p. ISBN 0-8262-0329-9. $22. LC 80-26064.

1795

Material culture and settlement history of a distinctive central Missouri region, surveyed by the Director of the Missouri Cultural Heritage Center. Based largely on field work; documents pre-1930 house and barn types with photos, sketches, and plans. Extensive reference and explanatory notes; lengthy bibliography; many citations to works by geographers.

Martin, Lawrence. The Physical geography of Wisconsin. 3rd ed. Madison, WI: University of Wisconin Press, 1965. 608 p. (1st ed., 1916. 549 p.). ISBN 0-299-03472-0. $30. ISBN 0-299-03475-5. $10.95. (pbk). LC 65-14538.

1796

Numerous maps, diagrams, photos, tables, and citations. One table shows altitudes of cities and villages. Many distinguished geographers aided in the preperation of this edition.

Mather, E. Cotton (Eugene); Hart, John Fraser; and Johnson, Hildegard Binder. Upper Coulee country. Prescott, WI: Trimbelle Press, 1975. 101 p. LC 75-13058.   1797
   Small gem of regional writing analyzing a 'regionette' in southwestern Wisconsin along the northern border of the Wisconsin Driftless region. Place data, landcape description, historical background, trend analysis, and commentary by three distinguished cultural geographers and an environmentally oriented planner trained in geography. Maps, graphs, photos, sketches, tables; reference and explanatory footnotes.

Mayer, Harold M. and Wade, Richard C. Chicago: growth of a metropolis. Chicago, IL: University of Chicago Press, 1973. 510 p. ISBN 0-226-51274-6. $17.50. (pbk). LC 68-54054.   1798
   Massive, authoritative work by an urban geographer (Mayer) and an urban historian (Wade). Organized chronologically by six periods of development commencing with Prairie Seaport, 1830-1851. Illustrations include hundreds of black and white photos and numerous maps and sketches. Statistical and non-statistical appendices; extensive bibliography.

McManis, Douglas R. The Initial evaluation and utilization of the Illinois prairie, 1815-1840. (Research paper no. 94). Chicago, IL: University of Chicago, Department of Geography, 1964. 109 p. LC 64-23588   1799
   An elaborately documented study by a historical geographer. Examines how settlers from the forested East viewed certain prairies favorably and utilized them on a small-scale trial and error basis before John Deere's sod-breaking plow appeared.

Nelson, Ronald E., ed. Illinois: land and life in the Prairie State. Dubuque, IA: Kendall-Hunt, 1978. 359 p. ISBN 0-8403-1831-6. $13.95. LC 77-91584.   1800
   Authored by nine Illinois geographers on behalf of the Illinois Geographical Society; introductory chapter by the editor; other chapters cover physical environment, historical geography, population and social geography, agriculture, mining and manufacturing, Metro East (East St. Louis area), and the Chicago Metropolitan Area. Clear writing; useful black and white maps and photos; bibliography; and an appendix of Illinois chronology.

Rafferty, Milton D. Missouri: a geography. Boulder, CO: Westview Press, 1983. 250 p. ISBN 0-86531-068-8. $36. ISBN 0-86531-435-7. $18. (pbk). LC 82-23817.   1801
   Basic geographical information presented effectively under traditional systematic headings with long concluding chapters on metropolitan centers and geographic regions. Strong historical focus. A straight-forward account, the only comprehensive geography of Missouri in print, it contains a wealth of maps, photos, a good bibliography, and excellent statistical data.

Rafferty, Milton D. The Ozarks: land and life. Norman, OK: University of Oklahoma Press, 1980. 294 p. ISBN 0-8061-1582-3. $21.95. LC 79-4738.   1802
   A topically organized geography by a leading authority on the Ozark region (southern Missouri, northern Arkansas, and portions of Kansas and Oklahoma). Contains much historical and cultural detail. Topics include Ozark landforms and geology, climate, Indians, settlement, the Civil War and its consequences, transportation and communication, mining, agriculture, lumbering, recreation, non-material culture traits, and the cultural landscape. It is written in very readable style with numerous maps, photos, and references. A basic work that covers all facets of Ozark regional geography that could also double as an excellent text on the history and development of the Ozarks.

Santer, Richard A. Michigan: heart of the Great Lakes. Dubuque, IA: Kendall/Hunt, 1977. 364 p. ISBN 0-8403-1698-4. $10.95. LC 76-53268.   1803
   A general geographical introduction to the state, organized topically and by fourteen planning and development regions. Largely descriptive, offering a potpourri of many kinds of information in liberally subheaded text, tables, maps, graphs, and photos. Lengthy list of sources.

Schroeder, Walter A. Bibliography of Missouri geography: a guide to written material on places and regions of Missouri. Columbia, MO: University of Missouri-Columbia Extension Division, 1977. 260 p. NUC 79-126781.　　　　　　　　　　　　　　1804

Has 1550 annotated entries arranged topically in 28 sections with author and place-name index. Includes articles, books, atlases, dissertations, theses, government documents, and selected unpublished materials. Most literature cited is from fields other than geography, but all of it has some geographical perspective.

Schroeder, Walter A. Missouri water atlas. Jefferson City, MO: Missouri Department of Natural Resources, 1982. 97 p. LC Map 82-620017.　　　　　　　　　　　1805

Detailed atlas of black-and-white maps by a geographer with short textual discussions. Personnel from 26 government agencies reviewed parts or all of this atlas. Maps of the entire state and major drainage basins show physical background, drainage systems, hydrologic cycle, water quality and quantity, and use. Tables; list of references cited.

Schroeder, Walter A. Presettlement prairie of Missouri. (Natural history series no. 2). Jefferson City, MO: Missouri Department of Conservation, 1981. 37 p.　　　　1806

Scholarly monograph providing explanatory text for author's map published separately in 1981 under the same title, and by the same publisher, at a scale of 1:500,000. Monograph reproduces the map by sections on a 1:1,000,000 scale. Map was built up by archival and field research over more than a decade, with main reliance on field notes of U.S. government land surveyors. Bibliography of 115 items. (Volume listed in Monthly checklist of state publications, April, 1982, page 637, entry 6899).

Self, Huber. Environment and man in Kansas: a geographical analysis. Lawrence, KS: The Regents Press of Kansas, 1978. 288 p. ISBN 0-7006-0162-7. $19.95. LC 77-5867.　1807

Intended as a college-level textbook; filled with facts, maps, tables, including time-series. Traditional systematic organization: historical background, physical elements and regions, population, resources, economic activities, urban development, including thumbnail sketches of towns and cities; short chapter on health and education. Many photos; long bibliography. A useful reference for libraries.

Sommers, Lawrence M. et al. Michigan: a geography. (State geography series). Boulder, CO: Westview Press, 1984. 254 p. ISBN 0-86531-093-9. $35. ISBN 0-86531-435-7. $18. (pbk). LC 83-19791.　　　　　　　　　　　　　　　　　　　　1808

A volume done with competence and authority. Clear writing; useful maps, graphs, tables, photos, bibliography.

Sublett, Michael D. Farmers on the road: interfarm migration and the farming of noncontiguous land in three midwestern townships, 1939-1969. (Research paper no. 168). Chicago, IL: University of Chicago, Department of Geography, 1975. 214 p. ISBN 0-89065-075-6. $10. LC 75-8522.　　　　　　　　　　　　　　　　　　1809

A geographical field analysis that throws much light on the nature and problems of midwestern family farmers. Based on hundreds of interviews and an extensive examination of published and unpublished public records and other literature. Examines two kinds of responses to the contemporary cost-price squeeze in agriculture: moving to another farm and farm enlargement, and the farming of noncontiguous land. Tables, diagrams, full documentation, selected bibliography.

Weaver, John E. Prairie plants and their environment: a fifty-year study in the Midwest. Lincoln, NE: University of Nebraska Press, 1968. 276 p. LC 67-19160.　　1810

Wide-ranging summation of prairie topics by a major plant ecologist: nature and environment of prairie; ecological studies; soil relations; plant competition; woodlands; effects of drought and grazing; methods of study. Bibliography of literature cited; seperate bibliography of author's own work.

U.S. South

Webb, Walter Prescott. The Great Plains. Boston: Ginn and Co., 1931. 525 p. Reprinted: Lincoln, NE. University of Nebraska Press, 1981. 525 p. ISBN 0-8032-9702-5. $9.95. LC 81-1821.

Classic book by a geographically and ecologically attuned historian; has evoked much subsequent work and discussion by scholars, including geographers. Conceives of the Great Plains not as commonly delimited on physiographic maps, but as a more extensive environment wherein settlers from the humid East had to modify their institutions drastically in response to a combination of relatively level terrain, treelessness, and moisture deficiencies.

1811

Worster, Donald. Dust Bowl: the southern plains in the 1930s. New York: Oxford University Press, 1979. 277 p. ISBN 0-19-502550-4. $22. ISBN 0-19-503212-8. $8.95. (pbk). LC 78-27018.

A generalized account and personal commentary by an environmentally oriented historian including case studies of Cimarron County, Oklahoma, and Haskell County, Kansas. Evocative photos; some maps; extensive reference notes.

1812

J. THE SOUTH   CHARLES S. AIKEN

Southeastern geographer. (Association of American Geographers. Southeastern Division). 1- (1961-). 2 per annum. Department of Geography, University of North Carolina at Charlotte, Charlotte, NC, 28223. ISSN 0038-366X.

Focus on the American South with articles and reviews.

1813

Tall timbers ecology and management conference procedings. (1969-). Annual. Tall Timbers Research Station, Route 1, Box 160, Tallahassee, FL, 32312. ISSN 0082-1527.

Volume 16 (1982, 206 p.) is a volume of original articles on the southern plantation by a group of geographers including Aiken, Anderson, Hart, Hilliard, and Prunty. The studies treat various regions and themes and employ historical, economic, and humanistic approaches. One of a continuing series.

1814

Anderson, James R. A Geography of agriculture in the United States' Southeast. (Geography of world agriculture, vol. 2). Budapest, Hungary: Akadémiai Kiadó, 1973. 135 p. $8.30. LC 74-153998.

A detailed discussion of recent changes and contemporary patterns of agriculture in the southeastern states by one thoroughly familiar with the area and its agricultural geography.

1815

Eller, Ronald D. Miners, millhands, and mountaineers: industrialization of the Appalachian South, 1880-1930. Knoxville, TN: University of Tennessee Press, 1982. 336 p. ISBN 0-87049-340-X. $23.50. ISBN 0-87049-341-8. $12.50. (pbk). LC 81-16020.

A provocative analysis of the economic and social revolution in southern Appalachia between 1880 and the Great Depression that integrated the region into the national economy. The basic premise is that the socioeconomic conditions of the region are products of the modernization of American life rather than the stagnation of a pre-twentieth century mountain culture.

1816

Ford, Thomas R., ed. The Southern Appalachian region: a survey. Lexington, KY: University of Kentucky Press, 1962. 308 p. LC 62-13456.

A symposium on southern Appalachian population, society, and folk arts with emphasis on changes between 1935 and 1960. A dated but thorough and still useful study of this major 'problem area' of the nation.

1817

Goldfield, David R. Cotton fields and skyscrapers: southern city and region, 1607-1980. Baton Rouge, LA: Louisiana State University Press, 1982. 232 p. ISBN 0-8071-1029-9. $20. LC 82-6582.

An interpretation of southern urbanization from the perspective that rural lifestyle, race, and a colonial economy are distinctive aspects of the South's history that contributed to the development of distinctive cities in the region.

1818

Hart, John Fraser. The South. 2nd. ed. New York: Van Nostrand, 1976. 166 p. (1st ed., The Southeastern United States, 1967. 106 p.). ISBN 0-442-29754-8. 75-16550.   1819
  Brief, semipopular treatment, basically regional in approach, and containing much up-to-date and original material.

Kniffen, Fred B. Louisiana: its land and people. Baton Rouge, LA: Louisiana State University Press, 1968. 196 p. ISBN 0-8071-0549-X. $16.95. LC 68-13448.   1820
  A brief illustrated account of Louisiana and the many features of physical and human geography that make it unique.

Lewis, Peirce F. New Orleans: the making of an urban landscape. (Comparative metropolitan analysis project). Cambridge, MA: Ballinger, 1976. 115 p. ISBN 0-88410-433-8. $8.95. LC 76-4797.   1821
  A well-written, original interpretation of the geography of one of the South's oldest cities employing historical perspective.

Vance, Rupert B. Human geography in the South: a study in regional resources and human adequacy. 2nd. ed. Chapel Hill, NC: University of North Carolina Press, 1935. 596 p. Reprinted: New York: Russell and Russell, 1968. 596 p. (1st ed., 1932. 596 p.) LC 68-25051.   1822
  Written by a sociologist, this treatment of the society, culture, and economy of the south is a dated but important source on the region prior to World War II.

K. THE WEST   JAMES W. SCOTT

Association of Pacific Coast Geographers. Yearbook. 1- (1935-). Annual. Oregon State University Press, 101 Waldo, Corvallis, OR. 97331. ISSN 0066-9628.   1823
  Articles on the Pacific Coast. Selected papers delivered at the annual meetings of the Association.

California geographer. (California Council for Geographic Education). 1- (1960-). Annual. Place of publication varies. LC 61-31909.   1824
  Articles on California or by California geographers.

Brown, Robert Harold. Wyoming: a geography. Boulder, CO: Westview Press, 1980. 375 p. ISBN 0-89158-560-5. $38.50. ISBN 0-686-96923-5. $20. (pbk). LC 80-16823.   1825
  A comprehensive treatment that emphasizes the historical; political; cultural; economic; and environmental aspects of the state's geography.

Comeaux, Malcolm L. Arizona: a geography. Boulder, CO: Westview Press, 1981. 336 p. ISBN 0-89158-563-X. $37.50. LC 80-13119.   1826
  A comprehensive text that emphasizes the cultural/historical and economic geography of Arizona.

Dicken, Samuel N., and Dicken, Emily F. Two centuries of Oregon geography. Portland, OR: Oregon Historical Society.
  Vol. 1: **The Making of Oregon: a study in historical geography.** 1979. 208 p. ISBN 0-87595-081-7. $19.95. ISBN 0-87595-061-7. $12.50. (pbk). LC 79-89087.
  Vol. 2: **Oregon divided: a regional geography.** 1982. 176 p. ISBN 0-87595-082-5. $16.95. ISBN 0-87595-064-7. $10.95. (pbk). LC 80-84480.   1827
  A meticulously researched and finely organized, if somewhat traditional, regional set that covers all aspects of Oregon's geography.

Easterbrook, Don J., and Rahm, David A. Landforms of Washington; the geologic environment. Bellingham, WA: Union Printing Co., 1970. 156 p. LC 72-15719.   1828
  A profusely illustrated volume that describes and analyzes the state's major physiographic regions.

Gibson, James R. Imperial Russia in frontier America: the changing geography of supply of Russian America, 1784-1867. New York, Oxford University Press, 1976. 257 p. ISBN 0-19-501875-3. $9.95. LC 74-21820.   1829

A major historical-geographical study of Russian Alaska soundly based on American and Russian archival sources.

Gregor, Howard F. An Agricultural typology of California. (Geography of world agriculture, v. 4). Budapest, Hungary: Akadémiai Kiadó, 1974. 107 p. ISBN 963-05-0174-0. $6.60. LC 75-311527.   1830

An authorative study of the major farm types of California agriculture.

Lantis, David W.; Steiner, Rodney; and Karinen, Arthur E. California: land of contrasts. 3rd rev. ed. Dubuque, IA: Kendall-Hunt, 1977. 486 p. (1st ed., 1963. 509 p.). ISBN 0-8403-0768-6. $15.95. LC 77-356160.   1831

A detailed standard work on the regional geography of California.

Liebman, Ellen. California farmland: a history of large agricultural landholdings. Totowa, NJ: Rowman and Allanheld, 1983. 226 p. ISBN 0-86598-107-8. $29.50. LC 82-20759.   1832

A valuable, objective study by a geographer of one of the more controversial aspects of California agriculture.

McGregor, Alexander Campbell. Counting sheep: from open range to agribusiness on the Columbia Plateau. Seattle, WA: University of Washington Press, 1983. 482 p. ISBN 0-295-95894-4. $25. LC 82-15903.   1833

An important contribution to the history and historical geography of grazing in the Columbia Basin during the past century. Heavily dependent on original sources.

McKee, Bates. Cascadia: the geologic evolution of the Pacific Northwest. New York: McGraw-Hill, 1972. 394 p. LC 74-169022.   1834

The standard work on the geology and geomorphology of the Pacific Northwest.

Miller, Crane S. and Hyslop, Richard S. California: a geography of diversity. Palo Alto, CA: Mayfield Publishing Company, 1983. 255 p. ISBN 0-87484-441-X. $16.95. LC 82-73744.   1835

A topical approach to the geography of the state. Stronger on the physical than the cultural aspects.

Miller, Orlano W. The Frontier in Alaska and the Matanuska Colony. (Western America series no. 26). New Haven, CT: Yale University Press, 1975. 329 p. ISBN 0-300-01638-7. $30. LC 74-82747.   1836

A detailed account of the establishment and the changing fortunes of the Matanuska Colony, and a valuable commentary on frontier settlement in high latitudes.

Nelson, Howard J. The Los Angeles metropolis. Dubuque, IA: Kendall-Hunt, 1983. 344 p. ISBN 0-8403-2939-3. $19.95. (pbk). LC 82-84490.   1837

Informative geographic monograph on the second most populous metropolitan region in the U.S. Includes suggested readings at the end of each chapter and a bibliography.

Reno, Philip. Mother earth, father sky, and economic resources: Navajo resources and their use. Alburquerque, NM: University of New Mexico Press, 1981. 183 p. ISBN 0-8263-0653-5. $7.95. (pbk). LC 80-54859.   1838

Discussion of the Navajo resource base and how the tribe is utilizing these resources in becoming a part of the overall American economy.

Robbins, William G.; Frank, Robert J.; and Ross, Richard E., eds. Regionalism and the Pacific Northwest. Corvallis, OR: Oregon State University Press, 1983. 256 p. ISBN 0-87071-337-X. $16.95. ISBN 0-87071-338-8. $9.95. (pbk). LC 83-2416.   1839

An essay collection that provides perspective on regionalism and explores the nature of the environments and the regional identity of the Pacific Northwest.

Seckler, David, ed. California water: a study in resource management. Berkeley, CA: University of California Press, 1975. 348 p. ISBN 0-520-01884-2. $48.50. LC 76-139773. 1840
A somewhat outdated, but still valuable treatment of water and water use in a sub-humid climate.

Unruh, John D., Jr. The Plains across: the overland emigrants and the Trans-Mississippi West, 1840-1860. Urbana, IL: University of Illinois Press, 1979. 565 p. ISBN 0-252-00698-4. $27.50. LC 78-9781. 1841
A major monograph that challenges the many myths about the overland passage, and provides comprehensive analyses of two decades of emigration to the American West.

White, Richard. Land use, environment, and social change: the shaping of Island County, Washington. Seattle, WA: University of Washington Press, 1980. 234 p. ISBN 0-295-95691-7. $17.50. LC 79-4845. 1842
A valuable regional study of land use and misuse in a small Washington county from Indian times to the present, and of the social consequences of environmental change.

4. CANADA*

R. Cole Harris and J. Keith Fraser

*In this section the editorial committee increased the number of entries recommended by the contributors.

A. SERIALS

Cahiers de géographie du Québec. 1- (1956-). 3 per annum. Les Presses de l'Université Laval, Boîte postal 2447, Québec, PQ G1K 7R4, Canada. ISSN 0007-9766. 1843
Devoted particularly to Quebec, Canada, and to cultural, economic, and regional geography. In French or English. Abstracts in English and French precede each article. Cumulative indexes: nos. 1-42 (1956-1973), 43-66 (1974-1981).

Canadian geographic. 1- (1930-). Bimonthly. The Royal Canadian Geographical Society, 488 Wilbrod Street, Ottawa, Ontario K1N 6M8, Canada. ISSN 0706-2168. 1844
Well-illustrated popular geographic articles on Canada and other parts of the world. Book reviews. Cumulative index: 1-99 (1930-1979).

See also: Canadian geographer (31).
        Cartographica (251).
        Géographie physique et quaternaire (437).

B. ATLASES AND MAPS

Nicholson, Norman L. and Selbert, L.M. The Maps of Canada: a guide to official Canadian maps, charts, atlases and gazetteers. Folkestone, U.K.; Mississauga, Ontario: William Dawson, 1981; Hamden, CT: Archon, 1982. 251 p. ISBN 0-208-01782-8. $35. LC 79-41118. 1845

Tessier, Yves. Carto-03: répertoire cartobibliographique sur la région de Québec. Québec: Cartothèque. Bibliothèque de l'Université Laval, 1983. 269 p. ISBN 2-920310-01-1. Can $15. LC 84-120601. 1846

Canada

Thompson, M.D.; Howarth, P.J.; Reyerson, R.A.; and Bonn, F.J. Landsat for monitoring the changing geography of Canada. Ottawa: Canadian Centre for Remote Sensing, 1982. 84 p. Free.   1847

Alberta. University, Edmonton. Dept. of Geography. Atlas of Alberta. Edmonton, Alberta: University of Alberta Press, in association with University of Toronto Press, 1969. 158 p. LC 78-653863.   1848•
Beautifully executed thematic atlas divided into the following subject categories: relief and geology; climate; water; vegetation; soil; wildlife; history; population; land use; agriculture; forestry; minerals; power; manufacturing; service; settlement patterns; and administration.

Canada. Surveys and Mapping Branch. Geography Division. The National atlas of Canada. 4th revised ed. Toronto: Macmillan Company of Canada, 1974. 254 p. ISBN 0-7705-1198-8. Can $ 56. LC 76-351950.   1849•
Indispensable thematic atlas on the physical, demographic, social, and economic characteristics of Canada containing double-page and smaller maps of the country printed on 254 plates. Based on the Canadian censuses of 1961 and 1971. Atlas is also available in French language edition.

Canada gazetteer atlas. Toronto: Macmillan of Canada in cooperation with Energy, Mines, and Resources Canada, and the Canadian Government Publishing Centre; Chicago, IL: University of Chicago Press, 1980. 164 p. ISBN 0-7705-1873-7. Can $39.95. ISBN 0-7705-1873-7. $65. LC 81-675339. (Also available in French: Canada atlas toponymique. Montréal: Guérin, 1980. 164 p. LC 80-136035).   1850
Consists of 48 regional maps and a detailed index, with population data collected for the 1976 Census on Canada. Particularly valuable for location of smaller places in Canada.

Dean, William G., ed., and Matthews, Geoffrey J., cartographer. Economic atlas of Ontario. Toronto: University of Toronto Press for the Government of Ontario, 1969. 113 plates. Bilingual. ISBN 0-8020-3235-4. Can $100. LC 73-653512.   1851•
The finest atlas of a Canadian province. The beautifully executed maps are organized into the following ten sections: aggregate economy; population; manufacturing; resource industries; wholesale and consumer trade; agriculture; recreation; transportation and communications; administration; and reference maps. Table of contents, text, and captions are written in English and French.

Farley, Albert L. Atlas of British Columbia: people, environment, and resource use. Vancouver, BC: University of British Columbia Press, 1979. 135 p. ISBN 0-7748-0092-5. Can $24.95. LC 79-689772.   1852•
Excellent thematic atlas divided into three sections entitled "People," "Environment," and "Resource Use."

Newfoundland. Federal-Provincial Task Force on Forestry. Land Capability-Land use sub-committee. Resource atlas, Island of Newfoundland. Ottawa: Surveys and Mapping Branch, Department of Energy, Mines, and Resources, 1974. 14 colored maps. LC 80-675543.   1853•
The atlas consists of a series of maps (at a scale of 1:1,000,000) indicating the known capability for forestry, recreation, wildlife (ungulates), and agriculture; a land capability analysis map; a series of resource use maps; and a resource use composite map.

Nova Scotia. Department of Lands and Forests. A book of maps: Land use and natural resources of Nova Scotia. Halifax, NS: Department of Lands and Forests, 1977. 48 p. LC 83-675180.   1854•
Twenty-four thematic maps drawn on a uniform scale (ca. 1:1,720,000) by the Maritime Resource Management Service, Amherst, Nova Scotia.

Richards, J. Howard, editor and director, and Fung, Ka Iu, cartographic editor. Atlas of Saskatchewan. Saskatoon, Saskatchewan: University of Saskatchewan, 1969. 236 p. LC 76-654088.
1855 •
Excellent thematic atlas arranged in 12 sections: introductory physical geography; historical geography; population geography; physical geography; zoogeography; forest geography; agricultural geography; minerals; commercial fish and fur; recreation; industry, services and circulation; urban geography and administrative areas; and statistical summary.

C. BOOKS

Bird, J. Brian (John). The Natural landscapes of Canada: a study in regional earth science. 2nd ed. Toronto: Wiley Canada, 1980. 260 p. (1st ed., 1972. 191 p.). ISBN 0-471-79952-1. Can. $14.95. ISBN 0-471-998-10-9. Can. $9.95. (pbk). C 79-94725-0
1856 •
A revealing overview of the formational factors that have influenced the areal characteristics of the Canadian landscapes.

Bird, J. Brian (John). The Physiography of Arctic Canada. Baltimore, MD: The Johns Hopkins University Press, 1967. 336 p. LC 67-16232.
1857
Background, major episodes, geomorphic processes, and special elements in landform development in the Canadian Arctic.

Bocking, Richard C. Canada's water: for sale? Toronto: James Lewis and Samuel (available through James Lorimer, Toronto), 1972. 188 p. ISBN 0-88862-028-4. Can. $12.95. ISBN 0-88862-044-6. Can. $3.75. (pbk). LC 72-90591.
1858 •
Water-resource development in the United States cannot be divorced from that in Canada, yet Americans eye Canadian water as if it were Yankee property. Bocking, a Canadian journalist, conveys the mood and concern in Canada regarding this perception problem in his incisive look into water management in both the U.S. and Canada.

Brody, Hugh. Maps and dreams: Indians and the British Columbia frontier. Vancouver, BC: Douglas and Mcintyre, 1981; London: Norman and Hobhouse, 1982; Penguin, 1983; New York: Pantheon, 1982. 297 p. ISBN 0-88894-338-5. Can. $19.95. ISBN 0-394-52104-8. $16.50. ISBN 0-394-74871-9. $7.95. (pbk). C 81-91269-4. LC 83-11368.
1859 •
Alternate descriptive and analytical chapters present a portrait of the Indians of northeastern British Columbia and their continuing struggles for survival.

Clayton, J.S. et al. Soils of Canada. Ottawa, Ontario: Canadian Government, 1977. 2 v. LC 78-321981.
Vol. 1: **Soil report.** ISBN 0-660-00502-6. Can. $25. (pbk).
Vol. 2: **Inventory.** ISBN 0-660-00503-4. Can. $25. (pbk).
1860
These volumes are the result of a cooperative project sponsored by the Canadian Soil Survey Committee and the Soil Research Institute of Canada.

Gentilcore, R. Louis (Rocco), and Head, C. Grant. Ontario's history in maps. Toronto; Buffalo, NY: University of Toronto Press, 1983. 284 p. ISBN 0-8020-3400-4. $65.
1861 •
In nine introductory essays the major aspects of Ontario's development are related to maps which reflect the province's history and the art of cartography.

Hamelin, Louis-Edmond. Canadian nordicity: it's your north, too. Translated from French by William Barr. Montreal, Québec: Harvest House, 1978. 373 p. ISBN 0-88772-175-3. Can. $12.95. (pbk). LC 80-466639. (Nordicité canadienne, Montréal: Hurtubise HMH, 1975. 458 p.).
1862 •
Portraits of the Canadian North; the diversity of its peoples and environments. Examines the ways in which the region has been integrated into the Canadian political economy and their implications.

Canada

Hare, F. Kenneth (Frederick), and Thomas, Morley K. Climate Canada. 2nd ed. Toronto: Wiley Canada, 1979; New York: Wiley, 1980. 230 p. (1st ed., 1974. 256 p.). ISBN 0- 471-99796-X. Can. $14.95. ISBN 0-471-08326-7. $29.95. (U.S.). LC 81-151093. C 79-94180-5.
    A good general overview of the various climates of Canada and their effects on the physical landscape.

1863

Harris, R. Cole (Richard Colebrook), and Warkentin, John. Canada before confederation: a study in historical geography. London; New York: Oxford University Press, 1974. 338 p. ISBN 0-19-501791-9. $9.55. (pbk). LC 73-87622.
    A regional survey of the historical geography of Canada before 1867.

1864 •

Hosie, Robert C. Native trees of Canada. 8th ed. Don Mills, Ontario: Fitzhenry and Whiteside; Ottawa, Ontario: Canadian Forestry Service, 1979. 380 p. (1st ed. by B.R. Morton). ISBN 0-88902-572-X. Can. $19.95. ISBN 0-88902-558-4. Can. $12.95. (pbk). C 80-12887 LC 80-466588 rev.
    An excellent reference concerning the trees and forests of Canada.

1865

Krueger, Ralph R. and Mitchell, Bruce, eds. Managing Canada's renewable resources. Toronto: Methuen, 1977. 333 p. ISBN 0-458-92350-8. Can. $14.95. (pbk). C 77-001128-4. LC 77-371033.
    A series of essays which provide an analytical framework for, and a number of specific case studies of, resource development.

1866 •

Lawrence, R.D. Canada's national parks. Toronto: Collins, 1983. ISBN 0-00-216890-1. Can. $44.95.
    Following a brief introduction to the parks system and its objectives, the landscapes, flora, fauna, and policies of each of the 29 parks (1983) are described.

1867

McBoyle, Geoffrey R. and Sommerville, Edward, eds. Canada's natural environment: essays in applied geography. Toronto: Methuen, 1976. 264 p. ISBN 0-458-91930-6. Can. $10.50. (pbk). C 76-017017-7. LC 76-376603.
    A series of essays concerned with the role of geographers in studying and managing the environment.

1868 •

McCann, Lawrence D., ed. Heartland and hinterland: a geography of Canada. Scarborough, Ont.: Prentice-Hall of Canada, 1982. 500 p. ISBN 0-13-385146-X. Can. $28.95. C 82-94100-X.
    A set of essays dealing with the evolving pattern and present character of the Canadian spatial economy. The collection is built around the analysis of core-periphery relationships; within this framework the organization is regional.

1869 •

Mitchell, Bruce, and Sewell, W.R. Derrick. Canadian resource policies: problems and prospects. Toronto: Methuen, 1981. 294 p. ISBN 0-458-94970-1. Can. $13.95. (pbk). C 81-94648-3.
    A series of essays which examine specific aspects of resources development and management in Canada.

1870 •

Nader, George A. Cities of Canada. Toronto: Macmillan of Canada, 1975-1976. 2 v. ISBN 0-77051-029-9. C 75-8440.
    Vol. 1: **Theoretical, historical, and planning perspectives.** 388 p.
    Vol. 2: **Profiles of fifteen metropolitan centres.** 453 p. ISBN 0-7705-1239-9. Can. $11.95. LC 74-10295.
    A theoretical and empirical survey of Canadian urbanization; focuses on major metropolitan centers.

1871

Nelson, James Gordon. Man's impact on the Western Canadian landscape. (Carleton Library no. 90). Toronto: McClelland and Stewart, 1976. 205 p. ISBN 0-7710-9790-5. Can. $4.95. (pbk). C 76-2987-3. LC 77-359147.

1872 •

A series of essays examining historical and contemporary processes shaping the human landscape in Western Canada.

Ray, Arthur J. Indians in the fur trade: their role as trappers, hunters, and middlemen in the lands southwest of Hudson Bay, 1660-1860. Toronto; Buffalo, NY: University of Toronto Press, 1974; 1978. 249 p. ISBN 0-8020-6226-1. Can. $10.95. (pbk). U.S.$ 10.95. (pbk). LC 73-89848.
    An account of the changing spatial pattern and human ecology of Indian involvement in the fur trade in Western Canada.

1873•

Ritchie, J.C. Past and present vegetation of the far Northwest of Canada. Toronto; Buffalo, NY: University of Toronto Press, 1984. 251 p. ISBN 0-8020-2523-4. Can. $35. U.S. $35.
    An excellent description of the physical and biological setting of the northern Yukon and Mackenzie River Delta region in the past and modern times. Covers an area largely unglaciated during the last ice age.

1874

Robinson, J. Lewis (John). Concepts and themes in the regional geography of Canada. Vancouver, BC: Talonbooks, 1983. 342 p. ISBN 0-88922-205-3. ISBN 0-88922-204-5. Can. $14.95. (pbk). LC 83-238512.
    A comprehensive regional geography, largely but not entirely contemporary in focus, providing much information and a general interpretation of the main areas of Canadian settlement.

1875•

Ross, Eric. Beyond the river and the bay: the Canadian Northwest in 1811. Toronto; Buffalo, NY: University of Toronto Press, 1970. 190 p. ISBN 0-8020-6188-5. $6.95. (pbk). LC 71-486954.
    A graceful, meticulously researched account, written in the style of the early 19th century, of the state of the Canadian Northwest in 1811.

1876•

Rowe, J.S. Forest regions of Canada. (Canadian Forestry publication no. 1300). Ottawa, Ontario: Information Canada, 1972. 172 p. (Based on W.E.D. Halliday's "A forest classification for Canada"). (Available from Falconiforme, Can. $2.50). LC 77-375359.
    Classifies the forest regions of Canada and portrays them on an accompanying map.

1877•

Simpson-Lewis, Wendy; McKechnie, R.; and Neimanis, V., eds. Stress on land in Canada. Ottawa, Ontario: Environment Canada, 1984. 323 p. ISBN 0-660-11367-8. Can. $ 18. (Also available in French: Les terres du Canada: stress et impacts. 1983. 349 p.).
    The impact of human activities on land is described in nine chapters ranging from localized effects of airports to whole industries such as agriculture and forestry. The approach is thematic with special attention to particular examples, supported by statistics.

1878•

Trotier, Louis, ed. Studies in Canadian geography. Toronto; Buffalo, NY: University of Toronto Press, 1972. (Published for the 22nd International Geographical Congress). 6 v. NUC 73-118895.
    Vol. 1: **The Atlantic provinces.** Macpherson, Alan, ed. 182 p. ISBN 0-8020-6158-3. $6. (pbk). LC 72-196413.
    Vol. 2: **Québec.** Grenier, Fernand, ed. 110 p. ISBN 0-8020-6159-1. $6. (pbk). LC 72-223945.
    Vol. 3: **Ontario.** Gentilcore, R. Louis, ed. 126 p. ISBN 0-8020-6160-5. $6. (pbk). LC 72-196549.
    Vol. 4: **The Prairie provinces.** Smith, P.J., ed. 141 p. ISBN 0-8020-6161-3. $6. (pbk). LC 72-196554.
    Vol. 5: **British Columbia.** Robinson, J. Lewis, ed. 139 p. ISBN 0-8020-6162-1. $6. (pbk). LC 72-197301.
    Vol. 6: **The North.** Wonders, William C., ed. 151 p. ISBN 0-8020-6164-8.
    Studies in the physical, economic, population, and urban geography of six regions in Canada by leading geographers of the country. Bibliographies.

1879•

Latin America

Troughton, Michael J. Canadian agriculture. (Geography of world agriculture, v. 10). Budapest, Hungary: Akadémiai Kiadó, 1982. 355 p. ISSN 963-05-2653-0. $38.
One of an excellent series of volumes that discuss all aspects of agriculture.

1880

Warkentin, John, ed. Canada: a geographical interpretation. Toronto; London: Methuen, 1968. 608 p. ISBN 0-458-95360-1. £14. C 68-1984. LC 68-82888. (Also in French: Le Canada: une interprétation géographique. 1970. 645 p.).
Prepared under the auspices of the Canadian Association of Geographers for the centennial of 1967, this is still a useful collection of topical and regional essays on Canada.

1881 •

Wynn, Graeme. Timber colony: a historical geography of early nineteenth century New Brunswick. Toronto; Buffalo, NY: University of Toronto Press, 1981. 224 p. ISBN 0-8020-5513-3. $25. ISBN 0-8020-6407-8. $10.95. (pbk). C 81-94014-0. LC 81-188246.
A richly textured study of the spatial patterns, regional economy, and settlements associated with a staple trade.

1882

Yeates, Maurice H. Main Street: Windsor to Quebec City. Toronto: Macmillan of Canada; Agincourt, Ontario: Gage, (in association with the Ministry of State for Urban Affairs and Information Canada), 1975. 431 p. ISBN 0-7705-1304-2. Can. $17.50. ISBN 0-7705-1305-0. Can. $5.95. (pbk). LC 76-358141.
A detailed study of the principal urbanized region of Canada, the principal locus of social and economic power in the country.

1883 •

5. LATIN AMERICA

Tom L. Martinson

A. BIBLIOGRAPHIES AND GEOGRAPHICAL REFERENCES

Bibliographic guide to Latin American studies; Guía bibliográfica para los estudios lationamericanos. 1- (1978-). Annual. G.K. Hall, 70 Lincoln St. Boston, MA, 02111. ISSN 0162-5314. LC 79-643128.
Annual continuation of the Catalog of the Latin American Collection of the Library of the University of Texas (entries 1902-1906).

1884

Cordeiro, Daniel R. A Bibliography of Latin American bibliographies: Social Sciences and Humanities. Metuchen, NJ: Scarecrow Press, 1979. ISBN 0-8108-1170-7. $17.50. LC 78-11935.
One of a basic series of bibliographies, preceded by Jones, Gropp, and Gropp supplement and succeeded by Piedracueva (entries 1893, 1888, 1889, 1897).

1885

Davidson, William V., comp. Geographical research on Latin America: a cartographic guide and bibliography of theses and dissertations, 1909-1978. Muncie, IN: Conference of Latin Americanist Geographers, 1980. 52 p. LC 80-80557.
An unannotated list of 800 theses and dissertations on Latin America, including a chart on graduate degree production in Latin American geography.

1886

Gerhard, Peter. A Guide to the historical geography of New Spain. (Latin American studies no. 14). Cambridge, U.K.; New York: Cambridge University Press, 1972. 476 p. ISBN 0-521-08073-8. $90. LC 72-163058.
Coverage is central and southern Mexico in the era of the Viceroy. Based on extensive archival research.

1887

Gropp, Arthur E., comp. A Bibliography of Latin American bibliographies. Metuchen, NJ: Scarecrow Press, 1968. 515 p. LC 68-9330.　　1888
An effort to update Jones' 1942 monumental bibliography by adding 4000 new references, all published before 1965. Subject arrangement; geographical arrangement by country within subject. Supplements have been provided by Gropp, 1971; Cordeiro, 1979; and Piedracueva, 1982 (entries 1893, 1889, 1885, 1897).

Gropp, Arthur E., comp. A Bibliography of Latin American bibliographies; supplement. Metuchen, NJ: Scarecrow Press, 1971. 277 p. ISBN 0-8108-0350-X. LC 68-9330.　1889
Updates the 1968 volume in the same format (entry 1888).

Gropp, Arthur E. A Bibliography of Latin American bibliographies published in periodicals. Metuchen, NJ: Scarecrow Press, 1976. 2 v., 1031 p. ISBN 0-8108-08382. LC 75-32552.　　1890
A companion work to his Bibliography of Latin American bibliographies.

Handbook of Latin American studies. 1- (1935-). Annual. University of Texas Press, Box 7819, Austin, TX, 78712. ISSN 0072-9833. LC 36-32633.　　1891
Prepared by a number of scholars for the Hispanic Division of the Library of Congress. Relatively comprehensive coverage of the literature on Latin America. Since volume 26 (1964), divided into two volumes, one for the Humanities, the other for the Social Sciences (including geography), each published in alternate years.

Hispanic American periodicals index (HAPI), 1976-1980. Los Angeles, CA: University of California at Los Angeles Latin American Center Publications and the Faxon Press, 1981. ISBN 0-87903-408-4. $200. LC 75-642408.　　1892
Contains current bibliographical information on events, trends, people, and culture in Latin America and Hispanic populations in the United States.

Jones, Cecil K. A Bibliography of Latin American bibliographies. (U.S. Library of Congress Latin American series no. 2). 2nd ed. Revised and enlarged by the author with the assistance of James A. Granier. Washington, DC: Government Printing Office, 1942. 311 p. Reprinted: Westport, CT: Greenwood Press, 1969. 307 p. (1st ed. as Hispanic American bibliographies, 1922. 200 p.). ISBN 0-8371-1160-9. $21. LC 69-13955.　1893
The monumental first effort to develop a comprehensive Latin American bibliography. Organized by country. Most entries are annotated.

Matos, Antonio, comp. and ed. Guía a las reseñas de libros de y sobre Hispanoamérica. A Guide to reviews from and about Hispanic America, 1960/64. Annual. Text in English, Spanish, and Portuguese. Detroit, MI: Blaine-Ethridge Books, 1960/64-. LC 66-96537.　　1894
Annual guide indexes reviews from about 580 of the principal review media in the Social Sciences and Humanities.

Mikesell, Marvin W., ed. Geographers abroad; essays on the problems and prospects of research in foreign areas. (Research paper no. 152). Chicago, IL: University of Chicago, Department of Geography, 1973. ISBN 0-89065-059-4. $10. LC 73-87829.　　1895
Current research by geographers on Latin America is discussed by James J. Parsons, "Latin America," p. 16-46 and in the West Indies by David Lowenthal, "The Caribbean region," p. 47-64.

Pan American Union. Department of Economic Affairs. Indice anotado de los trabajos aerofotográficos y mapas topográficos y de recursos naturales realizados en los países de la América Latina miembros de la OEA. Annotated index of aerial photographic coverage and mapping of topography and natural resources undertaken in the Latin American member countries of the OAS. Washington, DC: Pan American Union, 1964-1966. 19 v. Colored maps 45 x 56 cm. Text in Spanish and English except for Haiti, which is in French and English.

    Argentina, 1965. 28 p. LC Map 67-670.
    Bolivia, 1964. 13 p. LC Map 67-668.
    Brasil, 1965. 34 p. LC Map 67-639.
    Chile, 1964. 20 p. LC Map 67-656.
    Colombia, 1966. 31 p. LC Map 67-656.
    Costa Rica, 1965. 18 p. LC Map 67-636.
    Ecuador, 1964. 17 p. LC Map 67-635.
    Guatemala, 1965. 13 p. LC Map 67-671.
    Haiti, 1964. 9 p. LC Map 67-652.
    Honduras, 1965. 11 p. LC Map 67-657.
    México, 1965. 33 p. LC Map 67-650.
    Nicaragua, 1965. 13 p. LC Map 67-649.
    Panama, 1965. 13 p. LC Map 67-648.
    Paraguay, 1967. 9 p. LC Map 67-647.
    Peru, 1964. 26 p. LC Map 67-646.
    República Dominicana, 1964. 9 p. LC Map 67-645.
    El Salvador, 1965. 11 p. LC Map 67-644.
    Uruguay, 1964. 13 p. LC Map 67-643.
    Venezuela, 1964. 28 p. LC Map 67-637.   **1896**

Inventories all known aerial photography and topographic and thematic mapping of natural resources.

Piedracueva, Haydee, ed. A Bibliography of Latin American bibliographies, 1975 - 1979; Social Sciences and Humanities. Metuchen, NJ: Scarecrow Press, 1982. 313 p. ISBN 0-8108-1524-9. $25. LC 82-651.   **1897**

Includes monographs and periodical literature. See entries 1893, 1888, 1889, 1885. Third update of Gropp (1968).

The South American handbook including Caribbean, Mexico, and Central America. 1- (1924-). Annual. Trade and Travel Publications, Bath, BA1 1EN, U.K.; Rand McNally, 8255 N. Central Park, Box 728, Skokie, IL, 60076. ISSN 0081-2579. LC 25-514.   **1898**

Invaluable compendium for the geographer and the general traveler.

United Nations. Economic Commission for Latin America. Anuario estadístico de América Latina. Statistical yearbook for Latin America, 1973- . Annual. United Nations Sales Section, Rm. A-3315, New York, New York, 10017. ISSN 0041-6401. LC 75-645451.   **1899**

Statistics in this annual publication cover most Latin American countries and most topics, but the data often lag two or three years behind the publication date.

University of California at Los Angeles. Latin American Center. Statistical abstract of Latin America, 1- (1955-). Annual. University of California, Los Angeles, Latin American Center, Los Angeles, CA, 90024. ISSN 0081-4687. LC 56-63569.   **1900**

A convenient reference with a wealth of statistical data effectively arranged for both scholar and layman.

University of Texas Library, Austin. Catalog of the Latin American collection. Boston, MA: G.K. Hall, 1969. 31 v. ISBN 0-8161-08153. LC 70-10540.   **1902**

A printed card catalog arranged by author. Some 175,000 works represented.

University of Texas Library, Austin. Catalog of the Latin American collection, first supplement. Boston, MA: G.K. Hall, 1971. 5 v. ISBN 0-8161-1281-9. $430. (microfilm only). LC 75-646157.

Supplement to the 1969 volumes containing some 27,000 further entries in the same format.

1903

University of Texas Library, Austin. Catalog of the Latin American collection, second supplement. Boston, MA: G.K. Hall, 1973. 3 v. ISBN 0-8161-0979-6. ISBN 0-8161-1325-4. $340. (microfilm).

Updates the original volumes and first supplement.

1904

University of Texas Library, Austin. Catalog of the Latin American collection, third supplement. Boston, MA: G.K. Hall, 1975. 8 v. ISBN 0-8161-1107-3. ISBN 0-8161-1360-2. $930. (microfilm).

Updates earlier volumes and supplements.

1905

University of Texas Library, Austin. Catalog of the Latin American collection, fourth supplement. Boston, MA: G.K. Hall, 1977. 3 v. ISBN 0-8161-1156-1. $330. (microfilm).

Fourth and last supplement under the same title as the original (1969) volumes. Continued by Bibliographic guide to Latin Amerian Studies (entry 1884).

1906

Veliz, Claudio. Latin America and the Caribbean: a handbook. New York: Praeger, 1968. 840 p. LC 68-14143.

Thumbnail sketches of each nation with special reference to their political affairs, economic affairs, social background, contemporary arts, and history; with a fine set of maps.

1907

Wagley, Charles, ed. Social science research in Latin America. New York: Columbia University Press, 1964. LC 65-11971.

Includes a masterful resumé contributed by James J. Parsons, of important geographical research works in the 1950's and 1960's, "The contributions of geography to Latin American studies," p. 33-85.

1908

B. SERIALS

Anuario de geografía. 1- (1961-). Annual. Facultad de Filosofía y Letras, Colegio de Geografía, Cuidad Universitaria, México 20, DF, México. ISSN 0570-4073.

Large annual volume in Spanish of articles on Mexico.

1909

Caribbean geography. 1- (1983). Semiannual. Longman Jamaica, P.O,. Box 489, 6 Montrose Road, Kingston 10, Jamaica. LC 84-645680.

Devoted to the Caribbean region with some emphasis on Anglophone countries but not limited to them. Reviews and reports. In English.

1910

Ibero Americana. 1- (1932-). Irregular. University of California Press, 2223 Fulton St. Berkeley, CA 94720. ISSN 0073-4349.

Wide-ranging monographs on Latin American cultural history. At least 53 publications.

1911

Inter-American economic affairs. 1- (1947-). Quarterly. Inter-American Affairs Press, P.O. Box 181, Washington, DC 20044. ISSN 0020-4943. LC 50-1412.

Valuable contributions on economic issues from several disciplines.

1912

Journal of developing areas. 1- (1966). Quarterly. Western Illinois University, Macomb, IL 61455. ISSN 0022-037X.

Contributions from several disciplines with an emphasis on development issues.

1913

Latin America

Latin American research review. 1- (1965-). Three issues per year. University of New Mexico Press, 801 Yale NE. University of New Mexico, Albuquerque, NM 87106. ISSN 0023-8791. LC 65-9960.   1914
The leading interdisciplinary journal. Lengthy research articles and news of the Latin American Studies Association.

Revista Brasileira de geografía. 1- (1939-). Quarterly. Fundacao Instituto Brasileiro de Geografía e Estatistica, Av. Brasil 15671, Lucas ZC 91, Rio de Janiero, RJ, Brazil. ISSN 0034-723X.   1915
The leading geographical serial in Brazil. Current and prestigious. In Portuguese. Summaries in English and French.

Revista geográfica. (Instituto Panamericano de Geografía e Historia). 1- (1941-). 2 per annum. Servicios Bibliográficos, Secretaria General del IPGH, Ex-Arzobispado 29, México 18, DF, México. ISSN 0556-6630.   1916
The leading geographical periodical devoted to the Americas. Articles, bibliographies, notes, revi ws. Notes on the state of geography in American countries. Official reports and resolutions. In Spanish, English, or Portuguese. Cumulative indexes 1-63 (1941-1965), 64-77 (1966-1972), 78-90 (1973-1979).

Revista geográfica de América Central. 1- (1974-). Semiannual. Escuela de Ciencias Geográficas, Facultad de Ciencias de la Tierra y el Mar, Universidad Nacional, Heredia, Costa Rica.   1917
Articles particularly on Costa Rica and Central America. In Spanish. Abstracts in English and Spanish precede each article.

C. ATLASES

Arbingast, Stanley, and Hezlep, William. Atlas of Central America. Austin, TX: Bureau of Business Research, University of Texas, 1979. 62 p. ISBN 0-87755-262-2. $18. (pbk). LC 78-64336.   1918
Useful atlas of Central America containing separate maps of each country indicating administration, population, economic activity, transportation, and geology.

Arbingast, Stanley A., et al. Atlas of Mexico. Rev. ed. Austin, TX: Bureau of Business Research, University of Texas, 1975. 165 p. ISBN 0-87755-187-1. $20. LC 75-11269.   1919
Updates and expands the original 1970 edition with more thematic maps, statistics, and bibliographies.

Atlas Regional del Caribe, realizado por el departamento de Geografía Económica del Instituto de Geografía de la Academia de Ciencias de Cuba. La Habana, Cuba: Editorial Científico-Técnica. 1979. 69 p.   1920
"Good atlas defining the Caribbean region and displaying socioeconomic conditions." (Handbook of Latin American Studies, v. 43, 1981, entry 5430).

Brazil. Conselho Nacional de Geografía. Atlas nacional do Brasil. Rio de Janeiro, Brasil: Instituto Brasileiro de Geografía e Estátistica, 1966. unpaged. LC MAP67-771.   1921
Valuable overview provided by maps and text.

Centro Editor de América Latina. Atlas total de la República Argentina. Buenos Aires: Centro Editor de América Latina, 1982. 4 v. ISBN 950-25-1001-1. LC 84-227413.   1922
Multi-volume work containing text, graphs, tables, photographs, and maps. Volumes subdivided into four parts: physical, political, demographic, and economic.

Centro Editor de América Latina. Atlas total de la República Argentina. Buenos Aires: Centro Editor de América Latina, 1983. 145 parts. LC 83-169096.   1923
So far, 145 parts have been issued of this thematic atlas.

Chile. Instituto Geográfico Militar. Atlas de la República de Chile. 2nd ed. Santiago: Instituto Geográfico Militar, 1983. 349 p. (1st ed., 1970. 244 p.). LC 84-177761.　1924
Modern thematic atlas divided into five parts: general background, physical map of Chile, national thematic maps, regional thematic maps, and index.

Colombia. Instituto Geográfico Agustin Codazzi. Atlas de Colombia. 3rd rev. ed. Bogota: Instituto Geográfico Agustin Codazzi, 1977. 286 p. (1st ed., 1967. 206 p.). LC 79-354218.　1925
Excellent atlas, with many thematic maps and photos.

Cuba. Instituto Cubano de Geodesia y Cartografía. Atlas de Cuba: XX anniversario del triunfo de la Revolución Cubana. La Habana, Cuba: Instituto Cubano de Geodesia y Cartografía, 1978. 143 p. LC 79-689730.　1926
Attractively prepared thematic atlas containing physical maps, economic maps, population and cultural maps, history maps, and general maps. The general map of Cuba is in 18 sheets at a scale of 1:300,000, with an enlarged plan of the vicinity of Havana at 1:75,000.

Cuba. Instituto de Geografía de la Academia de Ciencias de Cuba y el Instituto de Geografía de la Academia de Ciencias de la URSS. Atlas nacional de Cuba, en el décimo aniversario de la Revolución. La Habana, Cuba: Instituto de Geografía, Academia de Ciencias de Cuba, 1970. 132 p. LC 72-654620.　1927
A very useful reference, with many colored maps and commentary.

Daroczi, Isabel; Garcia, Elena; and Liguera, Miguel. Atlas para la República Oriental del Uruguay. Montevideo: Ediciones Raschetti and Editiones de Montevideo, 1983. 24 p. LC 84-171665.　1928
Thematic atlas containing 27 maps, 18 graphs, and many colored photographs.

Delavaud, Anne Collin. Atlas del Ecuador. (Atlas del Mundo). Paris: Les Éditions J.A.; Banco Central del Ecuador, 1982. 80 p. ISBN 2-85258-250-3. LC 83-174392.　1929
Small but useful thematic atlas.

Ecuador. Instituto Geográfico Militar. Atlas geográfico de la República del Ecuador. Quito: Instituto Geográfico Militar, 1977. 82 p. LC 78-373185.　1930
Very fine thematic atlas divided into three parts: general geography; geography of Ecuador; and regional geography. Scale of principal maps 1:2,000,000.

El Salvador. Ministerio de Obras Públicas. Instituto Geográfico Nacional. Atlas de El Salvador. 3rd ed. San Salvador: Instituto Geográfico Nacional, 1979. 89 p. LC 81-147359.　1931
Good introductory atlas with a wide range of maps.

Garcia de Miranda, Enriqueta, and Falcon de Gyves, Zaida, eds. Atlas: nuevo atlas Porrua de la República Mexicana. 5th ed. Mexico, DF: Editorial Porrua, 1980. 197 p. ISBN 9-684-32200-3. LC 81-675750.　1932
Detailed thematic atlas.

Guatemala. Instituto Geográfico Militar. Atlas nacional de Guatemala. Ciudad de Guatemala : Instituto Geográfico Nacional, 1972. Unpaged. LC 74-650019.　1933
Very attractive atlas with wide range of maps and photos.

Jamaica. Ministry of Finance and Planning, Town Planning Department. National atlas of Jamaica. Kingston: Town Planning Department, 1971. 79 p. LC 72-189766　1934
Primarily uncolored thematic maps with text.

Panama. Instituto Geográfico Nacional Tommy Guardia, Departamentos de Geografía y Cartografía. Atlas nacional de Panamá. Ciudad de Panamá: Instituto Geográfico Nacional Tommy Guardia, 1975. 116 p. LC 78-379326.　1935
National thematic atlas containing a wealth of information.

Latin America

Perú. Instituto Nacional de Planificación. Atlas histórico geográfico y de paisajes Peruanos. Lima: Instituto Nacional de Planificación, 1970. 737 p. LC 71-654234.   1936
Exhaustive and detailed maps and text.

Sampedro V., Francisco. Atlas geográfico del Ecuador "SAM" con los básicas nociones históricos de la nacionalidad. Quito: Gráficos Claridad, 1980. 87 p. LC 81-147613.   1937
An introductory work.

U. S. Department of Agriculture. Agricultural geography of Latin America. (USDA misc. pub. no. 743). Washington, DC: Government Printing Office, 1958. 96 p. LC AGR 58-194.   1938
A highly useful atlas that offers distribution maps of major crops.

Venezuela. Dirección General de Cartografía Nacional. Atlas de Venezuela. 2nd ed. Caracas: Dirección General de Cartografía Nacional, 1979. 331 p. (1st ed., 1969. 216 p.). LC 80-675307.   1939
Update of 1969 edition with new thematic maps, text, and illustrations.

D. GENERAL WORKS

Blakemore, Harold, and Smith, Clifford T., eds. Latin America: geographical perspectives. (University paperbacks UP 511). 2nd ed. London; New York: Methuen, 1983. 557 p. (1st ed., 1974. 600 p.). ISBN 0-416-32830-X. $19.95. (pbk). LC 82-20877.   1940
Seven authors share writing this region-by-region text that emphasizes contemporary development issues.

Blouet, Brian W. and Blouet, Olwyn M., eds. Latin America: an introductory survey. New York: Wiley, 1982. 300 p. ISBN 0-471-08385-2. $21.95. LC 81-7451.   1941
Seven contributors offer thematic chapters on environmental and social patterns, with strong historical-cultural bent.

Blume, Helmut. The Caribbean islands. Translated from German by Ann Norton. London; New York: Longman, 1974. 464 p. ISBN 0-582-48164-3. LC 74-174848. (Die Westindischen Inseln. Braunschweig: G. Westermann, 1968. 352 p.).   1942
Encyclopedic and detailed survey of the Caribbean realm including an island-by-island summary, with many maps and photos.

Butland, Gilbert J. Latin America: a regional geography. 3rd ed. London: Longman, 1973. 391 p. (1st ed., 1960. 373 p.). ISBN 0-582-31076-8. LC 73-180522.   1943
A general introduction to Latin America following a region-by-region and then a country-by-country approach, with many maps and photographs.

Cole, John P. Latin America: an economic and social geography. 2nd ed. London: Butterworth, 1975. 470 p. (1st ed., 1970. 487 p.). ISBN 4-408-70653-8 £19.80. LC 75-328418.   1944
Topical chapters followed by chapters on major countries. Innovative maps and graphics.

Humbolt, Alexander von, and Bonpland, Aimé. Personal narrative of travels to the equinoctial regions of America during the years 1799-1804. Translated from French by Thomasina Ross. London: HG Bohn, 1852-53. 3 v Reprinted: New York: Ayer Co., 1969. 3 v. ISBN 0-405-08642-3. $60. LC 69-13241. (Voyages aux régions équinoxiales du nouveau continent fait en 1799, 1800, 1801, 1802, 1803, et 1804 Paris: Dufour, 1814-25. 3 v.).   1945
A classic work; greatest of the early travel accounts of Latin America. This is an abridged translation. Also available in fuller edition, translated by Helen Maria Williams. London: Longman, 1818-29. 7 v. Reprinted New York: AMS Press, 1966. 7 v. in 6. ISBN 0-404-03440-3. $215. LC 1-20872.

James, Preston E. Latin America. 4th ed. New York: Odyssey House, 1969. 949 p. (1st
ed., 1942. 908 p.). LC 69-10222.   1946
 The classic "bible" of Latin American geography.

Lentnek, Barry; Carmin, Robert L.; and Martinson, Tom L., eds. Geographic research on Latin America: benchmark 1970; proceedings of the conference of Latin
Americanist Geographers, vol. 1. Muncie, IN: Ball State University, 1971. 438 p. LC 73-
622963.   1947
 *Tour-de-force by leading Latin Americanist geographers who summarize contem-
 porary research themes as of 1970. Extensive bibliographies. A key work.*

Davidson, William V., ed. Geographic research on Latin America: benchmark 1980;
proceedings of the conference of Latin Americanist Geographers, vol. 8. Muncie, IN:
Conference of Latin Americanist Geographers, 1981. 482 p. LC 80-80557.   1948
 *A review of major research themes in Latin American geography for the decade of
 the 1970's by leading scholars. Extensive bibliographies. A key work. (Cf. no. 1947).*

Morris, Arthur S. Latin America: economic development and regional differentiation.
Totowa, NJ: Barnes and Noble, 1981. 256 p. ISBN 0-389-20194-4. $26.50. ISBN 0-389-
20195-2. $13.75. (pbk). LC 81-178645.   1949
 *Offers a historical analysis and a geographical investigation of the movement
 toward intra-regional development in Latin America.*

Odell, Peter R. and Preston, David A. Economies and societies in Latin America: a
geographical interpretation. 2nd ed. Chichester, U.K.; New York: Wiley, 1978. 289 p.
(1st ed., London: Macmillan, 1973. 265 p.). ISBN 0-041-99588-6. $44.95. ISBN 0-471-
99636-X. $23.95. (pbk). LC 77-12400.   1950
 *Joint authorship is maintained throughout this thematic text with Preston elaborat-
 ing major themes in social geography and Odell tackling the economic geography
 issues.*

Platt, Robert S. Latin America: countrysides and united regions. New York; London:
McGraw-Hill, 1942. 564 p. LC 43-74.   1951
 *Product of a lifetime of field experiences in Latin America. Illustrative of unique
 methodology.*

Sauer, Carl O. The Early Spanish main. Berkeley, CA: University of California Press,
1969. 306 p. ISBN 0-520-01415-4. $8.95. LC 66-15044.   1952
 *Analyzes the implications of European discovery, conquest, and colonization in the
 West Indies and adjacent Caribbean rimlands. Based on painstaking historical research into original sources.*

Wauchope, Robert., gen ed. Handbook of Middle American Indians. Austin, TX:
University of Texas Press, 1964-1976. 16 v. LC 64-10316.   1953
   Vol. 1: **Natural environment and early cultures.** West, Robert C., ed. 1964.
     578 p. ISBN 0-292-73259-7. $40.
   Vols. 2 & 3: **Archeology of southern Mesoamerica.** Willey, Gordon R., ed.
     1965. Vol. 2: 560 p.; vol. 3: 531 p. ISBN 0-292-73260-0. $85.
   Vol. 4: **Archeological frontiers and external connections.** Willey, Gordon
     R., and Ekholm, Gordon F., eds. 1966. 375 p. ISBN 0-292-73632-0. $35.
   Vol. 5: **Linguistics.** McQuown, Norman A., ed. 1967. 410 p. ISBN 0-292-
     73665-7. $35.
   Vol. 6: **Social anthropology.** Nash, Manning, ed. 1967. 605 p. ISBN 0-292-
     73666-5. $40.
   Vols. 7 & 8: **Ethnology.** Vogt, Egon Z., ed. 1969. Vol. 7: 584 p.; Vol 8: 388 p.
     ISBN 0-292-78419-8. $70.
   Vol. 9: **Physical anthropology.** Stewart, T. Dale, ed. 1970. 304 p. ISBN 0-
     292-70014-8. $45.

(Continued on next page)

Latin America

Vols. 10 & 11: **Archeology of northern Mesoamerica.** Bernal, Ignacio, and Ekholm, Gordon F., eds. 1971. 466 p. (v.10); 454 p. (v. 11). ISBN 0-292-70150-0. $65.
Vol. 12: **Guide to ethnohistorical sources, pt. 1.** Cline, Howard F., ed. 1972. 476 p. ISBN 0-292-70152-7. $35.
Vol. 13: **Guide to ethnohistorical sources, pt. 2.** Cline, Howard F. and Glass, John B., eds. 1973. 439 p. ISBN 0-292-70153-5. $35.
Vols. 14 & 15: **Guide to ethnohistorical sources, pts. 3 and 4.** Cline, Howard F., et al, eds. 1975. 410 p. (v. 14); 425 p. (v. 15). ISBN 0-292-70154-3. $65.
Vol. 16: **Sources cited and artifacts illustrated.** Harrison, Margaret A., ed. 1976. 332 p. ISBN 0-292-73004-7. $30.
Indispensible monumental guide to the environment, archeology, ethnology, social anthropology, ethnohistory, linguistics, and physical anthropology of the native peoples of Mexico and Central America.

Webb, Kempton, E. Geography of Latin America: a regional analysis. Englewood Cliffs, NJ: Prentice-Hall, 1972. 126 p. ISBN 0-13-351452-8. LC 71-144096.   1955
Topical and regional survey offers only a taste of Latin America's diversity.

West, Robert C. and Augelli, John P. Middle America: its lands and peoples. 2nd ed. Englewood Cliffs, NJ: Prentice-Hall, 1976. 494 p. (1st ed., 1966. 482 p.). ISBN 0-13-581546-0. LC 75-14417.   1956
Authoritative and comprehensive review of historical-cultural patterns in the West Indies, Mexico, and Central America.

E. SPECIAL TOPICS

Adams, Richard N. Cultural surveys of Panama-Nicaragua-Guatemala-El Salvador-Honduras. Washington, DC: Pan American Sanitary Bureau, 1957. 669 p. LC 58-3112.   1957
Coordinated regional survey based on field work and developed for general understanding of common cultural traits.

Brookfield, Harold C. Interdependent development. Pittsburgh, PA: University of Pittsburgh Press, 1975. 248 p. ISBN 0-8229-1118-3. $11.95. (pbk). LC 74-18752.   1958
Postulates a type of development in which all participants gain.

Caviedes, Cesar. The Politics of Chile: a sociogeographical assessment. (Special studies on Latin America). Boulder, CO: Westview Press, 1979. 357 p. ISBN 0-89158-311-4. $34.50. LC 78-23843.   1959
Expert interweaving of politics and environment using the thesis that Chilean political responses are based primarily on the people's place in society and the influence of their environment.

Cockroft, James D.; Frank, Andre Gunder; and Johnson, Dale L. Dependence and underdevelopment: Latin America's political economy. Garden City, NY: Anchor Books, 1972. 448 p. LC 79-171396.   1960
A broad-ranging survey of the dependency thesis applied to Latin America. Each author contributes chapters under broad subject headings: dependence, imperialism, and underdevelopment; class and politics; and social science and strategies of development. Some contributions have been reprinted from their original sources.

Crosby, Alfred W., Jr. The Columbian exchange: biological and cultural consequences of 1492. Westport, CT: Greenwood Press, 1972. 268 p. ISBN 0-8371-5821-4. $27.50. ISBN 0-8371-7228-4. $3.95. (pbk). LC 73-140916.   1961
Reveals (and revels in) the complex interconnections and resulting transformations of the contact between the old and new worlds, particularly in exchange of crops, animals, and diseases.

Denevan, William M., ed. The Native population of the Americas in 1492. Madison, WI: University of Wisconsin Press, 1976. 353 p. ISBN 0-299-07050-6. $29.50. LC 75-32071.   1962
Latest techniques and evidence allow for new estimates of American aboriginal population at the time of European contact. Eight authors summarize and debate the results.

Dorner, Peter, ed. Land reform in Latin America: issues and cases. Madison, WI: University of Wisconsin Land Tenure Center, 1971. 276 p. LC 70-188047.   1963
Case studies and introductory works by noted scholars, most of whom are affiliated with the Wisconsin Land Tenure Center.

Fittkau, E.J., and others, eds. Biogeography and ecology in South America. (Monographiae biologicae, v. 18). The Hague, Netherlands: W. Junk, 1969. 2 v. V. 2: 498 p. ISBN 90-6193-071-5. $48. LC 78-388001.   1964
The principal source of ecological information about South America.

Furtado, Celso. Economic development of Latin America: historical background and contemporary problems. Translated from Portuguese by Suzette Macedo. (Latin American studies, no. 8). 2nd rev. ed. Cambridge, U.K.; New York: Cambridge University Press, 1976. 317 p. (1st ed., 1970. 271 p.). ISBN 0-521-21197-2. $47. ISBN 0-521-29070-8. $12.95. (pbk). LC 76-14914. (A economia Latino-Americana: formação historico e problemas contemporâneos. São Paulo: Companhia Editora Nacional, 1976. 339 p. [1st ed., 1969. 365 p.]).   1965
Brazilian-born economist traces the growth of several Latin American economies from Conquest to the present emphasizing the influence of external dependence.

Gilbert, Alan. Latin American development: a geographical perspective. Harmondsworth, U.K.; Baltimore, MD: Penguin, 1974. 366 p. ISBN 0-14-021792-4. LC 75-308388.   1966
Author ably focuses on spatial adjustment, or changes that modify existing distributions of social and economic activity in Latin America.

Grunwald, Joseph, and Musgrove, Philip. Natural resources in Latin American development. Baltimore, MD: Johns Hopkins Press for Resources for the Future, 1970. 494 p. LC 77-108381.   1967
Provides information on the resource base, production, consumption, and trade of the principal resource commodity exports of Latin America as well as sections on the ecomomic history, major resource issues, and a statistical summary.

Hardoy, Jorge E. Urbanization in Latin America: approaches and issues. Garden City, NY: Anchor Books, 1975. 456 p. ISBN 0-385-08240-1. LC 76-2307.   1968
Detailed introduction to past, present, and the projected future growth of the Latin American city in a region already predominantly urbanized.

Harrison, Peter D. and Turner, Billie L., II., eds. Pre-Hispanic Maya agriculture. Albuquerque, NM: University of New Mexico Press, 1978. 414 p. ISBN 0-8263-0483-4. $20. LC 78-55703.   1969
Seventeen sections illustrate the diversity of emphasis and understanding of early Maya agriculture. New findings challenge old assumptions of a maize-oriented, swidden-dependent economy among the pre-Hispanic Maya.

Herrera, Ligia, and Pecht, Waldomiro. Crecimiento urbano de Américana Latina. Washington, DC: Banco Interamericano de Desarrollo; Santiago, Chile: Centro Latinoamericano de Demografía, no. 22. 1976. 2 v. 549 p. LC 82-131684.   1970
Volume in two parts first explores cities in their economic and environmental settings concentrating on major urbanized nations, then analyzes and projects their growth to 1980. A very useful compendium, but based on 1960 data.

Latin America

Higgins, Benjamin. Economic development, principles, problems, and policies. Rev. ed. New York: Norton, 1968. 918 p. (1st ed., 1959. 803 p.). ISBN 0-393-09714-5. $17.95. LC-67-11081. 1971
A standard reference work for various theories of economic development.

Honour, Hugh. The European vision of America. Cleveland, OH: Cleveland Museum of Art, 1975. 389 p. ISBN 0-910386-26-9. LC 75-35892. 1972
A compendium of images of America by early artists offers a unique geographical record and a stimulus for study of ancient landscapes.

Johnson, Edgar A.J. The Organization of space in developing countries. Cambridge, MA: Harvard University Press, 1970. ISBN 0-0-674-64338-0. $27.50. 452 p. LC 74-122216. 1973
Illustrates the alternatives to growth using the core-region model and proposes a development model promoting spatial equality.

Johnson, John J. Continuity and change in Latin America. Stanford, CA: Stanford University Press, 1964. 282 p. ISBN 0-8047-0184-9. $25. ISBN 0-8047-0185-7. $7.95. (pbk). LC 64-17001. 1974
Provides an analysis of social, economic, and political change in Latin America.

McPherson, Woodrow W., ed. Economic development of tropical agriculture: theory, policy, strategy, and organization. Gainsville, FL: University of Florida Press, 1968. 328 p. ISBN 0-8130-0161-7. $10. ISBN 0-8130-0330-X. $3.50. (pbk). LC 68-24368. 1975
Thirteen prominent agricultural economists discuss the importance of agriculture in economic development introducing models and study methods.

Morner, Magnus, ed. Race and class in Latin America. (Conference on Race and Class in Latin America, New York, 1965). New York: Columbia University Press, 1970. 309 p. ISBN 0-0-231-03295-5; ISBN 0-231-08661-X. $10.50. (pbk). LC 79-118357. 1976
Thirteen chapters in sequence on slavery, immigration, stratification, race relations, change in Indo-America in the nineteenth and twentieth centuries, the present state of knowledge, and the interdisciplinary tasks ahead.

Robinson, David J. Social fabric and spatial structure in colonial Latin America. Ann Arbor, MI: University Microfilms International for Syracuse University, Dept. of Geography, 1979. 478 p. ISBN 0-8357-0419-X. LC 79-15744. 1977
Eleven solid historical studies on the West Indies, Mexico, and continental South America inaugurate the Dellplain Latin America series with this volume.

Rodriguez Monegal, Emir, ed. and Colchie, Thomas. The Borzoi anthology of Latin American literature, from the time of Columbus to the 20th century. New York: Knopf, 1977. 2 v. 982 p. LC 76-19126.
    Vol. 1: ISBN 0-394-73301-0. $10.95. (pbk).
    Vol. 2: ISBN 0-394-73366-5. $10.95. (pbk). 1978
Often insights gained from literature complement or stimulate geographical field research.

Sawatzky, Harry L. They sought a country: Mennonite colonization in Mexico; with an appendix on Mennonite colonization in British Honduras. Berkeley, CA: University of California Press, 1971. 387 p. ISBN 0-520-01704-8. $38.50. LC 78-92673. 1979
Cultural and environmental background of the Chihuahua colonies founded by the Mennonites who moved primarily from Canada.

Steward, Julian H., ed. Handbook of South American Indians. (Smithsonian Institution. Bureau of American Ethnology. Bulletin 143). Washington, D.C.: U.S. Government Printing Office, 1946-1959. 7 v. Reprinted: New York: Cooper Square, 1963. 7 v. LC 63-17285.   1980

    Vol. 1: **The Marginal tribes.** 1946. 624 p.
    Vol. 2: **The Andean civilization.** 1946. 1035 p.
    Vol. 3: **The Tropical forest tribes.** 1948. 986 p.
    Vol. 4: **The Circum-Caribbean tribes.** 1948. 609 p.
    Vol. 5: **The Comparative ethnology of South American Indians.** 1949. 818 p.
    Vol. 6: **Physical anthropology, linguistics, and cultural geography of South American Indians.** 1950. 715 p.
    Vol. 7: **The Index.** 1959. 286 p.

The indispensible reference work on South American Indians.

Steward, Julian H. and Faron, Louis C. Native peoples of South America. New York: McGraw-Hill, 1959. 481 p. LC 58-10010.   1981

Summarizes the material in The Handbook of South American Indians *(entry 1980).*

Thomas, Robert N. and Hunter, John M., eds. Internal migration systems in the developing world; with special reference to Latin America. Cambridge, MA: Schenkman Publishing Co, 1980. 176 p. ISBN 0-87073-931-X. $9.95. LC 79-21875.   1982

Six of the nine chapters deal specifically with Latin America, including country studies of Colombia, Guatemala, Peru, Venezuela, Argentina, and one summary on Latin American national policies as related to spatial strategies.

Wharton, Clifton R. Jr., ed. Subsistence agriculture and economic development. Chicago, IL: Aldine, 1969. 481 p. LC 68-8163.   1983

Results of a conference organized to determine research priorities and set an agenda for the investigation of research issues in subsistence agriculture. Wide-ranging and useful despite its heavy reliance on the views of agricultural economists.

F. WEST INDIES AND GUIANAS

Alexander, Charles S. The Geography of Margarita and adjacent islands, Venezuela. Berkeley, CA: University of California Press, 1958. 108 p. LC 58-9195.   1984

Early Spanish discovery and exploitation made this area unique.

Eyre, L. Alan. Geographic aspects of population dynamics in Jamaica. Boca Raton, FL: Florida Atlantic University Press, 1972. 172 p. ISBN 0-8130-0440-3. $7.50. LC 76-151084.   1985

Overview of population patterns and an investigation of the relations between populations and social, economic, and ecologic factors present in Jamaica.

Guerra y Sanchez, Ramiro. Sugar and society in the Caribbean; an economic history of Cuban agriculture. (Caribbean series no. 7). Translated from Spanish by Marjory M. Urquidi. New Haven, CT: Yale University Press, 1964. 218 p.   1986

Explains the role of sugar in shaping the society and landscape of Cuba. (Volume cited in Handbook of Latin American studies, *vol. 27, page 108, entry no. 1045)*

Harris, David R. Plants, animals, and man in the outer Leeward Islands, West Indies: an ecological study of Antigua, Barbuda, and Anguilla. Berkeley, CA: University of California Press, 1965. 164 p. LC 66-63387.   1987

The modifications of these islands by people.

Jones, Clarence F. and Pico, Rafael, eds. Symposium on the geography of Puerto Rico. Rio Piedras, PR: University of Puerto Rico Press, 1955. 503 p. LC 55-43948.   1988

Contributions by 18 authors resulting from detailed field work on land tenure and use in Puerto Rico. Perhaps the most ambitious field research project ever undertaken in Latin American geography.

Latin America

Lowenthal, David. West Indian societies. London; New York: Oxford University Press, 1972. 385 p. ISBN 0-19-501559-2. $9.95. (pbk). LC 72-181627. 1989

An effective account of cultural complexity in an area where territorial fragmentation, social stratification, color, and colonial experience have created an exceptionally difficult challenge for cultural geography.

MacPherson, John. Caribbean lands. London; New York: Longman, 1980. 200 p. ISBN 0-582-76565-X. $8.95. (pbk). LC 80-131947. 1990

Developed for school use, text traces background information on environment and settlement patterns followed by brief chapters on individual islands.

Marrero y Artiles, Levi. Geografía de Cuba. 3rd ed. Havana: Editorial Selecta, 1957. 707 p. (1st ed. 1950. 736 p.). LC 59-52095. 1991

The standard text on pre-Castro Cuba.

May, Jacques, M. and McLellan, Donna L. The Ecology of malnutrition in the Caribbean: the Bahamas, Cuba, Jamaica, Hispaniola, Puerto Rico, the Lesser Antilles, and Trinidad and Tobago. (Studies in medical geography v. 12). New York: Hafner Press, 1973. 395 p. ISBN 0-02-848920-9. $21.95. LC 73-76140. 1992

A treasury of food and diet information, by country, with ample data.

Ortiz Fernández, Fernando. Cuban counterpoint; tobacco and sugar. Translated from Spanish by Harriet de Onís. New York: Knopf, 1947. 312 p. LC 47-1035. (Contrapunteo cubano del tabaco y azúcar. 1940. 475 p.). 1993

Influence of the two diametrically opposed crops that have shaped the landscape of Cuba.

Pico, Rafael. The Geography of Puerto Rico. Chicago, IL: Aldine, 1974. 439 p. ISBN 0-202-10056-1. LC 72-182916. 1994

Comprehensive volume treats the physical, social, and economic geography of the island, regionally and topically.

Schuchert, Charles. Historical geology of the Antillean-Caribbean region; or the lands bordering the Gulf of Mexico and the Caribbean Sea. New York: John Wiley; London: Chapman and Hall, 1935. 811 p. Reprinted: New York: Hafner, 1968. 811 p. LC 68-209664. 1995

Comprehensive review of the structural and stratigraphic history of the Caribbean area.

Smith, Raymond T. British Guiana. London; New York: Oxford University Press, 1962. 218 p. Reprinted: Westport, CT: Greenwood, 1980. 218 p. ISBN 0-313-22142-1. $24.75. LC 80-10964. 1996

Major aspects of Guyana described and analyzed by one of the country's keenest interpreters.

Steward, Julian H. ed. The People of Puerto Rico: a study in social anthropology. Urbana, IL: University of Illinois Press, 1956. 540 p. LC 56-5682. 1997

Several prominent anthropologists outline Puerto Rican culture.

Wood, Harold A. Northern Haiti; land use and settlements, a geographic investigation of the Département du Nord. Toronto: University of Toronto Press, 1963. 168 p. LC 63-6162. 1998

Based on detailed studies of the man-land relationship.

G. MEXICO AND CENTRAL AMERICA.

Aschmann, H. Homer. The Central desert of Baja California; demography and ecology. Riverside, CA: Manessier, 1967. 315 p. ISBN 0-910950-01-6. $69.50. LC 66-29636.
A model study of human occupance of a diversified environment.
1999

Brand, Donald D. Mexico: land of sunshine and shadow. Princeton, NJ: D. Van Nostrand, 1966. 159 p. LC 66-4405.
Brief but penetrating summary of Mexico's principal geographical, anthropological, and historical schema by a respected scholar.
2000

Browning, David G. El Salvador: landscape and society. Oxford, U.K.: Clarendon Press, 1971. 329 p. ISBN 0-19-823208-X. LC 71-855356.
Traces man-land relations since colonial times. A key work in understanding current problems in Central America.
2001

Chevalier, François. Land and society in colonial Mexico; the great hacienda. Translated from French by Alvin Eustis. Berkeley, CA: University of California Press, 1963. 334 p. ISBN 0-520-00229-6. $32. ISBN 0-520-04653-6. $8.95. (pbk). LC 63-20579. (La formation des grands domaines au Mexiqué. Paris: Institut d'ethnologie, 1952. 480 p. Travaux et mémoires, 56).
Examination of the growth of estates in Mexico in the 16th and 17th centuries.
2002

Coe, Michael D. The Maya. (Ancient peoples and places series). 3rd rev. ed. London: Thames and Hudson; New York: W.W. Norton, 1984. 207 p. (1st ed., 1966. 252 p.). ISBN 0-500-27327-8. $9.95. (pbk). LC 83-72969.
Valuable as an introduction to the rich Maya civilization.
2003

Davidson, William V. Historical geography of the Bay Islands, Honduras: Anglo-Hispanic conflict in the western Caribbean. Birmingham, AL: Southern University Press, 1974. 199 p. ISBN 0-87651-207-4. $16.95. ISBN 0-686-96890-5. $9.50. (pbk). LC 74-81225.
Eight diverse cultures have added their influences to the overall culture of the Bay Islands, but the principal cause of landscape change has been the conflict between Spanish and English interests.
2004

Diaz del Castillo, Bernal. The Discovery and conquest of Mexico, 1517-1521. Translated by Alfred P. Maudslay. New York: Farrar, Straus, and Cuhady, 1956. 478 p. ISBN 0-374-50384-2. $10.95. LC 56-5758. (Historia verdadera de la conquista de la Nueva España).
Contemporary account of the Conquest by one of Cortes's soldiers has not lost its appeal. A classic.
2005

Gierloff-Emden, Hans G. Mexico: eine Landeskunde. Berlin, W. Germany: Walter de Gruyter, 1970. 634 p. ISBN 3-11-002708-9. DM 154. LC 70-524263.
Detailed geography of Mexico accompanied by many maps, photographs, and line drawings, as well as a good bibliography.
2006

Guerra Borges, Alfredo. Geografía económica de Guatemala. Ciudad Guatemala: Editorial Universitaria, Universidad de San Carlos, 1969. 2 v. 751 p. V.1: 1969. 416 p. LC 77-471373.
Exhaustive compilation of the environmental background and primary economy of Guatemala, especially its agriculture. (Volume 1 cited in Handbook of Latin American studies, vol. 33 (1971), page 322, item 5117.)
2007

Johannessen, Carl L. Savannas of interior Honduras. (Ibero-Americana, 46). Berkeley, CA: University of California Press, 1963. 173 p. LC 63-64567.
Excellent example of field work on an ecological topic.
2008

Latin America

Jones, Chester L. Guatemala: past and present. Minneapolis, MN: University of Minnesota Press, 1940. 420 p. Reprinted: New York: Gordon, 1976. 420 p. ISBN 0-8490-1914-1. $59.95. LC 66-24713.　　　　　　　　　　　　　　　　　　　　　　2009
Classic text consists of three major sections: political development; economic advance; and social life. Extensive notes and bibliography.

McBryde, Felix Webster. Cultural and historical geography of southwest Guatemala. (Smithsonian Institution. Institute of Social Anthropology publication no. 4). Washington, DC: Government Printing Office, 1947. 184 p. LC 47-30636. Reprinted: Westport, CT: Greenwood Press, 1971. 184 p. ISBN 0-8371-3123-5. $36.25. LC 74-89017.　　2010
Extensive and detailed fieldwork make this a classic.

Nietschmann, Bernard Q. Caribbean edge: the coming of modern times to isolated people and wildlife. Indianapolis, IN: Bobbs-Merrill, 1979. 280 p. ISBN 0-672-52556-9. LC 78-11211.　　　　　　　　　　　　　　　　　　　　　　　　　　　　2011
Personal account of changes among the Miskito Indians of Nicaragua, especially their dependence on sea turtles for livelihood. Based on two years of fieldwork.

Price, John A. Tijuana: urbanization in a border culture. South Bend, IN: University of Notre Dame Press, 1973. 195 p. ISBN 0-268-00477-3; ISBN 0-268-00495-1. (pbk). LC 72-12641.　　　　　　　　　　　　　　　　　　　　　　　　　　　　2012
Description of the city within its kaleidoscopic milieu, concentrating on contemporary change along a unique border.

Romney, D.H., ed. Land in British Honduras: report of the British Honduras Land Use Survey Team. (England. Colonial Office. Research Department. Land Use Survey Team. Colonial research publication no. 24). London: Her Majesty's Stationary Office, 1959. 326 p.　　　　　　　　　　　　　　　　　　　　　　　　　　2013
Detailed description of land use with environmental and cultural references. A separate map folio accompanies the book.

Sandner, Gerhard. Agrar-Kolonisation in Costa Rica: Siedlung, Wirtschaft und Sozialgefüge in Costa Rica. (Schriften des Geographischen Instituts der Universität Kiel. Band 19, Heft 3). Kiel, W. Germany: Schmidt and Klaunig, 1961. 199 p. LC 64-54862.　　2014
Excellent maps accompany this detailed account of agricultural settlement.

Simpson, Eyler N. The Ejido: Mexico's way out. Chapel Hill, NC: 1937. 849 p. LC 37-18116.　　　　　　　　　　　　　　　　　　　　　　　　　　　　　2015
Classic investigation of the ejido (community landholding) that was at the core of Mexico's land reform program 50 years ago.

Tamayo, Jorge L. Geografía general de México. 2nd ed. Mexico, DF: Instituto Mexicano de Investigaciones Económicas, 1962. 4 v. (1st ed., 1949. 2 v. plus accompanying atlas). LC 63-34853.　　　　　　　　　　　　　　　　　　　　　　　　2016
The standard textbook on Mexico in Spanish.

Tax, Sol. Penny capitalism: a Guatemalan Indian economy. (Smithsonian Institution. Institute of Social Anthropology publication no. 16). Washington, DC: Government Printing Office, 1953. 230 p. Reprinted: New York: Octagon, 1971. 230 p. ISBN 0-374-97785-2. $21.50. LC 78-159254.　　　　　　　　　　　　　　　　　　　2017
Indispensable guide to the change in highland Indian societies from a traditional to a modern economy. Many unique maps and graphics of Panajachel and vicinity.

Whetten, Nathan L. Rural Mexico. Chicago, IL: University of Chicago Press, 1948. 671 p. LC 48-8023.　　　　　　　　　　　　　　　　　　　　　　　　　2018
Classic survey of rural life in Mexico, with many insights.

H. THE ANDEAN COUNTRIES

Bowman, Isaiah. Desert trails of Atacama. (American Geographical Society special publication no. 5). New York: American Geographical Society, 1924. 362 p. Reprinted: New York: AMS Press, 1976. ISBN 0-404-00964-6. $18.50. LC 76-111776.   2019
   A classic tale of exploration and discovery introducing the desert of northern Chile as a geographical laboratory.

Bowman, Isaiah. The Andes of southern Peru: geographical reconnaissance along the seventy-third meridian. New York: Published for the American Geographical Society by Henry Holt, 1916. 336 p. Reprinted: New York: Greenwood, 1968. 336 p. LC 68-23277.   2020
   Field study under rigorous conditions yielded a masterpiece.

Crist, Raymond E. and Leahy, Edward P. Venezuela: search for a middle ground. (Searchlight book no. 43). New York: Van Nostrand Reinhold, 1969. 128 p. LC 75-10875.   2021
   Background information on resource base and strategic position for understanding Venezuela's role in Latin American affairs.

Ford, Thomas R. Men and land in Peru. Gainsville, FL: University of Florida Press, 1955. 176 p. Reprinted: New York: Russell, 1971. 176 p. ISBN 0-8462-1611-6. $11. LC 76-152538.   2022
   Sociological study of economic development with special attention paid to land use and agricultural problems.

Guhl, Ernesto. Colombia: bosquejo de su geografía tropical. (Panamerican Institute of Geography and History. Publicación no. 273). Rio de Janeiro, Brazil: Instituto Panamericano de Geografica e Historia, 1967. 266 p. NUC 70-10297.   2023
   An introductory text on the geography of Colombia by the old master.

McBride, George M. Chile: land and society. (American Geographical Society, research series no. 19). New York: American Geographical Society, 1936. 408 p. Reprinted: New York: Octagon, 1971. 408 p. ISBN 0-374-95429-1. $29. LC 71-154618.   2024
   Study of the agricultural heartland of Chile recounts the development of its distinctive land use and tenure, and the resulting problems.

Muñoz Reyes, Jorge. Geografía de Bolivia. 2nd ed. La Paz: Academia Nacional de Ciencias de Bolivia, 1980. 514 p. (1st ed., 1977. 478 p.). LC 80-128897.   2025
   Textbook on Bolivia is an encyclopedic survey of environmental and cultural patterns containing information not readily available elsewhere.

Parsons, James J. Antioqueño colonization in western Colombia. Rev. ed. Berkeley, CA: University of California Press, 1968. 233 p. (1st ed., 1949. 225 p. [Ibero-Americana, 32]). ISBN 0-520-01464-2. $31. LC 68-58002.   2026
   Fascinating account of the rise of Antioquia's dominance in Colombia's economic affairs.

Prescott, William H. History of the conquest of Peru; with a preliminary view of the civilization of the Incas. New rev. ed. Philadelphia PA: J.B. Lippincott, 1874. 2 v. (1st ed., 1847). Frequently reprinted. Available in The conquest of Mexico and the conquest of Peru. New York: Random House, Modern Library, 1931. ISBN 0-394-60471-7. $10.95. LC 36-27495.   2027
   Classic account of the European discovery and destruction of the Inca empire.

Rodwin, Lloyd, et al. Planning urban growth and regional development; the experience of the Guayana Program of Venezuela. Cambridge, MA: MIT Press, 1969. 524 p. LC 68-18240.   2028
   The Guayana Program is one of the largest economic development projects in Latin America, and so deserves special notice.

Latin America

Smole, William J. The Yanoama Indians; a cultural geography. (Texas Pan- American series). Austin TX: University of Texas Press, 1976. 272 p. LC 75-16167.   2029
Concerned with spatial patterns and ecology of these contemporary "neolithic" peoples.

West, Robert C. Colonial placer mining in Colombia. (Louisiana State University studies. Social science series no. 2.). Baton Rouge, LA: Louisiana State University Press, 1952. 159 p. LC 52-14234.   2030
Northwest Colombia is the site of this integrated investigation of the physical and historical geography of a primary economic activity.

I. BRAZIL

Comissão Nacional do Brasil da União Geografica Internacional. Excursion guidebooks for the 18th International Geographical Congress. (Guia da excursão). Rio de Janeiro, Brazil: Edição do Conselho Nacional de Geografia, 1956. 9 v. LC 59-49385. LC 59-34528 rev.   2031
These nine volumes remain among the best introductory works on the geography of Brazil.

Cunha, Euclydes R. da. Rebellion in the backlands. Translated from Portuguese by Samuel Putnam. (Phoenix books, P. 22). Chicago, IL: University of Chicago Press, 1944, 1957. 532 p. ISBN 0-226-12444-4. $5.95. (pbk). LC 57-4329; LC 44-346. (Os sertões. 3rd ed. Rio de Janeiro: Laemmert, 1905. 618 p).   2032
Literary masterpiece relating the interrelations between people and environment in northeast Brazil.

Henshall, Janet D. and Momsen, Richard P., Jr. A Geography of Brazilian development. London: Bell, 1974. 305 p. ISBN 0-7135-1812-X. £9.50. ISBN 0-7135-1832-4. £6.95. (pbk). LC 75-307569.   2033
Welcome new text on the economic development aspects of Brazilian geography that stresses regional integration.

Meggers, Betty J. Amazonia: man and culture in a counterfeit paradise. (Worlds of man series). Chicago, IL: Aldine-Atherton, 1971. 182 p. (Now distributed by: Arlington Heights, IL: Harlan Davidson, Inc.). ISBN 0-88295-608-6. $17.95. ISBN 0-88295-609-4. $7.95. (pbk).   2034
Demonstrates how native people have perceived and used the tropical rainforest ecosystem.

Momsen, Richard P., Jr. Routes over the Serra do Mar; the evolution of transportation in the highlands of Rio de Janeiro and São Paulo. Muncie, IN: Ball State University Bookstore, 1965. 173 p. NUC 65-110373.   2035
Much of the economic development of Brazil can be traced to the opening of Brazil's interior by modern transportation along routes such as this. Volume, which includes a packet of maps, is cited in Handbook of Latin American studies, *vol. 23, page 348, item 2952a.*

Moran, Emilio F. Developing the Amazon. Bloomington, IN: Indiana University Press, 1981. 292 p. ISBN 0-253-14564-3. $22.50. LC 80-8382.   2036
Investigates how the tropical rainforest ecosystem and population survives technological change.

Smith, T. Lynn (Thomas). Brazil: people and institutions. Rev. ed. Baton Rouge, LA: Louisiana State University Press, 1963. 667 p. (1st ed., 1946. 843 p.). LC 71-146267.   2037
Review of major social patterns by a rural sociologist.

Wagley, Charles. An Introduction to Brazil. Rev. ed. New York: Columbia University Press, 1971. 341 p. (1st ed., 1963. 322 p.). ISBN 0-231-03542-X. $32. ISBN 0-231-03543-8. $12. (pbk). LC 63-15051.　　　　　　　　　　　　　　　　　　　　　　　　　2038
　　Social anthropologist reviews patterns of social structure and its regional variation.

Webb, Kempton E. The Changing face of northeast Brazil. New York: Columbia University Press, 1974. 205 p. ISBN 0-231-03767-8. $27.50. LC 74-1029.　　　　　　　2039
　　Analysis of the process of landscape evolution in the largest underdeveloped area of the largest developing country in the Western Hemisphere.

J. ARGENTINA, URUGUAY, AND PARAGUAY.

Aparacio, Francisco de, and Difrieri, Horacio. La Argentina; suma de geografía. Buenos Aires: Ediciones Peuser, 1958-1963. 9 v. LC 60-22647 rev.　　　　　　　2040
　　Massive and monumental, the ultimate geographical compendium for Argentina. Volume cited in the Handbook of Latin American studies, *vol. 29, page 405, item 5125.*

Eidt, Robert C. Pioneer settlement in northeast Argentina. Madison, WI: University of Wisconsin Press, 1971. 277 p. ISBN 0-299-05920-0. $27.50. LC 71-138058.　　　　2041
　　Investigates the scope and intensity of settlement and management of a marginal zone.

Jefferson, Mark. Peopling the Argentine pampa. (Latin American history and culture series). (American Geographical Society research series no. 16.) New York: American Geographical Society, 1926. 211 p. Reprinted: Port Washington, NY: Kennikut Press, 1971. 211 p. ISBN 0-8046-1378-8. $22.50. LC 71-123491.　　　　　　　　　　　2042
　　This study of European colonization in Argentina has become a classic.

Sargent, Charles S. The Spatial evolution of greater Buenos Aires, Argentina, 1870-1930. Tempe, AZ: Arizona State University, Center for Latin American Studies, 1974. 164 p.ISBN 0-87918-013-7. $12.95. LC 74-620000.　　　　　　　　　　　　　　　2043
　　Illustrates the growth of modern Buenos Aires resulting from improvements in transportation.

Scobie, James R. Argentina: a city and a nation. (Latin American history series). 2nd ed. New York: Oxford University Press, 1971. 323 p. (1st ed., 1964. 294 p.). ISBN 0-19-501480-4. $8.95. (pbk). LC 78-166005.　　　　　　　　　　　　　　　　　　2044
　　The best illustration of how Argentina's development has revolved around the growth of its primate city.

6. EUROPE

George W. Hoffman

Akademie der Wissenschaften der Deutschen Demokratischen Republik, durch die Kommission zur Herausgabe des Atlas DDR. Atlas Deutsche Demokratische Republik. Gotha/Leipzig, East Germany: VEB Hermann Haack, 1976-1981. 2 v. LC 78-381451. NUC 79-123089.　　　　　　　　　　　　　　　　　　　　　　　2045
Important thematic atlas of East Germany containing tables of contents and extensive map legends in German, English, French, Russian, and Spanish.

Akademie der Wissenschaften, Vienna. Kommission für Raumforschung und Wiederaufbau. Atlas der Republic Österreich. Vienna: Freytag-Berndt und Artaria, 1961- . LC Map 62-194.　　　　　　　　　　　　　　　　　　　　　　　　2046
Forty-eight sheets are planned. Thirty-five had been issued as of 1983.

Atlas ČSSR. 6th ed. Bratislava, Czechoslovakia: Slovenská Kartografia, 1982. 14 p., with 42 p. of plates. LC 83-243765.　　　　　　　　　　　　　　　　　　2047
Small but useful school atlas.

Atlas of Central Europe. London: Murray, 1963. 52 p. LC Map 64-539.　　2048
A collection of large-scale maps of the area; compiled and printed by the Bertelsmann Cartographical Institute.

Atlas: Republica Socialistă România. Bucureşti: Editura Academiei Republicii Socialiste România, 1974-1979. 13 v. LC Map 75-400012.　　　　　　　　　　2049
Text and legends in Romanian, English, French, and Russian. Issued in 13 sections, with several maps per section.

Atlas van Nederland. (Stichting wetenschappelijke Atlas van Nederland). 's-Gravenhage: Staatsuitgeverij, 1984- . 4 v. issued as of 1984. ISBN 90-12-05000-6. LC 84-219888.　2050
A new national atlas of the Netherlands combining text, photographs, graphs, and finely drawn maps. When completed the atlas will consist of 20 volumes.

Beaujeu-Garnier, Jacqueline, and Bastié, Jean. Atlas de Paris et de la région Parisienne. 2nd ed. Paris: Éditions Berger-Levrault, 1972. 50 p. (1st ed., 1967. 2 v.). LC Map 72-650139.　　　　　　　　　　　　　　　　　　　　　　　　　　　2051
Paris and its environs minutely dissected. Maps of all aspects of its geography-physical, urban, demographic, transport, agricultural, industrial, and commercial.

Bundesforschungsanstalt für Landeskunde und Raumordnung. Atlas zur Raumentwicklung. Bonn: Bundesforschungsanstalt, 1976-1982. 9 v. LC 77-564340.　2052
Thematic atlas of West Germany. Thus far, nine of a projected eleven volumes have been issued. Map titles are in German, English, and French. Text and statistical data presented in German only.

Browne, G.S., ed. Atlas of Europe: a profile of Western Europe. Edinburgh: John Bartholomew; New York: Charles Scribner's, 1974. 128 p. ISBN 0-684-13806-9. LC 73-20985.　　　　　　　　　　　　　　　　　　　　　　　　　　　　2053 •
Excellent introductory atlas giving an overview of Western Europe.

Clayton, Keith M. and Kormoss, I.B.F., eds. Western Europe. (Oxford regional economic atlas series). London: Oxford University Press, 1971. 160 p. ISBN 0-19-894306. LC 74-654598.　　　　　　　　　　　　　　　　　　　　　　2054 •
Part of an excellent series of thematic atlases that depict the economic landscape of world regions.

Comité National de Géographie (Belgique). Atlas de Belgique. Atlas van België. Brussels: Institut Géographique Militaire, 1950-1964. LC Map 55-350. 2055
  Invaluable source of information, beautifully produced. Cartography and geophysics, general geography, human geography, economic geography, regional geography, and administration.

Comité National Français de Géographie. Atlas de France. 2nd ed. Paris: Éditions Géographiques de France, 1951-58. (1st ed., 1933-46). LC Map 54-1163. 2056
  An excellent, serially-issued regional atlas, covering geomorphology, climatology, hydrography, biogeography, agriculture, industry, and human and political geography.

Finland. Maanmittaushallitus. Suomen kartasto. Atlas över Finland. Atlas of Finland. 5th ed. Helsinki: Maanmittaushallitus; Suomen Maantieteellinen Seura, 1976- . 7 v. as of 1983. ISBN 951-46-2570-6. (1st ed., 1920). LC 82-235050. 2057
  The fifth edition of the Atlas of Finland was begun in 1976 and is to be completed in 1986. More comprehensive than earlier editions, it will include more than 3000 thematic maps and diagrams with related text in Finnish, Swedish, and English, and will be issued in specific subject folios to be published as they are completed.

Germany. (Federal Republic, 1949-) Statistisches Bundesamt. Die 2058
Bundesrepublik Deutschland in Karten. Herausgeber: Statistisches Bundesamt, Institut für Landeskunde, Institut für Raumforschung. Mainz: W. Kohlhammer, 1965- . LC Map 66-259.

Irish National Committee for Geography. Atlas of Ireland. Dublin: Royal Irish Academy, 1979. 104 p. ISBN 0-901714-13-5. LC 80-675280. 2059
  Excellent national atlas containing approximately 250 maps divided into 13 categories of subject matter. These include general reference, geology and geophysics, geomorphology and hydrology, soils, climate, flora and fauna, settlement, population, primary production, manufacturing, tertiary activities, society and culture, and the Irish landscape and its representation on topographic maps.

Kartográfiai Vállalat, Budapest. National atlas of Hungary. (Prepared with the cooperation of the Geographical Committee of the Hungarian Academy of Sciences. Ed. in chief: Sándor Radó). Budapest: 1967. 112 p. LC 77-653067. 2060

Mardešić, Petar, and Dugački, Zvonimir. Geografski atlas Jugoslavije. Zagreb: Znanje, 1961. 256 p. LC Map 62-189. 2061

Nielsen, Niels, ed. The Atlas of Denmark; text and photographs. Translated from Danish by W.E. Calvert. Copenhagen: H. Hagerup, 1949-1961. 2 v. LC 51-3841. Published by the Royal Danish Geographical Society. 2062
  Covers the major aspects of the physical geography of Denmark with emphasis upon glacial landforms.

Otremba, Erich, ed. Atlas der Deutschen Agrarlandschaft. Wiesbaden: Franz Steiner, 1965- . LC Map 66-1135. 2063
  An ambitious project to map German farming through its crop systems, field patterns, social structure, history, and landscapes. Emphasis on small sample areas.

Oxford University Press. The Atlas of Britain and Northern Ireland. Oxford, U.K.: Clarendon Press, 1963. 200 p. LC 63-436. 2064 •
  Although now more than 20 years old, this is still an excellent thematic atlas tracing the resources of Great Britain. Most of the maps are at one of two scales, 1:2,000,000 or 1:1,000,000.

Europe

Polska Akademia Nauk. Instytut Geografii. Narodowy atlas Polski. Wrocław: Zakład Narodowy Imienia Ossolińskich, Wydawnictwo Polskiej Akademii Nauk, 1973-1978. 127 plates. LC 75-572481.
   Important national atlas of Poland containing 634 maps and 110 graphs on 127 sheets. The atlas is accompanied by a separate volume entitled Narodowy Atlas Polski; National atlas of Poland: texts and map keys in English *(185 p.).*

2065

Reader's Digest Association. The Reader's Digest complete atlas of the British Isles. London; Montreal: Reader's Digest Association, 1965. 229 p. LC 66-181.
   Attractive atlas containing complete coverage of Great Britain in 38 sheets at the scale of 1:380,160 and Ireland in eight sheets at 1:443,520. Physical aspects of the British Isles are covered in the section entitled "the fabric of the land" and the thematic topics in the section "the fabric of a nation." The final section is a gazetteer listing 32,000 place names in the British Isles.

2066 •

Spain. Instituto Geográfico y Catastral. Atlas nacional de España. Madrid: Instituto Geográfico y Catastral, 1965- . LC 66-892.
   Important thematic atlas issued in parts. Atlas accompanied by Reseña geográfica *(227 p.) and* Indice toponimico *(176 p.), both dated 1965.*

2067 •

Svenska Sällskapet för Antropologi och Geografi. Atlas över Sverige. National atlas of Sweden. Stockholm: Generalstabens litografiska anstalts forlag, 1953-1971. LC 54-1462.
   An excellent, exhaustive representation of various physical and cultural elements of the geography of Sweden. The atlas was issued in parts between 1953 and 1971 and consists of 520 maps. Text is in Swedish with table of contents, captions, and summaries in English.

2068 •

Switzerland. Landestopographie. Atlas der Schweiz. Edited by Eduard Imhof. Wabern-Bern: Verlag der Eidgenössischen Landestopographie, 1965- . LC Map 66-717.

2069

Thran, P., and Broekhuizen, Simon, eds. Agro-climatic atlas of Europe. New York: American Elsevier Publishing Co., 1965- . 3 v. LC Map 65-667.
   Volume one is an agro-ecological atlas of cereal growing in Europe. Volume two deals with the same subject and volume three discusses ecological aspects of European cereals.

2070

Tufescu, Victor. Atlas geografie: Republica Socialistă România. Bucureşti: Editura Didactică şi Pedagogică, 1965. 110 p. LC Map 66-1035.

2071

B. GENERAL WORKS

East, William Gordon. An Historical geography of Europe. 5th ed. London: Methuen; New York: Dutton (Barnes and Noble), 1966. 492 p. (1st ed., 1935. 480 p.). LC 54-36779.
   Reconstructs past geographies of Europe from the time of the Roman Empire to the late nineteenth century.

2072

Economic Commission for Europe, Geneva. Human settlements in Europe: postwar trends and policies. New York: United Nations, 1976. 141 p. Sale no.: E.76.II.E.9. LC 77-351220.
   A review prepared by the secretariat of the ECE as a contribution to the United Nations Conference on Human Settlements in Vancouver, Canada, June 1976. Study outlines certain problems and summarizes the policies pursued by ECE governments to solve them. Bibliography.

2073

Gutkind, Erwin. A. International history of city development. New York: Free Press, 1964-1972. 8 v. ISBN 0-02-913340-8. $300. (for the complete set). LC 64-13231.   2074
    Vol. 1: **Urban development in Central Europe.** 1964.
    Vol. 2: **Urban development in alpine and Scandinavian countries.** 1965. ISBN 0-02-913260-6. $50.
    Vol. 3: **Urban development in Southern Europe: Spain and Portugal.** 1966. ISBN 0-02-913270-3. $50.
    Vol. 4: **Urban development in Southern Europe: Italy and Greece.** 1969. ISBN 0-913280-0. $50.
    Vol. 5: **Urban development in Western Europe:France and Belgium.** 1970. ISBN 0-02-913300-9. $50.
    Vol. 6: **Urban Development in Western Europe: The Netherlands and Great Britain.** 1971. ISBN 0-02-913310-6. $50.
    Vol. 7: **Urban development in East-Central Europe: Poland, Czechoslovakia, and Hungary.** 1972. ISBN 0-02-913320-3. $40.
    Vol. 8: **Urban development in Eastern Europe: Bulgaria, Romania, and The U.S.S.R.** 1972. ISBN 0-02-913330-0. $50.
    Based on "cross-disciplinary studies and a comparative juxtaposition of urban development in individual countries." Richly illustrated, detailed analysis.

Hall, Peter, ed. Europe 2000. Final results of Plan 2000. London: Duckworth, 1977. 274 p. ISBN 0-7156-0987-4. LC 78-320635; New York: Columbia University Press, 1977. 274 p. ISBN 0-231-04462-3. $25. LC 77-9479.   2075
    A brief summary of the more than 20 reports in the project Europe 2000. Consists of main conclusions in education, industrialization, urbanization, and agriculture. Each topic has separate papers published between 1967-1975 by consulting with more than 200 experts from ten different countries.

Hoffman, George W., ed. A Geography of Europe: problems and prospects. 5th ed. New York; Chichester, U.K.: John Wiley, 1983. 647 p. (1st ed., New York: Ronald Press; London: Methuen, 1953. 775 p.). ISBN 0-471-89708-6. $31.95. LC 83-6964.   2076
    A multi-authored book, richly illustrated with maps and photographs; detailed in the discussions of the historical, physical, social, and economic geography of all regions of Europe, including the Soviet Union. Both systematic and regional analysis. Each regional chapter includes problem-oriented case studies with recommended solutions. Up-to-date statistical compilations and maps. Bibliography at end of each chapter.

Höll, Ottmar. Small states in Europe and dependence. Vienna: Wilhelm Braumüller (for Austrian Institute for International Affairs), 1983. 341 p. ISBN 3-7003-0545-1. LC 84-161794.   2077
    A useful study with a series of contributions based on conference papers in which the specific problems of small societies are discussed. The papers are generally of high quality and emphasize the close relationship of external and internal structural constraints of small nations. Valuable bibliographies at end of each chapter. Papers based on conference held in Laxenburg near Vienna June 10-12, 1981.

Jordan, Terry G. The European culture area: a systematic geography. (Harper's series in geography). New York: Harper and Row, 1973. 381 p. ISBN 0-06-043448-1. $29.95. LC 72-8624.   2078
    A useful systematic or topical analysis of Europe's human problems as opposed to region/subregion treatment. Within individual chapters emphasis is often on a regional analysis of individual problems. Well illustrated with an extensive bibliography, but much of the latter is outdated.

McNeill, William H. Europe's steppe frontier, 1500-1800. Chicago, IL: University of Chicago Press, 1964. 252 p. ISBN 0-226-56152-6. $13. (pbk). LC 64-22248. (Midway reprint service).   2079
    Settlement on the plains of Eastern Europe and the rivalries of the Austrian, Russian, and Turkish Empires from about 1500 to 1800.

Mitchell, Brian R. European historical statistics 1750-1975. (Cultural atlas series). 2nd rev. ed. London: Macmillan; New York: Facts on File, 1980. 868 p. (1st ed., London: Macmillan; New York: Columbia University Press, 1975. 827 p.). ISBN 0-87196-329-9. LC 80-67014. $85.   2080

A most valuable statistical compilation for an extended period on climate, popula-tion, labor force, agriculture, industry, external trade, transportation and communi-cation, finance, prices, and national accounts. Data is up-to-date to 1975.

Postan, Michael M. and Habakkuk, H.J., eds. The Cambridge economic history of Europe. Cambridge, U.K.; New York: Cambridge University Press, 1963-1982. 7 v.

> Vol. 1: **Agrarian life of the middle ages.** Postan, Michael M., ed. 1966. 872 p. ISBN 0-521-04505-3. $90.
>
> Vol. 2: **Trade and industry in the middle ages.** (date and price not set). ISBN 0-521-08709-0.
>
> Vol. 3: **Economic organization and policies in the middle ages.** Postan, Michael M., ed. 1963. 696 p. ISBN 0-521-04506-1. $69.50.
>
> Vol. 4: **The Economy of expanding Europe in the 16th and 17th centu-ries.** Rich, E.E. and Wilson, C.H., eds. 1977. 642 p. ISBN 0-521-04507-X. $62.50.
>
> Vol. 5: **The Economic organization of early modern Europe.** Rich. E.E. and Wilson, C.H., eds. 1977. 749 p. ISBN 0-521-08710-4. $79.50.
>
> Vol. 6: **The Industrial revolutions and after.** Postan, Michael M. and Ha-bakkuk, H.J., eds. 1965. Pt. 1: 602 p.; pt. 2: 438 p. ISBN 0-04508-8. $89.50.
>
> Vol. 7: **The Industrial economies: capital, labour, and enterprise, part 1: Britain, France, Germany, and Scandinavia.** Postan, Michael M. and Mathias, Peter, eds. 1978. 650 p. ISBN 0-521-21590-0. $72.50.
>
> Vol. 7: **The Industrial economies: capital, labour, and enterprise, part 2: the United States, Japan, and Russia.** Postan, Michael M. and Mathias, Peter, eds. 1978. 600 p. ISBN 0-521-21591-9. $72.50.   2081

Much first-class material on the background of agriculture, industry, and other matters in the regional growth of Europe.

Pounds, Norman J.G. An Historical geography of Europe 450 B.C. to 1330 A.D. Lon-don; New York: Cambridge University Press, 1973. 475 p. ISBN 0-521-08563-2. $64.50. ISBN 0-521-29126-7. $24.95. (pbk). 72-75299.   2082

Valuable work provides much backgroud.

Pounds, Norman J.G. An Historical geography of Europe 1500 to 1840. Cambridge; New York: Cambridge University Press, 1979. 438 p. ISBN 0-521-22379-2. $57.50. LC 79-11528.   2083

Excellent work that puts modern Europe into proper perspective.

C. SPECIAL TOPICS

Adams, John S.; Fricke, Werner; and Herden, Wolfgang, eds. Geography and re-gional policy: resource management by complex political systems. (Heidelberger Geo-graphische Arbeiten, Heft 73). Heidelberg, W. Germany: Geographisches Institut der Universität Heidelberg, 1983. 382 p.   2084

Papers of the second American-German International Seminar with German and English summaries of all papers. Papers stress the interface of geography and region-al policy in both political systems. This is a valuable book with maps, statistical com-pilations, and detailed bibliographies.

Berg, Leo Van Den; Drewett, Roy; Klassen, Leo H.; Rossi, Angelo; and Vijver-berg, Cornelius H.T. Urban Europe. Oxford U.K.; New York: Pergamon Press (for the European Coordination Centre for Research and Documentation in Social Sci-ences), 1982. 162 p. ISBN 0-08-023156-X. $44. LC 1-81233.   2085

The Urban Europe *series is the result of the collaboration of numerous social scientists and encompasses a cross-national comparative study in the costs of urban growth. Aimed to evaluate the costs associated with urban change and the financing of the urban systems. The first volume contains studies in two stages with numerous tables.*

Demko, George, ed. Regional development: problems and policies in Eastern and Western Europe. London: Croom Helm; New York: St. Martin's Press, 1984. 283 p. ISBN 0-312-66905-4. $27.95. LC 84-40039.   2086

A series of papers by specialists from Europe and the United States stressing both theoretical aspects of regional development and case study. Papers published have been carefully edited and the result is a most readable and valuable book. Excellent maps, tables, and bibliographies.

European Cultural Foundation, Central Scientific Committee. Chairman: Henri Janne. The Future of tomorrow. Prospective studies. The Hague, Netherlands: Martinus Nijhoff, 1972. 2 v. 585 p. (1st ed.: Amsterdam, 1970). NUC 74-149478.   2087

The books in this set are directed at the European public and their decision makers to make them aware of certain crucial problems for fundamental consideration: human relations in factories and offices, the situation of the working-person in the post-industrial age; education and teaching methods, how should Europeans be educated in the 21st century?; how can urban growth be controlled?; and old agricultural systems, the technological revolution which is transforming life in the countryside and the movements to towns.

European Cultural Foundation, Scientific Committee. Project 3, Chairman: Peter Hall, Director: Michel van Hulten. Fears and hopes for European urbanization: ten perspective papers and three evaluations. The Hague: Martinus Nijhoff, 1972. (originally published Amsterdam: European Cultural Foundation, 1970). 256 p. ISBN 9-02-471211-4. Dfl. 45.00, or $19.50. LC 73-160218.   2088

Part 3 of the Plan Europe 2000 Project, *this volume contains contributions by 13 authors from various European countries under overall themes of urbanization and planning human environments in Europe. An excellent introduction by Michel Van Hulten and Torsten Hägerstrand.*

Foster, Charles R., ed. Nations without a state: Ethnic minorities in western Europe. (Praeger special studies). New York: Praeger, 1980. 215 p. ISBN 0-03-056807-2. $29.95. LC 80-20900.   2089

This study of nine linguistic territorial regions which could be defined as nations without states emphasizes the cause and effect of political mobilization using a political culture approach, according to the editor. The different authors made a thorough analysis of these linguistic territorial regions and succeed in presenting a valuable scholarly study. Well footnoted with one appendix.

Franklin, S.H. The European peasantry: the final phase. London: Methuen, 1969. 256 p. LC 75-396472.   2090

A classic discussion of the evolution of the European peasantry since WW II. The discussions stress the economic and social consequences, the regional aspects of their evolution, and the role of the state in helping to solve their problems. Richly illustrated, maps, tables. Extensive bibliography, glossary.

Hall, Peter, and Hay, Dennis. Growth centers in the European urban system. London: Heinemann Education Books, 1980. 278 p. ISBN 0-435-3588-0. £15.50. LC 81-111138.   2091

Research into urban systems in the main developed regions of Europe. Study is part of a world-wide assessment. Tables, maps.

Europe

Hoffman, George W. The Changing European energy challenge: East and West. Durham, NC: Duke University Press, 1985. 230 p. ISBN 0-08223-0575-5. $34.75. LC 84-2474-8.   2092

An all-European analysis of the energy economy of both the industrialized and developed Western European countries and the Eastern European countries in the light of the price increases (and decreases of the last years) and oil supply disruption of the 1970s and their precarious energy situation. Special attention is given in the discussions on the proposed structural changes and diversification of supplies, which ultimately it is hoped will reduce their oil-energy supply vulnerability. Well illustrated with maps and statistical tables and detailed bibliographic references.

Miller, Mark J. Foreign workers in western Europe: an emerging political force. (Praeger special studies). New York: Praeger, 1981. 228 p. ISBN 0-03-059299-2. $34.95. LC 81-5853.   2093

The author raises the important question "what has become of all the minority aliens?" More than 30 million foreign workers and illegal aliens and 15 million political refugees all over the world have become a serious political problem. The literature on guest workers in western Europe has greatly increased. A valuable contribution to this literature and specifically to the discussion of the political impact of foreign labor policy. Selected bibliography.

Scargill, David I., gen ed. Problem regions of Europe. London; New York: Oxford University Press, 1973-1976. 16 v.

> Vol. 1: **The Eastern Alps.** Lichtenberger, Elisabeth. 1975. 48 p. ISBN 0-19-913106-6. $7.95. LC 75-326649.
> Vol. 2: **The Paris basin.** Thompson, Ian B. 1973. 48 p. ISBN 0-19-913278-X. £2.75. LC 73-158155.
> Vol. 3: **Andalusia.** Naylon, John. 1975. 48 p. ISBN 0-19-913108-2. £2.75. LC 75-319066.
> Vol. 4: **The Massif Central.** Clout, Hugh. 1973. 48 p. ISBN 0-19-913297-6. £2.50. LC 73-158145.
> Vol. 5: **The Mezzogiorno.** Mountjoy, Alan B. 1973. 48 p. ISBN 0-19-913100-7. £2.75. LC 73-158029.
> Vol. 6: **The Scandinavian northlands.** Mead, William R. 1974. 48 p. ISBN 0-19-913107-4. $6.95. LC 74-157847.
> Vol. 7: **London: metropolis and region.** Hall, John M. 1976. 48 p. ISBN 0-19-913194-5. £2.75. LC 77-355985.
> Vol. 8: **The Lower Rhône and Marseille.** Thompson, Ian B. 1975. 48 p. ISBN 0-19-913183-X. £2.75. LC 75-327222.
> Vol. 9: **The Franco-Belgian border region.** Clout, Hugh B. 1975. 48 p. ISBN 0-19-913182-1. £2.75. LC 75-323231.
> Vol. 10: **North-Rhine-Westphalia.** Hellen, John A. 1974. 48 p. ISBN 0-19-913290-9. £2.50. LC 75-315074.
> Vol. 11: **Poland's western and northern territories.** Hamilton, F.E. Ian. 1975. 48 p. ISBN 0-19-913105-8. £2.75. LC 75-323256.
> Vol. 12: **North East England.** Warren, Kenneth. 1973. 48 p. ISBN 0-19-913099-X. £2.75. LC 73-158932.
> Vol. 13: **Scotland's highlands and islands.** Turnock, David. 1974. 48 p. ISBN 0-19-913102-3. £2.50. LC 74-158450.
> Vol. 14: **Randstad, Holland.** Lawrence, George Richard Peter. 1973. 48 p. ISBN 0-19-913101-5. £2.75. LC 73-158153.
> Vol. 15: **Saar-Lorraine.** Burtenshaw, David. 1976. 48 p. ISBN 0-19-91319-7. £2.75. LC 76-367631.
> Vol. 16: **Northern Ireland.** Busteed, M.A. 1974. 48 p. ISBN 0-19-913103-1. LC 74-159344.   2094 •

Excellent series in paperbacks. Each author gives attention to one problem in a specific region and concentrates upon the tasks and difficulties of the area in the mid-1970's. Each book is well illustrated with maps and has a valuable bibliography and suggestions for further readings. The authors are all geographers.

D. NORTHERN AND WESTERN EUROPE: SERIALS

See also:
Annales de géographie. 27
Area. 28
Cahiers d'outre-mer. 30
Geoforum. 35
Geographical journal. 37
Geographical magazine. 38
Geography. 41
Institute of British Geographers. Transactions. 44
Norois. 47

Association de géographes français. Bulletin. 1- (1924-). Bi-monthly. Association de
Géographes Français, 191 rue Saint-Jacques, 75005 Paris, France. ISSN 0004-5322.   2095
*Papers presented at monthly meetings of the Association of French Geographers.
Often in the forefront in the development of geographical concepts or methods. In
French with abstracts in English and French. Cumulative indexes: 1924-1934; 1955-
1975.*

Cambria. A Welsh geographical review. 1- (1974-). Semi-annual. Department of Ge-
ography, University College of Swansea, Singleton Park, Swansea SA2 8PP, Wales,
U.K. ISSN 0306-9796.   2096
Specializes on Wales or on work by Welsh geographers.

East Midland geographer. 1- (1954-). Semi-annual. Department of Geography, University of Nottingham, Nottingham NG7 2RD, U.K. ISSN 0012-8481.   2097
*Articles on the East Midlands region of England. Book reviews. Cumulative index-
es at four-year intervals.*

Espace géographique: régions, environment, aménagement. 1- (1972-). Quarterly.
Doin Éditeurs, 8, Place de l'Odéon, F-75006 Paris, France. ISSN 0046-2497.   2098
*Articles on organization of space, on regions, on the environment and its percep-
tion, on planning, and on quantitative models and methods. In French. Abstracts in
French and English. Supplementary table of contents in English. Cumulative index:
1-5 (1972-1976), 6-10 (1977-1981).*

Fennia. 1- (1889-). Semi-annual. Academic Bookstore, Keskuskatu 1, SF-00100 Helsinki 10, Finland. ISSN 0015-0010.   2099
*Leading international periodical reporting on research on Finland and by Finnish
geographers. In English, with abstracts also in English. Cumulative index v 1-50
(1889-1928).*

Geografisk tidsskrift. 1- (1877-). Annual. Det Kongelige Danske Geografiske Selskab,
Haraldsgade 68, DK 2100 København, Denmark. ISSN 0016-7223.   2100
*Leading Danish geographical periodical. Articles, reviews, notes. In Danish or En-
glish. Abstracts in English. Cumulative indexes: v 1-20 (1887-1910), 21-33 (1911-1930),
34-50 (1931-1950), 51-59 (1951-1960), 60-69 (1961-1970).*

Hommes et terres du nord. 1963- . Semi-annual. Société de Géographie de Lille, 77,
rue Nationale, 59000 Lille, France. ISSN 0018-439X.   2101
*Articles on the north of France, the Low Countries, and other areas. Chronicle of
the north of France. Regional bibliography. In French. Abstracts in English and
French.*

Irish geography. 1- (1944-). Annual. The Geographical Society of Ireland. Place of
publication varies. ISSN 0075-0778.   2102
*Articles on all aspects of the geography of Ireland. Recent geographical literature
relating to Ireland.*

Europe

Norsk geografisk tidsskrift. Norwegian journal of geography. 1- (1926-). Quarterly. Order from Universitetsforlaget, P.O. Box 7508, Skillebekk, Oslo 2, Norway, or P.O. Box 258, Irvington-on-Hudson, New York, 10533. ISSN 0029-1951.　　　　2103
  Principal Norwegian geographical periodical. Annual Norwegian geographical bibliography. Mainly in English. Cumulative indexes v 1-10 (1926-1945), 11-24 (1946-1970).

Revue de géographie alpine. 1- (1913-). Quarterly. Institut de Géographie Alpine, rue Maurice-Gignoux, 38031 Grenoble, France. ISSN 0035-1121.　　　　2104
  Leading international journal devoted to geography of the Alps in all aspects both physical and human. Bibliography on the French Alps. In French. Abstracts in English and French. Cumulative indexes v 1-10 (1913-1922), 11-20 (1923-1932), 21-30 (1933-1942), 31-40 (1943-1952), 41-50 (1953-1962), 51-60 (1963-1972).

Revue de géographie de Lyon. 1- (1925-). Quarterly. 74 rue Pasteur, 69007 Lyon, France. ISSN 0035-113X.　　　　2105
  Devoted particularly to southeastern France. Chronicle for Rhône-Alpes. In French. Abstracts in English and French. Cumulative indexes v 1-10 (1925-1934), 11-20 (1935-1945), 21-30 (1946-1955).

Revue géographique de l'Est. 1- (1961-). Quarterly. Revue Géographique de l'Est, Presses universitaires de Nancy, 25 rue Baron-Louis, BP 454-5400 Nancy Cedex, France. ISSN 0035-3213.　　　　2106
  Articles, notes, and book reviews particulalrly on the east of France, the Rhine, Central, Eastern, and Southeastern Europe, and the Middle East. Chronicles. In French. Abstracts and tables of contents in French and English. Cumulative index 1-20 (1961-1980).

Revue géographique des Pyrénées et du Sud-Ouest. 1- (1930-). Quarterly. Service des Publications de l'Université, 56, rue du Taur, 31000 Toulouse, France. ISSN 0035-3221.　　　　2107
  Devoted to the Southwest of France and the Pyrenees. Chronicle. In French. Abstracts in English and French. Cumulative indexes 1930-1939, 1940-1949, 1950-1959, 1960-1969, 1970-1979.

Scottish geographical magazine. 1- (1885-). 3 per annum. Royal Scottish Geographical Society, 10 Randolph Crescent, Edinburgh EH3 7TU, Scotland, U.K. ISSN 0036-9225.　　　　2108
  Articles on Scotland. Recent geographical literature relating to Scotland. Book reviews. Cumulative indexes v 1-50 (1885-1934), 51-81 (1935-1965).

Société belge d'études géographiques. Bulletin. Belgische vereniging voor aardrijkskundige studies. Tijdschrift. 1- (1931-). 2 per annum. Belgische Vereniging voor Aardrijkskundige Studies, Celestijnenlaan 300, 3030 Leuven, Belgium. ISSN 0037-8925.　　2109
  Leading Belgian geographical journal. In French, Dutch, or English. Abstracts in English in front of each issue. Cumulative indexes 1931-1940, 1941-1950, 1951-1960, 1961-1970, 1971-1980.

Société languedocienne de géographie. Bulletin. 1-49 (1878-1929); s2 1-37 (1930-1966); s3 1 (90)- (1967-). Quarterly. Université Paul-Valery, B.P. 5043, 34032 Montpellier Cedex, France. ISSN 0373-3297.　　　　2110
  Devoted to Languedoc and nearby regions of southern France. In French. Abstracts in English and French. Cumulative indexes v 1-49 (1878-1929) and s 2 1-35 (1878-1964).

Svensk geografisk årsbok. Swedish geographical yearbook. 1- (1925-). Annual. Geografiska Institutionen, Sölvegatan 13, Lund, Sweden. ISSN 0081-9808.　　　　2111
  Articles on Sweden and general topics. Reviews. Annual Swedish geographical bibliography. Supplementary titles in table of contents and abstracts in English. Cumulative indexes 1925-1934, 1935-1950.

Anderson, Malcolm, ed. Frontier regions in Western Europe. London: Frank Cass; Totowa, NJ: Biblio Distribution Centre, 1983. 136 p. ISBN 0-7146-3217-1. $30. LC 83-135129.   2112
  A short book discussing frontier regions adjacent to international boundaries whose population is impacted by the political difficulties of the boundary, e.g., boundary disputes, subversive activities across international boundaries, problems of peripheral locations within the state. Valuable bibliography. The absence of maps is a disadvantage for the uninformed student of the political geography of Western Europe.

Beaujeu-Garnier, Jacqueline. France. (World's landscape series). London; New York: Longman, 1975. 132 p. ISBN 0-582-48178-3. $9.50. (pbk). LC 75-28288.   2113
  Masterful brief overview of France with discussions of the factors of diversity, rural landscapes, urban landscapes, and regional planning. References mainly in French.

Boal, Frederick W. and Douglas, J. Neville H., eds. Integration and division: geographical perspectives on the Northern Ireland problem. London; New York: Academic Press, 1982. 368 p. ISBN 0-12-108080-3. $40. LC 81-68978.   2114
  The authors reject the view of a Holy War between Protestants and Roman Catholics in Northern Ireland as a simplistic view and try in their discussions to set the Northern Irish problem in its proper perspective as a region with numerous complex cultural and political problems in a diverse community. The authors try to answer the question of what maintains the diversity and forces it toward conflict or constraints. Maps, tables, and excellent bibliography.

Burtenshaw, David; Bateman, M.; and Ashworth, Gregory J. The City in West Europe. Chichester, U.K.; New York: John Wiley, 1981. 340 p. ISBN 0-471-27929-3. $47.95. LC 80-41589.   2115
  A comparative view of Western Europe's urban development and planning. Well illustrated. Valuable bibliography.

Clemenson, Heather A. English country houses and landed estates. London: Croom Helm; New York: St. Martin's Press, 1982. 244 p. ISBN 0-312-25414-8. $30. LC 82-3298.   2116
  Examines the privately owned landed estates in the past and in the present day. An interesting historical geography. Insufficient illustrations and bibliography.

Clout, Hugh D. The Geography of post-war France: a social and economic approach. Oxford; New York; Toronto: Pergamon Press, 1972. 165 p. ISBN 0-08-016766-7. $10. (pbk). LC 70-172400.   2117
  Brief well-informed overview of the systematic and regional geography of France. Numerous informative maps.

Clout, Hugh D., ed. Regional development in Western Europe. 2nd ed. Chichester, U.K.; New York: John Wiley, 1981. 417 p. (1st ed., 1975. 328 p.). ISBN 0-471-27846-7. $49.95. ISBN 0-471-27845-9. $22.95. (pbk). LC 80-40852.   2118
  An in-depth examination of regional problems and planning programs by eleven geographers from Britain and the Continent. Second edition gives considerable attention to Europe's economic slowdown since the mid 1970s. Maps, tables.

Darby, H. Clifford (Henry). The Domesday geography of England. Cambridge, U.K.; New York: Cambridge University Press, 1954-1977. 7 v.   2119
    Vol. 1: **The Domesday geography of Eastern England.** 3rd ed. 1971. 400 p. ISBN 0-521-08022-3. £42. (U.K. only). LC 70-108106.
    Vol. 2: **The Domesday geography of Midland England.** (with Terrett, I.B.). 2nd ed. 1971. 490 p. LC 78-134626.
    Vol. 3: **The Domesday geography of South-East England.** (with Campbell, E.M.J.). 1962. 658 p. ISBN 0-521-04770-6. £50. (U.K. only). LC 62-6262.
    Vol. 4: **The Domesday geography of Northern England.** (with Maxwell, I.S.). 1962. 540 p. ISBN 0-521-04773-0. $102. LC 62-53452.
    Vol. 5: **The Domesday geography of South-West England.** (with Finn, R.W.). 1967. 469 p. LC 67-11519.
    Vol. 6: **Domesday gazetteer.** (with Versey, G.R.). 1975. 544 p. ISBN 0-521-20666-9. £60. (U.K. only). LC 75-19532.
    Vol. 7: **Domesday England.** 1977. 416 p. ISBN 0-521-21307-X. $85. LC 76-11485.
A classic study in historical geography, reconstructing the life and organization of a period from unique source materials.

Darby, H. Clifford (Henry), ed. A New historical geography of England. Cambridge, U.K.; New York: Cambridge University Press, 1976. 2 v.
    Vol. 1: 316 p. ISBN 0-521-22122-6. $59.50. LC 76-26141.
    Vol. 2: 460 p. ISBN 0-521-20116-0. $70. LC 76-26029.   2120
A completely new synthesis of the historical geography of England replacing the earlier pioneering volume entitled An historical geography of England before A.D. 1800 *(1936). This collected work by leading scholars in historical geography reflects much new research and the new viewpoints that developed over a forty-year period.*

Diem, Aubrey. Western Europe: a geographical analysis. New York; London: John Wiley, 1979. 534 p. ISBN 0-471-21400-0. $38.95. LC 77-24617.   2121
Geographic analysis of social and economic changes in Western Europe since World War II. Excellent illustrations and bibliography.

Dodgshon, Robert A. and Butlin, Robin A., eds. An Historical geography of England and Wales. London; New York: Academic Press, 1978. 450 p. ISBN 0-12-219250-8. $55. ISBN 0-12-219252-4. $24. (pbk). LC 78-18021.   2122
Provides interesting alternative approaches to the historical geography of England (cf. Darby, above) and Wales.

Dury, George H. The British Isles. 5th ed. London: Heinemann Educational Books; Totowa, NJ: Barnes and Noble Imports, 1973. 365 p. (1st ed. 1961. 503 p.). ISBN 0-435-35260-1. $19.50. LC 73-169618.   2123
A slightly revised edition of a standard text on the British Isles discussing the general geography in the first part and the regional geography in the second part of the book. Well illustrated. Detailed bibliography.

House, John W. France: an applied geography. London; New York: Methuen, 1978. 478 p. ISBN 0-416-15080-2. $49.95. LC 80-507208.   2124
A general overview.

Johnston, Ronald J. and Doornkamp, John C., eds. The Changing geography of the United Kingdom. London; New York: Methuen for the Institute of British Geographers, 1982. 430 p.ISBN 0-416-74800-7. $35. LC 82-12477.   2125
Geographic overview of the United Kingdom in a 50-year period since the founding of the Institute of British Geographers in 1933 with emphasis on recent changes for an understanding of the evolving economic, social, political, and environmental geography of Britain. The introduction and conclusion written by the editors serve to integrate the various contributions. This is a dynamic regional geography linking local, national, and global developments and their impact on the changing geography of the United Kingdom. Well illustrated with a valuable bibliography.

Lambert, Audrey M. The Making of the Dutch Landscape: a historical geography of the Netherlands. 2nd ed. London: Seminar Press; New York: Academic Press, 1971. 412 p. ISBN 0-12-785450-9. $44. LC 70-162378.   2126

In no other country in Europe has the hand of man had greater formative influence in shaping the landscape than the Netherlands. The discussions of the relentless battles to dike nearly one-half of the country lying beneath the waves is the story of the Netherlands. It is told in a most interesting, well-written and superbly illustrated book from the earliest time to present in a chronological approach.

Malmström, Vincent H. A Regional geography of Iceland. (National Research Council Publication no. 584; Division of Earth Sciences, Foreign Field Research Program Report no. 1). Washington, DC: National Academy of Sciences--National Research Council, 1958. 255 p. LC 58-60026.   2127

A basic work on Iceland's physical and cultural geography.

Manners, Ian R. North Sea oil and environmental planning: the United Kingdom experience. Austin, TX: University of Texas Press, 1982. 332 p. ISBN 0-292-76475-8. $37.50. LC 81-16170.   2128

Analysis of Britain's environmental policies in North Sea oil development; how planners in the U.K. dealt with major social and environmental impacts that accompanied the development of energy resources in the North Sea. A detailed study analyzing a wide range of problems influencing the formulation and implementation of environmental policies. Detailed bibliography and numerous diagrams and tables.

Martonne, Emmanuel de, and Demangeon, Albert. La France. (Géographie Universelle, Tome VI, parties 1 et 2). Paris: A Colin, 1946-1948. 3 v. LC 50-3334.   2129
    Part 1: **France physique.** Emmanuel de Martonne, 2nd ed., 1947.
    Part 2: **France économique et humaine.** Albert Demangeon, v 1, 1946; v 2, 1948.

A classic treatment of France by two of her most eminent geographers, one with physical, the other with human interests.

Mead, William R. An historical geography of Scandinavia. London; New York: Academic Press, 1981. 336 p. ISBN 0-12-48742-0. $39. LC 81-66377.   2130

A major contribution by the well-known British geographer who has spent a lifetime travelling and writing in and about the Scandinavian countries. This physical and human geography discusses the stepping stones of the human settlement pattern and takes a broad look at the surrounding scene. This is an extremely well-written and fascinating historical geography based on the concept of "Scandanavia." Charts, maps, detailed bibliography.

Noreng, Oysten. The Oil industry and government strategy in the North Sea. London: Croom Helm; Boulder, CO: Center for Energy and Economic Development, 1980. 268 p. ISBN 0-918714-12-8. $27.50. LC 80-81590.   2131

A distinguished economic analyst and planner discusses the question of the interdependence of government and oil companies in the extraction of a natural resource, focusing on the countries of the North Sea and the constraining factors of government policy as well as on relevant administrative patterns. Valuable bibliography and tables.

Nystrom, J. Warren (John), and Hoffman, George W. The Common Market. 2nd ed. New York: Van Nostrand Company, 1976. 147 p. (1st ed. by Nystrom, J. Warren, and Malof, Peter, 1962. 134 p.). ISBN 0-442-29756-4. LC 76-7100.   2132

A concise discussion and synthesis of various developments showing Western Europe's efforts since World War II to create an integrated community of nations. Maps, tables.

Europe

Parker, Geoffrey. The Logic of unity: a geography of the European Economic Community. 3rd ed. London; New York: Longman, 1981. 208 p. (1st ed., 1968. 178 p.). ISBN 0-582-30031-2. $11.95. (pbk). LC 80-40154.　　　　　　　　　　　　　　　　　2133

Third edition covers important developments in the European Community giving increased emphasis to the problems of energy conservation and an awareness of the dangers of environmental deterioration as a consequence of industrialization and heightened motorization and the growing importance of the regional fund for the poorer regions. The book is well illustrated and has an excellent up-to-date bibliography.

Stamp, Sir Laurence Dudley. The Land of Britain: its use and misuse. 3rd ed. London: Longman, Green, 1962. 546 p. LC 63-3030.　　　　　　　　　　　　　　　2134

Summarizes Stamp's pioneering work on land classification mapping in Britain. Also provides excellent coverage of Britain's geology, geomorphology, climate, soils, and types of farming.

Thompson, Ian B. Modern France: a social and economic geography. London; Toronto: Butterworths, 1970. 465 p. ISBN 0-408-70016-5. LC 77-532632.　　　　　2135

Treatise on the patterns of social development and of economic activity, with regional essays. Extensive tables, maps, references, and bibliography.

Watson, J. Wreford (James), and Sissons, J.B. The British Isles: a systematic geography. Edinburgh, U.K.: Nelson, 1964. 452 p. ISBN 0-17-711014-7. £10.25. LC 64-56444.　　　　　　　　　　　　　　　　　　　　　　　　　　　　　　　　2136

Twenty-two essays covering aspects of the physical, historical, cultural, economic, and political geography of the United Kingdom and Ireland. Also, a summary of British geography and geographers.

White, Paul. The West European city: a social geography. London; New York: Longman, 1984. 269 p. ISBN 0-582-30047-9. $13.50. (pbk). LC 83-9382.　　　　2137

The author clearly demonstrates that the cities of Western Europe differ basically from those of North America and other continents. He discusses in several chapters how this difference evolved over many centuries. A well-illustrated book with numerous tables and maps.

F. CENTRAL EUROPE: SERIALS

Berichte zur deutschen Landeskunde. 1- (1941-). Semi-annual. Zentralausschuss für deutsche Landeskunde, 1 Selbstverlag, D-5500, Trier, Federal Republic of Germany. ISSN 0005-9099.　　　　　　　　　　　　　　　　　　　　　　　　　　　2138

Articles, extensive book reviews, notes, and, through 1971, bibliographies on the geography of Germany. In German. Cumulative index v 1-35 (1941-1965).

Forschungen zur deutschen Landeskunde. 1- (1885-). Irregular. Zentralausschuss für deutsche Landeskunde, Selbstverlag, D-5500 Trier, Federal Republic of Germany.　　2139

Extensive research monographs on the geography of Germany. One of the longest series in the world of detailed research monographs on a single country. In German. Summaries in English.

Geographica helvetica. 1- (1946-). Quarterly. Fotorotar AG, Administration GH, Gewerbestrasse 18, 8132 Egg/ZH, Switzerland. ISSN 0016-7312.　　　　　　　2140

Principal geographical periodical of Switzerland with some concentration on Switzerland. In German. Some abstracts in English.

Neues Schrifttum zur deutschen Landeskunde. 1- (1979-). Semi-annual. Zentralausschuss für deutsche Landeskunde, Postfach 3825, D-5500 Trier, Federal Republic of Germany.　　　　　　　　　　　　　　　　　　　　　　　　　　2141

Detailed bibliography of German geography from 1979 (cf. entry 2138).

Geographischer Jahresbericht aus Österreich. 1- (1894-). Biennial. Geographisches Institut der Universität Wien, Universitätsstrasse 7, A-1010 Wien, Austria. ISSN 0029-9138.　2142
　　Detailed reports on dissertations, scientific publications, and lectures in geographical colloquia in each of the Austrian universities during each biennium: Wien, Graz, Innsbruck, Salzburg, Klagenfurt, and Wirtschaftsuniversität Wien. In German.

Österreichsche Geographische Gesellschaft. Mitteilungen. 1- (1857-). 2 per annum. Österreichische Geographische Gesellschaft, Karl Schweighofergasse 3, A-1070 Wien, Austria.　2143
　　Long established major scholarly geographical periodical. Includes annual bibliography of Austrian geography. In German. Summaries in English. Cumulative indexes: 1857-1907, 1908-1959, 1960-1980.

Note: See also Die Erde (33), Erdkunde (34), Geographische Zeitschrift (40), and GeoJournal (42).

G. CENTRAL EUROPE: BOOKS

Burghardt, Andrew F. Borderland: a historical and geographical study of Burgenland, Austria. Madison, WI: University of Wisconsin Press, 1962. (Available from Books on Demand, Ann Arbor, MI). $97.30. LC 62-15992.　2144
　　Study of a pivotal area in which stress is placed on the state-idea and on the assimilation of minorities.

Danz, Walter, and Henz, Hans-Rudolf. Integrated development of mountain areas. (Studies collection, regional policy series no. 20). Brussels, Belgium: Commission of the European Communities, 1981. 90 p. ISBN 9-282-52284-9. LC 81-190443.　2145
　　Useful short study of the environmental action program of the commission examining the varied environmental policy aspects of the hill and mountain areas of the EEC. Elements of common action in these areas are proposed in order to achieve these aims. Study analyzes main problems, needs, principles, and aims of an integrated community policy in these areas and proposes common action for problems of mountain regions in respect of urbanization.

Dickinson, Robert E. Germany: a general and regional geography. 2nd ed. London: Methuen; New York: Dutton, 1961. 716 p. (1st ed., 1953. 700 p.).　2146
　　A thorough, detailed, and well-written study in which the author succeeds in blending historic and cultural features with physical and economic to evoke the "personality" of the regions of Germany.

George, Pierre, and Tricart, Jean. L'Europe Centrale. Paris: Presses Universitaires de France, 1954. 2 v., 753 p. LC A55-2385.　2147
　　Comprehensive discussion, well illustrated, in traditional French manner. Volume I stresses the physical and human geography and volume II the economic geography.

Gutersohn, Heinrich. Geographie der Schweiz in drei bänden. Bern: Kümmerly and Frey, 1958-1969. 3 v. in 5 books. LC 59-28915. NUC 71-88213.　2148
　　Vol. 1: **Jura.** 1958. 260 p.
　　Vol. 2, Part 1: **The Alps.** 2nd ed. 1972. 486 p.
　　　　Part 2: **The Alps.** 1968. 440 p.
　　Vol. 3, Part 1: **Mittelland.** 1968. 292 p.
　　　　Part 2: **Mittelland.** 1969. 367 p.
　　The standard very comprehensive regional geography of Switzerland. An exhaustive study of each of the Swiss regions and the cantons, stressing both the physical and cultural geography. Very well illustrated.

Hoffman, George W., ed. Federalism and regional development: case studies on the experience in the United States and the Federal Republic of Germany. Austin, TX: University of Texas Press, 1981. 754 p. ISBN 0-292-73825-0. $40. LC 80-53735.　　　2149
Papers of first American-German seminar by American and German geographers with German and English abstracts. The theme of the papers was the impact of the federal system on regional development in the two countries. The papers focus on various political geographic issues which have had an impact on federal and sectoral spatial planning activities and which have become pressing societal issues. The volume presents theoretical and empirical studies of spatial activities and constraints of various federal and regional authorities. Some of the papers are of superior quality. Detailed bibliographies and numerous tables.

Mellor, Roy E.H. The Two Germanies: a modern geography. London: Harper and Row; Totowa, NJ: Barnes and Noble Imports, 1978. 461 p. ISBN 0-06-494778-5. $24.50. ISBN 0-06-494779-3. $9.95. (pbk). LC 79-100281.　　　2150
A regional geographic study of the two Germanys emphasizing patterns of their territorial space inherited from the past and the establishment of new ones. Detailed bibliography.

Mutton, Alice F.A. Central Europe: a regional and human geography. (Praeger advanced geographies). 2nd ed. London: Longman; New York: Praeger, 1968. 488 p. (1st ed., 1961. 475 p.). LC 68-15652.　　　2151
A richly illustrated regional text on the physical and human geography of Benelux, the two Germanys, Switzerland, Austria, and Czechoslovakia.

Pounds, Norman J.G. Ruhr: a study in historical and economic geography. Bloomington, IN: Indiana University Press, 1952. 283 p. Reprinted: Westport, CT: Greenwood, 1968. 283 p. ISBN 0-8371-0621-4. $15.75. LC 68-55636.　　　2152
The economic development of the Ruhr is analyzed at three separate periods-- 1800, 1850, and 1900--as a prelude to presenting the geographic picture in the mid-twentieth century. A final chapter assesses the significance of the Ruhr in the early post-war economy of Europe.

Sanguin, André-Louis. La Suisse, essai de géographie politique. Gap, France: Editions Ophrys, 1983. 363 p. ISBN 2-7080-0519-7. FFr 200. LC 83-187233.　　　2153
The study emphasizes the physical and human geography of Switzerland with the theme of the basic raison d'être of the Swiss national idea, the organization of its regions, the electoral system, and a spatial analysis of Switzerland's political frontiers and their international perspectives. Detailed bibliography and some maps.

Siegfried, André. Switzerland: a democratic way of life. Translated from French by Edward Fitzgerald. London: Cape; New York: Duell, Sloan, and Pearce, 1950. 223 p. Reprinted: Westport, CT: Hyperion Press, 1979. 283 p. ISBN 0-8371-0621-4. $15.75. LC 68-55636. (La Suisse: démocratie-témoin. Neuchâtel: La Baconnière, 1948. 238 p.).　　　2154
A classic discussion by the well-known French political geographer stressing the human geography, including economic and political developments.

Wild, Martin Trevor. West Germany: a geography of its people. Totowa, NJ: Barnes and Noble, 1980. 255 p. ISBN 0-06-497658-0. $23.50. LC 79-55698.　　　2155
This excellent short volume discusses socio-geographic responses to West Germany's postwar economic development and its influences upon contemporary human geography. The discussions are based largely on German sources and have numerous maps and a good bibliography.

H. EASTERN EUROPE: SERIALS

Geographia Polonica. (Polish Academy of Sciences. Institute of Geography and Spatial Organization). 1- (1964-). Irregular. Orders: Foreign Trade Enterprise Ars Polonia, Krakowskie Przedmieście 7, Warszawa, Poland. ISSN 0016-7282.   2156

Articles on the development of geography in Poland. Summaries or translations of the most important original research studies completed in Poland. Reports by Polish geographers on international congresses or conferences. Proceedings of many international conferences held in Poland or with substantial Polish participation. In some ways the most accessible and valuable geographical publication in east-central or Eastern Europe. Mainly in English. Cumulative index: 1-32 (1964-1975).

Studies in geography in Hungary. (Hungarian Academy of Sciences. Geographical Research Institute). 1- (1964-). Irregular. Budapest: Akadémiai Kiadó. Distributor: Kultura, Hungarian Foreign Trading Co. P.O. Box 149, H-1389, Budapest, Hungary. ISSN 0081-7961.   2157

Well-edited substantial monographs and collected works presenting the results of geographical research in the Institute of Geography of the Hungarian Academy of Sciences, or papers given at international meetings or symposia. In English.

Note: See also entry 48.

I. EASTERN EUROPE: BOOKS

Blanc, André. L'Europe Socialiste. (Europe de demain, 4). Paris: Presses Universitaires de France, 1974. 263 p. ISBN 3-528-353. FFr 177. LC 74-17196.   2158

A systematic analysis of the human and economic geography of six COMECON countries (excluding Yugoslavia and Albania) by a well-known French geographer. Numerous tables and maps.

Braun, Aurel. Small state security in the Balkans. London: Macmillan; Totowa, NJ: Barnes and Noble, 1983. 334 p. ISBN 0-389-20288-6. $23.50. LC 82-6639.   2159

Discussion of external security of the six Balkan states (including Turkey), but also stresses the impact of salient domestic factors such as nationalism and national minorities. The various influences interact with domestic factors which help determine the quality of regional security. This is a unique study in the field of political geography. Students of the Balkan countries will find the author's approach of interest and valuable for an understanding of small state behavior.

Carter, Francis W., ed. An Historical geography of the Balkans. London; New York: Academic Press, 1977. 599 p. ISBN 0-12-161750-5. $93. LC 77-77367.   2160

Contributions by an international team of social scientists. Some contributions cover the whole peninsula, some are only on the local level, and some contributions are on specific topics for one country of the region. Numerous maps, bibliography.

Enyedi, György. Hungary: an economic geography. Translated from Hungarian by Elek Helvei, translation revised and edited by Mary Völgyes. Boulder, CO: Westview Press, 1976. 289 p. ISBN 0-89158-030-1. $38. LC 76-3742.   2161

A useful overall view of the land and its people with a description of its economic development and regional economic patterns. Extensive bibliography of works mostly in Hungarian. Numerous maps and tables.

Faber, Bernard Lewis. The Social structure of Eastern Europe: transition and progress in Czechoslovakia, Hungary, Poland, Romania and Yugoslavia. New York: Praeger, 1976. 423 p. ISBN 0-275-55590-9. LC 75-23961.   2162

Seventeen papers presented at a conference, mostly by sociologists and anthropologists emphasizing the dynamics at work in society of Eastern Europe with the system of social stratification as a key factor in the process of societal transformation. Probably the most thorough analysis. Based on a careful selection of contributions.

Europe

Fischer-Galati, Stephen, ed. Man, state, and society in East European history. New York: Praeger, 1970. 343 p. LC 69-10516.
A series of papers written by historians discussing the interaction of the forces of nationalism and imperialism in an arena dominated by societies in slow transition from the feudal to the industrial stage. A careful selection of well-written papers makes this volume imperative for any serious student of Eastern Europe's past developments, which are basic for an understanding of present-day developments.
2163

Fisher, Jack. Yugoslavia, a multinational state: regional difference and administrative response. San Francisco, CA: Chandler Publishing Co., 1966. 244 p. LC 66-10335.
A detailed and very useful study about the regional differences and administrative response emphasizing the multinational character of the country and the social, economic, and political problems.
2164

French, R. A. (Richard Antony), and Hamilton, F.E. Ian. The Socialist city: spatial structure and urban policy. Chichester, U.K.; New York: Wiley, 1979. 541 p. ISBN 0-471-99689-0. $68.95. LC 78-16828.
The basic theme of all contributions deals with the question "is there a socialist city?" The contributions are organized in two distinct parts, each combining the thematic analysis with case studies. The first part examines selected aspects of the Soviet city, the second part consists of rather uneven case studies of Eastern European cities. Hamilton's initial discussions on Eastern Europe stress the importance of diverse national urbanization milieux for understanding growth and changes in particular cities. Well illustrated with tables and maps and detailed bibliographies after each chapter.
2165

Gianaris, Nicholas V. The economics of the Balkan countries: Albania, Bulgaria, Greece, Romania, Turkey, and Yugoslavia. New York: Praeger, 1982. 188 p. ISBN 0-03-060232-7. $29.95. LC 81-10748.
Discusses under three headings the diversity of the economies: socioeconomic background; organizational and developmental aspects; and foreign trade and economic cooperation. He stresses the difficulties of comparative analysis of countries with different economic systems. The main trend of these discussions is the author's feeling that closer cooperation has become a necessity. Numerous footnotes, but no bibliography or maps.
2166

Graham, Lawrence S. Romania: a developing Socialist state. Boulder, CO: Westview Press, 1982. 136 p. ISBN 0-89158-925-2. $20. LC 81-16493.
An excellent study by a political scientist trained in comparative analysis utilizing the concepts and perception of developmental politics and administration in a European setting, emphasizing Romania's status as a bridge country in both the East-West and North-South axis. The author's analysis explores Romania's current socioeconomic development and the country's unique relationship to the world, the larger European community, and the non-aligned Third World countries.
2167

Grothusen, Klaus-Detlev, ed. Jugoslawien. Yugoslavia. (Südosteuropa-Handbuch vol. I). Göttingen, W. Germany: Vandenhoeck and Ruprecht, 1975. 566 p. In German and English. ISBN 3-525-36200-5. LC 76-459121. (Deutsche Forschungsgemeinshaft. Arbeitskreis Südosteuropa-Forschung).
By German and American specialists. Covers the state and politics, economy, society and societal structure, culture, and science. Detailed appendices with a most useful bibliography. Contributions are of uneven character. Colored map.
2168

Grothusen, Klaus-Detlev, ed. Rumänien. Romania. (Südosteuropa-Handbuch vol. II). Göttingen, W. Germany: Vandenhoeck and Ruprecht, 1977. 711 p. In German and English. ISBN 3-525-36201-3. DM 175. LC 78-363627. (Deutsche Forschungsgemeinshaft. Arbeitskreis Südosteuropa-Forschung).
By German and American specialists. Covers state and politics, economy, society and societal structure, culture, and science. Valuable appendices. Colored map. Contributions are of uneven character, but factual information is of value.
2169

Hamilton, F.E. Ian. Yugoslavia: patterns of economic activity. (Praeger surveys in economic geography). London: Bell; New York: Praeger, 1968. 384 p. LC 68-19850.　　2170
An economic geography oriented toward analysis of industries and planning.

Hoffman, George W. and Neal, Fred. Yugoslavia and the new Communism. New York: Twentieth Century Fund, 1962. 546 p. Reprinted: Millwood, NY: Kraus Reprints. ISBN 0-527-02822-3. $10. LC 62-13485.　　2171
A valuable study of a country which has attempted to develop its own brand of socialism and to solve the problems of minority integration and federal administration, all on a restricted agricultural and industrial resource base.

Hoffman, George W., ed. Eastern Europe: essays in geographical problems. London: Methuen; New York: Praeger, 1971. 502 p. ISBN 0-416-15990. LC 75-588353. (U.K.); LC 70-134447. (U.S.). (Conference on East-Central and Southeast European geography, University of Texas at Austin, April 18-20, 1969).　　2172
The essays in this book present a wide range of problems in the area, showing contrasting social, demographic, and economic processes, and a variety of research techniques. The essays are introduced by a broad regional synthesis. Contributors are from both the United States and Europe. The book includes an extensive bibliography on the region as of 1970, over 40 maps, and numerous statistical compilations. Each chapter is followed by a brief summary of the discussions of each paper.

Hoffman, George W. Regional development strategy in southeast Europe: a comparative analysis of Albania, Bulgaria, Greece, Romania, and Yugoslavia. (Praeger special studies in international economics and development). New York: Praeger, 1972. 322 p. LC 75-18196.　　2173
Study is based on longterm field research showing the great spatial contrasts in this region, the continuous interaction of natural and social forces, the relationship between national and regional interests, and the rapid socioeconomic changes in the postwar period. A critical analysis of regional cooperation closes the discussions. Detailed statistical appendix and bibliography. Numerous maps.

Hoffman, George W. The Balkans in transition. Princeton, NJ: Van Nostrand, 1963. 124 p. Reprinted: Westport, CT: Greenwood Press, 1983. 124 p. ISBN 0-313-24288-7. $27.50. LC 83-18538.　　2174
Brief analysis of changing conditions in Albania, Bulgaria, and Yugoslavia.

Kosiński, Leszek A., ed. Demographic developments in Eastern Europe. New York: Praeger, 1977. 343 p. ISBN 0-275-56180-1. LC 76-12858.　　2175
The papers in this study are those presented in the social sciences at the International Slavic Conference in Banff in 1974.

Kostanick, H. Louis (Huey), ed. Population and migration trends in Eastern Europe. (Conference held on Demography and Urbanization in Eastern Europe, Los Angeles, February 5-9, 1976). Boulder, CO: Westview Press, 1977. 247 p. ISBN 0-0891-58147-2. $24. LC 77-1905.　　2176
Interdisciplinary analysis by 13 authors from the United States, United Kingdom, and certain Eastern European countries stressing demographic, urbanization, and migration developments in Eastern Europe. Numerous statistical compilations.

Paul, David. W. Czechoslovakia: profile of a Socialist republic at the crossroads of Europe. Boulder, CO: Westview Press, 1981. 196 p. ISBN 0- 89158-861-2. $22.50. ISBN 0-86531-506-X. $9.95. (pbk). LC 80-19333.　　2177
Up-to-date broad analysis of Czechoslovakian developments in the postwar period. Annotated bibliography, tables, charts.

Pounds, Norman J.G. Eastern Europe. Harlow, U.K.: Longman; Chicago, IL: Aldine, 1969. 912 p. SBN 582-48143-0. LC 70-409634.
2178
A comprehensive survey of the countries of Eastern Europe (East-Central and Southeast Europe excluding Greece). Both text and illustrations are of high quality. Six general chapters, eight covering the individual countries, and a concluding discussion.

Pounds, Norman J.G. Poland between East and West. Princeton, NJ: D. Van Nostrand, 1964. 132 p. LC 64-4686.
2179
Emphasizes the changes in Poland's boundaries and the continuity of the spirit of Polish nationalism.

Rugg, Dean S. The Geography of Eastern Europe. Lincoln, NE: Cliff's Notes, Inc., 1978. 115 p. ISBN 0-8220-1917-5. LC 78-67785.
2180
A synthesis of how cultures associated with the Eastern European countries have been organized in space. Emphasis is on historical interaction and economic modernization, but no attempt is made to cover all systematically. Numerous maps and tables and extensive bibliography.

Shoup, Paul. Communism and the Yugoslav national question. New York: Columbia University Press, 1968. 308 p. ISBN 0-231-03125-4. $28. LC 68-19759.
2181
The best and most thorough analysis of the national question of Yugoslavia since 1919. Detailed bibliography and valuable census data analysis.

Stefanov, Ivan; Dinev, Ljubomir; and Koev, Zdravko. Bulgarien: Land, Volk, Wirtschaft in Stichworten. Wien: Ferdinand Hirt, 1975. 127 p. ISBN 3-7019-8020-9. LC 76-452136.
2182
A concise geography of Bulgaria written by Bulgarian geographers discussing in a traditional way the physical geography, historical and cultural development, demographic aspects, and settlements. Excellent illustrations and literature references.

Sugar, Peter F. and Lederer, Ivo J., eds. Nationalism in Eastern Europe. Seattle WA; London: University of Washington Press, 1969. 465 p. LC 74-93026.
2183
A very valuable contribution to a most serious problem in all Eastern European countries. Emphasis in the eight country essays is on the history of nationalism. Fills in a very great gap in the literature. Explains the development of national identities of Eastern Europe, both past and present. Of special value is the introductory chapter by Professor Sugar on external and domestic roots of Eastern European nationalism. Heavily footnoted but no separate bibliography. The absence of maps showing the exact locations of the various nationalities is a disadvantage in this otherwise useful book.

Turnock, David. An Economic geography of Romania. London: G. Bell and Sons, 1974. 319 p. ISBN 0-7135-1628-3. £11.95. LC 74-180589.
2184
Survey of Romania's economic geography at the end of the Five Year Plan 1966-1970. While data on agriculture and industrial production are outdated, the book provides a good analysis of the basic strength and weakness of the country's economic development. Valuable bibliography, maps, and statistical tables.

Turnock, David. Eastern Europe. Folkstone, U.K.: Dawson; Boulder, CO: Westview, 1978. 273 p. ISBN 0-7129-0795-5. $32. LC 78-314844.
2185
A solid discussion of the industrial growth of the eight countries of Eastern Europe (excluding Yugoslavia) which vary in population size, resource base, and volume of industrial production. Discussions clearly indicate the great variations between and within individual countries. Numerous maps and tables.

Wilkinson, Henry R. Maps and politics: a review of the ethnographic cartography of Macedonia. Liverpool, U.K.: University of Liverpool Press, 1951. 366 p. LC 53-427.
2186
A classic book discussing the history of Macedonia largely based on its maps, which provide an important historical source and provide the evidence of growth and origins of the ethnographic disputes which have been at the heart of the Macedonian problem. Superb maps.

Estudios geográficos. 1- (1940-). Quarterly. Orders: Distribución de Publicaciones del Consejo Superior de Investigaciones Científicas, Vitrubio 8, Madrid 6, Spain. ISSN 0014-1496.　　2187
Leading geographical periodical of Spain. In Spanish. Abstracts in English and Spanish. Cumulative indexes v 1-10, no 1-37 (1940-1949), no 78-118 (1960-1969), no 118-157 (1970-1979).

Finisterra. Revista portuguesa de geografia (Lisboa. Universidade. Centro de Estudos Geográficos). 1- (1966-). Semi-annual. Orders: Livraria Portugal, Rua do Carmo 70, 1200 Lisboa, Portugal.　　2188
Articles on Portugal and former Portuguese territories in Africa. Notes, reviews. In Portuguese. Abstracts in English and French. Cumulative index 1966-1975.

Méditerranée: revue géographique des pays méditerranéens. 1-10 (1960-1969); ns 1- (1970-). Quarterly. Institut de Géographie, Université d'Aix-Marseille, 29, Avenue Robert Schuman, 13100 Aix-en-Provence, France. ISSN 0025-8296.　　2189
Devoted to the Mediterranean region. In French. Abstracts in English and French. Cumulative index 1960-1969.

Revista de geografía. 1- (1967-). Semi-Annual. Departamento de Geografía, Facultad de Letras. Universidad, Barcelona 7, Spain. ISSN 0048-7708.　　2190
Articles, information, documentation, and bibliography, especially on the geography of Catalonia, Spain, and the Mediterranean. In Spanish. Abstracts in English and French.

Rivista geografica italiana (Società di studi geografici). 1- (1893-). Quarterly. Pacini editore s.r.l. 56014, Ospedaletto (Pisa), Italy. ISSN 0035-6697.　　2191
Serious scholarly Italian journal of high quality and international reputation. Extensive reviews. Bibliographic notes. In Italian. Abstracts in English. Cumulative indexes v. 1-50 (1894-1943), 51-60 (1944-1953).

Società geografica italiana. Bollettino. 1- (1868-). Monthly with 2-3 numbers often combined into one issue. Villa Celimontana, Via della Navicella 12, 00184 Roma, Italy. ISSN 0037-8755.　　2192
Leading Italian geographical journal of international standing. Extensive reviews. In Italian. Supplementary English titles in table of contents and English abstracts. Cumulative indexes v. 1-12 (1868-1875), 13-24 (1876-1887), 25-36 (1888-1899), 37-48 (1900-1911), 49-60 (1912-1923), 61-72 (1924-1935).

Beckinsale, Monica, and Beckinsale, Robert. Southern Europe: a systematic geographical study. London: University of London Press; New York: Holmes and Meier, 1975. 334 p. ISBN 0-8419-0178-3. $55. LC 74-14940.　　2193
This valuable geographical analysis was written to aid students who have available only a basic textbook of the region. The book contains detailed regional sample studies within a political framework. Detailed bibliography.

Grothusen, Klaus-Detlev, ed. Griechenland. Greece. (Südosteuropa-Handbuch vol. III). Göttingen, W. Germany; Zurich, Switzerland: Vandenhoeck and Ruprecht, 1980, 1983. 770 p. In German and English. ISBN 3-525-36202-1. LC 80-153507 rev. (Deutsche Forschungsgemeinshaft. Arbeitskreis Südosteuropa-Forschung).　　2194
By German, Canadian, American, and Greek scholars. Topics covered include the state and politics, economy, society and social structures, education, sciences and arts, and a series of documentary appendices. Detailed bibliography. Of great value for reference purposes even though individual contributions are uneven in quality.

Houston, James M. The Western Mediterranean world: an introduction to its regional landscapes. London: Longmans; New York: Praeger, 1964. 800 p. ISBN 0-582-48133-3. £25. LC 67-25037.　　　　　　　　　　　　　　　　　　　　　　　　　2195

Analyzes landscapes in terms of landforms, ecological cover of soils and plants, and the cultural legacies imposed by man. Well illustrated with maps. Special contributions by J.R. Roglić on the Yugoslav littoral and by J.I. Clark on the Maghreb. A timely and timeless book.

Italy: a geographic survey. Edited by Mario Pinna and Domenico Ruocco. (Associazione dei Geografi Italiani). (Published for the 24th International Geographical Congress, Tokyo, 1980). Pisa, Italy: Pacini, 1980. 567 p.　　　　　　　　　　2196

The most authoratative treatment of the geography of Italy. Written by the leading geographers of Italy.

Kayser, Bernard. Géographie humaine de la Grèce: éléments pour l'étude de l'urbanisation. Paris: Presses Universitaires de France, 1964. 147 p. NUC 67-47451.　　2197

A comprehensive useful study of the demographic structure, distribution, migration, and emigration of the people of Greece, historically organized.

Semple, Ellen Churchill. The Geography of the Mediterranean world: its relation to ancient history. New York: Henry Holt, 1931; London: Constable, 1932. 737 p. Reprinted: New York: AMS Press, 1971. 737 p. ISBN 0-404-05751-9. $34.50. LC 70-137267.　2198

The tightly phrased, vivid, and even poetic sentences of this volume are documented by a total of more than 2700 references. This book is considered by many a classic volume in both style and content.

Stanislawski, Dan. The Individuality of Portugal: a study in historical-political geography. Austin, TX: University of Texas Press, 1959. 248 p. Reprinted: Westport, CT: Greenwood, 1969. 248 p. ISBN 0-8371-2120-5. $17.75. LC 71-88945.　　　　2199

A political and historical survey of the Portuguese culture area, with particular emphasis upon economic and historical developments prior to the sixteenth century.

Vita-Finzi, Claudio. The Mediterranean valleys: geological changes in historical times. Cambridge, U.K.; New York: Cambridge University Press, 1969. 140 p. ISBN 0-521-07355-3. $34.50. LC 69-10341.　　　　　　　　　　　　　　　　　　2200

Uses archeological evidence and radiocarbon dates to trace the geological changes that have taken place in the valleys of the Mediterranean basin. A study of interest to historical geographers. Well illustrated.

Walker, Donald S. A Geography of Italy. (Advanced geographies). 2nd ed. London: Methuen, 1967. 296 p. ISBN 0-416-42590-9. £3.50. LC 67-99883.　　　　　　2201

A comprehensive textbook, covering regional, economic, physical, and historical aspects of the area.

7. UNION OF SOVIET SOCIALIST REPUBLICS

Chauncy D. Harris

A. BIBLIOGRAPHIES

American bibliography of Slavic and East European studies. 1956- . Annual. Stanford University, CA: (Indiana University 1956-1966; Ohio State University Press, 1967). American Association for the Advancement of Slavic Studies. ISSN 0569-3497.   2202
> *Annual bibliography of studies by American scholars on the society and culture of the Soviet Union and Eastern Europe. Best source for following publications in English by Americans on the USSR and Eastern Europe.*

Harris, Chauncy D. Guide to geographical bibliographies and reference works in Russian or on the Soviet Union. (Research paper no. 164). Chicago, IL: University of Chicago, Department of Geography, 1975. 478 p. ISBN 0-89065-071-3. $10. LC 74-84784.   2203
> *Annotated list of 2660 bibliographies or reference aids, mainly in Russian or other languages of the Soviet Union, including general bibliographic aids, reference works, systematic fields of geography, and regions of the U.S.S.R.; also bibliographies and reference works on the Soviet Union in Western languages.*

Horak, Stephan M., comp., and Neiswender, Rosemary, ed. Russia, the U.S.S.R., and Eastern Europe: a bibliographic guide to English language publications, 1964-1974. Littleton, CO: Libraries Unlimited, 1978. 488 p. ISBN 0-87287-178-9. $30. LC 77-20696.   2204
> *Annotated classified inventory of 1611 publications in English on Russia, the Soviet Union, and Eastern Europe 1964-1974. The annotations are particularly well done being both informative and evaluative, often excerped or abstracted from reviews published in major American and British Slavic journals, especially the* Slavic Review, *with identification of reviewer and place of review. Supplements and updates Paul L. Horecky,* Russia and the Soviet Union: a Bibliographic Guide to Western Language Publications *(Chicago, University of Chicago Press, 1965, 473 p.).*

Jones, David Lewis. Books in English on the Soviet Union 1917-73: a bibliography. (Garland reference library of social sciences v. 3). London; New York: Garland Publishing Co., 1975. 331 p. ISBN 0-8240-1061-2. $43. LC 75-6887.   2205
> *Unannotated listing of 4587 books arranged by subject fields. Author index.*

B. SERIALS

Akademiia Nauk SSSR. Izvestia. Seriia geograficheskaia. 1- (1951-). Bi-monthly. Order from: Mezhdunarodnaya Kniga, Moscow C-200, U.S.S.R. or its authorized agents abroad. ISSN 0373-2444.   2206
> *The most comprehensive Soviet geographical periodical. Scholarly articles on all phases of Soviet geography, especially physical and economic geography and natural resources and their utilization. Scientific notes. Methods of research. Reports on conferences and on geography abroad. In Russian with supplementary table of contents in English. Cumulative index 1951-1966.*

Geograficheskoe obshchestvo S.S.S.R. Izvestia. 1- (1865-). Bi-monthly. Order from: Mezhdunarodnaya Kniga, Moscow G-200, U.S.S.R. or its authorized agents abroad. ISSN 0373-353X.   2207
> *Oldest of the Russian geographical periodicals. Articles especially on the physical geography of the Soviet Union. Scientific notes. News of the Society and its branches and sections. In Russian with supplementary table of contents in English. Cumulative indexes 1846-1875, 1876-1885, 1886-1895, 1896-1905.*

U.S.S.R.

Moskva (Moscow). Universitet. Vestnik. Geografiia. 1956- . Quarterly. Order from Mezhdunarodnaya Kniga, Moscow G-200, U.S.S.R. or from authorized agents abroad. ISSN 0027-1381 (0201-7385; 0579-9414). 2208
The leading university geographical periodical in the Soviet Union. Articles on all phases of geography, especially in the Soviet Union. Geographical notes. In Russian with supplementary English table of contents and abstracts.

Soviet geography. 1960- . 10 nos. a year. V.H. Winston and Sons, 7961 Eastern Avenue, Silver Spring, Maryland 20901 U.S.A. ISSN 0038-5417. 2209
The key source in English for following the significant articles published in the Soviet Union. Includes both translations of Russian articles and original articles in English. Valuable news notes. Lists of contents of Soviet geographical journals.

Voprosy geografii (Problems of geography). (Geograficheskoe Obshchestvo SSSR. Moskovskii Filial. Nauchnye Sborniki). 1- (1948-). Irregular. Izdatel'stvo "Mysl'," Leninskii Prospekt 15, 117071 Moskva V-71, U.S.S.R. 2210
The most interesting and valuable geographic series in the Soviet Union. Collections of articles on related themes. Each number also has a distinctive title as a monograph. In Russian with supplementary table of contents and abstracts in English.

C. ATLASES

Atlas SSSR. V.V. Tochenov, chairman of the editorial board. Moskva: Glavnoe Upravlenie Geodezii i Kartografii, 1983. 259 p. 10 Rubles. 2211
One-hundred and fifty-one large-format colored maps, including series of general physical maps of regions (hypsometric), elements of the physical environment, branches of the economy, and regional economic maps. Index of 30,000 names. A new edition of atlases of the same name, published in 1962 and 1969, edited by A.N. Baranov, but with more skilled use of color, more attractive format, and color coding of names in the index.

Atlas sel'skogo khoziaistva SSSR. (Atlas of Agriculture of the USSR.) A.I. Tulupnikov, M.I. Nikishov, and others, eds. Moskva: Glavnoe Upravlenie Geodezii i Kartografii, 1960. 308 p. 2212
An exemplary atlas with large detailed maps depicting distribution of natural conditions, agricultural organization, crops, and livestock for the Soviet Union as a whole, agriculture and land use for individual union republics, and value of production or yields of crops.

Geograficheskii atlas dlia uchitelei srednei shkoly. (Geographic atlas for teachers in secondary schools). 4th ed. Moskva: Glavnoe Upravlenie Geodezii i Kartografii, 1983. (1st ed., 1954; 2nd ed., 1959; 3rd ed., 1967). 238 p. LC 81-147080. 2213 •
World-wide atlas for teachers of geography in the USSR. Contains a section on the Soviet Union with maps of elements in the physical environment, and branches of the economy of the USSR as a whole. Also includes two sets of regional maps, one general physical, the other economic. A handy and useful atlas.

Dewdney, John C. USSR in maps. London; Toronto: Hodder and Stoughton Educational; New York: Holmes and Meier, 1982. 117 p. ISBN 0-8419-0760-9. $32.50. LC 52-1242. 2214 •
Effective black-and-white maps of the physical environment, human geography, economic geography, and regions of the Soviet Union. Each map accompanied on facing page by text which elucidates the significance of the topic and the most important features of the maps themselves.

Howe, G. Melvyn (George). The USSR. London: Hulton Educational Publications, 1972. 110 p. ISBN 0-7175-0585-5. LC 72-186451. 2215
Simple but effective black-and-white maps of the main features of the physical environment, history, population, and economy of the Soviet Union as a whole and of significant features of the major regions of the country. Photographs, Index.

Kish, George, et al. Economic atlas of the Soviet Union. 2nd ed. Ann Arbor, MI: University of Michigan Press, 1971. 90 p. (1st ed. 1960). LC 72-177958. 2216
A regional atlas with four maps (agriculture and land use; mining and minerals; industry; and transportation and cities) for each of 15 regions with five general maps for the country as a whole, with accompanying text. Index.

Landscape atlas of the U.S.S.R. Plummer, Thomas F., Jr.; Hanne, William G.; Bruner, Edward F.; and Thudium, Christian C., Jr. West Point, NY: United States Military Academy. Department of Earth, Space, and Graphic Sciences, 1971. 197 p. LC 73-176083. 2217
Seventy topographic maps with explanatory textual material.

U.S.S.R. agricultural atlas. Washington DC: Central Intelligence Agency, 1974. 58 p. LC 76-354161. 2218
Maps on the environment, technology, the system, and production of crops and livestock.

D. ENCYCLOPEDIAS AND GAZETTEERS

Cambridge Encyclopedia of Russia and the Soviet Union. Brown, Archie; Fennell, John; Kaser, Michael; and Williams, H.T., gen. eds.(Cambridge regional encyclopedias). London; New York: Cambridge University Press, 1982. 492 p. ISBN 0-521-23169-8. $37.50. LC 81-9965. 2219
Large well-illustrated volume with articles by many specialists. Includes sections on the physical environment, population, and the economy.

Great Soviet encyclopedia. London: Collier Macmillan; New York: Macmillan, 1973-1983. 31 v. plus index v. ISBN 0-686-81752-4. $1700. ($62 per individual volume). LC 73-10680. 2220
An English translation of Bol'shaia Sovetskaia Entsiklopediia (3rd ed., 1970-1978), providing articles on places, regions, systematic fields of learning, such as geography, and many other aspects of nature, society, economy, and culture of the Soviet Union.

Kratkaia geograficheskaia entsiklopediia. (Short geographical encyclopedia). A.A. Grigor'ev, main editor. Moskva: Sovetskaia Entsiklopediia, 1960-1966. 5 v. 2221
About 18,000 articles mostly on specific places or areas and on physical features, with detailed coverage of the Soviet Union. A major geographical reference work on the Soviet Union, combining a gazetteer, an encyclopedia of fields in geography, a dictionary of geographical terms, a biographical directory, and a bibliography.

E. STATISTICS

Mickiewicz, Ellen. Handbook of Soviet social science data. London: Collier-Macmillan; New York: The Free Press, 1973. 225 p. ISBN 0-02-921190-5. $30. LC 72-86510. 2222
Statistical tables for the Soviet Union and its union republics on demography, agriculture, production, housing, and other topics, with introductory text for each section by a specialist with informative foreword and introduction.

Narodnoe khoziaistvo SSSR: statisticheskii ezhegodnik (Economy of the USSR: statistical yearbook). Moskva: Finansy i Statistika, 1956- . Annual. 2223
The basic source of statistics for the Soviet Union. In Russian.

U.S.S.R.

U.S.S.R.: facts and figures annual. Scherer, John L., ed. 1- (1977-). Annual. Academic International Press, Box 1111, Gulf Breeze, FL, 32561. ISSN 0148-7760.
Handy compilation of statistics on the Soviet Union, including demography, economy, energy, industrial production by union republics, industries, agriculture, foreign aid and trade, transportation, and other topics.

2224

F. DISSERTATION

Dossick, Jesse J. Doctoral research on Russia and the Soviet Union, 1960-1975: a classified list of 3150 American, Canadian, and British dissertations with some critical and statistical analysis. New York; London: Garland Publishing, 1976. 345 p. ISBN 0-8240-1079-5. $45. LC 75-5115.
Dissertations listed by fields. Author index. Supplements the author's Doctoral Research on Russia and the Soviet Union *(New York: New York University Press, 1960, 248 p.).*

2225

G. GENERAL TEXTUAL TREATISES

Demko, George J. and Fuchs, Ronald J., eds. Geographical studies on the Soviet Union: essays in honor of Chauncy D. Harris. (Research paper no. 211). Chicago IL: University of Chicago. Department of Geography, 1984. 294 p. ISBN 0-89065-116-7. $10. LC 84-2498.
Leading Western scholars specializing on the Soviet Union contribute papers on Soviet geographical thought; urban networks, policies, and attitudes; administrative structure; demographic and ethnic processes; Siberian development; weather and agriculture; forests; and water quality.

2226

Demko, George J. and Fuchs, Roland J., eds. and translators. Geographical perspectives in the Soviet Union: a selection of readings. Columbus, OH: Ohio State University Press, 1974. 742 p. ISBN 0-8142-0196-2. $30. LC 74-9853.
Carefully selected articles by Soviet scholars, translated into English, on such fields as economic regionalization, resource management, agricultural, industrial, transportation, population, urban, and historical geography, and history, philosophy, and methodology of geography in the USSR.

2227

Gregory, James S. Russian land Soviet people: a geographical approach to the U.S.S.R. London; Toronto: George G. Harrap, 1968. 947 p. LC 68-17549.
Massive compendium on geography, especially historical geography, climate, natural vegetation, agriculture, and the regions of the Soviet Union.

2228

Kalesnik, S.V. and Pavlenko, V.G., eds. Soviet Union: a geographical survey. Moskva: Progress Publishers, 1976. 278 p. LC 77-372497. (English translation of revised, condensed text of Sovetskii Soiuz: Obshchii Obzor, Moskva, 1972, 813 p.).
Soviet presentation of natural conditions, population, and the economy of the Soviet Union as a whole. Beautifully illustrated with colored and black-and-white photographs and maps. Written by leading Soviet specialists from an official point of view.

2229

H. INTRODUCTORY TEXTBOOKS

Hooson, David J.M. The Soviet Union: people and regions. London: University of London Press (now Hodder and Stoughton); Belmont, CA: Wadsworth, 1966. 376 p. LC 66-25892.
Well-written systematic and regional geography.

2230

Howe, G. Melvyn (George). Soviet Union: a geographical survey. 2nd ed. Plymouth, U.K.: Macdonald and Evans, 1983. 501 p. (1st ed., 1968). ISBN 0-7121-1989-2. £15. ISBN 0-7121-1947-7. £10.95. (pbk). LC 83-126451. 2231•
    Organized in four parts: the physical framework; the economic framework; the human resources; and regional geography.

Lydolph, Paul E. Geography of the U.S.S.R.: topical analysis. Elkhart Lake, WI: Misty Valley Publishing, 1979 (1984). 522 p. plus 1984 supplement, 37 p. LC 78-66325. 2232
    Solid systematic treatment of the physical features, resources, population, and economic geography of the Soviet Union with numerous tables, diagrams, and maps. Updated by a 1984 supplement.

Lydolph, Paul E. Geography of the U.S.S.R. 3rd ed. London; New York; Toronto: John Wiley and Sons, 1977. (1st ed. 1964). 495 p. ISBN 0-471-55724-2. LC 76-26657. 2233
    A regional analysis of the Soviet Union based on 19 regions (18 economic regions plus Moldavia) with 2-color maps (black and brown), tables, illustrations, and bibliographies. The regional syntheses include physical setting and natural resources, population and cities, branches of the economy, and particular problems.

Parker, William H. The Soviet Union. 2nd ed. London; New York: Longman, 1983. 207 p. (1st ed. 1969). ISBN 0-582-30111-4. $13.95. (pbk). LC 82-20360. 2234•
    Brief introduction to the principal natural and cultural landscapes of the Soviet Union grouped by great vegetation belts: tundra, forest, steppe, desert, and mountains.

Symons, Leslie, ed. The Soviet Union: a systematic geography. London: Hodder and Stoughton Educational; Totowa, NJ: Barnes and Noble, 1983. 266 p. ISBN 0-389-20309-2. $32.50. ISBN 0-389-20310-6. $21.50. (pbk). LC 82-6683. 2235•
    Examination of the principal physical conditions, population, and major branches of the economy of the Soviet Union on a systematic basis.

I. PHYSICAL GEOGRAPHY

Lydolph, Paul E. Climates of the Soviet Union. (World survey of climatology, v.7). Amsterdam, Netherlands; Oxford, U.K.; New York: Elsevier Scientific Publishing Co., 1977. 443 p. ISBN 0-444-41516-5. $121.25. LC 76-46298. 2236
    Most detailed one-volume treatment of the climates of the Soviet Union, providing both a regional survey and systematic treatment of the thermal factor, the moisture factor, and distribution of climates. Numerous maps and 126 tables of climatic data.

J. ECONOMIC GEOGRAPHY: GENERAL

Cole, John P. Geography of the Soviet Union. London; Boston; Toronto: Butterworths, 1984. 452 p. ISBN 0-408-49752-1. $29.95. LC 83-14458. 2237
    Original and stimulating analysis of many aspects of the geography of the Soviet Union, particularly its economic geography. Numerous maps, diagrams, and tables, often with data presented in new ways. Supersedes J.P. Cole and F.C. German, A Geography of the U.S.S.R., 1st ed., 1961, 2nd ed., 1970.

Dewdney, John C. A Geography of the Soviet Union. 3rd ed. Oxford; New York; Toronto: Pergamon, 1979. 175 p. (1st ed., 1965). ISBN 0-08-023739-8. $34. ISBN 0-08-023738-X. $13.25 (pbk). LC 78-0992. 2238•
    Introductory textbook on the economic geography of the Soviet Union, systematically organized.

U.S.S.R.

Mathieson, Raymond S. The Soviet Union: an economic geography. London; Heinemann Educational Books; New York: Barnes and Noble, 1975. 342 p. ISBN 0-435-35601-1 £ 10.50. ISBN 0-06-494647-9. LC 75-10018.　　　　　　　　　　2239
   Systematic economic geography of the Soviet Union combined with descriptions of 11 economic regions.

Mellor, Roy E.H. The Soviet Union and its geographical problems. London; New York: Macmillan, 1982. 207 p. ISBN 0-333-7662-0. $27.95. ISBN 0-333-27663-9. $12.95. (pbk). LC 82-184376.　　　　　　　　　　2240 •
   Introductory systematic textbook examination of the economic geography of the Soviet Union. Replaces the author's Geography of the U.S.S.R. *(1964).*

K. ECONOMIC GEOGRAPHY: SPECIAL TOPICS

Barr, Brenton M. The Soviet wood-processing industry: a linear programming analysis of the role of transportation costs in location and flow patterns. (University of Toronto. Department of Geography. Research publication no. 5). Toronto: University of Toronto Press, 1970. 135 p. ISBN 0-8020-3259-1. Can $6. (pbk). LC 70-18546.　　　　2241
   Analysis of the wood-processing industry in relation to the forest resource, the market, the locational orientation of wood-processing and associated flow patterns, and transportation costs and the location of wood-processing.

Bergson, Abram, and Levine, Herbert S., eds. The Soviet economy: toward the year 2000. London; Winchester, MA: George Allen and Unwin, 1983. 452 p. ISBN 0-04-335045-3. $37.50. LC 82-15611.　　　　　　　　　　2242
   Valuable analyses by leading experts of important aspects of the Soviet economy, including population and labor force, agriculture, industry, energy, and regional economic development.

Dewdney, John C. The USSR: studies in industrial geography. Folkestone, U.K.: William Dawson and Sons; Boulder: Westview Press, 1976. 262 p. ISBN 0-0-89158-616-4. $31.50. LC 76-16744.　　　　　　　　　　2243
   Study of the factors in, major sectors of, and regional contrasts in manufacturing industries in the Soviet Union.

Dienes, Leslie. Locational factors and locational developments in the Soviet chemical industry. (Research paper no. 119). Chicago, IL: University of Chicago, Department of Geography, 1969. 262 p. ISBN 0-89065-27-6. $10. LC 69-18023.　　　　2244
   Analysis of locational factors in the development and regional distribution of the chemical industry of the Soviet Union with special attention to new petrochemical industries and centers.

Dienes, Leslie, and Shabad, Theodore. The Soviet energy system: resource use and policies. New York; London; Toronto: John Wiley and Sons, a Halsted Press book, for V.H. Winston and Sons, 1979. 298 p. ISBN 0-470-26629-5. $21.95. LC 78-20814.　　　2245
   Thorough analysis of patterns of energy resource production and of policy issues involved in the development of alternative sources and of final allocation. Effectively combines penetrating analysis with detailed presentation of basic data on production and distribution of energy sources in the Soviet Union.

Nove, Alec. The Soviet economic system. 2nd ed. London; Winchester, MA: Allen and Unwin, 1980. 406 p. (1st ed., 1977). ISBN 0-04-33504-9. $13.50. (pbk). LC 80-41221.　　2246
   Effective overview of the Soviet economy with some emphasis on planning and management.

Shabad, Theodore. Basic industrial resources of the U.S.S.R. New York; London: Columbia University Press, 1969. 393 p. ISBN 0-231-03077-0. $45. LC 75-101133.   2247

Informative presentation of general production trends in the fuel, electric power, metal, and chemical industries; and region by region survey of the development and status of these industries. Also available from Theodore Shabad, 145 East 84th St, New York, NY. 10028.

Stern, Jonathan P. Soviet natural gas development to 1990: the implications for the CMEA and the West. Lexington MA: Lexington Books; Toronto: D.C. Heath and Co., 1980. 190 p. ISBN 0-669-03233-6. $26. LC 79-2705.   2248

Treats the Soviet natural gas industry (the most rapidly expanding energy base in the Soviet Union), its development, present reserves, production, transportation, and use, and prospects for 1990. Extensive discussion of Soviet trade in natural gas with countries outside the CMEA, and options and motivations in that trade.

Symons, Leslie. Russian agriculture: a geographic survey. London: G. Bell and Sons; New York: John Wiley and Sons, 1972. 348 p. ISBN 0-7135-1627-5. $35. LC 72-197364.   2249

Evolution of agricultural patterns, economic and political factors, land and climate, labor, kolkhoz and sovkhoz, grain, industrial crops, fruit and vegetables, livestock, regions, and farm improvement and conservation.

Symons, Leslie, and White, Colin, eds. Russian transport: a historical and geographical survey. London: G. Bell and Sons, 1975. 192 p. ISBN 0-7135-1895-2. ISBN 0-7135-1911-8. (pbk). LC 75-33228.   2250

Chapters by separate authors on selected aspects of the development and effect of the transportation system of Russia and the Soviet Union: railways and the market for grain in the 1860s and 1870s and the economic development of Turkestan before the revolution; their contemporary role; the merchant marine; the northern sea route; and Soviet air transport.

L. POPULATION AND ETHNIC GROUPS

Katz, Zev; Rogers, Rosemarie; and Harned, Frederic, eds. Handbook of major Soviet nationalities. London: Collier Macmillan; New York: The Free Press, a division of Macmillan, 1975. 481 p. ISBN 0-02-917090-7. $25. LC 74-10458.   2251•

Statistical tables, text, and references for each of the nationalities (ethnic groups) of the Soviet Union forming the basis for the 15 union republics, plus Jews and Tatars.

Lewis, Robert A. and Rowland, Richard H. Population redistribution in the USSR: its impact on society, 1897-1977. New York: Praeger, 1979. 485 p. ISBN 0-03-050641-7. $49.95. LC 79-18076.   2252

Analyzes changes in regional patterns of urbanization, urban growth, city size, urban regions, and rural population 1897-1977 based on extensive adjustments to census data to achieve comparability of areas over time.

Lewis, Robert A.; Rowland, Richard H.; and Clem, Ralph S. Nationality and population change in Russia and the USSR: an evaluation of census data, 1897-1970. New York: Praeger, 1976. 456 p. ISBN 0-275-56480-0. LC 76-6942.   2253

Analysis of nationality and population change 1897-1970 with special reference to urbanization and regional distribution based on extensive adjustments to census data to achieve comparability of areas over time.

Wixman, Ronald. Language aspects of ethnic patterns and processes in the North Caucasus. (Research paper no. 191). Chicago, IL: University of Chicago, Department of Geography, 1980. 243 p. ISBN 0-089-65-098-5. $10. LC 78-31304.   2254

Original monographic investigation of the ethnic structure of the North Caucasus and of ethnic processes, nationality policies, and ethnic assimilation among the diverse minorities of this ethnically complex region.

U.S.S.R.

Wixman, Ronald. The Peoples of the USSR: an ethnographic handbook. Armonk, NY: M.E. Sharpe, 1984. 246 p. ISBN 0-87332-203-7. $35. LC 83-18433.　　　2255

Alphabetic list of ethnic groups of the Soviet Union, including information on alternative names, numbers according to the censuses of 1926, 1959, 1970, and 1979, language, religious affiliation, and location. Especially useful in correctly identifying groups which have been the subject of confusion and errors. Interesting in inclusion of self identifications, Russian nomenclature, and other designations (i.e. the Georgians are called Kartveli by themselves, Gruzin by the Russians, and Gurcu by their Turkik-speaking neighbors).

M. URBAN GEOGRAPHY

Harris, Chauncy D. Cities of the Soviet Union: studies in their functions, size, density, and growth. (AAG monograph series, no. 5). 2nd printing. Washington, DC: Association of American Geographers, 1972. (1st printing, 1970, Chicago: Rand McNally). 484 p. ISBN 0-89291-084-4. $4.95. LC 72-98437.　　　2256

Analysis of 1247 cities and towns in the Soviet Union. Treats Soviet urbanization in a world-wide setting; Soviet contributions to the study of Soviet cities; a functional classification of Soviet cities; relationships between size, central-place functions, and the administrative hierarchy of the country; spacing of cities; growth trends over the last century and a half; and growth patterns during the Soviet period. Tables, graphs, maps, and extensive bibliography.

See also entry 2165.

N. NATURAL RESOURCES, THE ENVIRONMENT, AND ENVIRONMENTAL POLLUTION

Gerasimov, I.P.; Armand, D.L.; and Yefron, K.M., eds. Jackson, W.A. Douglas, editor of English edition. Natural resources of the Soviet Union: their use and renewal. Translated form Russian by Jacek I. Romanowski. San Francisco, CA: W.H. Freeman and Co., 1971. 349 p. (Prirodnye Resursy Sovetskogo Soiuza: ikh ispol'zovanie i vosproizvodstvo. 1963. 241 p.). ISBN 0-7167-0248-7. $38.95. LC 74-138667.　　　2257

Use of natural resources in the U.S.S.R., especially water, climate, land, vegetation, and fish and game.

Jackson, W.A. Douglas, ed. Soviet resource management and the environment. Columbus, OH: American Association for the Advancement of Slavic Studies. (now at Stanford University, California), 1978. 239 p. NUC 79-126364.　　　2258

Papers presented at a conference in Seattle in 1974 on resource and environmental topics: manpower; water resources; electric power; energy; agriculture; the atmosphere; pollution; property rights; and policy.

Jensen, Robert G.; Shabad, Theodore; and Wright, Arthur W., eds. Soviet natural resources in the world economy. Chicago, IL: University of Chicago Press, 1983. 700 p. ISBN 0-226-39831-5. $100. LC 82-17317.　　　2259

This multi-authored volume provides a thoroughly researched, detailed, scholarly investigation and analysis of Soviet raw materials, their distribution, reserves, potential production, utilization, transportation, and possible role in Soviet foreign trade and the world economy. Includes 240 tables, 109 illustrations (82 of them maps), 2548 bibliographical and notational footnotes, and some emphasis on the special problems of Siberia.

Pryde, Philip R. Conservation in the Soviet Union. London; New York: Cambridge University Press, 1972. 301 p. ISBN 0-521-08432-6. LC 72-182025.　　　2260

Conservation practices and policy for land and soil resources, preserves, fisheries and wildlife, timber, minerals, water resources, environmental quality, and pollution.

O. PLANNING

Bater, James H. The Soviet city: ideal and reality. (Explorations in urban analysis, v. 2). London: Edward Arnold; Beverly Hills, CA: Sage Publications, 1980. 196 p. ISBN 0-8039-1466-0. $24. ISBN 0-8039-1467-9. $12. (pbk). LC 80-51193.

Effectively combines analysis of Soviet urban policy with evaluation of the actual results through discussion of ideology and the city, decision making and town planning, patterns of city growth, spatial organization of the Soviet city, and life in the Soviet city. Notes particularly three types of problems that have beset Soviet town planning: inadequate financial resources for city planning agencies; division of authority for decision making in each city; and lack of jurisdiction for city agencies beyond city boundaries.

2261

Pallot, Judith, and Shaw, Denis J.B. Planning in the Soviet Union. London: Croom Helm; Athens GA: University of Georgia Press, 1981. 303 p. ISBN 0-8203-0550-2. $27.50. LC 80-24723.

Regional and spatial aspects of Soviet planning for industry, agriculture, consumers, and urban settlements.

2262

P. HISTORICAL GEOGRAPHY

Bater, James H. St. Petersburg: industrialization and change. London: Edward Arnold; Toronto: McGill-Queen's University Press; New York: Holmes and Meier, 1976. 469 p. ISBN 0-7735-0266-1. $49.50. LC CN76-7171.

Describes and analyzes the transformation of St. Petersburg from a court administrative capital of the Russian Empire to a rapidly growing industrial and commercial center between the mid-1800s and the end of World War I. Extensive discussion of social conditions, including housing and disease, weak planning, and inadequete urban services. Richly illustrated with plates, tables, figures, and maps. Excellent study based on primary sources.

2263

Bater, James H. and French, R.A. (Richard Antony), eds. Studies in Russian historical geography. London; New York: Academic Press, 1983. 2v.

Vol. 1: ISBN 0-12-081201-0. $35.

Vol. 2: ISBN 0-12-081202-9. $35. LC 83-71080.

A collection of original studies in the historical geography of Russia by ten authors around four main themes: humans and the land; the frontier; Rus in Urbe; and the last century of the old regime. Papers characterized by high quality, foundation on primary sources, originality of contribution, and effectiveness of presentation. The most important publication on Russian historical geography in a Western language.

2264

Fedor, Thomas Stanley. Patterns of urban growth in the Russian Empire during the nineteenth century. (Research paper no. 163). Chicago, IL: University of Chicago. Department of Geography, 1975. 245 p. ISBN 0-89065-070-5. $10. LC 74-84783.

Study of urban growth patterns particularly in relation to industrial development in the Russian Empire 1811-1910.

2265

Gibson, James R. Feeding the Russian fur trade: provisionment of the Okhotsk seaboard and the Kamchatka Peninsula, 1639-1856. Madison, WI; London: University of Wisconsin Press, 1969. 337 p. ISBN 0-299-05230-3. $27.50. LC 79-81319.

Detailed careful study of the problem of supplying the remote Okhotsk seaboard and Kamchatka before the Russian acquisition of the Amur country to the south. Extensive utilization of primary and secondary sources in Russian and in Western languages.

2266

U.S.S.R.

Hamm, Michael F., ed. The City in Russian history. Lexington, KY: University Press of Kentucky, 1976. 349 p. ISBN 0-8131-1328-8. $35. LC 75-3544.
Historical perspectives on early Russian towns, the 19th century city, and growth and planning of cities in the 20th century, in studies by 16 authors.

2267

Parker, W.H. (William). A Historical geography of Russia. London: University of London Press Ltd., 1968. 416 p. ISBN 0-340-06940-6. LC 79-391284.
An attempt at an over-all synthesis of the full sweep of historical geography of Russia based on extensive utilization of Western European sources and some Russian sources. Numerous maps, diagrams, and plates.

2268

Rozman, Gilbert. Urban networks in Russia 1750-1800 and premodern periodization. Princeton NJ: Princeton University Press, 1976. 337 p. ISBN 0-691-09364-4. $35. LC 75-3472.
Study of the establishment and structure of the urban network of Russia, particularly 1750-1800, in view of central place theory and urban hierarchy, social structure, and comparisons with England and France.

2269

Q. REGIONS

Demko, George J. The Russian colonization of Kazakhstan: 1896-1916. (Indiana University Publications, Uralic and Altaic series, v. 99). Bloomington IN: Indiana University Research Center, 1969. 271 p. Reprinted: Atlantic Highlands, NJ: Research Center for Language and Semiotic Studies. 271 p. ISBN 0-87750-082-7. $10.50. (pbk). LC 67-66166.
An original monographic study of the peasant migration and the colonization of Kazakhstan and of the effects of this in-migration.

2270

Kirby, E. Stuart. The Soviet Far East. London: Macmillan, 1971. 268 p. ISBN 0-333-12955-5. LC 72-175857.
General review of the character, population, livelihood, and economic development of the Soviet Far East, followed by separate treatment of each of the eight major regions.

2271

North, Robert N. Transport in western Siberia: Tsarist and Soviet development. Vancouver, BC., University of British Columbia Press and Centre for transportation studies, 1979. 364 p. ISBN 0-7748-0097-6. Can $22. LC C78-002149-5.
Role of transportation in economic development of western Siberia from before the construction of the Siberian Railway and up to 1975, arranged by periods, with an introduction to the setting and a conclusion on continentality and economic development.

2272

Shabad, Theodore, and Mote, Victor L. Gateway to Siberian resources (the BAM). New York; London; Toronto: John Wiley and Sons. A Halsted Press Book, for Scripta Publishing Co., 1977. 189 p. ISBN 0-470-99040-6. LC 77-63.
Background material on the Baykal-Amur Mainline Railroad, the largest current construction project in the Soviet Union, extending 2000 miles through eastern Siberia, north of the Trans-Siberian Railway and north of Lake Baykal, in difficult terrain hitherto little developed and virtually uninhabited.

2273

8. ASIA: GENERAL

Clifton W. Pannell

A. BIBLIOGRAPHIES

Bibliography of Asian studies. (Association for Asian Studies). 1956- . Annual. 1
Lane Hall, University of Michigan, Ann Arbor, MI, 48109. ISSN 0067-7159.   2274
The main annual current bibliography on Monsoon Asia published in English. In-
cludes search of more than 600 periodicals as well as books. Arranged by regions and
country (i.e. East Asia/China) and topics such as geography, economics, etc. within
each region and country. An excellent research tool. Predecessor entitled Far Eastern bibliography, *1936-1955 (with slight variations in title) appeared in the* Far Eastern Quarterly. *See also* Cumulative Bibliography of Asian Studies, *next entry.*

Cumulative bibliography of Asian studies. 1941-1965. Plus supplements 1966- .
Boston, MA: G.K. Hall.   2275
     1941-1965. Author bibliography. 1969. 4 v. ISBN 0-8161-0805-6. $405;
     1941-1965. Subject bibliography. 1970. 4 v. ISBN 0-8161-0127-2. $415;
     Supplement. 1966-1970. Author bibliography. 3 v. ISBN 0-8161-0959-1.
     $415;
     Supplement. 1966-1970. Subject bibliography. 3 v. ISBN 0-8161-0235-X.
     $415.
Valuable research tools for larger libraries.

B. SERIALS

Asian geographer. Asian Geographical Association. 1- (1982-). Semi-annual. c/o
Department of Geography and Geology, University of Hong Kong, Hong Kong.   2276
Articles on East, Southeast, Southern, and Southwest Asia with some emphasis on
the People's Republic of China and Hong Kong. Lists of contents of geographical
journals published in Asia.

Asian survey. 1- (1960-). Monthly. University of California Press, 2120 Berkeley Way,
Berkeley, CA, 94720. ISSN 0004-4687.   2277
A leading periodical that focuses on social science research on all Asian countries.
Especially strong on politics and international affairs.

Asia yearbook. 1960- . Annual. Far Eastern Economic Review, Ltd. Box 160, Hong
Kong. ISSN 0071-3821.   2278•
Provides an excellent annual survey of political and economic matters for all the
countries of Monsoon Asia (Afghanistan through Japan). Includes descriptive narra-
tives as well as statistical data.

Journal for Asian studies. 1- (1941-). Quarterly. Association for Asian Studies, 1
Lane Hall, University of Michigan, Ann Arbor, MI, 48109. ISSN 0021-9118.   2279
The main scholarly periodical that deals with Asia. Covers all disciplines. Contains
an excellent book review section. The Bibliography of Asian Studies *is a separately*
published supplement.

C. GENERAL WORKS

Murphey, Rhoads. The Outsiders, the Western experiences in India and China.
(Michigan studies on China). Ann Arbor, MI: University of Michigan Press, 1977. 299 p.
ISBN 0-472-08679-0. LC 76-27279.   2280
An interesting and challenging thesis about the role of the West in the recent his-
torical development of India and China.

Spencer, Joseph E. Oriental Asia: themes toward a geography. (Foundations of world regional geography). Englewood Cliffs, NJ: Prentice-Hall, 1973. 146 p. LC 73-5645.　　2281
　　A brief and worthwhile outline of the main physical, environmental, and cultural patterns of Monsoon Asia.

Spencer, Joseph E. and Thomas, William L. Asia, east by south: a cultural geography. 2nd ed. New York: John Wiley, 1971. 669 p. (1st ed., 1954. 458 p.). (available from Books on Demand, Ann Arbor, MI, $160). LC 74-138920.　　2282
　　One of the best known general geographies of Monsoon Asia. Includes 11 topical and 16 regional chapters. The approach is cultural.

9. EAST ASIA

Clifton W. Pannell

A. GENERAL WORKS

Kolb, Albert. East Asia: China, Japan, and Korea: geography of a cultural region. Translated from German by C.A.M. Sym. London: Methuen, 1977. 591 p. ISBN 0-416-08420-6. ISBN 0-416-70780-7. £7. (pbk). LC 78-323679. (Ostasien: Geographie eines Kulturerdteiles. Heidelberg, W. Germany: Quelle and Meyer, 1963. 608 p.).　　2283
　　A somewhat dated but nonetheless magnificent regional geography with a focus on historical and cultural development of what the author describes as the Chinese culture region. Includes 25 maps in color and a number of black-and-white illustrations and tables.

Pannell, Clifton W., ed. East Asia: geographical and historical approaches to foreign area studies. Dubuque, IA: Kendall/Hunt, 1983. 253 p. ISBN 0-8403-2875-3. $13.95. (pbk). LC 82-83653.　　2284
　　A collection of readings that include physical, historical, cultural, economic, and political topics. Focus is mainly on China followed by Japan and Korea. Also includes a chapter on map and imagery coverage of East Asia. A useful although not definitive survey of the region.

B. JAPAN: BIBLIOGRAPHIES　　　　　　　　　　　　　　　　　CHAUNCY D. HARRIS

Hall, Robert B. and Noh, Toshio. Japanese geography: a guide to Japanese reference and research materials. (Center for Japanese Studies, bibliographical series no. 6). Rev. ed. Ann Arbor, MI: University of Michigan Press, 1970. 128 p. Reprinted: Westport, CT: Greenwood Press, 1978. 128 p. (1st ed., 1956. 233 p.). ISBN 0-313-20434-9. $21.75. LC 78-5578.　　2285
　　Contains 1486 entries to the Japanese literature on geography and on Japan. Titles are given in Romanized form, in Japanese script, and in English translation. A valuable window into Japanese research and writing.

Kokusai Bunka Shinkōkai. K.B.S.bibliography of standard reference books for Japanese studies with descriptive notes. Vol. II: geography and travel. Rev. ed. Tokyo: University of Tokyo Press, 1973. (1st ed. in 1962). 224 p. NUC 73-4747.　　2286
　　An annotated inventory of Japanese reference books on geography. Titles are given in Romanized form, in Japanese script, and in English translation. Highly useful.

Kornhauser, David H. Studies of Japan in Western languages of special interest to geographers. Tokyo: Kokon-Shoin, 1984. 99 p. ISBN 4-7722-1137-3. 4000 yen. (Revision of A selected list of writings on Japan pertinent to geography in Western languages, with emphasis on the work of Japan specialists. [Special publications no. 6]. Hiroshima, Japan: University of Hiroshima. Research and Sources Unit for Regional geography, 1979. 71 p.).　　2287
　　About 1200 entries on the physical, human, and regional geography of Japan.

Chirigaku hyoron. Geographical review of Japan. Series A. (Nippon Chiri Gakkai Association of Japanese Geographers). 1- (1925-). Monthly. Subscription Department, Business Center for Academic Societies of Japan. 2-4-16 Yayoi, Bunkyo-ku, Tokyo 113, Japan. ISSN 0016-7444.　　2288
The leading Japanese geographical periodical. Covers the entire range of geography. Articles mainly on Japan. Reviews. In Japanese. Supplementary table of contents and abstracts in English.

Geographical review of Japan. Series B. (The Association of Japanese Geographers). 57 (Ser. B)- (1984-). Semi-annual. Subscription Department,Business Center for Academic Societies of Japan, 2-4-16 Yayoi, Bunyo-ku, Tokyo 113, Japan. ISSN 0289-600.　　2289
New periodical in English to provide an international medium for Japanese geographers for original articles, subject review articles, and short notes. Valuable in providing a window into some of the more important scholarly contributions from Japan.

Jimbun chiri. Human geography. (Jimbun Chiri Gakkai. The Human Geographical Society of Japan). 1- (1948-). Bimonthly. Jimbun Chiri Gakkai, Geographical Institute, Faculty of Literature, Kyoto University, Kyoto, Japan. ISSN 0018-7216.　　2290
Devoted to human, historical, economic, and urban geography of Japan. Reviews. Notes. In Japanese. Supplementary table of contents and abstracts in English.

Tohoku University. Science reports. series 7. Geography. 1- (1952-). Semiannual. Department of Geography, Faculty of Science. Tohoku University, Aobayama, Sendai 980, Japan. ISSN 0375-7854.　　2291
Research articles on Japan. In English.

Tokyo Metropolitan University (Tokyo Toritsu Daigaku). Geographical reports. 1- (1966-). Annual. Department of Geography, Tokyo Metropolitan University, 2-1-1 Fukazawa, Setagaya-ku, Tokyo, 158, Japan. ISSN 0386-8710.　　2292
Scientific articles especially in physical geography. In English.

Tokyo. University. Department of geography. Bulletin. 1- (1969-). Annual. Department of Geography, Faculty of Science, University of Tokyo, Hongo, Tokyo 113, Japan. ISSN 0082-478X.　　2293
Research reports mainly in physical geography from the University of Tokyo. In English.

University of Tsukuba. Institute of Geosciences. Science reports. section a. Geographical sciences. 1- (1980-). Annual. Institute of Geoscience, University of Tsukuba, Sakura-mura, Niihari-gun, Ibaraki 305, Japan. ISSN 0388-6174.　　2294
Research reports from the University of Tsukuba mainly on Japan. In English.

Kokudo Chiriin. The National atlas of Japan. (Geographical Survey Institute). Tokyo: The Japan Map Center, 1977. 366 p. LC 77-365835.　　2295
A magnificent achievement of Japanese cartography that displays in 83 map folios and accompanying text the physical and human geography of modern Japan. Available also in Japanese.

Kokusai Kyoiku Joho Senta. Atlas of Japan: physical, economic, and social. 2nd ed. Tokyo: International Society for Educational Information, 1974. 64 p. (1st ed., 1970). LC 314450.　　2296
An informative well-presented series of thematic maps for Japan as a whole. In English.

East Asia

Teikoku's complete atlas of Japan. 6th ed. Tokyo: Teikoku-Shoin, 1977. 55 p. (1st ed., 1964). LC 77-378546.
    Inexpensive small atlas of Japan in English. Both thematic and regional maps. In the United States available from Denoyer-Geppert, Chicago, IL.

2297 •

Dempster, Prue. Japan advances: a geographical study. 2nd ed. London: Methuen; New York: Barnes and Noble, 1969. 332 p. (1st ed., 1967). ISBN 0-416-13290-1. £4.95. LC 75-473975.
    Systematic geography of Japan as a whole with discussion of the physical features, economic geography, population patterns, and cities. Bibliography.

2298

Geography of Japan. Edited by the Association of Japanese Geographers (Special publication no. 4). Tokyo: Teikoku-Shoin Co., Ltd., 1980. 440 p. ISBN 0-87040-494-6. $64. LC 81-114376.
    An excellent collection of essays on the geography of Japan contributed by leading Japanese specialists and published on the occasion of the 24th International Geographical Congress in Tokyo in 1980.

2299

Hall, Robert B. Jr. Japan: industrial power of Asia. 2nd ed. New York: D. Van Nostrand, 1976. 150 p. (1st ed., 1963. 127 p.). ISBN 0-442-29752-1. LC 75-16553.
    A brief and useful survey of contemporary Japan that covers most of the topics of interest to the geographer.

2300 •

Japanese cities: a geographical approach. (AJG special publication no. 2). Tokyo: Association of Japanese Geographers, 1970. 264 p. LC 70-597909.
    Original articles by Japanese geographers on the rapid urbanization of Japan up to 1970: historical aspects; urbanization and urban population; urban land use and townscape; urban structure; functional activities; urban transportation; and urban problems and planning. Tables, maps, and references.

2301

Kiuchi, Shinzo, ed. Geography in Japan. (AJG special publication no. 3). Tokyo: University of Tokyo Press, 1976. 294 p. ISBN 0-86008-159-1. $39.50. LC 76-382541.
    Survey in English of the work by Japanese geographers in the systematic and regional fields of geography. The overviews of each of the main subdivisions of physical and human geography include numerous studies in these fields on Japan itself.

2302

Kornhauser, David. Japan: geographic background to urban/industrial development. 2nd ed. London; New York: Longman, 1982. 189 p. (1st ed., Urban Japan: its foundations and growth, 1976. 180 p.). ISBN 0-582-30081-9. $11.95. (pbk). LC 81-19394.
    An excellent general introduction to the geography of Japan. Emphasis is on historical and urban landscapes. An informative overview by an experienced interpreter.

2303 •

Murata, Kiyoji, and Ota, Isamu, eds. An Industrial geography of Japan. London: Bell and Hyman; New York: St. Martin's Press, 1980. 205 p. ISBN 0-312-41428-5. $35. LC 80-13404.
    Studies by Japanese geographers on Japanese industries: industrial development and characteristics; the three major industrial regions; other industrial regions; the main manufacturing industries; and industrial location policy and environmentl issues.

2304

Noh, Toshio, and Gordon, Douglas H., eds. Modern Japan; land and man. Tokyo: Teikoku-Shoin, 1974. 146 p. LC 74-174834.
    Brief introduction to the regions of Japan from a Japanese viewpoint. Extensively illustrated with numerous two-color maps and diagrams. Based on a Japanese text by Toshio Noh, Taiji Yazawa, Ken-ichi Tanabe, and Hisashi Sato.

2305

Pezeu-Massabuau, Jacques. The Japanese islands: a physical and social geography. Translated from French by Paul C. Blum. Rutland, VT: Charles E. Tuttle, 1978. 283 p. ISBN 0-8048-1184-9. $12.50. LC 77-82140. (Géographie du Japon. 2nd ed. Paris: Presses universitaires de France, 1973. 127 p.).
2306
Lively, brief, and well-informed overview of the systematic geography of Japan. Emphasizes the inhospitable environment, natural hazards, the economy, and their interrelations.

Trewartha, Glenn T. Japan, a geography. Rev. ed. Madison, WI: The University of Wisconsin Press, 1965. 652 p. (1st ed., 1945. 607 p.). ISBN 0-299-03440-2. $30. ISBN 0-299-03444-5. $14.95. LC 65-11200.
2307
Although somewhat dated, this book remains an excellent detailed topical and regional geography of Japan.

F. KOREA

Bartz, Patricia McBride. South Korea. Oxford, U.K.: Clarendon Press, 1972. 203 p. ISBN 0-19-874008-5. £17.50. LC 72-196784.
2308
The most recent general and regional geography of South Korea. A good survey of the country.

Bunge, Frederica, ed. North Korea, a country study. 3rd ed. Washington, DC: Government Printing Office, 1981. 308 p. Stock number 008-020-00908-3. $11. LC 81-22915.
2309 •
A useful reference volume that includes chapters on history, society and geography, economy, government and national security; valuable because of of the paucity of good source material.

Bunge, Frederica, ed. South Korea, a country study. 3rd ed. Washington DC: Government Printing Office, 1982. 306 p. Stock number 008-001-00129-1. $4.50. LC 82-11385.
2310 •
A good basic reference volume with chapters on history, society and geography, economy, government and national security.

McCune, Shannon. Korea: land of broken calm. Princeton, NJ: D. Van Nostrand, 1966. 221 p. LC 66-16903.
2311
A nice overview of Korea viewed historically and from the perspective of geography and economic development. Material is current up to 1965.

G. CHINA: BIBLIOGRAPHY

Skinner, G. William, ed. Modern Chinese society, an analytical bibliography: vol. 1, Publications in Western languages - 1644-1972. Stanford CA: Stanford University Press, 1973. 802 p. ISBN 0-8047-0751-0. $55. LC 70-130831.
2312
A comprehensive bibliography of materials on Chinese society. A valuable sourcebook for students and specialists. Other volumes are available on Japanese and Chinese materials.

Williams, Jack F. China in maps, 1890-1960: a selective and annotated cartobibliography. (East Asia series, occasional paper no. 4). East Lansing, MI: Asian Studies Center, 1974. 365 p. LC 75-621921.
2313
An extremely valuable aid to the specialist or other user of general or specialized maps (including topographic) of China. A very useful reference tool.

East Asia

Acta geographica sinica (Dili xuebao). 1-32 no. 2 (1934-1966); 33- (1978-). Quarterly. Science Publishing House, Chegongzhuang Xilu 21, P.O. Box 339, Beijing, People's Republic of China.　　　2314
The principal geographic periodical of the PRC. In Chinese. Table of contents and abstracts in English.

China geographer. Edited by Clifton W. Pannell and Christopher L. Salter. 1- (1975-). Westview Press, 5500 Central Avenue, Boulder, CO, 80301.　　　2315
China geographer *was a periodical, nos. 1-10 (1975-1978). With no. 11 (1981) it became an irregular serial: No. 11 (Agriculture. 1981, 143 p.) and no. 12 (Environment. 1984, 183 p.) focus on a theme, with original articles by specialists including geographers from the People's Republic. An important collection of original articles on various aspects of China's geography.*

China quarterly. 1- (1960-). Quarterly. Contemporary China Institute, School of Oriental and African Studies, Malet St., London WC1E 7HP, U.K. ISSN 0009-4439.　　　2316
The leading scholarly periodical that deals with modern China. Emphasizes politics and economics.

Modern China, an international quarterly of history and social science. 1- (1974-). Quarterly. Sage Publications, 275 S. Beverly Dr., Beverly Hills, CA, 90212. ISSN 0097-7004.　　　2317
A comparatively new journal that especially welcomes views and interpretations on contemporary China.

Blunden, Caroline, and Elvin, Mark. Cultural atlas of China. New York: Facts on File (Equinox), 1983. 237 p. ISBN 0-87196-132-6. $35. LC 82-675304.　　　2318•
A historically oriented atlas composed more of photos and text than maps. Quite useful for cultural and historical studies.

Catchpole, Brian. A Map history of modern China. London; Portsmouth, NH: Heinemann Educational Books, 1976. 145 p. ISBN 0-435-31095-X. $7. (pbk). LC 77-372874.　　　2319•
An inexpensive paperback atlas that focuses on historical and political topics. Useful for school and undergraduate teaching.

Central Intelligence Agency. People's Republic of China, atlas. Washington, DC: Government Printing Office, 1971. 82 p. LC 72-178811.　　　2320•
A fine basic atlas of China prepared for President Nixon's visit to China in early 1972. Includes six regional maps, ten topical sections, and a variety of charts, tables, graphs, and text. Topics include population, transportation, administration, climate, and aspects of economy.

Fullard, Harold, ed. China in maps. London: George Philip, 1968. 25 p.　　　2321
Although some of the maps are dated, this remains the most widely available reasonably priced atlas of China for school and undergraduate use. Topical maps and discussion of location, history, physical, population and economic geography of China.

Geelan, P.J.M. and Twitchett, D.C., eds. The Times atlas of China. New York: D. Van Nostrand, 1984. 144 p. (1st ed., Tokyo: 1973). ISBN 0-7230-0118-9. $13.50. LC 75-313716.　　　2322•
An excellent atlas that contains a variety of topical maps, some of which originally appeared in the 1971 CIA atlas, and a series of physical, provincial, and city maps and accompanying notes. A handsome and useful atlas.

Hsieh, Chiao-min. Atlas of China. New York: McGraw-Hill, 1973. 282 p. ISBN 0-07-030628-1. $22.95. LC 72-8717.
> A valuable general atlas of China generously supplemented with textual discussion. Main topics include physical, historical, cultural, economic, and regional geography. Useful for secondary and college students as well as specialists.

2323

J. CHINA: REFERENCE WORKS

Buck, John Lossing. Land utilization in China. Shanghai, China: The Commercial Press, 1937. 494 p. Reprint: New York: Garland, 1982. 476 p. ISBN 0-8240-4683-8. $55. LC 80-8828.
> A classic study based on field surveys of 16,786 farms conducted in 22 of China's provinces 1929-1933. A good beginning point for any study of China's agriculture.

2324

Bunge, Frederica, and Shinn, Rinn-Sup, eds. China, a country study. (Area handbook series). 3rd ed. Washington, DC: Government Printing Office, 1981. 590 p. LC 81-12878.
> A useful basic reference volume with chapters on the history, land, society, economy, politics, education, science, and national defense.

2325•

Eckstein, Alexander, ed. Quantitative measures of China's economic output. (Sponsored by Social Sciences Research Council). Ann Arbor, MI: University of Michigan Press, 1980. 443 p. ISBN 0-472-08754-1. $26.50. LC 79-17822.
> An important source book for statistical data on China's economy prior to 1980. Contains sections on agriculture, industry, capital formation, national product, and general economy. Origin and evaluation of the quality of the economic data are discussed.

2326

Goldstein, Steven M. and Sears, Kathrin, eds. The People's Republic of China: a basic handbook. 4th ed. Croton-on-Hudson, NY: Learning Resources in International Studies, Council on International and Public Affairs, 1984. 160 p. ISBN 0-936876-17-4. $7.50.
> Inexpensive, easy-to-use reference book on contemporary China. Topics include the land and people, history, foreign relations and national defense, economic development, trade, welfare, health, education, and daily life. A very useful sourcebook with a number of maps.

2327•

Hook, Brian, ed. The Cambridge encyclopedia of China. Cambridge, U.K.; New York: Cambridge University Press, 1982. 492 p. ISBN 0-521-23099-3. $37.50. LC 81-9927.
> A compendium of topics written by leading specialists in the field of Chinese studies. Includes a good overview of the geography of China.

2328•

Posner, Arlene, and deKeijzer, Arne J., eds. China, a resource and curriculum guide. 2nd rev. ed. Chicago, IL: University of Chicago Press, 1976. 317 p. ISBN 0-226-67559-9. $17. ISBN 0-226-67560-2. $5. (pbk). LC 75-9061.
> A valuable guide to materials and teaching on China. An essential resource guide for teachers.

2329•

The World Bank (International Bank for Reconstruction and Development). China: socialist economic development. Washington, DC: The World Bank, 1983. 464 p. 3 v. ISBN 0-8213-0245-0. $40. LC 83-14698.
> Detailed World Bank country study of China's economy, statistical system, economic sectors, and social sectors. Includes well-written narrative discussions as well as considerable statistical data on China's economy and society drawn from official Chinese sources.

2330

Xue, Muqiao, ed. Almanac of China's economy, 1981. (Economics Research Center and State Council of the People's Republic of China). New York: Eurasia Press, 1982. 1143 p. ISBN 0-88410-894-5. $155.　　　　　　　　　　　　　　　　　　2331
   An important reference book compiled in the PRC from official sources. Covers all sectors of the economy and includes provincial surveys. A valuable sourcebook.

K. CHINA: GENERAL WORKS

Buchanan, Keith; Fitzgerald, Charles P.; and Ronan, Colin A. China: the land and people. New York: Crown Publishers, 1981. 520 p. ISBN 0-517-54494-6. $19.95. LC 81-2275.　　　　　　　　　　　　　　　　　　　　　　　　　　　　　2332•
   A useful general interest volume with sections on the land and people, history, impact of the West, and science and technology.

Buchanan, Keith. The Transformation of the Chinese earth. London: G. Bell; New York: Praeger, 1970. 336 p. ISBN 0-7135-1549-X. LC 76-100908.　　　　　　2333
   A sympathetic description of China, its land, people, spatial organization, and general geography up through the late 1960's. Although out of date, it offers an interesting view of China during the heyday of the Maoist period.

China Handbook Editorial Committee. Geography (of China). (China handbook series). Beijing: Foreign Language Press, 1983. 260 p. ISBN 0-8351-0984-4. $4.95.　　2334•
   A useful recent descriptive geography from China that includes four chapters on physical geography and lengthy regional descriptions.

deCrespigny, Rafe R.C. China: the land and its people. Melbourne, Australia: Nelson; New York: St. Martin's Press, 1971. 235 p. ISBN 0-17-004636-2. LC 76-180738.　2335
   A brief descriptive geography of China that follows a regional approach.

Dwyer, Denis J. China now. Harlow, U.K.; New York: Longman, 1974. 510 p. (Available from Books on Demand, Ann Arbor, Michigan, $130). LC 73-87225.　　　　2336
   A useful collection of articles that focus on topics of interest in geography--the land, farming, population, transportation, industry, cities, politics, and approaches to development.

Ginsburg, Norton, and Lalor, Bernard, eds. China: the 80's era. Boulder, CO: Westview Press, 1984. 378 p. ISBN 0-86531-668-6. $28.50. LC 84-50347.　　　　2337
   An important collection of papers that focus on recent changes in China and the likely direction of future trends. Topics include politics, society, economy, population, education, science, and the arts.

Greer, Charles. Water management in the People's Republic of China. Austin TX:, University of Texas Press, 1979. 174 p. ISBN 0-292-79011-2. $12.50. LC 78-15303.　2338
   A good summary and review of river management with emphasis on the Yellow and Yangtze Rivers.

Knapp, Ronald, ed. China's island frontier. Honolulu, HI: University of Hawaii Press, 1980. 296 p. ISBN 0-8248-0705-7. $20. LC 80-18578.　　　　　　　　2339•
   A fine collection of essays on the historical development of Taiwan as a frontier of China. Of interest to historians and other social scientists as well as geographers.

Leung, Chi-Keung. China: railway patterns and national goals. (Research paper no. 195). Chicago, IL: Univerity of Chicago. Department of Geography, 1980. 243 p. ISBN 0-89065-102-7. $10. LC 80-17030.　　　　　　　　　　　　　　　　　2340
   Examines the historical interrelationship of politics and policy, railway development, and national goals of China.

Leung, Chi-Keung, and Ginsburg, Norton, eds. China: urbanization and national development. (Research paper no. 196). University of Chicago. Department of Geography, 1980. 283 p. ISBN 0-89065-103-5. $10. LC 80-29142.　　　　　　2341
　　A collection of recent essays by geographers who specialize on China and its development problems. Topics focus on cities and on urban and transportation development.

Liang, Ernest P. China: railways and agricultural development. (Research paper no. 203). Chicago, IL: University of Chicago. Department of Geography, 1982. 186 p. ISBN 0-89065-109-4. $10. LC 82-4749.　　　　　　2342
　　An analytical study of the role of railroads in the commercialization of agriculture and the economic development of China.

Ma, Laurence J.C. and Hanten, Edward W., eds. Urban development in modern China. Boulder, CO: Westview Press, 1981. 264 p. ISBN 0-86531-120-X. $23.50. LC 81-10336.　　　　　　2343
　　A fine collection of papers on Chinese cities and urbanization by specialists in the field. An important contribution to knowledge on the cities of China.

Ma, Laurence J.C. and Noble, Allen G., eds. The Environment: Chinese and American views. New York: Methuen (for the Ohio Academy of Science), 1981. 397 p. ISBN 0-416-32050-3. $32. LC 81-215147.　　　　　　2344
　　A collection of papers by Chinese and American geographers on the environment. Important mainly for the Chinese papers which give a view of the state of geography in China in the late 1970's.

Murphey, Rhoads. The Fading of the Maoist vision: city and country in China's development. New York: Methuen, 1980. 169 p. ISBN 0-416-60201-0. $13.95. LC 80-10115.　　　　　　2345•
　　An interesting short volume that advances the idea of China's retreat from the more ideological trends associated with Mao Zedong to a more pragmatic direction focused on ecomonic growth.

Nickum, James E., ed. Water management organization in the People's Republic of China. Armonk, NY: M.E. Sharp, 1980. ISBN 0-87332-140-5. $30. LC 80-5458.　　　　　　2346
　　A carefully selected and edited set of translated readings from key Chinese sources on the topic of irrigation and water management. Includes a good discursive introduction by the editor.

Pannell, Clifton W. and Ma, Laurence J.C. China: the geography of development and modernization. New York: Halsted (John Wiley) and V.H. Winston, 1983. 342 p. ISBN 0-470-27376-3. $49.95. ISBN 0-470-27377-1. $19.95. (pbk). LC 83-113355.　　　　　　2347•
　　A recent geographical study that focuses on China's modernization and development in the context of human/environment relationships and spatial organization. Topics include physical and historical geography, population, resources, agriculture, and other areas of economic geography. Also includes chapters on Taiwan and Hong Kong. A useful and recent addition to the geographic literature on China.

Pannell, Clifton W. Taichung, Taiwan: structure and function. (Research paper no. 144). Chicago, IL: University of Chicago. Department of Geography, 1973. 199 p. ISBN 0-89065-051-9. $10. LC 72-91223.　　　　　　2348
　　A study of urban form and function in a middle size city in Taiwan. The focus is on land use, which is compared with existing models of Western city structure. Of interest to comparative urbanists.

Samuels, Marwyn S. Contest for the South China Sea. New York: Methuen, 1982. 203 p. ISBN 0-416-33140-8. $24.95. LC 81-18868.　　　　　　2349
　　A carefully researched monograph on a sensitive topic in political geography, the territorial claims and control of the South China Sea. The author relies heavily on Chinese sources and emphasizes the Chinese viewpoint.

Shabad, Theodore. China's changing map: national and regional development. Rev. ed. New York: Praeger, 1972. 370 p. (1st ed., 1956. 295 p.). LC 71-178868.   2350 •
An informative geography of China up to 1970. Includes a topical section on physical, political, and economic geography as well as good regional descriptions of the entire country.

Skinner, G. William, ed. The City in late imperial China. Stanford, CA: Stanford University Press, 1977. 820 p. ISBN 0-8047-0892-4. $55. LC 75-184.   2351
Valuable collection of essays on urbanization and cities in China by specialists in the field.

Smil, Vaclav. The Bad earth: environmental degradation in China. Armonk, NY: M.E. Sharpe, 1984. 245 p. ISBN 0-87332-230-4. $25. ISBN 0-87332-253-3. $13.95. (pbk). LC 83-14821.   2352 •
An interesting and readable account of the serious environmental problems that China is experiencing as the country industrializes and the population continues to grow.

Tregear, T.R. (Thomas). China: a geographical survey. London: Hodder and Stoughton; New York: Halsted Press, John Wiley, 1980. 372 p. (1st ed. A geography of China. London: University of London Press; Chicago, IL: Aldine, 1965. 342 p.). ISBN 0-470-26925-1. $34.95. (U.S.). ISBN 0-340-23739-2. (U.K.). LC 82-140081.   2353
A good general geography of China especially strong on physical and historical geography. It draws heavily on Tregear's previous works and little updating was done. Therefore, statistical data on population and economy are drawn mainly from the early 1960's. Important recent policy changes are not discussed.

U.S. Congress. Joint Economic Committee. China under the four modernizations. Washington, DC: Government Printing Office, 1982. 2 v.: 610 p.; 381 p. Stock number: 052-070-05758-1. LC 82-603023.   2354
Every three or four years the Joint Economic Committee publishes a report on China's economy and modernization with chapter contributions from leading analysts in government and academia. Topics include policy, science and technology, military, and population as well as various aspects of the economy.

U.S. Congress. Joint Economic Committee. Chinese economy: post Mao. Washington, DC: Government Printing Office, 1978. 880 p. LC 79-600609.   2355
The 1978 version of the Joint Economic Committee report on China's policy for modernization, economic development, population, and labor supply. Although supplanted in part by the 1982 report, some of the essays in this volume are still useful.

10. SOUTHEAST ASIA

Alvar W. Carlson

A. BIBLIOGRAPHIES

Bixler, Paul H. Southeast Asia; bibliographic directions in a complex area. (Choice bibliographical essay series, no. 2). Middletown, CT: Choice, 1974. 98 p. ISBN 0-914492-01-2. LC 74-2049.
    2356
Excellent bibliography organized by country. Includes category of "overseas Chinese". Very informative discussions of materials by academic discipline before the entries listed by country. A publication of the Association of College and Research Libraries, a division of the American Library Association.

Breese, Gerald, ed. Urban Southeast Asia; a selected bibliography of accessible research, reports, and related materials on urbanism and urbanization. New York: Southeast Asia Development Advisory Group (SEADAG) of the Asia Society, 1973. 165 p. LC 73-161125.
    2357
Extensive bibliography on urbanism and urbanization in Hong Kong, Indonesia, Malaysia, the Philippines, Singapore, Thailand, and Vietnam by ten major contributors, including three geographers. Unannotated entries by country followed by entries on major cities. Excellent reference for background research on urbanization.

Carlson, Alvar W. A Bibliography of the geographical literature on Southeast Asia, 1920-1972. (Exchange bibliography nos. 598, 599, and 600). Monticello, IL: Council of Planning Librarians, June 1974. 127 p.
    2358
An unannotated bibliography of 1312 entries by country authored by geographers or published in geographical periodicals. Includes entries for 88 bibliographies.

Chen, John H.M. Vietnam: a comprehensive bibliography. Metuchen, NJ: Scarecrow Press, 1973. 314 p. ISBN 0-8108-0562-6. LC 72-10549.
    2359
Very comprehensive unannotated bibliography of 2331 entries, largely French and English. Entries are fully indexed by both title and subject. Brief background account is given on the history and geography of Vietnam.

Chen, Virginia, comp. The Economic conditions of East and Southeast Asia: a bibliography of English-language materials, 1965-1977. Westport, CT: Greenwood Press, 1978. 788 p. ISBN 0-313-20565-5. $75. LC 78-57762.
    2360
More than 225 pages are devoted to Southeast Asian countries, except for those within Indochina. Includes entries of other bibliographies.

Chiang, Tao-Chang. Singapore in thematic maps: an annotated carto-bibliography. (Research project series no. 10). Singapore: Nanyang University, Institute for Humanities and Social Sciences, February 1979. 177 p.
    2361
Compiled by a geographer, this is an exhaustive annotated bibliography of maps which show aspects of the physical and cultural geography of Singapore. References are to maps found in English language sources. A total of 899 entries and an appendix with annotations on 57 topographical maps. No index.

Cotter, Michael. Vietnam: a guide to reference sources. Boston, MA: G.K. Hall, 1977. 272 p. ISBN 0-8161-8050-4. LC 77-22448.
    2362
More than 1400 entries on all aspects of Vietnam. Some are in French, others translated. Essential for research on Vietnam.

Heussler, Robert. British Malaya: a bibliographical and biographical compendium. (Garland reference library of social sciences, vol. 79; themes in European expansion, vol. 1). New York: Garland Publishing Company, 1981. 193 p. ISBN 0-8240-9369-0. $33. LC 80-8968.
    2363
Part 1 is an annotated bibliography of 499 entries of which many would interest cultural geographers. Part 2 lists biographies. Both parts cover the colonial period to 1957.

Southeast Asia

Pelzer, Karl J. West Malaysia and Singapore: a selected bibliography. New Haven, CT: Human Relations Area Files Press, 1971. 394 p. ISBN 0-87536-235-4. $25. LC 72-87853.

    Comprehensive, unannotated compilation of approximately 4000 entries. Organized largely by three usual interests in geography: physical, cultural, and economic. Confusing author index.

2364

Sardesai, D.R. and Sardesai, Bhanu D. Theses and dissertations on Southeast Asia: an international bibliography in social sciences, education, and fine arts. (Bibliotheca Asiatica no. 6). Zug, Switzerland: Inter Documentation Company, 1970. 176 p. Swiss francs 40. LC 77-869967.

    Entries are listed by eight disciplines, including geography, which has nearly 100 entries. Valuable for theses and dissertations from many countries on Southeast Asia.

2365

Tanger, Frank N. Burma: a selected and annotated bibliography. New Haven, CT: Human Relations Area Files Press, 1973. 356 p. ISBN 0-87536-227-3. $25. LC 72-90939.

    An updated version of the 1956 Annotated bibliography of Burma. *Total of 2086 entries, mostly in the English language, including bibliographies, books, journal articles, documents, and dissertations. Entries also from sources in Burmese, Russian, and Eastern European languages. All entries are annotated except for the journal articles and Russian and East European sources. Many entries are of a geographical nature.*

2366

Tregonning, Kennedy G. Southeast Asia; a critical bibliography. Tucson, AZ: University of Arizona Press, 1969. 103 p. ISBN 0-8165-0271-4. $4.95. (pbk). LC 68-9845.

    Annotated and carefully selected entries, mostly in English, on Southeast Asia as a region and by country. Entries date from 1945 onward and are organized by categories of geography, history, politics, economics, and social/cultural aspects.

2367

University of Wisconsin, Madison. Land Tenure Center Library. Land tenure and agrarian reform in East and Southeast Asia: an annotated bibliography. Boston, MA: G.K. Hall, 1980. 557 p. ISBN 0-8161-8221-3. LC 80-18049.

    Annotated entries for Southeast Asia are by the region and individual countries on p. 235-497. Has personal name index, subject index, and corporate name index. Many of the 116 entries are by geographers. Excellent reference.

2368

B. SERIALS

See also:
Malaysian journal of tropical geography, 2736
Singapore journal of tropical geography, 2737

Indonesian journal of geography. v. 1-5, no. 1-9 (1960-1965); v. 6- , no. 10/31- (1976-). Semiannual. Faculty of Geography, Gadjah Mada University, Bulaksumur, Yogyakarta, Indonesia. ISSN 0024-9521.

    Articles mainly on Indonesia and Southeast Asia. In English.

2369

Philippine geographical journal. 1-6 (1953-1958); 7- (1963-). Quarterly. Philippine Geographic Journal, P.O. Box 2116, Manilla, Philippines. ISSN 0031-7551.

    Articles mainly on the Philippines, developing areas, and geographical problems of concern to them. In English.

2370

Barton, Thomas F.; Kingsbury, Robert C.; and Showalter, Gerald R. Southeast Asia in maps. 2nd ed. Chicago, IL: Denoyer-Geppert Company, 1973. (1st ed., 1970). 96 p. LC 74-83474.

This atlas has 60 black-and-white maps with brief text on the ten countries of Southeast Asia. About one-half of the maps are historical and political in nature.

2371 •

Morgan, Joseph R. and Valencia, Mark J., eds. Atlas for marine policy in Southeast Asian seas. Berkeley, CA: University of California Press, 1983. 144 p. ISBN 0-520-05005-3. $125. LC 83-47891.

Very impressive, but expensive atlas published for the East-West Environment and Policy Institute, Honolulu. A must for any reference library. Emphasis on importance of the seas and management questions. 102 maps, including 26 in color, accompanied by textual material not found elsewhere, ranging from marine resources, economic exploitation, ports and commercial activities, tourism, and pollution.

2372

Prescott, J.R.V.; Collier, H.J.; and Prescott, D.F. Frontiers of Asia and Southeast Asia. Carlton, Victoria, Australia: Melbourne University Press, 1977. 106 p. ISBN 0-522-84116-3. $15. (Available from International Specialized Book Services, Beaverton, OR). LC 78-305222.

A basic gazetteer of 44 maps with a detailed historical/geographical discussion of each. Emphasis is on the boundaries between countries and related problems. 22 maps involve Southeast Asian countries.

2373 •

Ward, R. Gerard, and Lea, David A.M., eds. An Atlas of Papua and New Guinea. Glasgow, U.K.: Collins, Longman, and Department of Geography, University of Papua and New Guinea, 1970. 101 p. ISBN 0-582-00081-5. LC 70-654459.

A comprehensive atlas edited by two geographers, containing 51 maps with commentaries by more than 20 contributors. Covers all facets of the island including population, social, medical, political, physical, and economic aspects. An excellent basic reference on New Guinea.

2374

Abdul Rahim Mokhzani, Datuk B., ed. Rural development in Southeast Asia. New Delhi, India: Vikas Publishing House, 1979. 254 p. ISBN 0-7069-0762-0. $22.50. (Available from Advent Books, New York). LC 79-902701.

Good analysis of land development, the social and economic characteristics of sponsored migration and resettlement programs in Singapore, Malaya, Thailand, and the Philippines. Two chapters deal with resettlement schemes in Hong Kong. Chapters are result of Southeast Asian Social Science Association Conference on Rural Development in Southeast Asia, Kuala Lampur, 1975.

2375

Allen, J.; Golson, J.; and Jones, R., eds. Sunda and Sahul: prehistoric studies in Southeast Asia, Melanesia, and Australia. New York: Academic Press, 1977. 647 p. ISBN 0-12-051250-5. $75. LC 77-74808.

Consists of 18 chapters analyzing paleolithic peoples, their subsistence strategies, exchange systems, technology, and livelihoods during and after the Pleistocene period in the land bridge area of the continental shelves (Sunda and Sahul) between mainland Southeast Asia and Australia. Particular value is found in the discussion of Carl Sauer's hypothesis on agricultural origins.

2376

Banks, David J., ed. Changing identities in modern Southeast Asia. The Hague: Mouton Publishers, 1976. 357 p. ISBN 0-202-90046-0. DM 46. LC 77-370481.

Excellent book on ethnicity in Southeast Asia, consisting of 19 articles, discussing the different ethnic groups and their livelihoods in each country. Topics include culture contact and change as well as tourism. Result of the 9th International Congress of Anthropological and Ethnological Sciences, Chicago, 1973.

2377

Southeast Asia

Dobby, Ernest Henry George. Southeast Asia. 11th ed. London: University of London Press; Mystic, CT: Lawrence Verry, 1973. 429 p. (1st ed., 1950. 415 p.). ISBN 0-340-17385-8. $12.50.　　　　　　　　　　　　　　　　　　　　　　　　　　　2378
    A standard geography text divided into three parts: 1) physical environments (5 chapters, including one on soils); 2) countries and regions within countries (16 chapters); and 3) economic and political environments (4 chapters). Chapter bibliographies, 118 maps and diagrams.

Dutt, Ashok K., ed. Southeast Asia: realm of contrasts. 3rd rev. ed. Boulder, CO; London: Westview Press, 1985. 275 p. (1st ed., 1974. 194 p.). ISBN 0-86531-561-2. $37.50. ISBN 0-86531-562-0. $15.95 (pbk).　　　　　　　　　　　　　　　　2379•
    Intended to be a text, this book has 15 chapters written by 12 geographers. Part one (7 chapters) covers the usual physical, economic, political and cultural aspects of Southeast Asia. Part two includes eight case studies on the individual countries. Well illustrated with 69 figures and 21 tables.

Fisher, Charles A. South-East Asia: a social, economic, and political geography. 2nd ed. London; New York: Methuen, 1966. 831 p. (1st ed., 1964. 831 p.). ISBN 0-416-42480-5. $65. LC 66-72859.　　　　　　　　　　　　　　　　　　　　　　2380
    This book remains the best and most comprehensive geography text on Southeast Asia. Includes 108 tables, 110 maps, and an extensive bibliography. No photos. Each country and Southeast Asia as a region are discussed in detail.

Fryer, Donald W. Emerging Southeast Asia: a study in growth and stagnation. 2nd ed. London: George Phillips; New York: John Wiley and Sons, 1979. 540 p. (1st ed., 1970. 486 p.). ISBN 0-470-26298-2. LC 78-7804.　　　　　　　　　　　　　　　2381•
    Comprehensive text by a geographer who is recognized as an authority on the region. Revised from the 1970 edition, especially material on Indochina. Includes overview, analysis of countries of progress (Thailand, Philippines, Malaysia, and Singapore), and countries of stagnation (Vietnam, Laos, Cambodia, Burma, and Indonesia). Has 29 pages of bibliography and necessary maps.

Hainsworth, Geoffrey B., ed. Village-level modernization in Southeast Asia : the political economy of rice and water. Vancouver, BC: University of British Columbia Press, 1982. 411 p. ISBN 0-7748-0156-5. Can $34. LC 83-109477.　　　　　　2382
    Up-to-date and varied accounts of the changes and trends in the rural economies found in Indonesia, Philippines, Malaysia, and Thailand. Geographical topics include rice cultivation (green revolution), fishing, and tourism. Result of the First International Conference of the Canadian Council for Southeast Asian Studies, Vancouver, 1979.

Hanks, Lucien M. Rice and man: agricultural ecology in Southeast Asia. (Worlds of man series). Arlington Hts, IL: Davidson, Harlan, Inc., 1972. 174 p. ISBN 0-88295-606-X. $16.95. ISBN 0-88295-607-8. $8.95. (pbk). LC 78-169512.　　　　　　　2383
    An anthropological/historical study of rice production as the staple crop in the community of Bang Chan, near Bangkok, Thailand. Emphasis is on cultural ecology, ecosystems of rice, and modes of rice cultivation, typical of much of Southeast Asia. Includes two appendices, one of which is on numerical data regarding rice production by modes of cultivation in Southeast Asia. Monograph is of great significance to geographers.

Hill, Ronald David, ed. South-East Asia: a systematic geography. (Chia Lin Sien and others). Kuala Lumpur, Malaysia; New York: Oxford University Press, 1979. 214 p. ISBN 0-19-580393-0. ISBN 0-580398-1. (pbk). LC 79-941024.　　　　　　　2384•
    Intended as a text for secondary-level (high school) students, seven contributors (five geographers) provide ten chapters, largely on the physical, economic, political, and population geography of the region. Weak in cultural geography. Well-illustrated with 62 figures (many maps) and 68 tables.

Hill, Ronald David, and Bray, Jennifer M., eds. Geography and the environment in Southeast Asia. Hong Kong: Hong Kong University Press, 1978. 485 p. ISBN 9-622-09009-5. HK \$50. LC 79-103361.

Conference proceedings (Department of Geography and Geology Jubilee Symposium, University of Hong Kong) largely on the impact of urbanization, what it means to rural areas, and the role of planning. Includes 28 chapters (papers), largely by geographers, with 78 tables and 68 figures, including many maps.

2385

Hutterer, Karl L., ed. Economic exchange and social interaction in Southeast Asia: perspectives from prehistory, history, and ethnography. (Michigan papers on South and Southeast Asia, no. 13). Ann Arbor, MI: University of Michigan, Center for South and Southeast Asian Studies, 1977. 318 p. ISBN 0-89148-013-7. \$9.50. (pbk). LC 77-95147.

Anthropological and historical in nature, this book includes 14 essays by 14 contributors. Concentrates on the early trade strategies and exchange systems in many parts of Southeast Asia. Short on maps, but good photographs and detailed tables. Includes 43 pages of bibliography. Result of a seminar by the Center for South and Southeast Asian Studies, University of Michigan, Ann Arbor, 1976.

2386

Kantner, John F. and McCaffrey, Lee, eds. Population and development in Southeast Asia. Lexington, MA: Lexington Books, 1975. 323 p. ISBN 0-669-98145-1. LC 74-30885.

Fourteen of the fifteen contributed chapters were seminar papers given by authorities on the region's current demographics sponsored by the Population Panel of the Southeast Asia Development Advisory Group (SEADAG) of the Asia Society, New York. Invaluable for understanding the region's population geography.

2387

Keyes, Charles F. The Golden peninsula: culture and adaptation in mainland Southeast Asia. New York: Macmillan, 1977. 370 p. ISBN 0-02-364430-3. LC 76-15175.

Interestingly written, this ethnological study emphasizes the region's sociocultural diversity within rural and urban contexts. Has current account on Vietnam. Much material on the role of Buddhism. Could very well be used as a cultural geography text on the region.

2388 •

Leng, Lee Yong. Southeast Asia and the Law of the Sea: some preliminary observations on the political geography of Southeast Asian seas. 2nd ed. Singapore: Singapore University Press, 1980. 75 p. (1st ed. 1978). ISBN 0-8214-0466-0. \$4. (pbk). (Available from Ohio University Press, Athens, Ohio). LC 82-93029.

Monograph by a political geographer on issues discussed at the 3rd United Nations Conference on the Law of the Sea and how they would affect Southeast Asian countries, particularly Singapore. Issues and maps include economic zones, maritime and offshore boundaries, seabed resources, fishing problems, and military significance. Indicates growing interest in Southeast Asian seas.

2389

McGee, Terence G. and Yeung, Yue-man. Hawkers in Southeast Asian cities: planning for the bazaar economy. Ottawa, Canada: International Development Research Centre, 1977. 139 p. ISBN 0-88936-120-7. LC 79-302771.

First authoritative in-depth and comparative analysis of street and market hawkers and vendors in the urban space and retailing/marketing systems of six Southeast Asian cities: two each in Malaysia, Indonesia, and the Philippines. The authors are geographers who have included the necessary maps, diagrams, and photographs. Excellent bibliography.

2390

Milner, G.B., ed. Natural symbols in South East Asia. (Collected Papers in Oriental and African Studies). London: University of London. School of Oriental and African Studies, 1978. 182 p. ISBN 0-7286-0043-9. £4. LC 80-481508.

An exciting collection of eight anthropological essays including much cultural geography because of the heavy emphasis on perceptions of space (e.g. architectural and political), time, life, and attitudes toward plants and animals. A study of the natural environment of man in various parts of Southeast Asia. Contains 13 photos, five figures/diagrams and two tables.

2391

Southeast Asia

Pryor, Robin J. ed. Migration and development in South-East Asia: a demographic perspective. Kuala Lumpur, Malaysia; London; New York: Oxford University Press, 1979. 354 p. ISBN 0-19-580420-1. £22. LC 79-123125.

    2392 •

    Most valuable population geography of region (Thailand, Malaysia, Singapore, Indonesia, Philippines) consisting of 25 chapters. Extensive bibliography of up-to-date entries.

Smith, Ralph B. and Watson, William, eds. Early South East Asia: essays in archeology, history, and historical geography. (School of Oriental and African Studies). London; New York: Oxford University Press, 1979. 561 p. ISBN 0-19-713587-0. $55. LC 79-112301.

    2393

    Best background information and interpretation of data on prehistoric Southeast Asia. 36 chapters based upon papers given at Colloquy on Early South East Asia, School of Oriental and African Studies, London, England, 1973. Part one (18 chapters) is on prehistory of the region and part two is on first millennium A.D. Extensive bibliography and index.

Taylor, Alice. Focus on Southeast Asia. (American Geographical Society, Focus series). New York: Praeger, 1972. 229 p. LC 72-188045.

    2394 •

    A basic geographical account of each country with maps and photographs.

Wallace, Ben J. Village life in insular Southeast Asia. Boston, MA: Little, Brown, and Co., 1971. 146 p. LC 71-155319. .

    2395

    A brief, but fascinating undergraduate anthropology text/reader on contemporary peoples and cultures of Southeast Asia, including hunters and gatherers, sea fishermen, shifting cultivators, and wet rice agriculturalists. Two examples are given of each of these groups. Excellent photographs.

Wheatley, Paul. Nāgara and commandery: origins of the Southeast Asian urban traditions. (Research paper nos. 207-208). Chicago, IL: University of Chicago, Department of Geography, 1983. 472 p. ISBN 0-89065-113-2. $20. LC 83-18014.

    2396

    Very scholarly, historical geographical discussion of the establishment of the city, its origins, diversity, and resulting hierarchies. Greatly dependent upon archeological research findings from prehistoric Southeast Asia. 23 figures, including maps and diagrams of early cities.

Wolters, Oliver W. History, culture, and region in Southeast Asian perspectives. Singapore: Institute of Southeast Asian Studies, 1982. 120 p. ISBN 9971-902-42-7. $22.50. (pbk). (Available from Gower Publishing Co., Brookfield, Vermont). LC 83-940898.

    2397

    Excellent background information on the development of cultures in pre-European Southeast Asia. Contains much historical geography and geographical history. Discusses the Mandala as a concept of space organization. Provides insights into today's cultural diversity and regional literary cultures. Contains lengthy up-to-date bibliography.

Wood, William, ed. Cultural-ecological perspectives on Southeast Asia: a symposium. (Southeast Asia series, no. 41). Athens, OH: Ohio University Center for International Studies, 1977. 192 p. ISBN 0-8214-0322-2. LC 76-620062.

    2398

    All 11 articles provide valuable background on the region's contemporary and prehistoric ecological anthropology. Extensive bibliography of 19 pages.

Yeung, Yue-man, and Lo, C.P., eds. Changing South-East Asian cities: readings on urbanization. Singapore; New York: Oxford University Press, 1976. 245 p. ISBN 0-19-580316-7. $19.95. LC 77-359038.

    2399

    A collection of reprints of the best journal articles on urbanization in Southeast Asia published between 1955 and 1972. Topics include historical development of the cities, patterns of urbanization, and case studies.

E. SPECIAL TOPICS

Brown, Paula. Highland peoples of New Guinea. Cambridge, U.K.; New York: Cambridge University Press, 1978. 258 p. ISBN 0-521-21748-2. $37. ISBN 0-521-29249-2. $10.95. (pbk). LC 77-80830.
Offers much ethnological information on Papua New Guinea.

2400

Buchanan, Iain. Singapore in Southeast Asia: an economic and political appraisal. London: G. Bell and Sons, 1972. 336 p. ISBN 0-7135-1656-9. £9.50. LC 72-183727.
Most comprehensive geographical study by a geographer of the cultural, economic, and political aspects of Singapore. Emphasis is on the 1960's.

2401

Burley, T.M. The Philippines: an economic and social geography. London: G. Bell and Sons, 1973. 375 p. ISBN 0-7135-1706-9. LC 73-331109.
A basic text emphasizing the economic and social geography of the Philippines. Analyzes six types of urban settlements, several major crops and 14 regions of the country. Many maps.

2402

Cho, Lee-Jay; Suharto, Sam; McNicoll, Geoffrey; and Mamas, S.G. Made. Population growth of Indonesia:an analysis of fertility and mortality based upon the 1971 population census. (Monographs of the Center for Southeast Asian Studies, Kyoto University). Honolulu, HI: The University of Hawaii Press, 1980. 123 p. ISBN 0-8248-0691-3. $16. ISBN 0-8248-0696-4. $10. (pbk). LC 79-18275.
An indepth study of Indonesian demography. Emphasis on geographical subdivisions and rural vs. urban populations. Includes 13 maps and 29 tables.

2403

Clarke, William C. Place and people: an ecology of a New Guinean community. Berkeley, CA: University of California Press, 1971. 265 p. ISBN 0-520-01791-9. $39.50. LC 78-126764.
Study of the components of a typical New Guinea ecosystem and subsistence behavior. Provides a very informative account of people preparing swiddens/gardens. Good photos and diagrams. Four appendices, including two with data on cultivated plants and wild animals.

2404

Eder, James F. Who shall succeed? Agricultural development and social inequality on a Philippine frontier. Cambridge, U.K.; New York: Cambridge University Press, 1982. 264 p. ISBN 0-521-24218-5. $42.50. LC 81-10178.
An enlightening, localized account of migrants who move from Cuyo to Palawan, Philippines, their livelihoods, differences in agricultural production and origins of social/economic inequality. Gives insights into swiddening on a frontier. Contains much geographical material with statistical tables based largely upon field work.

2405

Fryer, Donald W. and Jackson, James C. Indonesia. Boulder, CO: Westview Press; London: E. Benn, 1977. 313 p. ISBN 0-89158-028-X. $38.50. LC 76-3707.
A broad geographical survey which emphasizes the human and political dimensions of Indonesia.

2406 •

Hansen, Gary E., ed. Agricultural and rural development in Indonesia. Boulder, CO: Westview Press, 1981. 312 p. ISBN 0-86531-124-2. $25. (pbk). LC 80-24414.
Up-to-date discussions in 16 chapters on the types and reasons for diverse agricultural production. Emphasis is on Java, the man/land ratio, and the green revolution. There are no discussions of plantations and swidden agriculture, but chapters on aquaculture and forestry. Micro-scale analysis of gardens and household economies.

2407

Hardjono, J.M. Transmigration in Indonesia. Kuala Lumpur, Malaysia: Oxford University Press, 1977. 116 p. ISBN 0-19-580344-2. LC 77-152740.
Very readable discussion of organized transfers and resettling of peoples because of overpopulation and the resulting infrastructures from the Dutch colonial period of the early 1900's to 1974. Written by an Indonesian geographer.

2408

Hayami, Yujiro, and Kikuchi, Masao. Asian village economy at the crossroads: an economic approach to institutional change. Tokyo, Japan: University of Tokyo Press; Baltimore, MD: Johns Hopkins University Press, 1982. 275 p. ISBN 0-8018-2774-4. $26.50. LC 81-83546.　　　　　　　　　　　　　　　　　　　　　　　　　　　　2409

Provides geographical insights into the agrarian structures, changes, and institutions of several villages in the Philippines and Indonesia in the 1970's. Excellent for its analyses of the different rice harvesting systems. Good maps and bibliography.

Hill, Ronald David. Rice in Malaysia: a study in historical geography. Kuala Lumpur, Malaysia; New York: Oxford University Press, 1977. 234 p. ISBN 0-19-580335-3. LC 78-940674.　　　　　　　　　　　　　　　　　　　　　　　　　　　　　　2410

Rice production in both prehistoric and historic times to 1910. Excellent analysis of agricultural development and land tenure by internal regions within Malaya. Extensive bibliography.

Ho, Robert, and Chapman, E.C., eds. Studies of contemporary Thailand. (Department of Human Geography, publication HG/8). Canberra, ACT, Australia: Australian National University. Research School of the Pacific, 1973. Reprinted: 1975. 416 p. ISBN 0-7081-0294-8. LC 72-97053.　　　　　　　　　　　　　　　　　　　　2411

All 17 chapters, including six by geographers, provide valuable information about Thailand. Numerous maps.

Ingram, James C. Economic change in Thailand, 1850-1970. 2nd ed. Stanford, CA: Stanford University Press, 1971. 352 p. (1st ed., 1955. 254 p.). ISBN 0-8047-0782-0. $25. LC 70-150325.　　　　　　　　　　　　　　　　　　　　　　　　　　　　2412

An economic history especially of agricultural production in Thailand. Chapters 1-10 are unrevised from the 1st ed. Chapters 11 and 12 provide data from 1950 onward.

LeBar, Frank M., ed. and comp. Ethnic groups of insular Southeast Asia. New Haven, CT: Human Relations Areas Files Press.
　　Vol. 1: **Indonesia, Andaman Islands, and Madagascar.** 1972. 226 p. ISBN 0-87536-403-9. $25. LC 72-90940.
　　Vol. 2: **Philippines and Formosa.** 1975. 167 p. ISBN 0-87536-405-5. $25. LC 74-19513.　　　　　　　　　　　　　　　　　　　　　　　　　　　　2413 •

Encyclopedic-type listings and ethnographic descriptions of the insular peoples/tribes of Indonesia and the Philippines. These volumes complement the survey Ethnic groups of mainland Southeast Asia *by Frank LeBar, Gerald C. Hickey, and John K. Musgrave (New Haven, CT: HRAF Press, 1964). Accompanying maps and bibliography.*

Lee, David. The Sinking ark: environmental problems in Malaysia and Southeast Asia. Kuala Lumpur, Malaysia; New York: Heinemann Educational Books, 1980. 85 p. ISBN 0-686-72742-8. $10. LC 81-142995.　　　　　　　　　　　　　　　　　　　　2414 •

Easily readable text-like monograph on ecological problems in Southeast Asia, especially in Malaysia. Discusses the plight of the tropical rainforests and problems of population growth. Excellent photos and diagrams.

Lewis, Henry T. Illocano rice farmers: a comparative study of two Philippine barrios. Honolulu, HI: University of Hawaii Press, 1971. 209 p. ISBN 0-87022-460-3. LC 70-127330.　　　　　　　　　　　　　　　　　　　　　　　　　　　　　2415

An anthropological study of two barrios engaged in wet-rice cultivation in Northern Luzon. Investigates technological and social adaptations of each barrio. Contains a good glossary of Filipino terms.

McKinnon, John, and Bhruksasri, Wanat, eds. Highlanders of Thailand. Kuala Lumpur, Malaysia; New York: Oxford University Press, 1983. 358 p. ISBN 0-19-580472-4. $49.50. LC 83-232171.　　　　　　　　　　　　　　　　　　　　　　2416

Highly interesting anthropological/geographical account comprised of 22 chapters (19 contributors) on the characteristics, land use, settlements, linguistics, and religions of ten northern Thai tribes and peoples (Yuan, Shan, Yunnanese, Lua, Karen, Hmong, Yao, Lisu, Lahu, and Akha). Includes chapter on tourism among the hill tribes. Well-illustrated by 35 photos, 14 tables, six figures, and 11 maps. Publication sponsored by the Tribal Research Centre, Chiang Mai, Thailand.

Missen, G.J. Viewpoint on Indonesia: a geographical study. Melbourne, Victoria, Australia: Thomas Nelson, 1972. 359 p. LC 72-172121.　　　　　　　　　　2417
The best comprehensive economic analysis by a geographer on Indonesia. Excellent accounts of the 'ladang' and 'sawah' agricultural systems. Considerable historical geographical material. Many maps and photos.

Morauta, Louise; Pernetta, John; and Heaney, William, eds. Traditional conservation in New Guinea: implications for today. (Monograph 16). Boroko, New Guinea: The Institute of Applied Social and Economic Research, 1982. 392 p. ISBN 0-7247-0264-4. Kina 1.50.　　　　　　　　　　2418
Published conference proceedings on conceptions of the environment and conservation practices. Total of 43 papers discuss land and water resources and are for the most part geographic. Conference held by publisher and Office of Environment and Conservation.

Neill, Wilfred T. Twentieth-century Indonesia. New York: Columbia University Press, 1973. 413 p. ISBN 0-231-03547-0. $35. ISBN 0-231-08316-5. $15. (pbk). LC 72-11718.　　　　　　　　　　2419
Very readable geographic account of the Indonesian peoples and their religious and colonial backgrounds. Much emphasis on biogeography as exemplified by the chapters on plant and animal life and their relationships to man. Bibliography of 14 pages of selected references.

Ooi, Jin-Bee. Peninsular Malaysia. Rev. ed. London; New York: Longman, 1976. 437 p. (1st ed: Land, people, and economy in Malaya, 1963. 426 p.). ISBN 0-582-48185-6. £8.95. LC 75-42166.　　　　　　　　　　2420 •
A broad geography text on Malaya divided into three parts: land; people; and economy. Many photos, maps, and a fairly extensive bibliography.

Ooi, Jin-Bee, and Sien, Chia Lin, eds. The Climate of West Malaysia and Singapore. Kuala Lumpur, Malaysia: Oxford University Press, 1974. 262 p. LC 75-300682.　　　　2421
Edited by two geographers, this is an authoritative account of the climate fashioned into 19 chapters by 12 contributors. Divided into three parts: the upper air; climatic elements; and applied climatology. Many maps. Readable and a good source for climatic data. Excellent bibliography.

Paijmans, K., ed. New Guinea vegetation. Canberra ACT, Australia: National University Press, 1977. 213 p. ISBN 0-7081-0572-6. A $17.50. (1982, Commonwealth Scientific and Industrial Research Organization. (Available from State Mutual Book and Periodical Service, New York. $49). LC 76-22206.　　　　　　　　　　2422
Authoritative account divided into three parts: phytogeography; vegetation; and ethnobotany. Valuable to both physical and cultural geographers interested in New Guinea. Lengthy bibliography. Well-illustrated with 53 photos, 12 figures, and 33 tables.

Pataki-Schweizer, K.J. A New Guinea landscape: community, space, and time in the eastern highlands. (Anthropological studies in the eastern highlands of New Guinea, vol. 4). Seattle, WA: University of Washington, 1980. 165 p. ISBN 0-295-95656-9. $40. LC 78-21211.　　　　　　　　　　2423
An excellent anthropological/geographical examination of how several landscapes/human ecosystems have been developed and used. Contains 24 maps, 25 tables, 16 diagrams, five appendices, numerous photographs and a lengthy up-to-date bibliography.

Southeast Asia

Pelzer, Karl J. Planter and peasant: colonial policy and agrarian struggle in East Sumatra, 1863-1947. (Verhandelingen van het Koninklijk Instituut voor Taal--, Land-en Volkenkunde, Vol. 84). 'S-Gravenhage, Netherlands: Martinus Nijhoff, 1978. 163 p. ISBN 90-247-2174-1. LC 79-322001. 2424

Detailed historical-geographical account of agrarian policies and problems concerning the establishment of plantations and their impact upon the indigenous peasants of eastern Sumatra. Well-illustrated with 13 maps and 35 photos in addition to 23 tables.

Pernia, Ernesto del Mar. Urbanization, population growth, and economic development in the Philippines. (Studies in population and urban demography, no. 3. International Population and Urban Research, University of California at Berkeley). Westport CT: Greenwood Press, Inc., 1977. 213 p. ISBN 0-8371-9721-X. $29.95. LC 77-24588. 2425

Very detailed demographic analysis of urbanization and internal migration in the Philippines. Of considerable importance to population geographers and others as a source book on population based largely upon the 1970 census. Contains 59 tables in the text, six annex tables, and an appendix of 25 tables plus a map of census regions.

Reed, Robert R. Colonial Manila: the context of Hispanic urbanism and process of morphogenesis. (Publications in geography, vol. 22). Berkeley, CA: University of California Press, 1978. 129 p. ISBN 0-520-09579-0. $18.50. LC 77-80476. 2426

Fills the gap for geographical literature on the colonial development and raison d'etre of Manila. Extensive bibliography and detailed maps.

Salita, Domingo C. Geography and natural resources of the Philippines. Quezon City, Philippines: University of the Philippines System, College of Arts and Sciences, 1974. 338 p. 2427

Most emphasis on economic geography and the country's natural resources. Well-illustrated with 30 figures, including maps. Contains 29 tables, appendices, selected bibliography, and index.

Salita, Domingo C. and Rosell, Dominador Z. Economic geography of the Philippines. Bicutan, Philippines: National Research Council of the Philippines, 1980. 341 p. LC 82-101390. 2428 •

Written by prominent Filipino geographers who provide a comprehensive analysis of all aspects of the Philippine's economic geography as well as some demographic factors. Contains good bibliography.

Simkins, Paul D. and Wernstedt, Frederick L. Philippine migration: the settlement of the Digos-Padada Valley, Davao Province. (Monograph series, no. 16). New Haven, CT: Yale University Southeast Asia Studies, 1971. 147 p. ISBN 0-938692-13-5. $8.25. LC 73-154010. 2429

Study made by geographers on the migrations and pioneering efforts on the frontier or undeveloped lands of Mindanao from the 1920's to the 1960's. Complements study by James Eder cited above.

Waddell, Eric. The Mound builders: agricultural practices, environment, and society in the central highlands of New Guinea. (American Ethnological Society, monograph 53). Seattle, WA: University of Washington Press, 1972. 253 p. ISBN 0-295-95169-9. LC 70-159437. 2430

An anthropological geographical account of the use of time and space evolving around agricultural systems. Well-illustrated with maps, photographs, and graphs.

Wernstedt, Frederick L. and Spencer, J. E. The Philippine island world: a physical, cultural, and regional geography. Berkeley, CA: University of California Press, 1967. Reprinted: 1978. 742 p. ISBN 0-520-03513-5. $49. LC 67-14001. 2431 •

Still the best analysis by geographers of the Philippines. Three parts: physical environment; cultural and economic environments; and regional environments. Lengthy statistical appendix of 37 tables and bibliography.

11. SOUTH ASIA

Joseph E. Schwartzberg

A. BIBLIOGRAPHY

Patterson, Maureen L.P. in collaboration with Alspaugh, William J. South Asian civilizations: a bibliographic synthesis. Chicago, IL; London: University of Chicago Press, 1981. 853 p. ISBN 0-226-64910-5. $60. LC 81-52518.   2432

More than 28,000 entries in Western languages arranged in an original three-dimensional classification of chronology, geographic area, and topic. Extensive author and subject indexes. Includes India, Sri Lanka, Pakistan, Bangladesh, Nepal, Sikkim, and Bhutan. Although selective based on language, accessibility, up-to-dateness, and length, the bibliography is massive and provides the most comprehensive bibliography of Southern Asia.

B. SERIALS

Deccan geographer. 1- (1962-). Semi-annual. Subhadra Bhawan, 120A, Nehru Nagar East, Secunderabad 500 026, Andhra Pradesh, India. ISSN 0011-7269.   2433
Articles on geographic topics in India.

Geographical review of India. 1- (1932-). Quarterly. Geographical Society of India, 35, Ballygunge Circular Road, Calcutta 700019, India. ISSN 0046-5690.   2434
Generally the most substantial geographical periodical of India. Articles on a wide range of aspects of the geography of India and its parts. Geographical notes.

Indian geographical journal. 1- (1926-). Semi-annual. The Indian Geographical Society, Department of Geography, University of Madras, Madras 600 005, India. ISSN 0019-4824.   2435
The oldest geographical periodical of India. Articles on the geography of India, particularly South India. Research notes. Book reviews.

National geographical journal of India. 1- (1955-). Quarterly. The National Geographical Society of India, Banaras Hindu University, Varanasi 221 005, India. ISSN 0027-9374.   2436
One of the most substantial and regular geographical periodicals of India. Geographical studies particularly of India but also on other parts of Asia. Book reviews. News and notes.

National geographer. 1-5 (1958-1962); 6- (1971-). Semi-annual. The Allahabad Geographical Society, Department of Geography, University of Allahabad, Allahabad 211 002, India. ISSN 0470-0929.   2437
Articles on India, particularly North India. Book reviews. News and views.

Oriental geographer. 1- (1957-). Semi-annual. Bangladesh Geographical Society, Department of Geography, University of Dhaka, Dhaka 2, Bangladesh. ISSN 0030-5308.   2438
Articles mainly on Bangladesh.

C. ATLAS

Schwartzberg, Joseph E., ed. An Historical atlas of South Asia. Chicago, IL: University of Chicago Press, 1978. 354 p. ISBN 0-226-74221-0. $200. LC 77-81713.   2439•
This large-format atlas provides an annotated cartographic record of the political, cultural, social, economic, and demographic history of South Asia from the Old Stone Age to the year of publication. It both synthesizes the historical geography of the region and relates it to its broader Asian or even global setting. The work includes 600 maps, hundreds of graphs, photos, drawings, and tables, a lengthy text, a bibliography of more than 4000 titles, and an index of some 15,000 entries.

South Asia

D. BOOKS

Bayliss-Smith, Tim P. and Wanmali, Sudhir, eds. Understanding green revolutions: agrarian change and development planning in South Asia. Cambridge, U.K.; New York: Cambridge University Press, 1984. 384 p. ISBN 0-521-24942-2. $59.50. LC 83-14434.

2440

This "festschrift", honoring B.H. Farmer, comprises a collection of incisive essays, mainly geographic, that look beyond the polemical rhetoric engendered by the Green Revolution at the realities experienced by agrarian communities in a wide range of environments in India, Sri Lanka, and Bangladesh. Particularly interesting are Bayliss-Smith's comparison of energy flows prior to and subsequent to the adoption of the new technology and Graham Chapman's structural analysis of the perceptual framework for decision making on two Bangladeshi farms.

Bhardwaj, Surinder Mohan. Hindu places of pilgrimage in India: a study in cultural geography. Berkeley, CA: University of California Press; Delhi: Thomson Press, 1973. 258 p. ISBN 0-520-02135-5. $40. LC 74-18631.

2441

Based on extensive field work and an examination of a wide range of religious texts, this study documents the importance of the institution of pilgrimage, both in contemporary India and over the last two millennia. Bhardwaj classifies places of pilgrimage into a regional hierarchy, from pan-Indian to local, and relates their characteristic functions to the "Great" and "Little Traditions" of Hinduism and their role in diffusing religious ideas and secular knowledge.

Blaikie, Piers M. Family planning in India: diffusion and policy. London: Edward Arnold, 1975. 168 p. ISBN 0-7131-5780-1. £20. LC 75-32973.

2442

Blaikie judiciously studies India's family planning program at four scales: the decision makers; the micro-regional; the District; and the national; and demonstrates the inter-relatedness among them. Underpinning the study are meticulously collected data on family planning motivation, knowledge, and practice collected from hundreds of married couples in backward regions of Bihar. The findings lead to a trenchant critique of classical spatial diffusion theory and to recommendations for a practical integrated family planning strategy.

Blair, Harry W. Voting, caste, community, society: explorations in aggregate analysis in India and Bangladesh. New Delhi, India: Young Asia Publications, 1979. 199 p. Rs. 64. LC 79-901696.

2443

Blair, a political scientist, seeks here to discover the determinants and correlates of voting behavior in Bihar and Bangladesh. His investigations, solidly rooted in fieldwork, are supported by aggregate data analysis, both through statistical methods and well executed maps. In particular, his mapping of caste and religious groups in Bihar is remarkably detailed and politically revealing.

Chandra Sekhar, A. and Burman, Roy B.K., eds. Economic and socio-cultural dimensions of regionalism: an Indo-U.S.S.R. collaborative study. (Census centenary monograph no. 7). New Delhi, India: Office of the Registrar General, India, 1972. 538 p. LC 73-907495.

2444

Most of the two dozen essays comprising this volume attempt, with greatly varying degrees of sophistication, to regionalize the whole of India or certain portions of it according to a wide variety of factors (demographic, economic, social, degree of urbanization, land-use, etc.) or to delineate, by other means, the country's spatial structure. Planning regions are suggested and a number of essays put foward specific planning recommendations.

Crane, Robert I., ed. Regions and regionalism in South Asian studies: an exploratory study. (Monograph and occasional papers series, monograph no. 5). Durham, NC: Duke University, Program in Comparative Studies on Southern Asia, 1966. 281 p. LC 67-8233.

2445

Stimulating views presented by historians, anthropologists, sociologists, geographers, and political scientists who seek to come to terms with the problems of regions and regionalism in the context of South Asia. Some of the empirical data presented (e.g. on regional differences in family structure) are of considerable non-theoretical interest.

De Silva, K.M. Sri Lanka: a survey. London: C. Hurst; Honolulu, HI: University of Hawaii Press, 1977. 496 p. ISBN 0-8248-0568-2. $25. LC 77-372752.　　　　2446
This comprehensive and scholarly reference work is the collaborative effort of eleven Sri Lankan scholars. The text abounds in material indispensible for a thorough geographic study of the island's plural society, economy, and politics. The presentation is blessedly dispassionate.

Dupree, Louis. Afghanistan. Princeton, NJ: Princeton University Press, 1973. 760 p. ISBN 0-691-03006-5. $60. ISBN 0-691-00023-9. $12.50. (pbk). LC 76-154993.　　　2447
Dupree, an anthropologist who has worked mainly in and on Afghanistan since 1949, offers a superb, but regrettably partially dated, introduction to that country's land, people, past, and "present." The author has a good feel for geography. Twenty maps and 103 illustrations supplement his authoritative text.

Farmer, Bertram Hughes. Agricultural colonization in India since independence. London; New York: Oxford University Press (for the Royal Institute of International Affairs), 1974. 372 p. LC 74-196689.　　　　2448
Systematically compares roughly sixty government-sponsored agricultural colonization schemes in thirteen different areas of India. Draws heavily on the author's own prior experience studying pioneer settlement in what was then Ceylon and on comparable studies on Indonesia. It also points to the causes of relative success or failure in planned settlement ventures, and indicates the not very extensive limits within which further settlement may be attempted.

Farmer, Bertram Hughes, ed. Green Revolution? Technology and change in rice-growing areas of Tamil Nadu and Sri Lanka. Boulder, CO: Westview Press; London: Macmillan, 1977. 429 p. ISBN 0-89158-709-8. LC 76-51278; ISBN 0-333-19679-1. LC 77-376981.　　　　2449
The fruit of collaboration by an international team of geographers and other scholars who investigated the diffusion and impact of the Green Revolution in selected non-deltaic rice-growing areas of Tamil Nadu and Sri Lanka. Provides a valuable corrective to numerous disciplinarily biased studies, including macroeconomic analyses whose sweeping generalizations are inapplicable over much of South Asia. Argues for more regionally sensitive research and development.

Fox, Richard G., ed. Realms and region in traditional India. (Monograph and Occasional Papers series, monograph no. 14). Durham, NC: Duke University, Program in Comparative Studies on Southern Asia, 1977. 307 p. ISBN 0-916-994-12-0. $12. LC 76-3151.　　　　2450
Eight of the essays in this volume deal, to a large degree, with the historical geography of India. Four argue for or against the applicability of Aidan Southall's "segmentary state" model - originally developed for parts of Africa - to pre-modern India. Another four, including two by geographers, are concerned with problems of measurment of historic and pre-historic states and of definition of their boundaries.

Johnson, Basil Leonard Clyde. Development in South Asia. Harmondsworth, Middlesex, U.K.; New York: Penguin Books, 1983. 250 p. ISBN 0-14-080460-9. $5.95. LC 83-238459.　　　2451•
Less regionally detailed than the author's own South Asia *(London: Heinemann Educational Books, 2nd ed., 1981), this brief work provides a remarkably balanced, country-by-country assessment of the physical, ideological, and institutional factors bearing on economic development, of actual development experience in the period since independence, of current development levels, and the prospects of the future.*

Karve, Irwati K. Kinship organization in India. 3rd ed. New York: Asian Publishing House, 1968. 304 p. (1st ed.: 1953, 304 p.). LC 79-7261. 2452
The only detailed regional analysis of Indian kinship systems, which lie at the root of many forms of social interaction. Treats the related regional differences in regard to the customary laws of property, succession, and inheritance.

Leshnik, Lawrence S. and Sontheimer, Gunther-Dietz, eds. Pastoralists and nomads in South Asia. (Monograph series of the South Asian Institute of the University of Heidelberg). Wiesbaden, West Germany: Otto Harrassowitz, 1975. 276 p. ISBN 3-447-01552-7. DM 112. LC 75-512060. 2453
Rich, multi-disciplinary essays. David Sopher's study of Indian pastoral castes and livestock ecologies is particularly stimulating. The essays range temporally from prehistoric to contemporary and relate to a wide diversity of environments. The groups studied include not only various pastoral nomadic societies, but also semi-nomadic transhumant groups, sedentary pastoralists and non-pastoral nomads.

Lodrick, Deryck O. Sacred cows, sacred places: origins and survivals of animal homes in India. Berkeley, CA: University of California Press, 1981. 307 p. ISBN 0-520-04109-7. $29.50. LC 80-51241. 2454
This fascinating study examines the origins and contemporary importance of "goshalas" (homes for useless cattle) and "pinjrapoles" (animal shelters/hospitals) that are found over most of India, but especially in Gujarat and Rajasthan, where the influence of Jainism has been strongest. The economic and ecological implications of the institutions are examined and a critical light is cast thereby on the controversial materialist views of the anthropologist, Marvin Harris, in respect to the sanctity of cattle in India.

Maloney, Clarence. Peoples of South Asia. New York: Holt, Rinehart, and Winston, 1974. 584 p. LC 73-17484. 2455•
Though written by an anthropologist, this excellent, well-illustrated text is the closest thing yet available to a systematic overview of the socio-cultural geography of South Asia, treating pre-history, race, language, religion, caste, tribes, village and urban life, social change, and demographics. Six regional chapters supplement the topical treatment.

Marriot, McKim. Caste ranking and community structure in five regions of India and Pakistan. 2nd ed. Poona, India: Deccan College Postgraduate and Research Institute, 1965. 111 p. (1st ed., 1960). Rs 7.50. SA 67-902. 2456
This seminal anthropological study is the first systematic attempt to describe, explain, and interpret regional differences in caste structure. It is based on comparative "transactional analyses" in the field in five localities of India and Pakistan. Regrettably, there is no way to specify over how broad a region each set of Marriott's localized findings is applicable.

Michel, Aloys A. The Indus Rivers, the effects of partition. New Haven, CT: Yale University Press, 1967. 595 p. LC 67-13444. 2457
This is a magisterial and exhaustive examination of the effects of partition on the irrigation-based economy of the Indus drainage basin and of the complicated, but on the whole successful, diplomatic and engineering efforts to mitigate the damage done by the interposition within that region of a new international boundary.

Miller, Barbara D. The Endangered sex, neglect of female children in rural north India. Ithaca, NY: Cornell University Press, 1981. 201 p. ISBN 0-8014-1371-0. $22.50. LC 81-3226. 2458
Miller, an anthropologist, has produced a remarkably geographic study probing the marked demographic imbalance of the sexes in India. Among the study's dozen well-made maps, for example, is one of district-wise residuals from regression of juvenile sex ratio on male-female disparity in labor participation rates. Throughout, the work displays a sensitivity to scale and spatial correlation, supplementing a comparative regional approach.

Misra, Rameshwar Prasad, ed. Million cities of India. New Delhi, India: Vikas Publishing House, 1978. 405 p. ISBN 0-7069-0554. Rs. 80. LC 78-904602.   2459
This collection of essays, mainly by geographers, focuses on nine Indian cities with populations exceeding one million at the 1971 census. It treats each such city in regard to its historical development, population patterns, economic bases, infrastructure, land use pattern, living conditions, and future prospects, and seeks to relate the cities collectively to trends in urbanization. Concluding essays consider global problems of metropolitan planning and administration and the development of an effective urbanization policy.

Misra, Rameshwar Prasad, and Sundaram, K.V. Multi-level planning and integrated rural development in India. New Delhi, India: Heritage Publishers, 1980. 234 p. LC 80-900626.   2460
Reviewing a wide range of experience with multi-level planning for integrated rural development, the authors stress the primary importance of efficient administration at the village, block, and district levels and regard rigorous application of geographic models as a secondary concern.

Misra, Rameshwar Prasad; Sundaram, K.V.; and Prakasa Rao, V.L.S. Regional development planning in India: a new strategy. Delhi, India: Vikas Publishing House, 1974. 398 p. LC 74-902980.   2461
This work explains the bases and mechanics of multilevel regional planning, limns the spatial structure of the Indian economy, and investigates the prospects for removing spatial imbalances therein through a growth pole and growth centre strategy. Detailed field studies of a tribal region, a region of intensive agriculture, and a region of heavy industry are summarized to demonstrate the anticipated impacts of suggested governmental investment in specified growth centres/poles.

Noble, Allen G. and Dutt, Ashok K. India: cultural patterns and processes. Boulder, CO: Westview Press, 1982. 378 p. ISBN 0-86531-237-0. $20. LC 82-70173.   2462
The nineteen essays, almost all by geographers, comprising this volume vary enormously in quality. Collectively they display a pronounced materialist bias in their approach to cultural geography, treating such topics as urban landscapes, rural and urban house types, folk arts and religious architecture. Various aspects of religion, caste, and medical geography also receive notice.

Noble, Allen G. and Dutt, Ashok K., eds. Indian urbanization and planning, vehicles of modernization. New Delhi, India: Tata McGraw-Hill Publishing Company Ltd., 1977. 366 p. Rs. 120.   2463
Eleven of the twenty essays in this volume are by geographers; the remainder are drawn from eight different disciplines. The collection combines retrospective views of urban development from pre-colonial times to the present, empirical analyses of contemporary conditions, and recommendations for more effective urban and regional planning.

Planalp, Jack M. Heat stress and culture in northern India. Washington, DC: U.S. Army Medical Research and Development Command, 1971. 557 p. LC 72-601763.   2464
This penetrating analysis of cultural adaptation to heat stress in the Gangetic Plain lies on the frontiers of geography, anthropology, and medicine. Topics considered include housing, dress, diet, daily and seasonal activity patterns, cooling devices, and the epidemiology and treatment of heat injuries.

Er-Rashid, Haroun. Geography of Bangladesh. Dacca: University Press; Boulder, CO: Westview Press, 1977. 579 p. ISBN 0-89158-356-4. $36.50. LC 78-19679.   2465•
Though dry and encyclopedic, this is a rather comprehensive and, on the whole, authoritative text. It demonstrates that, even in as small a deltaic region as Bangladesh, there is remarkable areal diversity that warrants the scrutiny of a geographer.

Schwartzberg, Joseph E. Occupational structure and level of economic development in India: a regional analysis. (Census of India, 1961, monograph no. 4). New Delhi, India: Office of the Registrar General, India, 1969. 268 p. SA 68-15842.　　　2466
Through a combination of census-based studies and field work in a dozen widely scattered areas of diverse levels and types of economic development, the author explores the strikingly different regional occupational structures within India. It derives an explanatory model based on the interplay of degree of commercialization of agriculture, level of affluence, and degree of socio-economic conservatism (i.e. adherance to the "jajmani" system) that fits most of the regions rather well.

Sen Gupta, P., and Sdasyuk, Galina. Economic regionalization of India: problems and approaches. (Census of India, 1961, monograph series v. 1, no. 8). Delhi, India: The Manager of Publications, 1968. 257 p. SA 68-15660.　　　2467
Part one of this monograph, mainly by Sdasyuk, sets forth "concepts, principles and methods of regionalization," with particular reference to India. Part two, by both authors, considers economic regionalization by specific components: natural resources; agriculture; industry; and transport. Part three, by Sen Gupta, presents comprehensive economic and planning regionalizations of India at the macro- and meso-scales.

Simoons, Frederick J. A Ceremonial ox of India: the mithan in nature, culture, and history - with notes on the domestication of common cattle. Madison, WI: University of Wisconsin Press, 1968. 323 p. ISBN 0-299-04980-9. $37.50. LC 68-9023.　　　2468
Squarely in the Sauerian tradition, this study considers the mithan (Bos Frontalis) culture complex of tribal regions of Bhutan, northeastern India, Bangladesh, and Burma. Though embodying little first-hand field work, it admirably synthesizes and evaluates the findings of scores of anthropologists, administrators, and other observers of the little known semi-feral ox that figures so predominantly in the ceremonial life of the areas involved.

Singh, R.L., ed. India, a regional geography. Varanasi, India: National Geographical Society of India, 1971. 992 p. LC 78-92087.　　　2469 •
A collaborative effort by several dozen authors from all major regions of India, this massive, rather encyclopedic, text inevitably displays considerable unevenness in depth and quality of exposition. Nevertheless, it contains a wealth of regional detail that is missing in the regional half of the Spate-Learmonth text, especially in regard to settlement patterns and land use, and, less consistently, for social geography.

Sopher, David E., ed. An Exploration of India: geographical perspectives on society and culture. Ithaca, NY: Cornell University Press, 1980. 334 p. ISBN 0-8014-1258-7. $27.50. LC 79-17033.　　　2470
Essays on one of the world's most complex and difficult culture areas by a competent team of cultural geographers.

Spate, O.H.K. (Oskar Hermann Khristian), and Learmonth, Andrew T.A. India and Pakistan: a general and regional geography. 3rd ed. London: Methuen, 1967. 877 p. (1st ed.: 1954, 827 p.). LC 68-86324.　　　2471
Although largely dated, especially in respect to economic matters, this vividly written text ranks among the classics of regional geography. Its descriptions of South Asia's physical environment still merit study. Its principal weakness is its insufficient attention to the region's complex social geography, especially in regard to caste in India.

The State of India's environment, a citizens report. 1982- . Annual. New Delhi, India: Centre for Science and Environment, 1982. 192 p. The 1982 report edited by Anil Agarwal, Ravi Chopra, and Kalpana Sharma. Available from The Centre for Science and Environment, 807 Vishal Bhaval, 95 Nehru Place, New Delhi, India, 110019. Rs 125. (Reference: Library of Congress. Accessions list of South Asia, vol. 3, no. 4, April 1983, p. 264). LC 82-913698.　　　2472 •

This landmark citizens' initiative, probably the first of its kind in South Asia, reports on the state of all major aspects of the Indian environment. Though presented in a popular style, the report generally strives for scientific objectivity. It pinpoints specific problems, seeks to quantify the nature and pace of environmental alteration, and suggests courses of constructive action. Future annual reports are anticipated.
Sections include: land; water; forests; dams; atmosphere; habitat; people; health; energy; wildlife; and government.

Sundaram, K.V. Urban and regional planning in India. 2nd ed. New Delhi, India: Vikas Publishing House, 1979. 432 p. (1st ed., 1977, 432 p.). ISBN 0-7069-0536-9. Rs. 125. LC 77-907131. 2473

Sundaram argues for a pragmatic planning policy, with due regard to what is economically and administratively feasible, in place of the elaboration of theoretically optimal models that cannot be followed in real-world situations.

Wanmali, Sudhir. Periodic markets and rural development in India. Delhi, India: B.R. Publishing Co., 1981. 236 p. Rs 90. LC 81-902406. 2474

The only detailed investigations of a complete local system of periodic markets yet made in India. It views markets from the perspectives of both consumers and traders and relates its findings to central place theory, to empirical studies from other areas, and to development theory and practice. Regrettably, the study area is a tribal region of South Bihar rather than a more representative locale of Hindu peasantry.

12. SOUTHWEST ASIA AND NORTH AFRICA

Ian R. Manners

A. BIBLIOGRAPHIES

Atiyeh, George N. The Contemporary Middle East: 1948-1973: a selective and annotated bibliography. Boston, MA: G.K. Hall, 1975. 664 p. ISBN 0-8161-1085-7. $57.50. LC 74-19247. 2475

Basically a social science bibliography. Annotated listing of significant publications.

Center for the Study of the Modern Arab World (CEMAM). Arab culture and society in change: a partially annotated bibliography of books and articles in English, French, German, and Italian. Beirut, Lebanon: Dar el-Mashreq Publishers, 1973. 318 p. L£ 65. LC 73-960590. 2476

An extensive bibliography covering a wide range of social issues, conditions of women, new political ideologies, and modern political codes.

Grimwood-Jones, Diana, and Hopwood, Derek, eds. Middle East and Islam: a bibliographical introduction. Rev. ed. Zug, Switzerland: Inter-Documentation Co., 1979. 429 p. (1st ed., 1972. 368 p.). ISBN 3-85750-032-8. SFr 75. SFr 18. (microfiche). LC 72-85349. 2477

Bibliographical listings arranged according to subject, reference, history, Islamic studies, region, language, and literature.

Grimwood-Jones, Diana; Hopwood, Derek; and Pearson, J.D. (James), eds. Arab Islamic bibliography: the Middle East Library Committee guide. Hassocks, U.K.: Harvester Press; Atlantic Highlands, NJ: Humanities Press, 1977. 292 p. ISBN 0-391-00691-1. LC 76-51397. 2478

Bibliographic listings of Middle East literature.

Littlefield, David W. The Islamic Near East and North Africa: an annotated guide to books in English for non-specialists. Littleton, CO: Libraries Unlimited, 1977. 375 p. ISBN 0-87287-159-2. LC 76-218. 2479 •

Annotated listings of general works, historical, political, social, and economic studies, as well as travel and general guidebooks on a country-by-country basis.

Southwest Asia and North Africa

Quarterly index Islamicus; a catalogue of articles on Islamic subjects in periodicals and other collective publications. 1- (1977-). Quarterly. Mansell Publishing Co., 6 All Saints St., London N1 9RL, England. ISSN 0308-7395.   2480
   Formerly called Index Islamicus *(1906-1955), this periodical is published quarterly and cumulated at 5-year intervals. Contains bibliographic listings from periodical literature on Islamic areas.*

B. SERIALS

International journal of Middle East studies. (Middle East Studies Association). 1- (1970-). Quarterly. Cambridge University Press, 32 East 57th St., New York, NY, 10022. ISSN 0020-7438.   2481
   Articles concerning the Middle East from the seventh century to modern times, with particular attention given to history, anthropology, economics, sociology, law, and philosophy.

Israel exploration journal. 1- (1950/51-). Quarterly. Israel Exploration Society, Box 7041, Jerusalem, Israel. ISSN 0021-2059.   2482
   Publishes many articles by Israeli geographers. Text in English and French.

The Middle East. 1- (1974-). Monthly. IC Publications, Ltd. 122 East 42nd St. Rm. 1121, New York, NY, 10168. ISSN 0305-0734.   2483
   In-depth political and regional articles pertaining to Middle Eastern countries.

The Middle East and North Africa. 1948- . Annual. Europa Publications, Ltd. 18 Bedford Sq., London WC1B 3JN, England. ISSN 0076-8502.   2484
   A basic reference. Contains general information on Middle Eastern and North African geography, history, economy, and education. Also contains a who's who in the Middle East, bibliographies, a listing of research institutes and statistical summaries.

Middle East economic digest. 1- (1957-). Weekly. MEED, Ltd., 551 5th Ave., New York, NY, 10176. ISSN 0047-7230.   2485
   News, analysis, and forecasts concerning current economic issues in Middle Eastern countries.

Middle East journal. 1- (1947-). Quarterly. Middle East Institute, 1761 N Street N.W., Washington, DC, 20006. ISSN 0026-3141.   2486
   The leading American periodical dealing with political, social, and economic issues in the region extending from Morocco to Pakistan. Its "chronology" and "review of the periodical literature" are especially valuable.

Middle East review. 1- (1968-). Quarterly. American Academic Association for Peace in the Middle East, 330 7th Ave., New York, NY, 10001. ISSN 0097-9791.   2487
   Articles on economic, political, and social aspects of the Middle East.

Revue de géographie du Maroc. 1-22 (1962-1972); n.s. 1- (1977-). Annual. Société de Géographie du Maroc, Faculté des Lettres, Université de Rabat, Rabat, Maroc. ISSN 0035-1156.   2488
   Articles on Morocco, in French or Arabic. Abstracts in English, French, and Arabic.

Revue tunisienne de géographie. 1- (1978-). Semi-annual. Faculté des Lettres et des Sciences Humaines de Tunis. Boulevard du 9 avril 1938, B.P. 1128, Tunis, Tunisia.   2489
   Articles on Tunisia. In French. Abstracts in English, French, and Arabic.

Société de Géographie d'Egypte. Bulletin. 1- (1875-). Annual. Société de Géographie d'Egypte, Sharia Kasr El-Aini, Cairo, Egypt.   2490
   Topical and regional geography of Egypt and the Middle East. Majority of articles are in English.

C. ATLASES

Amiran, David H.K.; Shachar, Arie; and Kimhi, Israel, eds. Atlas of Jerusalem. Berlin; New York: W. De Gruyter, 1973. 173 p. ISBN 3-11-003623-1. LC 72-650051.   2491
A detailed urban atlas covering physical geography, history, land use, demography, and economy.

Bindaqji, Husayn Hamzah. Atlas of Saudi Arabia. Oxford, U.K.: Oxford University Press, 1978. 61 p. ISBN 0-19-919101-8. LC 79-318291.   2492
Particularly detailed maps of the physical and human geography of Saudi Arabia.

Brice, William C., ed. An Historical atlas of Islam. Leiden, The Netherlands: E.J. Brill, 1981. 71 p. ISBN 9-00-406116-9. Dutch Guilders 120. LC 81-212100.   2493 •
An informative collection of both general and regional maps which provides a graphic view of the Islamic world from the rise of Islam to the onset of World War I.

Comité de Géographie du Maroc. Atlas du Maroc. Rabat, Morocco: Comité de Géographie du Maroc, 1954- . LC Map 56-593.   2494
A valuable serial publication which includes descriptive pamphlets as well as maps.

Djambatan Uitgeversbedrijf, N.W. Amsterdam. Atlas of the Arab world and the Middle East. London: Macmillan; New York: St. Martin's Press, 1960. 17 p. LC Map 60-313.   2495
Contains maps of physical and cultural characteristics as well as photographs and accompanying text.

German Research Council, and The University of Tübingen. Der Tübinger atlas des Vorderen Orients (TAVO). [The Tübingen atlas of the Middle East (TAVO)]. Wiesbaden, West Germany: Dr. Ludwig Reichert Verlag, 1977-. ISBN 3-88226-610-4. LC 79-689722.   2496
Perhaps the single most important academic project ever undertaken in Middle Eastern studies, the Tübingen atlas will eventually comprise a total of 350 maps on 300 sheets. Published in the form of large unbound sheets, the project is scheduled to be completed by the end of the 1980's. Detailed and highly informative maps on physical and human geography and history. Acccompanying scholarly monographs.

Israel, Survey Department. Atlas of Israel. 2nd ed. Jerusalem, Israel: Survey of Israel, Ministry of Labor; Amsterdam, The Netherlands: Elsevier Scientific Publishing Co., 1970. Various pagings. 296 p. (1st ed.: Israel. Mahlekat ha-medidot. Jerusalem: The Jewish Agency, 1956-1964. 296 p. In Hebrew). ISBN 0-444-40740-5. Dutch Guilders 650. LC 76-653754.   2497
Thorough coverage of physical geography, history, demography, economics and education.

Martin, Gilbert. Atlas of the Arab-Israeli conflict. New York: Macmillan, 1975. 103 p. ISBN 0-02-543370-9. LC 74-13917.   2498
A collection of maps depicting the Arab-Israeli conflict up to 1975.

Robinson, Francis. Atlas of the Islamic world since 1500. Oxford, U.K.: Phaidon; New York: Facts on File, 1982. 238 p. ISBN 0-87196-629-8. $35. LC 82-675002.   2499 •
A collection of maps tracing the history of Islam together with interpretive essays on religion and culture.

U.S. Central Intelligence Agency. Issues in the Middle East; atlas. Washington DC: Government Printing Office, 1973. 40 p. LC 74-187210.   2500 •
This atlas focuses on basic geographical, social, and economic issues in the Middle East through the use of maps, charts, photographs, and brief interpretive text.

Southwest Asia and North Africa

Vilnay, Zev. The New Israel atlas. Bible to present day. Translated from Hebrew by Moshe Aumann. Jerusalem, Israel: Israel Universities Press; London: H.A. Humphrey; New York: McGraw-Hill, 1968. 112 p. ISBN 0-250-80744-0. LC 79-653326. (Atlas tav shin kaf het).
2501 •
A valuable series of maps and explanatory texts covering the history and geography of Israel. Includes a gazetteer of settlements with foundation dates.

D. GENERAL WORKS

American University. Foreign Area Studies Division. Country Studies series. Washington, DC: Government Printing Office.
   Afghanistan. Smith, Harvey H., ed. 4th ed. 1973. (Reprinted 1978). 453 p. $13. LC 73-600084.
   Algeria. Nelson, Harold P., ed. 3rd ed. 1979. 370 p. $11. LC 79-13466.
   Egypt. Nyrop, Richard F., ed. 4th ed. 1983. 362 p. $8. LC 83-600110.
   Iran. Nyrop, Richard F., ed. 3rd ed. 1978. 492 p. $13. LC 78-11871.
   Israel. Nyrop, Richard F., ed. 2nd ed. 1979. 414 p. $12. LC 79-13733.
   Jordan. Nyrop, Richard F., ed. 3rd ed. 1980. 310 p. $11. LC 80-607127.
   Libya. Nelson, Harold P., ed. 3rd ed. 1979. 350 p. $11. LC 79-24183.
   Morocco. Nelson, Harold P., ed. 4th ed. 1978. 410 p. $12. LC 78-11528.
   Persian Gulf. Nyrop. Richard F., ed. 1977. 448 p. $12. LC 77-23854.
   Saudi Arabia. Nyrop, Richard F., ed. Rev. ed. 1982. 404 p. $13.
      LC 76-51268.
   Syria. Nyrop, Richard F., ed. 3rd ed. 1979. 268 p. $11. LC 79-607771.
   Tunisia. Reese, Howard C., ed. Rev. ed. 1979. 326 p. $11. LC 79-15707.
   Turkey. Nyrop, Richard F., ed. 3rd ed. 1980. 370 p. $12. LC 80-607042.
   Yemens. Nyrop, Richard F., ed. 1977. 266 p. $10. LC 77-608104.
2502 •
Written by teams of social scientists, these area handbooks examine economic, military, political, and geographic conditions of Middle Eastern countries.

Beaumont, Peter; Blake, Gerald H.; and Wagstaff, J. Malcolm. The Middle East: a geographical study. London; New York: John Wiley, 1976. 572 p. ISBN 0-471-06117-4. $59.95. ISBN 0-471-06119-0. $31.95. (pbk). LC 74-28284.
2503
Comprehensive geographical study of the region. Subject is approached both systematically and through thematic essays that focus on the potential for and the problems of development in individual countries.

Berger, Morroe. The Arab world today. Garden City, NY: Doubleday, 1962. Reprint: Octagon Press, 1980. 480 p. ISBN 0-374-90602-5. $34.50. LC 62-7601.
2504
A perceptive if rather dated sociological analysis of the Arab world prior to 1960 with examples drawn from Egypt, Syria, Lebanon, Jordan, and Iraq.

Beydoun, Ziad Rafiq, and Dunnington, H.V. The Petroleum geology and resources of the Middle East. Beaconsfield, U.K.: Scientific Press, 1975. 99 p. ISBN 0-901360-08-2. £20. LC 76-352984.
2505
A detailed summary of Middle East geology with discussion of the problems involved in accurately estimating oil reserves.

Birot, Pierre, and Gabert, Pierre. La Méditerranée et le Moyen-Orient. (ORBIS: introduction aux études de géographie). 2nd rev. ed. Paris: Presses Universitaires de France, v. 1. Généralités, Péninsule Ibérique Italie, 1964. 551 p. LC 74-208667. Only volume published (?). For North Africa and Southwest Asia see 1st ed. by Pierre Birot and Jean Dresch, 1953-1956. 2 v. LC 54-7204 rev.:
   Vol. 1: **La Mediterranée Occidental: Péninsule Ibérique, Italie, Afrique du Nord,** 1953. 549 p.
   Vol. 2: **La Mediterranée Oriental er le Moyen-Orient: les Balkans, lÁsie Mineure, le Moyen-Orient,** 1956. 526 p.
2506
An exhaustive presentation of the physical and human geography of the area in the French tradition. Special emphasis is placed on general discussions of the region's geology and climate, and the "human problems" encountered there. A fine geographic reference work.

Blake, Gerald H. and Lawless, Richard I., eds. The Changing Middle Eastern city. London: Croom Helm; New York: Barnes and Noble, 1980. 273 p. ISBN 0-85664-576-1. $28.50. LC 80-479247.
2507
Valuable collection of essays dealing with the changing structure and morphology of Middle Eastern cities and the problems resulting from rapid, unplanned growth.

Brice, William C., ed. The Environmental history of the Near and Middle East since the last Ice Age. London; New York: Academic Press, 1978. 384 p. ISBN 0-12-133850-9. $60. LC 77-81377.
2508
Detailed case studies by specialists intended to assist in interpreting the nature of past environments in the Middle East.

Brown, L. Carl, ed. From Madina to Metropolis: heritage and change in the Near Eastern city. Princeton, NJ: The Darwin Press, 1973. 343 p. ISBN 0-87850-006-5. $24.95. ISBN 0-87850-007-3. $17.95. (pbk). LC 76-161054.
2509
A collection of essays devoted to urban planning and prospects in the Near East and North Africa.

Clarke, John I. and Fisher, William B., eds. Populations of the Middle East and North Africa: a geographical approach. London: University of London Press; New York: Holmes and Meier, 1972. 432 p. ISBN 0-8419-0125-2. $44.50. LC 72-80410.
2510
A study of demographic patterns and population trends of the Middle East. Individual country chapters deal with these general themes as well as particular issues such as religious pluralism, population-resource relationships, and migration.

Clarke, John I. and Bowen-Jones, Howard, eds. Change and development in the Middle East. London; New York: Methuen, 1981. 322 p. ISBN 0-416-71080-8. $45. LC 80-41649.
2511
Essays pertaining to economic, social, and political conditions in the Near East.

Clawson, Marion; Landsberg, Hans H.; and Alexander, Lyle T. The Agricultural potential of the Middle East. New York: American Elsevier Publishing Co., 1971. 312 p. LC 79-135058.
2512
Detailed assessment of the region's agricultural potential. Coverage of the resource base (soils, water) as well as population growth, labor markets, capital requirements, and institutional and infrastructure needs.

Costello, Vincent F. Urbanization in the Middle East. Cambridge, U.K.; New York: Cambridge University Press, 1977. 121 p. ISBN 0-521-21324-X. $22.95. ISBN 0-521-29110-0. $9.95. (pbk). LC 76-11075.
2513
An interdisciplinary study of urban growth in the Middle East. Discussions range from the pre-industrial city to contemporary rural-urban migration and social change.

Fisher, William B. The Middle East: a physical, social, and regional geography. 7th ed. London; New York: Methuen, 1978. 615 p. (1st ed., 1950. 514 p.). ISBN 0-416-71510-9. $25. LC 79-301895.
2514•
Textbook providing systematic coverage of physical and human geography as well as detailed regional chapters that relate past and present human activities to the physical environment.

Gischler, Christiaan E. Water resources in the Arab Middle East and North Africa. Cambridge, U.K.: Middle East and North African Studies Press, 1979. 132 p. ISBN 0-906559-00-6. £11.5. ISBN 0-906559-01-4. £7.50. LC 80-48486.
2515
Technical assessment of the water resources available to Arab world countries and associated management problems.

Southwest Asia and North Africa

Ibn Battutah. The Travels of Ibn Battutah. Translated from Arabic by Samuel Lee. Cambridge U.K.: Oriental Translation Series, 1829. Reprinted: New York: Ben Franklin, 1971. 243 p. ISBN 0-8337-2051. $18.50. LC 74-172523. (Tuhfat al-nuzzār). 2516
A translation of the travels of Ibn Battutah in Africa and Asia, providing a valuable account of Islamic society in the fourteenth century.

Ibn Khaldun. The Muqqadimah: an introduction to history. Translated from Arabic by Franz Rosenthal. 2nd ed. Princeton, NJ: Princeton University Press, 1967. 3 v. (1st ed., New York: Pantheon, 1958. 3 v.). LC 74-186373. (Kitab al-ibar al-Muqqaddimah). 2517
One of the great works of fourteenth century Islamic scholarship. Intended as an introduction to the study of history, with particular reference to the social, cultural, and economic context within which human social organizations have developed. Discusses the influence of the physical environment on human activities and the origin and forms of social organization with particular attention to desert and urban ways of life.

Issawi, Charles P., ed. The Economic history of the Middle East, 1800-1914: a book of readings. Chicago, IL: University of Chicago Press, 1966, Reprinted 1976. 543 p. ISBN 0-226-38609-0. $19. LC 66-11883. 2518
An excellent symposium of papers on Middle Eastern economic history, many appearing in English for the first time. Deals with Turkey, the Arab East, and the Nile Valley.

Issawi, Charles P. An Economic history of the Middle East and North Africa. New York: Columbia University Press, 1982. 304 p. ISBN 0-231-03443-1. $28. LC 81-19518. 2519 •
An excellent study of the patterns and trends of development in the Middle East and North Africa over the last two hundred years.

Johnson, Douglas L. The Nature of nomadism: a comparative study of pastoral migrations in Southwestern Asia and Northern Africa. (Research paper no. 118). Chicago, IL: Department of Geography, University of Chicago, 1969. 200 p. ISBN 0-89065-025-X. $10. LC 69-18022. 2520
A comparative analysis of nomadic migration patterns in the Middle East and North Africa, with particular emphasis on the ecological basis of nomadism.

Kraeling, Carl H. and Adams, Robert M., eds. City invincible: a symposium on urbanization and cultural development in the ancient Near East. Chicago, IL: University of Chicago Press, 1960. 447 p. LC 60-13791. 2521
Valuable papers on the early culture history of the Near East and the Mediterranean region.

Lapides, Ira, ed. Middle Eastern cities: a symposium on ancient, Islamic, and contemporary Middle Eastern urbanism. Berkeley, CA: University of California Press, 1969. 206 p. LC 72-81939. 2522
A thoughtful discussion of research on Middle Eastern cities, past and present, with contributions by social scientists and historians.

Miquel, André. La géographie humaine du monde Musulman jusqu'au milieu du 11e siècle. 1. Géographie et géographie humaine dans la litterature arabe des origines à 1050. Paris; The Hague, The Netherlands: Mouton, 1973. 427 p. ISBN 2-7132-0044-X. ISBN 3538386. FFr 198. LC 68-76953. 2523
Scholarly treatment of Arab geographic thinking during the early years of Islam.

Planhol, Xavier de. The World of Islam; La monde Islamique: essai de géographie religieuse. Ithaca, NY: Cornell University Press, 1959. 142 p. LC 59-16313. 2524
Excellent, but controversial essay. Explores the morphology of religious landscapes and the geographical significance of religious institutions; confined almost entirely to the Middle Eastern Islamic world.

Planhol, Xavier de. Les Fondements géographiques de l'histoire de l'Islam. Paris: Flammarion, 1968. 443 p. ISBN 2215408. FFr 69.LC 68-108964. 2525
 Expansion and refinement of the theme of the author's World of Islam *(see above entry) exploring the morphology of religious landscapes and the geographical significance of religious institutions in the Moslem Middle East.*

Royal Institute of International Affairs, Information Department. Mansfield, Peter, ed. The Middle East: a political and economic survey. Oxford, U.K.; New York: Oxford University Press, 1980. 579 p. ISBN 0-19-215851-7. $35. LC 79-23699. 2526
 A useful encylopedic reference work. Usually revised every few years.

Serjeant, Robert Bertram. The Islamic city. Paris: UNESCO, 1980. 210 p. ISBN 9-231-01665-2. FFr 36. LC 81-106664. 2527
 A collection of 12 papers presented at a UNESCO colloquium dealing with the problems of preserving the architectural heritage of Islam.

E. SOUTHWEST ASIA

Adams, Robert M. Land behind Baghdad: a history of settlement on the Diyala Plains. Chicago, IL: University of Chicago Press, 1965. 187 p. ISBN 0-226-00425-2. $22. LC 65-17279. 2528
 The full sequence of broadly changing patterns of irrigation, agriculture, and urban settlement is traced and analyzed within the framework of a small but historically crucial region of the Near East.

Aharoni, Yohanan. The Land of the Bible: a historical geography. Translated from Hebrew by A.F. Rainey. 2nd rev. ed. Philadelphia, PA: Westminster, 1979. 481 p. (1st ed. London: Burns and Oates, 1967. 409 p.). ISBN 0-664-24266-9. $19.95. (pbk). LC 80-14168. (Erets-Yisrael bi-tekufat ha-Mikra.) 2529
 Erudite, well-documented historical geography of Palestine up to the end of the Judean Kingdom. Includes a table giving ancient place names and their modern Arab and Hebrew nams.

Amiran, David H.K. and Ben-Arieh, Y., eds. Geography in Israel. Jerusalem, Israel: Israel National Committee, International Gegraphical Union, 1976. 412 p. (Reference: Kiryat Sefer, a bibliograhical quarterly of the Jewish National and Hebrew University Library, vol. 52, no. 2. April, 1977. Entry 1992). 2530
 Papers presented at the 23rd International Geographical Congress in 1976 summarizing geographical research in Israel. Studies of geomorphological problems as well as of socioeconomic conditions.

Benedict, Peter; Tümertekin, Erol; and Mansur, Fatma, eds. Turkey: geographic and social perspectives. Leiden, The Netherlands: E.J. Brill, 1974. 446 p. ISBN 9-00-403889-2. Dutch guilders 148. LC 75-302611. 2531
 Collection of essays dealing with aspects of rural and urban change in Turkey. Themes include sedentarization, land tenure, rural economic development, and urbanization.

Cole, Donald Powell. Nomads of the nomads: the Al Murrah Bedouin of the Empty Quarter. Chicago, IL: Aldine Publishing Co., (now Arlington Heights, IL: Harlan Davidson), 1975. ISBN 0-88295-605-1. $8.95. (pbk). LC 74-18211. 2532•
 A study of the symbiotic relationship between Al Murrah and their environment in the Rub Al Khali of the Arabian Desert.

Donkin, R.A. Manna: an historical geography. (Biogeographica, v. 17). The Hague, The Netherlands: W. Junk; Hingham, MA: Kluwer Academic, 1980. 161 p. ISBN 90-6193-218-1. $47.50. LC 79-24368. 2533
 Comprehensive study of the conditions under which different species of trees and shrubs growing in the arid lands of Turkestan, Iran, and Afganistan produce manna and of its historical use as a food and for its therapeutic effects.

Southwest Asia and North Africa

El Mallakh, Ragae, and El Mallakh, Dorothea H., eds. Saudi Arabia: energy, developmental planning, and industrialization. Lexington, MA: Lexington Books, 1982. 204 p. ISBN 0-669-04801-1. $25. LC 81-47746.　　　　　　　　　　　　　　　　2534
　　A series of papers on the economic development of Saudi Arabia, including a review of the Third Development Plan (1980-1985), its objectives and ramifications.

English, Paul W. City and village in Iran: settlement and economy in the Kirman Basin. Madison, WI: University of Wisconsin Press, 1966. 204 p. LC 66-22856.　　　　2535
　　One of the best studies of the Middle East by a geographer. Covers the physical environment, pattern and morphology of settlement, socio-economic structure, urban dominance, and regional economy of an area in central Iran. Bibliography.

Fernea, Elizabeth. Guests of the Sheikh. Garden City, NY: Doubleday; London: Hale, 1965. 346 p. ISBN 0-385-01485-6. $5.95. (pbk). LC 65-13098.　　　　　　　2536
　　An ethnography of an Iraqi village, with emphasis on the role of women and social life and customs.

Fisher, William B., ed. The Land of Iran. (The Cambridge History of Iran, v. 1). London; New York: Cambridge University Press, 1968. 784 p. ISBN 0-521-06935-1. $74.50. LC 67-12845.　　　　　　　　　　　　　　　　　　　　　　　　　　　　　2537
　　A collection of essays, some highly specialized, on the physical geography, people, and economy of Iran.

Golany, Gideon, ed. Arid zone settlement planning: the Israeli experience. Oxford, U.K.; New York: Pergamon Press, 1979. 567 p. ISBN 0-08-023378-3. $63. LC 78-26517.　　2538
　　A collection of papers by architects, planners, and social scientists that focus on the problems of and approaches to arid land settlement in Israel.

Hooglund, Eric J. Land and revolution in Iran, 1960-1980. Austin, TX: University of Texas Press, 1982. 191 p. ISBN 0-292-74633-4. $19.95. LC 81-21959.　　　　　　2539
　　A study of land reform, politics and government in Iran.

Humlum, Johannes. La géographie de l'Afghanistan: étude d'un pays aride. Copenhagen, Denmark: Gyldendal, 1959. 421 p. LC 60-25143.　　　　　　　　　　　　　2540
　　The book is divided into two sections: general overview and topical geography. Considerable detail with excellent use of photographs, maps, drawings, and tables.

Karmon, Yehuda. Israel: a regional geography. London; New York: Wiley-Interscience, 1971. 345 p. ISBN 0-471-45870-8. LC 70-116162.　　　　　　　　　　　2541
　　A regional geography of Israel written in the classical tradition. Includes a systematic treatment of the physical and human geography of Israel as well as detailed regional descriptions.

Kolars, John F. Tradition, season, and change in a Turkish village. (Research paper no. 82). Chicago, IL: Department of Geography, University of Chicago, 1963. 205 p. LC 63-17961.　　　　　　　　　　　　　　　　　　　　　　　　　　　　　2542
　　Geographical study of fifteen settlements in the Anatalya region of the southern coast of Turkey.

Lambton, Ann Katherine. The Persian land reform, 1962-1966. Oxford, U.K.: Clarendon Press, 1969. 386 p. ISBN 0-19-828163-3. LC 79-441517.　　　　　　　　2543
　　A comprehensive review of the 1962 land reform law, including the changing nature of the reform program and its initial implementation.

Makki, M.S. Medina, Saudi Arabia: a geographic analysis of the city and region. Amersham, U.K.: Avebury Publishing, 1982. 231 p. ISBN 0-86127-301-X. £24. ISBN 0-86127-304-4. £16. (pbk). LC 81-195397.　　　　　　　　　　　　　　　　2544
　　A useful study of Medina with discussion of the city's physical setting and historic evolution.

Marx, Emanuel. Bedouin of the Negev. Manchester, U.K.: Manchester University Press; New York: Praeger, 1967. 260 p. LC 67-19214.   2545
   The ecology, administrative order, land ownership, movement cycles, and political organization of Bedouin in the Negev.

Orni, Ephraim, and Efrat, Elisha. Geography of Israel. 4th ed. Jerusalem, Israel: Israeli Universities Press, 1980. 556 p. (1st ed. 1964). ISBN 0-7065-1124-7. $12.50. (From Keter Publishing Co., Jerusalem). LC 71-178048. (ha-Geografyah shel artsenu.) .   2546•
   The first comprehensive study of the geography of Israel in the English language. Numerous maps and photographs. Extensive bibliography.

Salmanzadeh, Cyrus. Rural society in southern Iran. Cambridge, U.K.: MENA Press, 1980. 275 p. ISBN 0-906559-02-2. £15. ISBN 0-906559-03-0. £9.50. (pbk).   2547
   An impressive study of land reform and its aftermath in Khuzestan. Provides a comprehensive treatment of pre-land reform conditions as well as the emergence of agribusiness.

Shahrani, M. Nazif Mohib. The Kirghiz and Wakhi of Afghanistan: adaptation to closed frontiers. Seattle, WA: University of Washington Press, 1979. 264 p. ISBN 0-295-95669-0. $22.50. LC 79-11665.   2548
   Ethnographic treatment of a population adapted to life at high altitudes. Particular concern for the way in which the closing of frontiers has affected interrelationships between societies, habitat, and the central government.

Smith, George A. The Historical geography of the Holy Land. 25th ed. Reprinted New York: Harper and Row; London: Collins; Magnolia, MA: Peter Smith, 1966. 512 p. (1st ed.: London: Hodder and Stoughton, 1894. 692 p.; 25th ed., 1931). ISBN 0-8446-2956-1. $12. LC 66-19780.   2549
   A classic description of Palestine and the physical conditioning of its cultural and religious history.

Thalen, D.C.P. Ecology and utilization of desert shrub rangelands in Iraq. The Hague, The Netherlands: W. Junk, 1979. 448 p. ISBN 9-06-193593-8. Dutch Guilders 150. LC 80-450004.   2550
   Discussion of the impact of human activities and of livestock grazing on the ecology of the Iraqi plains. Includes recommendations for improved range management.

Thesiger, Wilfred. Arabian sands. London: Longmans, Green; New York: Dutton, 1959. 326 p. Reprinted New York: Viking/Penguin, 1984. 347 p. ISBN 0-670-13005-2. $4.95. (pbk). LC 83-50119.   2551•
   A good descriptive account of exploration of eastern Arabian deserts during 1945-1950.

Thesiger, Wilfred. The Marsh Arabs. London: Longmans; New York: Dutton; Harmondsworth, U.K.: Penguin, 1964. 242 p. ISBN 0-14-002573-1. £2.25. (pbk). LC 64-6741.   2552
   A thorough account of the habitat and activities of these inhabitants of southern Iraq.

Weulersse, Jacques. Paysans de Syrie et du Proche-Orient. Paris: Gallimard, 1946. 329 p. LC 47-2893.   2553
   Penetrating and original analysis of the economic and social geography of the sedentary agricultural communities of the Near East.

Wilkinson, John C. Water and tribal settlement in South-East Arabia: a study of the Aflaj of Oman. Oxford, U.K.: Clarendon Press, 1977. 276 p. ISBN 0-19-823217-9. £18.50. LC 77-7347.   2554
   Perceptive account of settlement history, water management, and social organization in the interior of Oman that provides a valuable baseline study of the pattern and process of change in a traditional society.

Southwest Asia and North Africa

Abu-Lughod, Janet. Cairo: 1001 years of the city victorious. Princeton, NJ: Princeton University Press, 1971. 284 p. ISBN 0-691-03085-5. $55. LC 73-112992.   2555
   An outstanding study of Cairo which describes the spatial and sociological charac-teristics of the city as these have evolved during the past millennium.

Abu-Lughod, Janet. Rabat: urban apartheid in Morocco. Princeton, NJ: Princeton University Press, 1980. 374 p. ISBN 0-691-05315-4. $40. ISBN 0-691-10098-5. $14.50. (pbk). LC 80-7508.   2556
   An incisive study of Rabat and of the impact of French colonial policies on the mor-phology and social structure of the city.

Briggs, Lloyd C. Tribes of the Sahara. Cambridge, MA: Harvard University Press, 1960. 295 p. (Available from Books on Demand, Ann Arbor, MI. $82). LC 60-7988.   2557
   Probably the best book in English on the peoples of the Sahara. In addition to ac-counts of the individual tribes of the Sahara, the text includes excellent discussions of the historical and ecological background of the area.

Butzer, Karl W. and Hansen, Carl L. Desert and river in Nubia: geomorphology and prehistoric environments at the Aswan Reservoir. Madison, WI: University of Wiscon-sin Press, 1968. 562 p. ISBN 0-299-04770-9. $45. LC 67-20761.   2558
   An impressive reconstruction of the physical history of the middle Nile Valley and the sequence of changing environments in which early societies lived.

Butzer, Karl W. Early hydraulic civilization in Egypt: a study in cultural ecology. Chi-cago; London: University of Chicago Press, 1976. 134 p. ISBN 0-226-08634-8. LC 75-36398.   2559
   Scholarly analysis of the complex relationship of irrigation technology and land use to environmental, demographic, and social change traced through five thousand years of riverine history.

Chiapuris, John. The Ait Ayash of the high Moulouya Plain: rural social organization in Morocco. (University of Michigan, Museum of Anthropology, Anthropological pa-pers no. 69). Ann Arbor, MI: University of Michigan Press, 1979. 186 p. ISBN 0-932206-83-2. $6. (pbk). LC 80-623508.   2560
   A study of the changing pattern of control over resources and of social organization among a Berber-speaking community of agriculturists farming in the high Moulouya Plain.

Despois, Jean, and Raynal, René. Géographie de l'Afrique du Nord-Ouest. Paris: Payot, 1967. 571 p. LC 67-88736.   2561
   A detailed regional geography of Morocco, Algeria, Tunisia, and the Sahara. Ex-haustive bibliography of French studies.

Johnson, Douglas L. Jabal al-Akhdar, Cyrenaica: an historical geography of settle-ment and livelihood. (Research paper no. 148). Chicago, IL: University of Chicago, De-partment of Geography, 1973. 240 p. ISBN 0-89065-055-1. $10. (pbk). LC 73-79883.   2562
   A valuable study of contemporary patterns of resource use and settlement among nomadic groups in the eastern Jabal al-Akhdar based on field investigation.

Mikesell, Marvin W. Northern Morocco: a cultural geography. (Publications in geog-raphy, v. 14). Berkeley, CA: University of California Press, 1961. 135 p. ISSN 0068-6441. $3. (pbk). LC 62-62686.   2563
   A survey of physical geography and culture history, followed by more detailed ac-counts of settlement patterns, livelihood, and the effects of man's activity on the land.

Troin, Jean-Françoise. Les Souks marocains: marchés ruraux et organisation de l'espace dans la moitié nord du Maroc. Aix-en-Provence, France: Edisud, 1975. 503 p. ISBN 2-85744-009-X. FFr 250. LC 76-458149.   2564
    A meticulous study of the function, evolution, and spatial characteristics of periodic rural markets in northern Morocco and of their relationship to permanent urban markets.

Waterbury, John. Egypt: burdens of the past, options of the future. Bloomington, IN: Indiana University Press, 1978. 318 p. ISBN 0-253-31943-9. $12.50. (pbk). LC 78-3248.   2565
    A collection of papers on key development issues confronting Egypt during the 1970's.

Waterbury, John. Hydropolitics of the Nile Valley. Syracuse, NY: Syracuse University Press, 1979. 301 p. ISBN 0-8156-2192-2. $20. LC 79-16246.   2566
    An excellent account of the interplay of hydrology and politics in the development of water resources of the Nile Basin.

13. AFRICA SOUTH OF THE SAHARA

Sanford H. Bederman

See also preceding Section 12. Southwest Asia and North Africa.

A. BIBLIOGRAPHIES

A Current bibliography on African affairs. 1-6 (1962-1967). New series. 1- (1968-). Quarterly. Baywood publishing Company, 120 Marine St., Box D, Farmingdale, NY, 11735. ISSN 0011-3255.   2567
    Contains feature articles, book reviews, and bibliographical sections for general subjects and regional studies.

Asamani, J. Index Africanus. Stanford, CA: Stanford University, Hoover Institution Press, 1975. 659 p. ISBN 0-8179-2531-7. $30. LC 76-187266.   2568
    Lists more than 200 journals concerned with Africa and 23,000 articles published between 1885 and 1965.

Bederman, Sanford H. Africa: a bibliography of geography and related disciplines. 3rd ed. Atlanta, GA: Georgia State University, School of Business Administration, Publishing Services Division, 1974. 334 p. (1st ed., 1970. 212 p.). ISBN 0-88406-089-6. $11. LC 74-22175.   2569
    Contains 3629 citations of articles and books organized by region. Included are useful regional, country, and author indexes.

International African bibliography. 1- (1971-). Quarterly. Mansell Publishers Ltd. 6 All Saints St., London, N1 9RL, England. ISSN 0020-5877.   2570
    Previously a section in the journal Africa. *Material is arranged geographically by country, and there is also a subject tracing at the end of each entry.*

O'Connor, Anthony M. Urbanization in tropical Africa: an annotated bibliography. Boston, MA: G.K. Hall, 1981. 381 p. ISBN 0-8161-8262-0. $42. LC 80-24853.   2571
    Covers publications in French and English from 1960 on. Arranged by area, it includes about 2500 citations.

Africa South of the Sahara

Panofsky, Hans. A Bibliography of Africana. Westport, CT: Greenwood Press, 1975. 350 p. ISBN 0-8371-6391-9. $29.95. LC 72-823.　　2572
　Contains a guide to resources by subject and discipline with relevant sections under sciences (agriculture) and social sciences (demography, urbanization, and economics).

U.S. Library of Congress, Africa section. Africa south of the Sahara: index to periodical literature 1900-1970. Boston, MA: G.K. Hall, 1971. 4 v. ISBN 0-8161-0892-7. $395. LC 74-170939.
　First Supplement: 1973, 521 p. ISBN 0-8161-1048-4. $110.
　Second Supplement, (June 1972-December, 1976): 1982, 3 v. ISBN 0-8161-0293-7. LC 82-167629.　　2573
　This bibliography contains more than 100,000 citations.

B. SERIALS

Africa: journal of the International African Institute. 1- (1928-). Quarterly. Journals Department, Manchester University Press, Oxford Road, Manchester M13 9PL, United Kingdom. ISSN 0001-9720.　　2574
　This high quality journal, emphasizing primarily social science topics, consistently publishes articles written by geographers. From 1985 will include a fifth volume each year: African bibliography.

Africa: the international business, economic, and political monthly. 1- (May, 1971-). Monthly. Africa Journal Ltd., Kirkman House, 54a Tottenham Ct. Rd., London, W1P 0B7, England. ISSN 0044-6475.　　2575
　A commercially produced monthly news magazine providing current analyses of political events and economic trends.

Africa report. 1- (1956-). Bi-monthly. African-American Institute, 833 United Nations Plaza, New York, NY, 10017. ISSN 0001-9836.　　2576
　Essential to a basic library on contemporary Africa. Specializing in current affairs and international relations, it presents current reports and clear political analyses.

African affairs: journal of the Royal African Society. 1- (1901-). Quarterly. Oxford University Press, 200 Madison Avenue, New York, NY, 10016. ISSN 0001-9909.　　2577
　In addition to the usual scholarly articles on all aspects of African studies, this venerable journal provides numerous book reviews and a useful current bibliography.

African studies review. (Originally African studies bulletin). 1- (April, 1958-). Quarterly. African Studies Association, 255 Kinsey Hall, University of California - Los Angeles, 405 Hilgard Ave., Los Angeles, CA, 90024. ISSN 0002-0206.　　2578
　Covers all aspects of African studies including literature and the arts.

The Journal of modern African studies: a quarterly survey of politics, economics, and related topics in contemporary Africa. 1- (1963-). Quarterly. Cambridge University Press, 32 E. 57th St., New York, NY, 10022. ISSN 0022-278X.　　2579
　A prestigious journal that consistently publishes high-quality research on a variety of African social science topics. It is particularly noted for its emphasis on economic life and political and geo-political relationships.

Kenyan geographer. (Geographical Society of Kenya). 1- (1975-). Semi-annual. Kenya Literature Bureau, P.O. Box 30022, Nairobi, Kenya.　　2580
　Articles on Kenya, East Africa, and Africa in general.

Madagascar. Revue de géographie. 1- (1962-). Semi-annual. Subscriptions: Fondation National de l'Enseignement Supérieur, Antananarivo, Madagascar. ISSN 0047-5416.　　2581

Articles on Madagascar and the islands of the Indian Ocean. In French. Abstracts in English and French. Cumulative index nos. 1-30 (1962-1977) in no. 30 (1977).

Savanna: a journal of environmental and social sciences. 1- (1972-). Semi-annual. Ahmadu Bello University Press (Savanna), P.M.B. 1094. Zaria, Nigeria.   2582
Articles particularly on Nigeria and the savanna areas of West Africa. Reviews. Bibliographies on Nigeria and its regions.

South African geographical journal. 1- (1917-). Annual. South African Geographical Journal, P.O. Box 31201, 2017 Braamfontein, Transvaal, South Africa.   2583
Articles on South Africa. Mainly in English with abstracts in English.

Suid-Afrikaanse geograaf. South African geographer. (Vereniging vir Geografie). 1- (1957-). Semi-annual. P.O. Box 2031, Dennesig 7601, South Africa. ISSN 0378-5327.   2584
Articles on South Africa. Section on geography in the schools. In Afrikaans or English. Abstracts in English.

See also entries 2488-2490.

C. ATLASES

Africana Publishing Corp. Graphic perspectives of a developing country series. London: University of London Press; New York: Holmes and Meier.
> **Liberia in maps.** Gnielinski, Stefan, ed. 1972. 111 p. ISBN 0-8419-0126-0. $35. LC 72-80411.
> **Malawi in maps.** Agnew, Swanzie, and Stubbs, Michael, eds. 1972. 143 p. ISBN 0-8419-0127-9. $35. LC 74-654433.
> **Nigeria in maps.** Barbour, Michael, et al., eds. 1982. 148 p. ISBN 0-8419-0763-3. $35. LC 82-675001.
> **Sierra Leone in maps.** Clarke, John I., ed. 2nd ed. 1972. 120 p. ISBN 0-8419-0070-1. $34.50. LC 79-654429.
> **Tanzania in maps.** Berry, Leonard, ed. 1972. 172 p. ISBN 0-8419-0076-0. $34.50. LC 70-654258.
> **Zambia in maps.** Davies, D.H., ed. 1972. 128 p. ISBN 0-8419-0081-7. $34.60. LC 70-654258.   2585

All the atlases include clear black-and-white maps and drawings to illustrate essays on a variety of individual topics which focus mostly on the physical setting and socio-economic development. All the volumes are carefully edited and contain accurate geographical data.

Davies, Harold Richard J. Tropical Africa: an atlas for rural development. Cardiff, U.K.: University of Wales Press; Mystic, CT: Verry, 1973. 81 p. ISBN 0-7083-0522-9. $25. LC 73-175199.   2586
Covers Africa from the Sahara to the southern margin of the Zaire Basin. Forty maps depict physical background, traditional life, and modernization of society.

Fage, John D. An Atlas of African history. 2nd ed. New York: Africana Publishing Co.; London: E. Arnold, 1978. 84 p. (1st ed. 1958. 64 p.). ISBN 0-8419-0429-4. $37.50. LC 78-16131.   2587•
Incorporates the results of recent historical and archeological research. Containing 71 maps, it is the most scholarly of the historical atlases.

Murray, Jocelyn, ed. Cultural atlas of Africa. New York: Facts on File, 1982. 240 p. ISBN 0-87196-558-5. $35. LC 80-27762.   2588•
The title is misleading because most of the maps are of individual countries showing locations of cities and major towns and where crops are grown. Part II contains useful historical and cultural maps. A total of 96 maps are included.

Reader's Digest atlas of Southern Africa. Cape Town, South Africa: Reader's Digest, 1984. 256 p. ISBN 0-947008-02-0. LC 84-245810.

2589•

An excellent, much needed, modern atlas of Southern Africa produced by Reader's Digest in conjunction with the Republic of South Africa's Directorate of Surveys and Mapping, Department of Community Development. The atlas is divided into three sections: "Anatomy of South Africa," (thematic maps and text); "Africa South of the Sahara;" and "Our Land in Close-up." The latter section includes 138 pages of six-color topographic maps ranging from 1:50,000 to 1:2,500,000 in scale and based largely on maps prepared by the South African government.

Van Chi-Bonnardel, Regine, director. The Atlas of Africa. (Grand atlas du continent africain). Paris: Jeune Afrique; New York: Hippocrene, 1973. 335 p. ISBN 0-903274-03-5. $100. LC 72-650052.

2590

Available in either French or English, it includes general continental maps at various scales, as well as maps for individual countries.

D. GENERAL WORKS

American University. Foreign Areas Studies Division. "Area Handbook" series replaced by "Country Studies" series. Washington, DC: Government Printing Office.

AREA HANDBOOK SERIES

Burundi. McDonald, G.C., ed. 1969. (Reprinted 1975). 203 p. $9.50. LC 70-605915.

Chad. Nelson, H.D. et al., eds. 1972. (Reprinted 1979). 261 p. $10. LC 72-600075.

Cameroon. Nelson, H.D. et al., eds. 1974. 335 p. $11. LC 73-600274.

Ghana. Kaplan, I. et al., eds. 1971. 449 p. $12. LC 74-611338.

Guinea. Nelson, H.D. et al., eds. 2nd ed. 1975. 386 p. $12. LC 75-26515.

Kenya. Nelson, H.D. ed. 3rd ed. 1983. 334 p. $14. LC 84-6420.

Liberia. Roberts, T.R. et al., eds. 1972. 387 p. $12. LC 72-600021.

Malagasy Republic. Nelson, H.D. et al, eds. 1973. 327 p. LC 73-600012.

Malawi. Nelson, H.D., ed. 1975. (Reprinted 1980). 353 p. $12. LC 75-619072.

Mozambique. Kaplan, I., ed. 2nd ed. 1977. 240 p. $10. LC 77-13475.

Rwanda. Nyrop, R.F. et al., eds. 1969. 212 p. LC 72-606089.

Senegal. Nelson, H.D., ed. 1974. (Reprinted 1977). 410 p. $11. LC 72-600061.

Sierra Leone. Kaplan, I., et al, eds. 1976. 400 p. $12. LC 76-49498.

COUNTRY STUDIES SERIES

Angola. Kaplan, I., ed. 2nd ed. 1979. 286 p. $11. LC 79-21789.

Ethiopia. Kaplan, I. and Nelson, H.D., eds. 3rd ed. 1981. 366 p. $12. LC 81-7928.

Nigeria. Nelson, H.D., ed. 4th ed. 1982. 358 p. $12. LC 82-6795.

Somalia. Nelson, H.D., ed. 3rd ed. 1982. 346 p. $13. LC 82-16401.

South Africa. Nelson, H.D., ed. 2nd ed. 1981. 464 p. $12. LC 81-19155.

Tanzania. Kaplan, I., ed. 2nd ed. 1978. 344 p. $11. LC 78-10304.

Zaire. Kaplan, I., ed. 3rd ed. 1979. 332 p. $11. LC 79-9987.

Zambia. Kaplan, I. ed. 3rd ed. 1979. 308 p. $11. LC 79-21324.

Zimbabwe. Nelson, H.D., ed. 2nd ed. 1983. 360 p. $11. LC 83-11946.

2591•

Originally designed for use by military and other personnel who needed a convenient compilation of basic facts about the social, economic, political, and military institutions and practices of various countries. The new series appeals to a broader group of people, especially students, scholars, and even tourists. All contain good maps, up-to-date data, and an excellent bibliography.

Best, Alan C.G. and de Blij, Harm J. African survey. New York: John Wiley and Sons, 1977. 626 p. ISBN 0-471-20063-8. $38.50. LC 76-44520.

2592

An expanded, updated revision of de Blij's A Geography of Sub-Saharan Africa (1964). Covering the whole continent, each chapter focuses on a specific aspect of the country in question.

Fage, J.D. and Oliver, Roland, gen. eds. The Cambridge history of Africa. Cambridge, U.K.; New York: Cambridge University Press, 1975 to date. 6 v. LC 76-2261 (set).

> Vol. i: **From the earliest times to c. 500 B.C.** Clark, J. Desmond, ed. 1982. 1157 p. ISBN 0-521-22215-X. $102.50.
>
> Vol. 2: **c. 500 B.C. to A.D. 1050.** Fage, J.D., ed. 1978. 840 p. ISBN 0-521-21592-7. $97.50.
>
> Vol. 3: **c. 1050 to c. 1600.** Oliver, Roland, ed. 1977. 803 p. ISBN 0-521-20981-0. $89.50.
>
> Vol. 4: **c. 1600 to c. 1790.** Gray, Richard, ed. 1975. 738 p. ISBN 0-521-20413-5. $84.50.
>
> Vol. 5: **c. 1790 to c. 1870.** Flint, John E., ed. 1976. 617 p. ISBN 0-521-20701-0. $84.50.
>
> Vol. 8: **c. 1940 to c. 1975.** Crowder, Michael, ed. 1984. 800 p. ISBN 0-521-22409-8. $84.50.   2593

The authoritative history of Africa covering the continent from the earliest times to the present day. Original essays prepared by outstanding scholars are included in this magnificent multi-volume work.

Clarke, John I. et al. An Advanced geography of Africa. Amersham, U.K.: Hulton Educational Publications, Ltd., 1975. 528 p. ISBN 0-7175-0600-2. £6.90. LC 75-324442.   2594
  An excellent systematic coverage of the continent. Twelve chapters cover the environment, the political map, society, and the economy.

de Blij, Harm J. and Martin, Esmond, eds. African perspectives. New York: Methuen, 1981. 264 p. ISBN 0-416-60231-2. $17.95. LC 80-13953.   2595
  A volume of essays by specialists interpreting the economic geography of nine African nations.

Gourou, Pierre. L'Afrique. Paris: Hachette, 1970. 488 p. FFr 187. LC 76-517688.   2596
  A masterfully written and superbly documented achievement by a great scholar who spent most of his career studying the world's tropics.

Grove, Alfred Thomas. Africa. 3rd ed. Oxford, U.K.; New York: Oxford University Press, 1978. 337 p. (1st ed., 1967. 275 p.). ISBN 0-19-913244-5. $21.95. LC 78-322859.   2597•
  This new edition has been expanded to cover the whole continent. It contains both systematic and regional chapters.

Hallett, Robin. Africa since 1875: a modern history. Ann Arbor, MI: University of Michigan Press, 1974. 807 p. ISBN 0-472-07170-X. $15. LC 72-91505.   2598
  The last century of Africa's history is presented regionally.

Hance, William A. The Geography of modern Africa. 2nd ed. New York: Columbia University Press, 1975. 657 p. (1st ed. 1964. 653 p.). ISBN 0-231-03869-0. $30. LC 75-2329.   2599
  A comprehensive and authoritative treatment of the continent that utilizes both the systematic and regional approach. More than anything else, however, economic activities are emphasized.

Harrison Church, Ronald J., et al. Africa and the islands. 4th ed. London; New York: Longman, 1977. 542 p. (1st ed., 1964. 494 p.). ISBN 0-582-35197-9. LC 78-304729.   2600•
  A factual, straight-foward presentation of physical and cultural topics for all of Africa's countries. New edition provides current information.

Knight, C. Gregory, and Newman, James L., eds. Contemporary Africa: geography and change. Englewood Cliffs, NJ: Prentice-Hall, 1976. 546 p. ISBN 0-13-170035-9. $34.95. LC 76-4902.   2601
  Thirty-two essays by well-known African specialists systematically discuss population, rural and urban change, rural-urban systems, and modernization. This is one of the best compendiums available on the geography of Africa.

Africa South of the Sahara

Paden, John W. and Soja, Edward W., eds. The African experience. Evanston, IL: Northwestern University Press; London: Heinemann, 1970. 3 v. ISBN 0-89771-005-3. (Vol. 3 available from State Mutual Books and Periodicals, New York, for $5.95). LC 70-98466.   2602

Volume one includes 31 original essays covering all aspects of the contemporary scene, whereas volume two presents a syllabus to aid in the teaching of Africa. Volume three contains more than 1100 pages of bibliography and a short guide to resources. One of the editors is a geographer.

Prothero, R. Mansell (Ralph), ed. A Geography of Africa: regional essays on fundamental characteristics, issues, and problems. 2nd rev. ed. London; Boston, MA: Routledge and Kegan Paul, 1973. 482 p. (1st ed. 1969. 480 p.). ISBN 0-7100-7669-X. £15.95. LC 73-179540.   2603

Contains nine original essays, eight of which focus on the various broad regions of Africa.

Schulthess, Emil, and Birrer, Emil. Africa. 2nd rev. English ed. New York: Simon and Schuster, 1969. 75 p. LC 75-87881.   2604 •

One of the world's finest photographers presents both grim and glorious Africa in 136 spectacular color and black-and-white pictures.

Scott, Earl P., ed. Life before the drought. London; Boston, MA: Allen and Unwin, 1984. 196 p. ISBN 0-04-910076-9. $24.95. LC 84-9235.   2605

Seven authors, all having done fieldwork in the drought-prone parts of Africa, generally show that traditional food production systems have well-developed adaptive mechanisms for coping with drought. The approaches are those of political economy and human ecology, and most of the studies are about West African locations.

Skinner, Elliott P., ed. Peoples and cultures of Africa: an anthropological reader. Garden City, NY: Doubleday/Natural History Press, 1973. 756 p. LC 72-77003.   2606

Thirty-six previously published articles and portions of books are organized to provide understanding of Africa's peoples, their social and political institutions, and their beliefs and religion.

E. SPECIAL TOPICS

Boateng, E.A. A Political geography of Africa. Cambridge, U.K.; New York: Cambridge University Press, 1978. 292 p. ISBN 0-521-21764-4. $42.50. ISBN 0-521-29269-7. $16.95. (pbk). LC 77-80828.   2607

A traditional, factual, non-theoretical work that draws mainly from historical rather than political sources.

Christopher, Anthony J. Colonial Africa. London: Croom Helm; Totowa, NJ: Barnes and Noble, 1984. 232 p. ISBN 0-389-20452-8. $28.50. LC 84-673289.   2608

A systematic historical geography, continental in scale, in which specific attention is given to the impress of explorers, politicians and administrators, traders, missionaries, planters, miners, and settlers. Emphasis is on the significant changes made on the African landscape by colonizers.

Clarke, John I. and Kosiński, Leszek A., eds. Redistribution of population in Africa. London; Exeter, NH: Heinemann Educational, 1982. 212 p. ISBN 0-435-95030-4. $58. ISBN 0-435-95031-2. $20. (pbk). LC 81-170774.   2609

A gold mine of information and current data on the distribution of people in Africa. The volume includes 27 papers covering all parts of the continent.

Dalby, David, and Harrison Church, Ronald J., eds. Drought in Africa 2. Rev. and expanded ed. London: International African Institute, 1977. 200 p. (1st ed. 1973. 124 p.). ISBN 0-85302-056-6. £6.95. LC 78-310180.   2610

Contains 20 original essays in English and French. The persistence of drought is one of the most serious problems Africans are contending with at the end of the 20th century.

Davidson, Basil. Can Africa survive? Arguments against growth without development. Boston, MA; Toronto: Little, Brown, and Company, 1974. 207 p. ISBN 0-316-17434-3. $5.95. (pbk). LC 74-6490.　　　　　　　　　　　　　　　　　　　　　　2611
　　A personal view of Africa's economic problems.

de Vos, Antoon. Africa, the devastated continent. The Hague, The Netherlands: Dr. W. Junk; Hingham, MA: Kluwer, 1975. 236 p. ISBN 90-6193-078-2. $39.50. LC 75-322501.　　　　　　　　　　　　　　　　　　　　　　　　　　　　2612•
　　An important synthesis of our knowledge of environmental problems in Africa.

Furon, Raymond. Geology of Africa. Translated from French by A. Hallam and L.A. Stevens. New York: Hafner, 1963. 377 p. LC 63-2498. (Géologie de l'Afrique. 2nd ed. Paris: Payot, 1960. 400 p.).　　　　　　　　　　　　　　　　　　2613
　　One of the first definitive reviews of contemporary approaches to the geology of Africa. Highly useful by scholars and advanced students. The volume, though dated, remains the best source available.

Griffith, J.F., ed. Climates of Africa. (World survey of climatology, v. 10.) Amsterdam, The Netherlands; New York: Elsevier Publishing Co., 1972. 604 p. ISBN 0-444-40893-2. $157.50. LC 72-135485.　　　　　　　　　　　　　　　　2614
　　An excellent compendium of regional climatologies. The volume contains dozens of maps, diagrams and tables.

Hance, William A. Black Africa develops. Waltham, MA: Crossroads Press, 1977. 158 p. ISBN 0-918456-17-7. $10. LC 77-155100.　　　　　　　　　　　2615
　　A brief topical presentation of the state of economic development in Africa south of the Sahara.

Hance, William A. Population, migration, and urbanization in Africa. New York: Columbia University Press, 1970. 450 p. LC 75-116378.　　　　　　　　2616
　　Dated but still good for its factual and straight-foward presentation. Written by a geographer who has traveled extensively throughout Africa. The section on African cities is particularly interesting.

Hoyle, Brian S. and Hilling, David, eds. Seaports and development in tropical Africa. London: Macmillan; New York: Praeger, 1970. 272 p. LC 71-112020.　　　2617
　　The important role of seaports in the economic growth of underdeveloped areas is emphasized in 15 original essays. Case studies of Dakar, Freetown, and Abidjan are found along with 36 maps and 44 tables.

Jones, William O. Marketing staple food crops in tropical Africa. Ithaca, NY: Cornell University Press, 1972. 293 p. ISBN 0-8014-0736-2. LC 72-4176.　　　　2618
　　A masterful study of one of Africa's most pressing problems--the adequacy of food supplies. Included are lengthy case studies of Sierra Leone and Kenya.

O'Connor, Anthony M. The African city. New York: Africana Publishing Co, 1983. 359 p. ISBN 0-8419-0881-8. $27.50. ISBN 0-8419-0882-6. $17.50. (pbk). LC 83-10648.　　2619
　　A systematic study of urbanization on a continental scale. Topics covered include rural-urban migration, the urban economy, housing, spatial structures, and urban systems.

O'Connor, Anthony M. The Geography of tropical African development. 2nd ed. Oxford, U.K.; New York: Pergamon Press, 1978. 229 p. (1st ed., 1971. 207 p.). ISBN 0-08-021847-4. $35. ISBN 0-08-021848-2. $11.25. (pbk). LC 77-30470.　　　　2620
　　A study of spatial patterns of economic change since independence. Sixty maps are included.

Ominde, Simeon H. and Ejiogu, C.N., eds. Population growth and economic development in Africa. London; Nairobi, Kenya; Ibadan, Nigeria: Heinemann (In association with the Population Council, New York), 1972. 421 p. ISBN 0-435-97470-X. £11. LC 73-161318.

Fifty-three essays by experts. Topics include population distribution and variable growth, migration, demographic data problems, and family planning programs. Included are many case studies of countries.

2621

Pritchard, J.M. Landform and landscape in Africa. London: Edward Arnold, 1979. 160 p. ISBN 0-7131-0204-7. £4.25. (pbk). LC 80-456479.

A geomorphology of the continent covering such topics as surface morphology and drainage, and riverine, glaciated, tectonic, volcanic, desert, and coastal landscapes.

2622

Rotberg, Robert I., ed. Africa and its explorers: motives, methods, and impact. Cambridge, MA: Harvard University Press, 1970. 351 p. ISBN 0-674-00777-8. $7.95. (pbk). LC 77-134327.

Original essays detailing the accomplishments of nine 19th-century European explorers. Those on Heinrich Barth and Gerhard Rohlfs are particularly well-done.

2623 •

Severin, Timothy. The African adventure: four hundred years of exploration in the dangerous continent. New York: E.P. Dutton; London: Hamilton, 1973. 288 p. ISBN 0-525-05110-4. LC 72-96378.

A well-written, heavily illustrated account of how Europeans first learned about the "Dark Continent." It begins with the search for Prester John and ends with Stanley and de Brazza in the Congo.

2624 •

Strage, Mark. Cape to Cairo: rape of a continent. New York: Harcourt, Brace, and Jovanovich; London: J. Cape, 1973. 278 p. LC 73-13958.

A well-written account of how Imperial Europe attempted to build a railway from South Africa to the Mediterranean.

2625 •

Thomas, Michael F. and Whittington, Graeme W., eds. Environment and land use in Africa. London: Methuen, 1969. 554 p. ISBN 0-416-10840-7. LC 74-410364.

One of the best compendiums available on people and their use of land in Africa. Eighteen essays comprise this volume, six of them are studies of the natural environment.

2626

Trimmingham, J. Spencer (John). The Influences of Islam upon Africa. (Arab background series). New York: Longman, 1980. 182 p. (1st ed., 1968). ISBN 0-582-78499-9. $27.

After a survey of the history and characteristics of Islamic culture zones in Africa, there are chapters on religious and cultural change, the religious life of African Muslims, the influence of Islam on social life, and the African Muslim in the era of change. A new chapter was written for the 2nd edition.

2627

United Nations. Economic Commission for Africa. African statistical yearbook, 1976. New York: The United Nations, 1977. 4 v. LC 76-643332.

Despite spotty geographical and chronological coverage, these volumes remain one of the best sources of comparable statistical data for the continent available at this time.

2628

The World Bank. Accelerated development in Sub-Saharan Africa: an agenda for action. Washington, DC: The World Bank, 1981. 198 p. $6. LC 81-16828.

The central theme of this valuable report is that more efficient use of scarce resources, both human and capital, is vital for improving economic conditions in most African countries.

2629

Dickson, Kwamina B. An Historical geography of Ghana. Cambridge, UK: Cambridge University Press, 1969. 379 p. LC 69-19375.
A well-documented, pioneer study of a major African nation.

2630

Dickson, Kwamina B. and Benneh, George. A New geography of Ghana. London: Longman, 1970. (Metricated ed., 1977). 173 p. ISBN 0-582-60343-9. £3.60. LC 78-317440.
Updated and more topical review of above entry.

2631

Gugler, Josef, and Flanagan, William G. Urbanization and social change in West Africa. Cambridge, U.K.; New York: Cambridge University Press, 1978. 235 p. ISBN 0-521-21348-7. $32.50. ISBN 0-521-29118-6. $10.95. (pbk). LC 76-9175.
A multidisciplinary approach to the topic and a good, but brief, summary of recent work on urbanization in West Africa. The volume emphasizes the dynamic nature of the dual rural-urban system that has emerged in the recent decades of this century.

2632

Harrison Church, Ronald J. West Africa: a study of the environment and of man's use of it. 8th ed. London; New York: Longman, 1980. 526 p. (1st ed., 1957. 547 p.) ISBN 0-582-30020-7. $26. (pbk). LC 79-41255.
Perhaps the most venerable geographical study of the most populous region of the continent. The first two parts contain systematic physical and cultural chapters, whereas the third part includes a chapter on each country.

2633 •

Mabogunje, Akin L. (Akinlawon Lapido). Regional mobility and resource development in West Africa. Montreal, PQ; London: McGill-Queens University Press, 1972. 154 p. ISBN 0-77350-120-7. LC 79-173461.
A well-rounded and balanced analysis of interregional population movements in West Africa. A short but important book.

2634

Morgan, W.T.W. (William Thomas Wilson). Nigeria. (World's landscapes series). London; New York: Longman, 1983. 179 p. ISBN 0-582-30003-7. £6.50. (pbk). LC 82-15265.
Although a new book, little attention is paid to the effects of the Civil War, the oil boom, and the economic recession of the 1980's. Otherwise a clearly written, well-illustrated study generally utilizing the theme of landscape.

2635

Ojo, Oyediran. The Climates of West Africa. London; Ibadan, Nigeria; Nairobi, Kenya: Heinemann Educational Books Ltd., 1977. 218 p. ISBN 0-435-95700-1. £5.95. LC 78-319464.
A systematic study including chapters on temperature, evaporation and water balance, winds, the distribution of rainfall, and regional climatic patterns. Contains excellent maps and diagrams.

2636

Riddell, J. Barry. The Spatial dynamics of modernization in Sierra Leone: structure, diffusion, and response. Evanston, IL: Northwestern University Press, 1970. 142 p. ISBN 0-8101-0309-5. $11.95. LC 75-108134.
Concerned with the areal incidence of development in Sierra Leone. A strong quantitative approach is taken in this study.

2637

Udo, Reuben. K. A Comprehensive geography of West Africa. New York: Africana Publishing Corporation, 1978. 304 p. ISBN 0-8419-0379-4. $35. ISBN 0-8419-0380-8. $17.50. (pbk). LC 78-2295.
Both the systematic and regional approaches are used. Demographic and economic data are current through the mid-1970's. A chapter for each country is included.

2638

Udo, Reuben K. Geographical regions of Nigeria. Berkeley, CA: University of California Press, 1970. 212 p. LC 70-94980.
Replaced Buchanan and Pugh's Land and people of Nigeria as the standard geography of the country. Includes 84 maps and diagrams, and 52 photographs.

2639

White, Henry Patrick, and Gleave, M.B. An Economic geography of West Africa. London: G. Bell, 1971. 322 p. ISBN 0-7135-1721-2. £5.95. LC 70-864176. 2640
Four themes prevail: continuing importance of the traditional economy; the modernization of the economy; regional variations; and conflicts between traditional and modernizing elements in the economy.

Zachariah, K.C. (Kunniparampil Curien) and Condé, Julien. Migration in West Africa. New York: Oxford University Press (for the World Bank), 1981. 130 p. ISBN 0-19-520187-6. $19.95. ISBN 0-19-520187-6. $8.95 (pbk). LC 80-21352. 2641
A detailed study of recent migration streams in West Africa. Twenty-two maps are included in this important study.

G. CENTRAL AFRICA

Balon, Eugene K. and Coche, A.G., eds. Lake Kariba: a man-made tropical ecosystem in Central Africa. The Hague, The Netherlands: Dr. W. Junk; Hingham MA: Kluwer, 1974. 767 p. ISBN 90-6193-076-6. $131.50. LC 75-300246. 2642
Particularly valuable for its assessment of the careless application of technology during the construction of the Kariba Dam on the Zambezi River in the late 1950's.

Kofele-Kale, Ndiva, ed. An African experiment in nation building: the bilingual Cameroon Republic since reunification. Boulder, CO: Westview Press, 1980. 369 p. ISBN 0-89158-685-7. $35. LC 79-5356. 2643
A variety of viewpoints are presented in 12 essays that attempt to explain the dynamic nature of one of the most fascinating countries in Africa. Considerable attention is paid to political and economic conditions.

Miracle, Marvin P. Agriculture in the Congo Basin: tradition and change in African rural economies. Madison, WI: University of Wisconsin Press, 1967. 355 p. LC 67-26628. 2644
An intensive survey and synthesis of literature on traditional shifting cultivation and animal husbandry in Central Africa, and a study of African response to new crops and techniques. Rejects oversimplification of agriculture and overemphasis on conservation.

West, Richard. Congo: an account of a century of European exploration and exploitation in the heart of Africa. New York: Holt, Rinehart, and Winston, 1972. 304 p. ISBN 0-03-09391-8. LC 79-182765. 2645
A seasoned traveler and journalist writes about the great names associated with this remote region. Among those highlighted are Du Chaillu, de Brazza, and Stanley.

H. EAST AFRICA

History of East Africa. Oxford, U.K.: Clarendon Press, 1963-1976. 3 v. LC 63-4375.
Vol. 1: Oliver, Roland A., and Mathew, Gervase, eds. 1963. 500 p.
Vol. 2: Harlow, Vincent; Chilver, E.M.; and Smith, Alison, eds. 1965. 768 p.
Vol. 3: Low, D.A. and Smith, Alison, eds. 1976. 691 p. 2646
A splendid history. Volume one goes from the Stone Age to the onset of colonial partition, with the greatest attention to the immediate pre-contact period. Volume two traces the establishment of British and German rule and the African response with separate treatment of Kenya, Uganda, Tanganyika, and Zanzibar. Volume three continues the story up to independence.

Hoyle, Brian S. Seaports and development: the experiences of Kenya and Tanzania. London; New York: Gordon and Breach, 1983. 334 p. ISBN 0-667-06030-0. $61.50. LC 82-12128. 2647
Contains an in-depth discussion of highly localized port activity set in the context of East Africa's space economy.

Knight, C. Gregory. Ecology and change: rural modernization in an African community. New York; London: Academic Press, 1974. 300 p. ISBN 0-12-785435-5. $46.50. LC 73-7446.

An excellent analysis of the process of change occurring among the rural Nyihah who reside in Southwestern Tanzania.

2648

Matthiessen, Peter, and Porter, Eliot, photographer. The Tree where man was born. New York: Crescent Books, 1972. Reprinted: New York: Dutton, 1983. 247 p. ISBN 0-525-48032-3. $8.95. (pbk). LC 73-157328.

Two first-rate naturalists take the reader on a tour of East Africa. This volume is graced by over 100 magnificent photographs.

2649•

Miller, Charles. The Lunatic express: an entertainment in imperialism. New York: Macmillan, 1971. 559 p. LC 71-153759.

A history of the construction of the Uganda Railway from Mombasa to Lake Victoria.

2650•

Morgan, W.T.W. (William Thomas Wilson). East Africa. London: Longman, 1973. 410 p. Reprinted ["New ed."], 1976. ISBN 0-582-48573-8. £10.50. (pbk). LC 74-161775.

A straight-foward systematic and regional geography of the area comprised by Kenya, Uganda, and Tanzania. Includes physical, economic, and cultural aspects.

2651

Morgan, W.T.W. (William Thomas Wilson), ed. East Africa: its people and resources. 2nd ed. Nairobi, Kenya; New York: Oxford University Press, 1969. 312 p. LC 75-17100.

The revised, expanded, and updated edition of Edward W. Russell, ed. Natural resources of East Africa (Nairobi, Kenya: D.A. Hawkins in association with East African Literature Bureau, 1962. 144 p.). This volume shifts emphasis towards human activities. Includes six colored maps of East Africa at a scale of 1:1,000,000.

2652

Morgan, W.T.W. (William Thomas Wilson), ed. Nairobi: city and region. 2nd ed. Nairobi, Kenya: Oxford University Press, 1971. 154 p. (1st ed. 1967. 154 p.). LC 67-6616.

A dozen essays, ranging from geology and climate to agricultural land use and manufacturing industries, detail the geographical personality of the most important urban area in East Africa.

2653

Naipaul, Shiva. North of south: an African journey. New York: Simon and Schuster; Penguin; London: Deutsch, 1979. 349 p. ISBN 0-671-24742-5. ISBN 0-14-004894-4. $5.95. (pbk). LC 78-20954.

An outstanding travel book written by a person of Indian heritage. A very sensitive report on modern life in Tanzania, Kenya, and Zambia.

2654

Newman, James L. The Ecological basis for subsistance change among the Sandawe of Tanzania. (Foreign field research program, 36). Washington, DC: National Academy of Science, 1970. 199 p. ISBN 0-309-01851-X. $12.75. LC 75-607171.

This volume presents the findings of a field geographer who spent over a year in a remote part of East Africa.

2655

I. SOUTH AFRICA

Christopher, Anthony J. South Africa. London; New York: Longman, 1982. 237 p. ISBN 0-582-47001-4. $15.95. LC 81-8254.

A systematic approach including material on urban activity, industry, and agriculture. The author plays down the negative aspects of South African Government racial policy.

2656

Christopher, Anthony J. Southern Africa. Folkestone, U.K.: Dawson; Hamden, CT: Archon Books, Shoe String Press, 1976. 292 p. ISBN 0-208-01620-1. $22.50. LC 76-21207.

2657

A chronologically arranged historical geography of settlement and economic activity. The period from the first Dutch settlement to 1960 is covered.

Fair, T.J.D. (Thomas). South Africa: spatial frameworks for development. Capetown, South Africa: Juta and Company, 1982. 93 p. ISBN 0-7021-1307-7. LC 82-214967.

2658

A short, non-controversial analysis of the South African space economy providing a framework for understanding the country's contemporary economic geography.

Smith, David M., ed. Living under apartheid: aspects of urbanization and social change in South Africa. London; Boston, MA: Allen and Unwin, 1982. 256 p. ISBN 0-04-309110-5. $35. LC 82-11605.

2659

Racial segregation in South African cities is carefully scrutinized by 16 scholars.

Western, John. Outcast Cape Town. Minneapolis, MN: University of Minnesota Press, 1981; London: Allen and Unwin, 1982. 372 p. ISBN 0-8166-1025-8. $22.50. LC 81-14640.

2660

A major contribution by a social geographer to the study of the impact of apartheid policies on a unique South African city. Emphasis is on the coloured population in Cape Town.

14. OCEANIA

Gordon R. Lewthwaite

A. BIBLIOGRAPHIES

Cammack, Floyd M. and Saito, Shiro. Pacific Island bibliography. New York: Scarecrow Press, 1962. 421 p. LC 62-10126.

2661

Designed largely as a supplement to C.H.R. Taylor Pacific bibliography (1st ed., 1951) cited below. Focused on the tropical Pacific.

Sachet, Marie Hélène, and Fosberg, F. Raymond (Frances). Island bibliographies: Micronesian botany, land environment, and ecology of coral atolls, vegetation of tropical Pacific islands. (National Research Council. Publication 335). Washington, DC: National Academy of Sciences, National Research Council, 1955. 577 p. LC 55-60007 rev.

2662

Broad coverage and perceptive annotations increase the value of this work for the geographer.

Sachet, Marie Hélène, and Fosberg, F. Raymond (Frances). Island bibliographies supplement: Micronesian botany, land environment, and ecology of coral atolls, vegetation of tropical Pacific islands. Washington, DC: National Academy of Sciences, 1971. 427 p.

2663

Carries Island bibliographies forward to 1970, with supplements on botany, environment and ecology, and vegetation.

Taylor, Clyde R.H. A Pacific bibliography: printed matter relating to the native peoples of Polynesia, Melanesia, and Micronesia. 2nd ed. Oxford, U.K.: Clarendon Press; New York: Oxford University Press, 1965. 692 p. (1st ed., Memoirs of the Polynesian Society, v. 24. Wellington, N.Z.: Polynesian Society, 1951. 492 p.). LC 66-1568.

2664

An invaluable work, ethnological in emphasis, with contents arranged in largely geographical order.

University of Hiroshima, Research and Sources Unit in Regional Geography. Information retrieval system for geographical literature and source material: database GEOGRA. (Special publication no. 5). Hiroshima, Japan: University of Hiroshima, 1978. 305 p. In Japanese and English. 2665

An ingenious geographical thesaurus and bibliography of Oceania utilizing an information retrieval system developed for geographers using Japanese and English. Books and articles with authors, titles, contents, key words and reviews are correlated.

B. SERIALS

Australian geographer. 1- (1928-). Semiannual. Geographical Society of New South Wales, P.O. Box 328, North Ryde, N.S.W. 2113, Australia. ISSN 0004-9182. 2666

Articles on a wide range of geographical problems but especially on Australia. Notes. Book reviews. Cumulative index every three years.

Australian geographical studies. 1- (1963-). Semiannual. Department of Human Geography, Research School of Pacific Studies, Australian National University, P.O. Box 4, Canberra, A.C.T. 2061, Australia. ISSN 0004-9190. 2667

A scholarly journal of academic geographers in Australia devoted particularly to Australia and neighboring regions. Notes. Reviews.

New Zealand geographer. 1- (1945-). Semiannual. New Zealand Geographical Society, Department of Geography, University of Canterbury, Christchurch, New Zealand. ISSN 0028-8144. 2668

Lively and valuable geographical periodical with some specialization on New Zealand. Geographic notebook. Reviews of articles on New Zealand in other journals. Book reviews. Cumulative index: 1-25 (1945-1969).

Pacific perspective. 1- (1972-). Semiannual. South Pacific Social Science Association, Box 5083, Suva, Fiji. ISSN 0379-525X. 2669

An interdisciplinary journal produced within the insular Pacific, with emphasis on history, culture, politics, and current issues. Some articles are of significance to human geography.

Pacific viewpoint. 1- (1960-). Semiannual. Department of Geography, Victoria University of Wellington, P.O. Box 196, Wellington, New Zealand. ISSN 0030-8978. 2670

Articles on Pacific and Asian areas. Notes and comments. Reviews. Cumulative indexes 1-5 (1960-1964), 6-10 (1965-1969), 11-22 (1970-1981).

C. ATLASES

Anderson, Allan Grant, ed. New Zealand in maps. London: Hodder and Stoughton, 1977; New York: Holmes and Meier, 1978. 141 p. ISBN 0-8419-0324-7. $42. LC 77-18023. 2671•

More than 60 pages of maps and graphs, often several to a page, each accompanied by more than 1000 words of text. Based partly on the 1971 census, but broad information is contributed by ten geographers on the physical and biological environments, population and settlement, and the economic and social structure.

Australia. Division of National Mapping. Atlas of Australian resources. 3rd series. Canberra, ACT: Australia, Division of National Mapping, 1980- . 3 volumes to date, 3 more in preparation. (2nd series, 1962). ISBN 0-642-51458-5. LC 82-675088.

Vol. 1: **Soils and land use.** 1980. 25 p.
Vol. 2: **Population.** 1980. 23 p.
Vol. 3: **Agriculture.** 1982. 24 p.
Vol. 4: **Climate.** (In preparation).
Vol. 5: **Geology and minerals.** (In preparation).
Vol. 6: **Water.** (In preparation) 2672

A magnificent series of colored maps, mostly at a scale of 1:5,000,000, with accompanying commentary and photographs.

Oceania

Duncan, J.S., ed. Atlas of Victoria. Melbourne, Victoria, Australia: Victorian Government, 1982. 239 p. ISBN 0-7241-8255-1. Australian $39.95.
2673 •
A very fine atlas with contemporary and factual emphasis, produced by more than 30 contributors. Some 170 colored maps and diagrams (including 20 maps at a scale of 1:2,000,000) and 209 photographs, with accompanying text, present a comprehensive and detailed view of the physical, human, and regional geography of the state.

Ford, Edgar, ed. Papua New Guinea resource atlas. Milton, Queensland, Australia: Jacaranda Press, 1974. 56 p. ISBN 0-7016-8214-0. LC 76-367879.
2674
A full-color atlas in large format with photographs, sketches, and maps accompanied by explanatory texts. Presentation progresses from physical, climatic, and biogeographical components to primary and secondary industry, infrastructure, population, societal elements and archeological sites.

King, David, and Ranck, Stephen, eds. Atlas of Papua New Guinea: a nation in transition. Bathurst, N.S.W., Australia: Robert Brown and Associates, Pty. Ltd., in conjunction with the University of Papua New Guinea, 1981. 109 p. ISBN 0-909197-14-8. Australian $10.
2675 •
A very serviceable atlas with 51 colored maps, coupled with relevant essays, bibliography, and 40 representative photographs. Primacy is given to historical, social, and economic aspects; components of the physical environment follow.

New Zealand Department of Lands and Survey. Atlas of the South Pacific. (New Zealand Mapping Service 295). Wellington, New Zealand: New Zealand Department of Lands and Survey, for External Intelligence Bureau, Prime Minister's Department, 1978, (1979). 47 p. ISBN 0-477-01500-X. NZ $12.95. LC 80-675087.
2676 •
Twenty plates with detailed maps of the South Pacific islands. Scales vary, and relief is shown by shading and spot heights. Each plate is accompanied by a brief text incorporating physical and human data.

ORSTOM. (Office de la Recherche Scientifique et Technique Outre-Mer). Atlas de la Nouvelle-Calédonie et dépendances. Paris: ORSTOM, 1982. 108 p. 53 numbered plates. ISBN 2-7099-0601-5. FFr 900.
2677
An exceptional atlas, detailed and comprehensive, with 53 colored plates commonly at a 1:1,000,000 scale, with extended commentaries in French and summaries in English. From the general setting it progresses to natural conditions, origins and settlement of populations, rural land use, economy and infrastructure, social services, and urban centers.

Reader's Digest Services. Reader's Digest atlas of Australia. Sydney, N.S.W., Australia: Reader's Digest Services, 1978. 288 p. ISBN 0-909486-54-9. Australian $37.95. LC 79-306687.
2678
The most comprehensive single-volume atlas and gazetteer of Australia, utilizing 1:1,000,000 topographic maps prepared by the Division of National Mapping, with explanation for the layman. The physical and climatic section is followed by presentation of Aboriginal and European peoples, urbanization, farming, mining, manufacturing and individual cities.

Wards, Ian, ed. New Zealand atlas. Wellington, New Zealand: A.R. Shearer, Government Printer, 1976. 291 p. ISBN 0-477-01000-8. LC 76-381751.
2679 •
The official national atlas, an inventory of land and people, and the factors influencing both, by 31 contributors. Maps, text, and photographs are combined to present cultural and especially physical features.

D. GENERAL WORKS

Bellwood, Peter S. Man's conquest of the Pacific: the prehistory of Southeast Asia and Oceania. Auckland, New Zealand: William Collins; New York: Oxford University Press, 1978. 462 p. ISBN 0-00-216911-8. NZ $35. ISBN 0-19-520103-5. LC 78-59765. 2680

A fine comprehensive work invaluable for cultural geographers. Beginning with Southeast Asia and advancing into the Pacific, it covers populations, cultural and linguistic history, and subsistence patterns, then it moves from mainland and insular Asia into Melanesia, Micronesia, tropical Polynesia and New Zealand.

McKnight, Thomas. Australia's corner of the world: a geographical summation. Englewood Cliffs, NJ: Prentice-Hall, 1970. 116 p. ISBN 0-13-053801. LC 73-104897. 2681

A general introductory work, this clear, direct text describes Australia's Pacific setting, unique land, process of peopling, primary industries, urban civilization, changing international relationships, and its partner, New Zealand.

White, J. Peter (John), and O'Connell, James F. A Prehistory of Australia, New Guinea, and Sahul. Sydney, N.S.W., Australia; New York: Academic Press, 1979, 1983. 286 p. ISBN 0-12-746-750-5. US $29.50. LC 81-71781. 2682

A survey of contemporary knowledge of the prehistory of once-united Australia and New Guinea (Sahul), in largely chronological order. The interaction of people with changing environments--shorelines, climates, vegetation, fauna--is often in view, and cultural complexes and regions are discussed.

E. AUSTRALIA

The Australian Heritage Commission. The Heritage of Australia:the illustrated register of the National Estate. South Melbourne, Victoria, Australia: The Macmillan Company of Australia (in association with the Australian Heritage Commission), 1981. 1164 p. ISBN 0-333-33750-6. LC 81-188161. 2683 •

A official stock-taking of the "national estate," including 6600 places of environmental, historic, and cultural value. Each entry is described and illustrated, arranged by states and territories with four sub-regions. Essays cover the heritages of landform, nature, aboriginal life, and colonial architecture. Perhaps the first single-volume inventory for any country.

Burnley, I.H. The Australian urban system: growth, change, and differentiation. Melbourne, Victoria, Australia: Longman Cheshire, 1980. 339 p. ISBN 0-582-71101-0. Australian $12.25. (pbk). LC 81-104131. 2684

A comprehensive work, with conceptual background, broad aspects of the urban system, and the historical background to Australian urbanization the focus of the three parts. Both the system of cities and their internal characteristics are considered.

Cardew, Richard V.; Langdale, John V.; and Rich, David C., eds. Why cities change: urban development and economic change in Sydney. North Sydney, N.S.W., Australia; Boston, MA: Allen and Unwin (in association with the Geographical Society of New South Wales), 1982. 307 p. ISBN 0-86861-252-9. US $37.50. LC 82-229404. 2685

A searching analysis of factors causing urban change in general and in Sydney in particular. Financial and technological aspects are highlighted, temporal and spatial changes analysed, community impact discussed, and governmental response is advocated.

Courtenay, Percy P. Northern Australia: patterns and problems of tropical development in an advanced country. Melbourne, Victoria, Australia: Longman Cheshire, 1982. 335 p. ISBN 0-582-71476-1. Australian $15.95. LC 83-106459. 2686 •

A broad survey of an underdeveloped tropical region in a developed country, including past and present settlement patterns, pastoral, agricultural and mineral activities, the infrastructural aspects, and problems and prospects of urbanization and development of the tertiary sector. Stresses peripheral location remote from the developed Australian core.

Oceania

Dury, George H. and Logan, M.I., eds. Studies in Australian geography. Melbourne, Victoria, Australia: Heinemann Educational Australia, 1968. 368 p. LC 70-371193.
2687
A fine if uneven collection of essays ranging from geomorphology and biogeography through agricultural settlement to urban-industrial geography.

Gentilli, Joseph, ed. Climates of Australia and New Zealand. (World survey of climatology, v. 13, edited by H.E. Landsberg). Amsterdam, The Netherlands; New York: Elsevier Publishing Company, 1971. 405 p. ISBN 9-444-40827-4. $134. LC 71-103354.
2688
A somewhat technical, well-mapped climatological work by U. Radok, J. Gentilli, and W.J. Maunder. The region is related to Southern Hemisphere circulation, Australian climatic factors, dynamics, elements, and fluctuations; New Zealand's climatic elements and areas are also described.

Gentilli, Joseph. Western landscapes. (Sesquicentennial publications series). Nedlands, Western Australia, Australia: University of Western Australia Press, 1979. 526 p. (distributed by International Scholarly Book Services, Forest Grove, OR). ISBN 0-85564-155-X. LC 79-670404.
2689
A coherent anthology of papers, with maps and illustrations, published for Western Australia's sesquicentenary. Sections on land and nature, land and people, and western metropolis involve a geographical coverage.

Hanley, Wayne, and Cooper, Malcolm, eds. Man and the Australian environment: current issues and viewpoints. Sydney, N.S.W., Australia; New York: McGraw-Hill, 1982. 362 p. ISBN 0-07-072952-2. US $22. LC 82-125157.
2690•
A wide ranging anthology by geographers and other specialists on Australia's physical, economic, social, and political environments, stressing their holistic interrelationships and advocating balanced planning to control emergent problems in a marginal setting.

Heathcote, R.L. (Ronald Leslie). Australia. London; New York: Longman, 1975. 246 p. ISBN 0-582-48179-1. £6.95. LC 75-22092.
2691•
Quiet Australia, 1770, is compared to unquiet Australia, 1970. Part one focuses on land, seasons, ecosystems, and indigenes; part two on landscapes created by maritime, pastoral, agricultural, mining, and urban activities; and part three on varying visions of Australia.

Holmes, John H., ed. Man and environment: regional perspectives. Hawthorn, Victoria, Australia: Longman Australia, 1976. 261 p. ISBN 0-582-68227-4. Australian $13.40. LC 77-352797.
2692•
Initiated as an educational aid, this volume examines man-environment relations in regional settings. Subdivided between case studies of man's impact and of conflict in man-environment relations, the focus is on urban and rural Australian examples with some from Asia. A concluding section suggests guidelines for local study.

Jeans, D.N. (Dennis Norman), ed. Australia: a geography. Sydney, N.S.W., Australia: Sydney University Press, 1977. 571 p. ISBN 0-424-00036-9. Australian $25. LC 78-310917; New York: St. Martins Press, 1978. ISBN 0-312-06116-1. LC 77-85122.
2693
A superb and comprehensive (if sometimes technical) coverage of Australia's physical and human geography in 24 expert contributions with many maps and illustrations. Systematic in approach, with physical and human aspects to be divided into two volumes in forthcoming revision.

Jeans, D.N. (Dennis Norman). An Historical geography of New South Wales to 1901. Sydney, N.S.W., Australia: Reed Education; Melbourne, Victoria, Australia: Longman Cheshire, 1972. 328 p. Australian $13.50. LC 73-150851.
2694
A fine analysis of early European response and adjustment to a strange land, and of successive locational shifts as pastoralism, gold, wheat, sugarcane, and dairying developed and contributed to regional formulation.

Learmonth, Nancy, and Learmonth, Andrew. Regional landscapes of Australia: form, function, and change. Sydney, N.S.W., Australia: Angus and Robertson; London: Heinemann, 1971. 493 p. LC 72-197955.
2695 •
Written to meet the need for a descriptive regional coverage of Australia, this analysis moves from region to region in non-technical language accompanied by telling illustrations and maps. With seven major regions subdivided into a total of 53 subregions, coverage is both panoramic and detailed.

Leeper, G.W., ed. The Australian environment. 4th ed. of CSIRO publication under the same title. Melbourne, Victoria, Australia: Commonwealth Scientific and Industrial Research Organization, 1970. 163 p. (1st ed., 1949). LC 76-23890.
2696
A substantially rewritten edition of an established work, covering most elements of the physical and bioclimatic environment, introduced pastures and crops, forestry, livestock, and native fauna. Many maps and illustrations.

Linge, G.J.R. (Godfrey). Industrial awakening: a geography of Australian manufacturing 1788 to 1890. Canberra, ACT, Australia: Australian National University Press, 1979. 846 p. (Distributed by Books Australia, Norwalk, CT). ISBN 0-7081-0419-3. $39.95. LC 78-73140.
2697
A thoroughly researched, almost encyclopedic account of the evolution of manufacturing as remote colonial Australia was progressively involved in global commerce and technology. Data are presented colony by colony and era by era, with accompanying maps, tables, and interlinking essays.

Lonsdale, Richard E. and Holmes, John H., eds. Settlement systems in sparsely populated regions: the United States and Australia. New York: Pergamon, 1981. 397 p. ISBN 0-08-023111-X. $44. LC 80-27278.
2698
Australian sparseland development is compared with the American frontier by means of demographic, functional, social, behaviorial, educational, and health characteristics outlined in 18 papers. Case studies are combined with broad principles and policy guidelines.

Meinig, Donald W. On the margins of the good earth: the South Australian wheat frontier, 1869-1884. (Association of American Geographers, monograph series, no. 2). Chicago, IL: Rand McNally, (Now Washington DC: Association of American Geographers); London: J. Murray, 1962. 231 p. US $3.95. LC 62-7266. NUC 65-45000.
2699
An historical geography which documents the resolution of the boundary problems between plowed land and grazing land in the colony of South Australia. A study of selected features of the colonization process.

Powell, J.M. (Joseph Michael), ed. The Making of rural Australia, environment, society, and economy: geographical readings. Melbourne, Victoria, Australia: Sorrett Publishing, 1974. 179 p. ISBN 0-909752-12-5. LC 75-325009.
2700 •
A collection of readings on the historical evolution and contemporary characteristics of rural Australia. Largely genetic and regional, these readings focus on pioneer settlement processes, central authority, drought, irrigation, economic farm structure, international trade, and resource conflicts. A companion volume to Urban and industrial Australia *(below).*

Powell, J.M. (Joseph Michael), ed. Urban and industrial Australia: readings in human geography. Melbourne, Victoria, Australia: Sorrett Publishing, 1974. 252 p. ISBN 0-909752-13-3. LC 75-329830.
2701 •
An anthology of significant papers selected to elucidate the processes of urban and industrial change, grouped according to political factors, city systems, city industry, internal structure, retailing, regional linkages, and continuing resilience. A companion volume to The making of rural Australia *(above).*

Oceania

Powell, J.M. (Joseph Michael), and Williams, Michael, eds. Australian space, Australian time: geographical perspectives. Melbourne, Victoria, Australia; New York: Oxford University Press, 1975. 256 p. ISBN 0-19-550456-9. Australian $25. ISBN 0-19-550463-1. Australian $14.95. (pbk). LC 75-328436.　　　　　　　　　　2702
　　Selected to illuminate goals and processes of geographical change during the development era. Seven contributors discuss the impress of central authority, conservation, rural settlement intensification, urbanization, industrialization, transportation, and federal-state relations.

Scott, Peter. Australian agriculture: resource development and spatial organization. (Geography of world agriculture, v. 9). Budapest, Hungary: Akadémiai Kiadó, 1981. 151 p. ISBN 9-630-52445-7. US $15. LC 81-190315.　　　　　　　　　　　2703
　　Perhaps the most detailed survey of Australian agricultural geography, elucidating and mapping distribution and economic organization of agriculture, the resource base, more than 20 different crop and livestock systems, and the potential for expansion of production and export.

Seddon, George, and Davis, Mari, eds. Man and landscape in Australia: towards an ecological vision. (UNESCO Programme on Man and the Biosphere, no. 2). Canberra, ACT, Australia: Australian Government Publishing Service, 1976. 373 p. ISBN 0-642-02119-8. Australian $10. LC 78-313249.　　　　　　　　　　　　　　　2704
　　Directed towards the themes of perception and remodelling of the landscape, 27 papers are clustered into five sections focusing on perceptual attitudes in the arts, the built environment, and changing perceptions of Australia's environments.

Spate, O.H.K. (Oscar Hermann Khristian). Australia. Melbourne, Victoria, Australia: Lothian; London: Benn; New York: Praeger, 1968. 328 p. LC 68-19862.　　2705
　　A geographer's witty and sympathetically critical survey of Australia with emphasis on social, economic, and political (as well as broadly geographical) aspects.

Williams, Michael. The Making of the South Australian landscape: a study in the historical geography of Australia. London; New York: Academic Press, 1974. 518 p. ISBN 0-12-785995-1. US $69.50. LC 73-9483.　　　　　　　　　　　　　2706•
　　A well-written and illustrated account of the formation of the humanized landscape, with sequential analyses of human impact on woodland, swamps, desert, and soils, and the building of townships and Adelaide.

Wilson, R.K. (Robert Kent). Australia's resources and their development. Sydney, N.S.W., Australia: University of Sydney, Department of Adult Education, 1980. 406 p. ISBN 0-8580-023-5. Australian $12. LC 79-3928.　　　　　　　　　　2707•
　　An enlarged successor to John Andrew Australian resources and their utilization (University of Sydney, Department of Tutorial Classes, 1953, 1965), this volume provides a panaoramic view of continental resources, development history, and social environment, then presents descriptions of rural industries, mining, manufacturing, services and trade. It concludes with post-1945 developments and comments on current issues.

F. NEW ZEALAND

Anderson, A. Grant (Allan), ed. The Land our future: essays on land use and conservation in New Zealand in honour of Kenneth Cumberland. (New Zealand Geographical Society special publications, miscellaneous series no. 7,). Auckland, New Zealand: Longman Paul, 1980. 324 p. ISBN 0-582-71774-4. NZ $ 22.95. (pbk). New Zealand Geographical Society, NZ $25.　　　　　　　　　　　　　　　　　　2708
　　An interdisciplinary but well-focused festschrift by 20 contributors. The content is both empirical and theoretical with emphasis on human-land linkages and conservation in relationship to soils, forestry, farming, water resources, mountains, the rural-urban fringe and the future of farming.

Bedford, Richard D. and Sturman, Andrew P., eds. Canterbury at the crossroads: issues for the eighties. (New Zealand Geographical Society special publications, miscellaneous series no. 8). Christchurch, New Zealand: New Zealand Geographical Society, 1983. 450 p. NZ $12.50.

A wide-ranging collection of papers by geographers and other specialists on the varied environments of Canterbury Province, the management of water, land, and air resources, changes in town and country, and the problems and prospects of regional stagnation or development.

2709

Cumberland, Kenneth B. Landmarks. Surry Hills, N.S.W., Australia: Reader's Digest Services, 1981. 304 p.

Lucidly written and tellingly illustrated, this popular but professional volume (produced to accompany a T.V. series), describes the formation of the New Zealand landscape from pre-European to present times. Ten chapters sequentially cover the peoples and technologies involved, mini-biographies highlight individual landscape makers, and a diversified future is prognosticated.

2710•

Cumberland, Kenneth B. and Whitelaw, James S. New Zealand. (World's landscapes series). Harlow, U.K.: Longmans; Chicago, IL: Aldine, 1970. 194 p. LC 73-110624.

Traces the New Zealand landscape from pre-human through Polynesian to European eras, places contemporary landscapes in a gradient from "wild" through agricultural to urban, and discusses future potentials.

2711

Franklin, S. Harvey (Samuel). Trade growth and anxiety: New Zealand beyond the welfare state. Wellington, New Zealand: Methuen New Zealand, 1978. 402 p. ISBN 0-456-02320-8. LC 79-313386.

A geographer's vigorously expressed view that approaches serving socio-economic egalitarianism inhibit essential economic restructuring. Mores, population, urbanization, trade, farming, manufacturing, and differential regional changes are examined.

2712

Hargreaves, R.P. (Raymond Philip), and Heenan, L.D.B. (Leonard). An Annotated bibliography of New Zealand population. Dunedin, New Zealand: University of Otago Press, 1972. 230 p. ISBN 0-908569-11-4. NZ$8.95. LC 72-195977.

A comprehensive bibliography of published material on multiple aspects of New Zealand's population, including historical changes, spatial distribution, demographic structure, and ethnic composition. A supplement is in preparation.

2713

Johnston, Ronald J., ed. Society and environment in New Zealand. Christchurch, New Zealand: Whitcombe and Tombs, 1974. 204 p. LC 75-329218.

A largely descriptive but still valuable account of aspects of New Zealand geography by ten contributors. Chapters cover landform systems, climatic resources, historical development, population, regionalization of agricultural systems, urban patterns, resource management, and landscape.

2714

Johnston, Ronald J., ed. Urbanization in New Zealand. Wellington, New Zealand: Reed Education; now Auckland: Longman Paul, 1973. 328 p. ISBN 0-589-04847-3. NZ $5.95. (pbk). LC 74-167119.

A useful if somewhat dated anthology of fifteen essays on national and regional urban systems, the roles of transport, migration, ethnic groups, and residential patterns. Emphasis is placed on mapping, spatial analysis, and environmental management.

2715

McLintock, Alexander H., ed. An Encyclopedia of New Zealand. Wellington, New Zealand: R.E. Owen, Government Printer; New York: International Publications Service, 1966. 3 vol., v.1: 928 p; v.2: 894 p.; v.3: 848 p. LC 67-4443.

An encyclopedia containing an unusual concentration of relevant data and geographical articles: economic geography, resources, regional studies, etc.

2716•

Oceania

Moran, W. (Warren), and Taylor, M.J., eds. Auckland and the Central North Island. Auckland, New Zealand: Longman, Paul, 1979. 257 p. ISBN 0-582-71767-1. NZ $11.95. (pbk).
    A collection of 13 papers originally designed to illuminate an I.G.U. geographical excursion but revised to gain wide use in education. The first five papers emphasize Auckland city and its environs: others deal with geomorphology, energy, soils, farming, and historical geography.

2717

Neville, R.J. Warwick, and O'Neill, C. James, eds. The Population of New Zealand: interdisciplinary perspectives. Auckland, New Zealand: Longman, Paul, 1980. 339 p. ISBN 0-582-71771-X. NZ $16.95. LC 80-490159.
    An interdisciplinary collection of 13 essays by social scientists, including three geographers, centered on population and the linkages between different factors. Migration, demographic structure, redistribution, labor force, and the connections between population and economy, social policy, environment, and the future are explored.

2718

Wilkes, C. and Shirley, I., eds. In the public interest: health, work, and housing in New Zealand society. Auckland, New Zealand: Benton Ross, 1984. 307 p.
    Considers whether the public interest is being served in relation to health, work, and housing. The conflict between the private and public interest is charted together with chapters on the history of policy developments and the effect of state activity or lack thereof on the consumer. Offering a variety of perspectives, this book illustrates the importance of analysis rather than simple description.

2719

G. PACIFIC ISLANDS

British Admiralty, Naval Intelligence Division. The Pacific islands. (Great Britain, Naval Intelligence Division, Geographical Handbook series). Cambridge, U.K.: Cambridge University Press for His Majesty's Stationary Office, 1945. 4 v.: v.1: 599 p.; v.2: 741 p.; v.3: 739 p.; v.4: 526 p.
    Prepared for use during the Second World War by an interdisciplinary team, this work still provides a very useful historical background and the most comprehensive coverage of the physical geography of the Pacific islands. The four volumes respectively center on a General Survey, the Eastern Pacific, the Western Pacific from Tonga to the Solomons, and the Western Pacific including New Guinea and Micronesia.

2720

Brookfield, Harold C., ed. The Pacific in transition: geographical pers ectives on adaptation and change. London: Edward Arnold; New York: St. Martin's Press, 1973. 332 p. LC 73-82818.
    A combination of papers from 12 contributors with extensive experience in Oceania and sometimes divergent viewpoints. Though half the contributions focus on aspects of Papua New Guinea, papers on Vanuatu (New Hebrides), Fiji, Tonga, and Pacific maritime economies are included.

2721

Brookfield, Harold C. and Hart, Doreen. Melanesia: a geographical interpretation of an island world. London: Methuen, 1971. 464 p. LC 71-579836.
    A splendid regional overview fusing relevant approaches from traditional and contemporary geography. A detailed synthesis of environment, peopling, human ecology, subsistance and cash economy is followed by an evaluation of the development process, with emphasis on the dual economy, trade, central places, and continuity and change.

2722

Bryan, Edwin H. Jr. Guide to place names in the Trust Territory of the Pacific islands. Honolulu, HI: Bishop Museum, 1971. LC 72-31246.
    A systematic identification of places in the Marshall, Caroline, and Mariana groups (excluding Guam), with relevant maps. For each unit, places are listed clockwise from the northernmost point. Alternative U.S., Japanese, and indigenous names are tabulated, and geographic coordinates given.

2723

Craig, Robert D. and King, Frank P., eds. Historical dictionary of Oceania. Westport, CT; London: Greenwood Press, 1981. 392 p. ISBN 0-313-21060-8. US $55. LC 80-24779.
2724•

Though few contributors are geographers, this encyclopedic volume assembles many otherwise scattered data on the archeology, anthropology, history, and geography of Oceania, especially Micronesia.

Doumenge, François. L'Homme dans le Pacific Sud: étude géographique. (Publications de la Société des Oceanistes no. 19). Paris: Musée de l'Homme, 1966. 633 p. FFr 100. LC 67-40610.
2725

An invaluable, well-illustrated and comprehensive topical survey of the physical and human geography of the South Pacific Islands, with particular strength on French Oceania.

Howlett, Diana. Papua New Guinea: geography and change. Revised, expanded, and metricated edition. Melbourne, Victoria, Australia: Thomas Nelson, 1973. 180 p. (1st ed., Geography of Papua and New Guinea, 1967, 159 p., reprinted 1971). ISBN 0-17-002169-6. Australian $5.95. LC 74-176821.
2726

Within the framework of the human-environment system, this study of Papua New Guinea on the eve of independence moves from traditional society and physical environment to resource use by subsistence economies, expatriates, transitional villagers, towns, and industries. The socio-technological interaction of old and new is emphasized.

Kissling, Christopher C., ed. Transport and communications for Pacific microstates: issues in organisation and management. Suva, Fiji: Institute of Pacific Studies, University of the South Pacific, 1984. 192 p. $8. $6. (pbk).
2727

Nowhere else in the world are populations so scattered and dependent on long-range international transport and communications. Eleven experts tell in simple language of the acheivements, the problems, and the potentials in this aspect of Pacific Islands development.

Levison, Michael; Ward, R. Gerard (Ralph); and Webb, John W. The Settlement of Polynesia: a computer simulation. Minneapolis, MN: The University of Minnesota Press, 1973. 137 p. LC 72-92337.
2728

A computer examination of divergent theories on the means whereby Polynesians reached the islands and established contact areas. Significant for review of literature, methodology, and results.

May, R.J. (Ronald James), and Nelson, Hank, eds. Melanesia: beyond diversity. Canberra, ACT, Australia: The Australian National University, Research School of Pacific Studies, 1982. 2 v. 690 p. (Distributed in the U.S. by Miami, FL: Books Australia). ISBN 0-86784-045-5. LC 82-73076.
2729

A broad-ranging, well-organized sequence of papers by experts. The first volume considers natural and cultural aspects, prehistory, and history; the second covers recent developments especially in agriculture, transportation, and socio-political life.

Mitton, Robert D. The Lost world of Irian Jaya. Melbourne, Victoria, Australia; New York: Oxford University Press, 1983. 234 p. ISBN 0-19-554368-8. Australian $50. LC 83-144617.
2730

A superb book on an area originally little known and now seldom visited. Magnificent photos and a sensitive and perceptive text by a mining geologist. Published posthumously from his papers.

Oliver, Douglas L. The Pacific islands. Cambridge, MA: Harvard University Press, 1962. 313 p. Reprinted Honolulu HI: University of Hawaii Press, 1975. ISBN 0-8248-0397-3. $8.95. LC 75-19352.
2731

Much is outdated by subsequent research in prehistory and recent post-colonial political changes, but the book stands as an anthropologist's historical overview of history, society, and economy.

The Tropics

Pacific Islands yearbook. 1- (1930-). Irregular. Pacific Publications Pty. Ltd., GPO
Box 3408, Sydney, N.S.W., 2001, Australia. ISSN 0078-7523. (15th ed.: 1984. 557 p.
$60).   2732•

*An essential reference, with up-dated information on all Pacific island countries
and territories. Coverage includes physical geography, socio-economic conditions,
history, demography, trade, and transport. Each area has a small-scale black-and-
white map; a back pocket holds a larger overview map. Revisions are intermittently
published.*

Ward, R. Gerard (Ralph), ed. Man in the Pacific islands: essays on geographical
change in the Pacific islands. Oxford, U.K.: Clarendon Press, 1972. 339 p. ISBN 0-19-
823210-1. LC 73-150468.   2733

*A collection of eleven essays focused on change with case studies. Topics include
prehistory, vegetation, the "beche-de-mer" trade, the labor trade, plantations, Ma-
katea phosphate, population growth, land tenure, horticulture, and urbanization.*

Ward, R. Gerard (Ralph), and Proctor, Andrew, eds. South Pacific agriculture:
choices and constraints. Canberra, ACT, Australia; Norwalk CT: Australian National
University Press, 1980. 525 p. ISBN 0-7081-1944-1. U.S. $21. LC 79-56229.   2734

*A comprehensive survey, by an interdisciplinary team, of development issues in the
tropical Pacific. Forestry, fishing, and particularly agricultural potentials and problems are examined with special reference to Kiribati, the Cook Islands, Tonga, Western Samoa, the Solomon Islands, Fiji, and Papua New Guinea.*

Winslow, John H., ed. The Melanesian environment. Canberra, ACT, Australia: Australian National University Press, 1977. 562 p. ISBN 0-7081-0824-5. Australian $12. LC
78-309200.   2735

*A useful collection of papers from the Ninth (1975) Waigani Seminar, largely on Pa-
pua New Guinea but including most of Melanisia, and Hawaii. Tensions between development and conservation are highlighted as environment is related to pre-history,
agriculture, fishing, mining, forestry, tourism, health, and planning.*

15. THE TROPICS

Connie Weil

A. SERIALS

Malaysian journal of tropical geography. 1- (1980-). Semiannual. University of Malaya, Department of Geography. Kuala Lumpur; 22-11, Malaysia. SN 82-21062.   2736

Supersedes in part Journal of Tropical Geography *1-49. International authorship.
In English.*

Singapore journal of tropical geography. 1- (1980-). Semiannual. National University of Singapore, Department of Geography, Kent Ridge, Singapore 0511, Singapore.
ISSN 0129-7619. SN 82-644401.   2737

*Substantial scholarly and well-illustrated articles on equatorial areas in Asia, Afri-
ca, Latin America, and the Pacific. International authorship of articles. In English.
Supersedes in part* Journal of Tropical Geography *1-49 (1953-1979). Cumulative indexes: 1-29 (1953-1969), 30-39 (1970-1974).*

B. BOOKS

Bates, Marston. Where winter never comes: a study of man and nature in the tropics.
New York: Charles Scribner's Sons, 1952. 310 p. LC 52-6464.   2738

*Although this natural history is dated in many ways, it remains a uniquely readable
and insightful work by a well trained naturalist in love with the tropics. It serves as an
antidote to the perjorative view of tropical lands and peoples that so often permeates
ostensibly objective, scientific writing.*

Bourlière, François, ed. Tropical savannas. (Ecosystems of the world, v. 13). Amsterdam, The Netherlands; New York: Elsevier Scientific Publishing Co., 1983. 730 p. ISBN 0-0444-42035-5. $197.75. LC 81-19415.　　　　　　　　　　　　　　　　　　2739

This is the best overall reference work on tropical savannas. It is one volume in an ambitious encyclopedia of terrestial and aquatic ecosystems. With a few exceptions, the authors are biologists. In addition to papers which survey tropical savanna paleogeography, climate, soils, and life forms, several papers examine the tropical savannas of particular parts of the world (Africa, India, Southeast Asia, Australia and Southwest Pacific, and Latin America).

Crowder, Loy V. and Chheda, H.R. Tropical grassland husbandry. London; New York: Longman, 1982. 562 p. ISBN 0-582-46677-6. $60. LC 82-15239.　　　　　2740

This textbook serves well as a reference work. The first chapters concisely survey the characteristics and distribution of tropical climates, soils, grasses, and legumes. The rest of the volume is a practical guide to tropical grassland management. Socioeconomic aspects receive only cursory treatment.

Ewusie, J. Yanney. Elements of tropical ecology, with reference to the African, Asian, Pacific and New World tropics. London; Portsmouth, NH: Heinemann, 1980. 205 p. ISBN 0-435-93700-6. $16.50. (pbk).　　　　　　　　　　　　　　　　　2741 •

This clearly organized and written introductory textbook provides a succinct survey of the physical and biological characteristics of the latitudinal tropics. Photographs and diagrams elucidate processes and relationships well. Examples from specific places are frequent and apt but not detailed. Socioeconomic issues receive minimal coverage.

Golley, Frank B. and Werger, M.J., eds. Tropical rain forest ecosystems: part A: structure and function. (Ecosystems of the world, v. 14A). Amsterdam, The Netherlands; New York: Elsevier Scientific Publishing Company, 1982. 382 p. ISBN 0-444-41986-1. $112.75. LC 81-7861.　　　　　　　　　　　　　　　　　　　　　　　　2742

This is the best general reference on the biological ecology of tropical rain forests. This volume focuses on environmental characteristics and ecosystemic processes. Volume 14B in this encyclopedic series on terrestrial and aquatic ecosystems also will be devoted to tropical rain forests, but it has yet to be published or even titled.

Goodland, Robert J.A.; Watson, Catharine; and Ledec, George. Environmental management in tropical agriculture. Boulder, CO: Westview Press, 1984. 237 p. ISBN 0-86531-715-1. $20. (pbk). LC 83-14807.　　　　　　　　　　　　　　　　　　2743

All three authors of this volume have been associated with the Office of Environmental Affairs at the World Bank, but this is not a World Bank publication. The book argues that economic development efforts must emphasize sustainable agriculture. A little more than half of the book discusses specific tropical crops. The remainder addresses such issues as integrated pest management and soil erosion.

Gourou, Pierre. The Tropical world: its social and economic conditions and its future status. Translated from French by S.H. Beaver. 5th ed. London: Longman, 1980. 190 p. (1st ed., 1953. 156 p.). ISBN 0-582-30012-6. £5.95. (pbk). LC 79-41083. (Pays Tropicaux. Paris: Presses Universitaires de France, 1947. 196 p.).　　　　　　　　　　2744 •

This updated version of Gourou's classic consideration of the humid tropics has 16 new maps and 40 photographs. Although population, disease ecology, industrial potential, and "problems due to European intervention" are treated, the emphasis is on agriculture, particularly traditional. A conversational style and fascinating details make this a highly readable introduction.

Grigg, David. The Harsh lands: a study in agricultural development. London: Macmillan; New York: St. Martin's Press, 1970. 321 p. LC 73-98070.　　　　　　　　2745

Grigg's Harsh Lands *include arid lands (with an emphasis on tropical and subtropical areas), the humid tropics, savannas, and the monsoon lands of South Asia. The book explores the reasons for inadequate food production in these places. Environmental and socioeconomic facets are treated equally well.*

Hames, Raymond B. and Vickers, William T., eds. Adaptive responses of native Amazonians. New York: Academic Press, 1983. 516 p. ISBN 0-12-321250-2. $49. LC 82-18399.　　　　　　　　　　　　　　　　　　　　　　　　　　　　2746

Although the contributors to this book are all ecological anthropologists and the subject matter is specialized, this volume deserves the attention of geographers interested in the tropics. It provides detailed generally rigorous case studies of traditional peoples' adaptation to a tropical environment. The major themes are shifting cultivation, hunting and fishing, nutrition, and settlement patterns.

Hanson, Haldore; Borlaug, Norman E.; and Anderson, R. Glenn. Wheat in the Third World. Boulder, CO: Westview Press, 1982. 174 p. ISBN 0-86531-357-1. $19. LC 82-8478.　　　　　　　　　　　　　　　　　　　　　　　　　　　　2747

This specialized monograph is included here because it deals with recent developments in "green revolution" technology, of which the authors are loyal supporters. The authors have worked with the Mexican Wheat Program jointly sponsored by the Mexican Government and the Rockefeller Foundation. Case study reports are included for India, Bangladesh, Brazil, and other countries in addition to Mexico.

Harris, David R., ed. Human ecology in savanna environments. London; New York: Academic Press, 1980. 522 p. ISBN 0-12-326550-9. $57.50. LC 80-40210.　　　　2748

This book is the product of a Wenner-Gren sponsored conference. Harris provides an overview of tropical savanna environments, then 22 papers of exceptional quality are presented under three major headings: resource use (e.g., variations in agriculture and pastoralism in West African savannas); management problems (with emphasis on adaptation to seasonality and on 'development' issues); and human biology.

Hodder, B.W. Economic development in the tropics. 3rd ed. London; New York: Methuen, 1980. 225 p. (1st ed., 1968. 258 p.). ISBN 0-416-74250-5. $9.95. ISBN 0-416-74260-2. $9.95. (pbk). LC 74-18546.　　　　　　　　　　　　　　　　　　2749•

This is a substantially revised version of Hodder's analysis of tropical "development." Hodder has not abandoned his faith in planning. He describes the climates, soils, and vegetation of the tropics as well as systems of tropical agriculture, but he emphasizes socioeconomic issues. He concludes this edition with case studies of Brazil, Nigeria, and India.

Jordan, Carl F., ed. Tropical ecology. Stroudsburg, PA: Hutchinson Ross Publishing Company, 1981. 356 p. ISBN 0-87933-398-7. $45. LC 81-4260.　　　　　　　2750

The papers in this volume represent many disciplines and have been reprinted from numerous sources. The volume is particularly appropriate for libraries which do not have a comprehensive collection of scientific periodicals. The major themes are the abundance and diversity of species and ecosystemic processes in the tropics.

Kamarck, Andrew M. The Tropics and economic development: a provocative inquiry into the poverty of nations. Baltimore, MD: Johns Hopkins University Press for the World Bank, 1976. 113 p. ISBN 0-8018-1891-5. $12.50. ISBN 0-8081-1903-2. $5. (pbk). LC 76-17242.　　　　　　　　　　　　　　　　　　　　　　　　　　2751

Dismissing the claim that Third World poverty is largely the result of exploitation, Kamarck argues that inherent features of tropical climate are obstacles to economic productivity and human health. For example, the diversity of species in the humid tropics maximizes crop pests and human disease agents while making lumber extraction costly.

Manshard, Walther. Tropical agriculture: a geographical introduction and appraisal. Translated from German by D.A.M. Naylon. London; New York: Longman, 1974. 226 p. ISBN 0-582-48187-2. $17.95. (pbk). LC 73-88902. (Agrargeographie der Tropen. Mannheim, W. Germany: Bibliographisches Institut, 1968. 307 p.).　　　　　2752

Manshard examines the natural ecological zones (ranging from rainforest to desert), dietary patterns, and socioeconomic "regions" of the tropics. He surveys major types of tropical agriculture, including a brief but interesting description of "game cropping" or wildlife management. Other topics in this wide-ranging synthesis include disease ecology, altitudinal variation in ecological zonation, and the "green revolution."

Moran, Emilio F., ed. The Dilemma of Amazonian development. Boulder, CO: Westview Press, 1983. 347 p. ISBN 0-86531-373-3. $20. LC 82-23693.   2753
Despite this book's specific focus on the Amazon Basin, its themes and insights are relevant for tropical "development" elsewhere. The papers by social scientists stress the impact of development policies on local communities and overall equity. Other contributors emphasize the diversity of climate, soils, and vegetation within Amazonia.

National Research Council. Committee on selected biological problems in the humid tropics. Ecological aspects of development in the humid tropics. Washington, DC: National Academy Press, 1982. 297 p. ISBN 0-309-03235-0. $14.50. (pbk). LC 82-3620.   2754
This is the first of four publications planned to review technical knowledge necessary for sound natural resource management. This volume concerns the humid tropics. Themes of specific chapters include environmental constraints on "development," characteristics of tropical ecosystems, conservation of genetic resources, tropical agriculture, and soil management. The sources cited are multi-disciplinary and up-to-date.

Owen, D.F. (Denis Frank). Man in tropical Africa: the environmental predicament. New York: Oxford University Press, 1973. 214 p. ISBN 0-19-519746-1. LC 73-85079.   2755
Although this volume treats only tropical Africa, the issues it addresses are relevant for tropical regions elsewhere. Contrasts between rural and urban life; agricultural adaptations to environment; "weeds, pests, and diseases of cultivation"; food supply; and disease ecology are among the topics examined with unusual sophistication.

Riehl, Herbert Climate and weather in the tropics. London; New York: Academic Press, 1979. 611 p. ISBN 0-12-588180-0. $67.50. LC 78-73890.   2756
The author and the intended audience for this comprehensive volume are atmospheric scientists, but geographers interested in factual information about tropical climate and weather will find it here. Although parts of the book are too technical for the non-specialist, the basic description is lucid, and examples of phenomena are reported for specific places.

Ruthenberg, Hans, with contributions by J.D. MacArthur, H.D. Zandstra, and M.P. Collinson. Farming systems in the tropics. Oxford, U.K.: Clarendon Press; New York: Oxford University Press, 1980. 424 p. Reprinted 1983. (1st ed., 1971. 313 p.). ISBN 1-19-859482-8. $39.95. LC 79-41135.   2757
This overview of types of tropical agriculture includes regulated ley and irrigation farming, traditional pastoralism, and ranching in addition to the predictable shifting and semi-permanent cultivation. An excellent introduction to the characteristics and distribution of each type of agriculture is illustrated with case studies from Africa and/or Asia.

Senior, Michael. Tropical lands: a human geography. London; New York: Longman, 1979. 316 p. ISBN 0-582-60353-6. $10.95. (pbk). LC 79-670352.   2758•
After a brief but competent survey of the tropical world's environmental characteristics, this introductory textbook describes its population and economy. Well chosen examples from Latin America, Africa, Asia, and the Pacific illustrate the book's themes, which include migration, urbanization, agriculture, and tourism among others. The book has numerous maps and photographs.

Thomas, Michael F. Tropical geomorphology: a study of weathering and landform development in warm climates. New York: John Wiley and Sons; London: Macmillan, 1974. 332 p. LC 73-13428.   2759
This volume analyzes the tropical morphogenic processes, then surveys the characteristic landforms of the tropics. The description is terse but comprehensive. Reflecting the author's fieldwork experiences, most of the examples are drawn from West Africa, Papua New Guinea, and Australia. Numerous photographs supplement the text.

Warren, Kenneth S., and Mahmoud, Adel A.F., eds. Tropical and geograpical medicine. New York; Maidenhead, Berks, U.K.: McGraw-Hill Book Company, 1984. 1175 p. ISBN 0-07-068327-1. $75. LC 83-11959.   2760
  This is the most comprehensive reference on tropical diseases which embodies an ecological perspective. Most of the papers explore a specific disease or group of related diseases. Nearly all address geographical facets of disease: distribution; environmental requisites for survival of pathogens; and, where appropriate, the impact of landscape alteration on health patterns.

16. ARID LANDS

Douglas L. Johnson

A. BIBLIOGRAPHIES

Hopkins, Stephen T. and Jones, Douglas E., with the technical assistance of Rodgers, John A. Research guide to the arid lands of the world. Phoenix, AZ: Oryx Press, 1983. 391 p. ISBN 0-98774-066-1. $74.50. LC 83-42500.   2761
  Entries are grouped by research topic and country, and estimates of dryland degradation risk are made for each continent and country. This guide's review of basic reference material makes it an important introductory resource.

Paylore, Patricia, comp. and ed. Desertification: a world bibliography. (Prepared for the 23rd International Geographical Congress, Moscow, 1976; Pre-conference meeting of the IGU Working Group on Desertification at the Desert Research Institute, Ashkhabad, Turkmen SSR, USSR). Some annotations in French and German; Russian references translated by Theodore Guerchon and Janice Chlopowicz. Tucson, AZ: Office of Arid Lands Studies, University of Arizona, 1976. 644 p. LC 76-376355.   2762
  Although much work on desertification has taken place since this massive compilation, it remains a basic introduction to the literature of the field. 1409 references in English, French, and German, many of them annotated, are organized by continent and country. These are supplemented by 237 references translated from Russian and grouped by topics. Cross-indexed by key words. A supplemental volume published in 1980 updates the earlier version.

Paylore, Patricia, and Mabbutt, J.A., comps. and eds. Desertification world bibliography update, 1976-1980. Tucson, AZ: University of Arizona. Office of Arid Land Studies, 1980. 196 p. $15 plus $4 postage (order through the University).   2763
  Updates the 1976 bibliography. Includes 402 annotated references. Takes account of the 1977 United Nations Conference on Desertification.

Paylore, Patricia, ed. Arid lands research institutions: a world directory. Revised and updated ed. Tucson, AZ: The University of Arizona Press, 1977. 317 p. (1st ed., 1967. 268 p.). ISBN 0-8165-0631-0. $12.50. (pbk). LC 77-8982.   2764
  The research interests and history of more than 160 academic and governmental institutions are organized by continent and country. This is the most comprehensive guide available, and its utility is enhanced by a staff name and subject index.

B. GENERAL WORKS

Bender, Gordon L., ed. Reference handbook on the deserts of North America. Westport, CT; London: Greenwood Press, 1982. 594 p. ISBN 0-313-21307-0. $75. LC 80-24791.   2765
  Detailed scholarly articles on desert regions and systematic topics are supplemented by massive bibliographies and useful appendices listing research areas and facilities.

Hills, E.S. (Edwin Sherbon), ed. Arid lands: a geographic appraisal. London: Methuen; Paris, France: UNESCO, 1966. 461 p. ISBN 0-412-20750-8. £4.40. LC 67-72563.   2766
Twenty essays by a superb cadre of scholars provides an indispensible baseline for new knowledge garnered during the last two decades.

C. DESERTIFICATION

Eckholm, Eric, and Brown, Lester R. Spreading deserts: the hand of man. (Worldwatch paper no. 13). Washington, DC: Worldwatch Institute, 1977. 40 p. ISBN 0-916468-12-7. $2. LC 77-81479.   2767•
This lively and provocative overview manages to avoid excessive oversimplification despite its brevity.

Glantz, Michael H., ed. Desertification: environmental degradation in and around arid lands. Boulder, CO: Westview Press, 1977. 346 p. ISBN 0-89158-115-4. $34.50. LC 77-3901.   2768
Twelve essays address the causes and consequences of, and possible solutions to, the degradation of the world's dryland ecosystems. One half of the essays deal with the African land use problems while the remainder present global and regional scale aspects of the problem.

Sheridan, David. Desertification of the United States. Washington, DC: Council on Environmental Quality, 1981. 142 p. $4.25. LC 81-602136.   2769•
Bad conservation practices characterize the use of much of the United States' arid zone. Sheridan's readable account systematically documents the evidence for overgrazing, soil erosion, ground water depletion, salinization, and mining disturbances in the American West.

Spooner, Brian, and Mann, H.S., eds. Desertification and development: dryland ecology in social perspective. London; New York: Academic Press, 1982. 407 p. ISBN 0-12-658050-2. $56.   2770
An extremely important contribution to the literature on dryland management, this study examines both the social causes and combative techniques associated with desertification. Several surprisingly counter-intuitive cautionary examples from Africa and elsewhere are counter-pointed with the Indian and Iranian experience of dryland resource exploitation.

United Nations Conference on Desertification, Secretariat. Desertification: its causes and consequences. New York; Oxford; Toronto: Pergamon Press, 1977. 448 p. ISBN 0-08-022023-1. $97. ISBN 0-08-022-395-8. $45. (pbk). LC 77-81423.   2771
Four state-of-the-art summaries of mid-seventies knowledge of desertificaton are supplemented by an overview essay. Global reviews of climate, ecological change, human consequences, and technological solutions provided the scientific basis for the anti-desertification plan of action developed at the United Nation's Conference on Desertification.

Walls, James. Land, man, and sand: desertification and its solution. New York: Macmillan, 1980. 336 p. ISBN 0-02-699810-6. $19.95. LC 79-7852.   2772•
Fifteen case studies prepared for the United Nations Conference on Desertification are summarized in approachable prose that makes available to the non-specialist the best documented examples of success and failure in arid land management.

Arid Lands

Cooke, Ronald U. and Warren, Andrew. Geomorphology in deserts. Berkeley; Los Angeles, CA: University of California Press, 1973. 394 p. ISBN 0-520-02280-7. $38.50. LC 72-82230.

    Now somewhat dated, this massive literature survey remains among the best systematic summaries of the surface processes that shape deserts.

2773

Cooke, Ronald U.; Brunsden, D.; Doornkamp, J.C.; and Jones, D.K.C., with contributions by Griffiths, J.; Knott, P.; Potter, R.; and Russell, R. Urban geomorphology in drylands. London; New York: Oxford University Press for the United Nations University, 1982. 324 p. ISBN 0-19-823239-X. $32. LC 82-3561.

    A heroic initial attempt to incorporate geomorphological processes into urban site selection and development. The volume uses case studies drawn from consulting experience to explore the impact of sand and dust movements, water and sediment problems, and salt action, and also to suggest planning solutions.

2774

Dregne, H.E. (Harold) Soils of arid regions. (Developments in soil science, no. 6). Amsterdam, The Netherlands; New York; London: Elsevier Scientific Publishing Co., 1976. 237 p. ISBN 0-444-41439-8. LC 76-21740.

    The physical, chemical, and biological properties of arid zone soils and their classification are examined systematically as well as in a regional, continent-by-continent survey.

2775

Goodall, D.W. and Perry, R.A., eds. with the assistance of Howes, K.M.W. Arid land ecosystems: structure, functioning, and management. Vol. 1. (International Biological Programme series no. 16). Cambridge, U.K.; London; New York: Cambridge University Press, 1979. 881 p. ISBN 0-521-21842-X. $135. LC 77-84810.

    Everything one could wish to know about the structure and atmospheric, soil, and plant and animal proceses of drylands is covered in 35 essays by internationally known participants in the International Biological Programme.

2776

Goudie, Andrew, and Wilkinson, John. The Warm desert environment. Cambridge, U.K.; London; New York: Cambridge University Press, 1977. 88 p. ISBN 0-521-213304-7. $16.95. ISBN 0-521-29105-4. $8.95. (pbk). LC 76-9731.

    As an overview of desert environments and the major traditional livelihood systems that extract a living from them, this brief monograph is particularly useful.

2777 •

Mabbutt, J.A. Desert landforms. (Introduction to systematic geomorphology, v. 2). Canberra, ACT, Australia: Australia University Press; Cambridge, MA; London: MIT Press, 1977. 340 p. ISBN 0-262-13131-5. $25. LC 77-76688.

    Copiously supplied with illustrations and line drawings and organized around landform types and the processes that shape them, this intelligent synthesis brings to life the Australian desert landscape.

2778 •

Petrov, Mikhail Platonovitch. Deserts of the world. Translated from Russian by the IPST staff. Jerusalem: Israel Program for Scientific Translations; New York; Toronto: Halsted Press, John Wiley and Sons, 1976. 447 p. ISBN 0-470-68447-X. $99.95. LC 75-12921. (Pustyni zemnogo shara. Leningrad: Nauka, 1973. 435 p.).

    This is an encyclopedic global survey of arid regions. Primary emphasis is placed on the characteristics of the physical environments of the deserts. However, an optimistic assessment of the development prospects of arid land natural resources pervades the concluding chapters.

2779

West, Neil E., ed. Temperate deserts and semi-deserts. (Ecosystems of the world, v. 5). Amsterdam, The Netherlands; New York; Oxford, U.K.: Elsevier Scientific Publishing Co., 1983. 522 p. ISBN 0-444-41931-4. $170.25. LC 80-28760.

2780

Very detailed, lavishly illustrated ecological descriptions of the cool xeric ecosystems of the mid-latitudes make this a basic introduction to North American, Eurasian, and South American deserts. Land use is only examined in the six North American chapters, but a useful concluding chapter attempts to develop comparative generalizations. This is a companion volume to number 12 (Hot desert and arid shrubland) in the series, which has not yet been published.

E. RESOURCE MANAGEMENT

Campos-Lopez, Enrique, and Anderson, Robert J., eds. Natural resources and development in arid regions. Boulder, CO: Westview Press, 1983. 362 p. ISBN 0-86531-418-7. $29. LC 82-63172.　　2781

This somewhat uneven collection of papers explores natural assessment systems, development options, and transfer of technology that is presently available but is often under utilized. The Mexican experience receives particular attention in this collection.

Golany, Gideon, ed. Urban planning for arid zones: American experiences and directions. New York; Chichester, U.K.; Toronto: John Wiley and Sons, 1978. 245 p. ISBN 0-471-02948-3. LC 77-10472.　　2782

Despite the title, examples from the Third World abound. While the lessons generated by this collection of papers do not apply to most developing societies, they do provide insight into the pattern of urban development characteristic of much of the American Southwest.

Gonzales, Nancie L., ed. Social and technological management in dry lands: past and present, indigenous and imposed. (AAAS Selected Symposium no. 10). Boulder, CO: Westview Press for the American Association for the Advancement of Science, 1978. 199 p. ISBN 0-89158-438-2. $24. LC 77-93023.　　2783

Eight essays of varying length, largely written from an anthropological perspective, explore the lessons to be learned from traditional and modern management systems in coping with drought and desertification.

Heathcote, R.L. (Ronald Leslie). The Arid lands: their use and abuse. London; New York: Longman, 1983. 323 p. ISBN 0-582-30048-7. £7.95. LC 82-17935.　　2784 •

This is the most comprehensive overview of arid land resource management presently available. The text emphasizes the natural resources of drylands and the resource-use systems that have evolved to exploit them. Environmental impacts and management problems are particularly well handled.

Kovda, Victor A. Land aridization and drought control. Translated from Russian. Boulder, CO: Westview Press, 1980. 277 p. ISBN 0-89158-259-2. $35. LC 79-19062. (Aridatsiia sushi i bor'ba s zasukhoi. Moskva: Nauka, 1977. 272 p.).　　2785

This review of food and drought problems and irrigation impacts by a noted Soviet soil scientist is an optimistic assessment of dryland productive potential if bad practices are avoided.

Martin, William E.; Ingram, Helen M.; Laney, Nancy K.; and Griffin, Adrian H. Saving water in a desert city. Washington, DC: Resources for the Future, 1984. 111 p. ISBN 0-915707-04-7. $10. LC 83-43263.　　2786

Living in a desert often means water is in short supply. How Tucson, Arizona, has confronted water conservation policy, and its difficulties in grappling with resource limits to growth, is the basis for a case study that is a microcosm of public participation in municipal water-use problems in the American Southwest.

Arid Lands

Matlock, W. Gerald. Realistic planning for arid lands: natural resource limitations to agricultural development. (Advances in desert and arid land technology and development, v. 2). Chur, Switzerland; London; New York: Harwood Academic Publishers, 1981. 261 p. ISBN 3-7186-0051-X. $49.50. LC 81-3817.   2787

The use of long-term planning methodology based on systems analysis and on a realistic appraisal of the resource base constraints of arid lands is articulately advocated as a solution to the degradation of drylands and as a basis for future development.

UNITAR. (United Nations Institute for Training and Research). Alternative strategies for desert development and management. Proceedings of an international conference held in Sacramento, California, May 31-June 10, 1977. (Environmental sciences and applications series, v. 3). New York; Oxford; Toronto: Pergamon Press, 1982. 4 v. 1396 p. ISBN 0-08-022401-6. $200. LC 81-23433.

       Vol. 1: **Energy and minerals.** ISBN 0-08-022402-4. $45.
       Vol. 2: **Agriculture.** ISBN 0-022403-2. $40.
       Vol. 3: **Water.** ISBN 0-022404-0. $70.
       Vol. 4: **Desert managment.** ISBN 0-022405-9. $65.   2788

The experience of industrialized states is compared with the Third World countries, and the overall tone of the conference is optimistic both about management options and development prospects.

White, Gilbert F. Science and the future of arid lands. Paris: UNESCO, 1960. 95 p. Reprinted: Westport, CT: Greenwood Press, 1976. 95 p. ISBN 0-8317-8786-9. $15. LC 76-4551.   2789•

A profoundly insightful overview of arid land resource management issues, as relevant today as when it was first penned 25 years ago.

Wilkinson, J.C. (John Craven). Water and tribal settlement in South-East Arabia: a study of the Aflaj of Oman. (Oxford research studies in geography). Oxford, U.K.: Clarendon Press, 1977. 276 p. ISBN 0-19-823217-9. £18.50. LC 77-7347.   2790

An extraordinarily satisfying multi-disciplinary study by a geographer. This work illustrates how detailed examination of the cultural ecology of an irrigation system can generate insight into both theory and pragmatic development issues.

Worthington, E. Barton, ed. Arid land irrigation in developing countries: environmental problems and effects. International symposium, 16-21 February 1976, Alexandria, Egypt. New York; Toronto; Oxford, U.K.: Pergamon Press, 1977. ISBN 0-08-021588-2. $115. LC 76-58394.   2791

An important collection of scholarly case studies by leading international specialists, this volume examines the impact of irrigation on water, land use, soil, biological balances, and human health. An introductory section on the main environmental problems and management issues associated with irrigation provides a helpful integrated context for the specialist studies that follow.

17. POLAR AREAS

Roger G. Barry

A. BIBLIOGRAPHIES

Antarctic bibliography. 1965- . Annual. United States National Science Foundation, Office of Polar Programs, 10 First Street, Washington, DC, 20550. ISSN 0066-4626. LC 65-61825.
Material arranged by 13 subject categories. Four indexes. Information is published monthly in Current Antarctic Literature *and cumulated at intervals of a year or eighteen months. The earlier period 1951-1961 was covered in an unnumbered volume published in 1970.*

Arctic bibliography. Arctic Institute of North America. Montreal, PQ: McGill-Queen's University Press, 1953-1975. 16 v. ISSN 0066-6947. LC 53-61783.
Comprehensive author listing, with abstracts, of material in all languages; indexed by subject and geographical location. Indispensable.

Library catalogue of the Scott Polar Research Institute, Cambridge, England. Boston, MA: G.K. Hall, 1976; Supplement 1981. 19 v.; Supplement, 5 v. ISBN 0-8161-1216-9. $1865; Supplement: ISBN 0-8161-0334-8. $680.
Offers author, subject, and regional access to the holdings of the Scott Polar Research Institute. The library, founded in 1920, is the largest single collection of its kind and covers the whole field of knowledge as it applies to the polar region.

Recent polar and glaciological literature. 1973- . 3 nos. per year. Scott Polar Research Institute, Lensfield Rd., Cambridge CB2 1ER, England. ISSN 0263-547X.
Selectively covers the recently published literature of the polar regions; the scientific study of snow and ice and its applied aspects. Separated from Polar Record *in 1973.*

B. SERIALS

Antarctic journal of the United States. 1- (1966-). 5 nos. per year. Orders: Superintendant of Documents, Washington, DC, 20402. ISSN 0003-5335.
Short reports on U.S. Antarctic research.

Arctic (Journal of the Arctic Institute of North America). 1- (1948-). Quarterly. Orders: Allen Press, Inc., Box 368, Lawrence, KS, 66044. Or University of Calgary Press, 2500 University Drive, NW, Calgary, AB, Canada T2N 1N4. ISSN 0004-0843.
High quality articles on the arctic environment, biology, archeology and anthropology; news, notes, and book reviews.

Arctic and alpine research. 1- (1969-). Quarterly. Institute of Arctic and Alpine Research, University of Colorado, Campus Box 450, Boulder, CO, 80309. ISSN 0004-0851.
High quality articles on all aspects of arctic and alpine environments: geography; geomorphology; glaciology; geology; climatology; biology; ecology; and soils. Book reviews. Cumulative index v. 1-10 (1969-1978).

Glaciological data, report. 1- (1977-). 1-2 issues per year. World Data Center-A for Glaciology, CIRES, University of Colorado, Campus Box 449, Boulder, CO, 80309. ISSN 0017-0712.
Reports on data-related topics for snow and ice cover variables; individual issues have included Arctic sea ice, ice cores, permafrost, and Antarctic climate data.

2792

2793

2794

2795

2796

2797

2798

2799

Polar Areas

Polar geography and geology. 1- (1977-). Quarterly. V.H. Winston and Sons, 7961 Eastern Avenue, Silver Springs, MD, 20901. ISSN 0273-8457. 2800
  English translations of current Soviet, European, and Japanese articles on the human geography, physical geography, and geology of the Arctic and Antarctic. Also original articles. Lists current polar literature and new notes on the northern Soviet Union.

Polar record. 1- (1931-). 3 nos. per year. Scott Polar Research Institute, Lensfield Rd., Cambridge CB2 1ER, England. ISSN 0032-2474. 2801
  High quality articles and reports of current polar research and field programs by all countries; news notes and book reviews; includes the bulletins of the Scientific Committee on Antarctic Research.

C. ATLASES

Bushnell, Vivian C., ed. Antarctic map folio series. New York: American Geographical Society, 1964-1975. 19 folios. LC Map 64-29. 2802
  Detailed map folios on geodetic, geophysical, glaciological, meteorological, and oceanographic aspects of Antarctica. Folio 19 treats the history of exploration and scientific investigation.

Drewry, David J., ed. Antarctica: glaciological and geophysical folio. Cambridge, U.K.: Scott Polar Research Institute, University of Cambridge, 1983. 9 sheets. ISBN 0-901021-04-0. £59. 2803
  Original maps, cross-sections, and commentary with data tables and references on the ice sheet surface, ice thickness, bedrock surface, and other geophysical-glaciological variables.

Gorshkov, Sergei G., editor-in-chief. World ocean atlas, v. 3: Arctic ocean. Translated from Russian by D.A. Brown. Oxford; New York: Pergamon Press, 1983. 184 plates. ISBN 0-0802-8735-2. $400. LC 78-40616. (Atlas okeanov. Tom 3. Severnyi Ledovityi Okean, Moskva: Glavnoe upravlenie navigatsii i okeanografii Ministerstva oborony SSSR; Voenno-morskoi flot, 1980. 184 plates. 4 p. index). 2804
  Maps on history of exploration, ocean floor, geomorphology, climate, atmospheric circulation, hydrology, ice hydrochemistry, and biogeography of the Arctic Ocean, also charts on magnetism, aurora, and astronomy. Introduction and index in English. Map plates in Russian.

United States. Central Intelligence Agency. Polar regions atlas. Washington, DC: Central Intelligence Agency, National Foreign Assessment Center, 1978. 66 p. LC 78-602630. 2805 •
  Valuable short illustrated accounts of exploration, physical environment, resources, transportation, fisheries, peoples, science projects, and sovereignity problems in the polar regions and a gazetteer including Antarctic research stations.

D. GENERAL WORKS

Armstrong, Terence E.; Rogers, George; and Rowley, Graham. The Circumpolar north. London; New York: Methuen, 1978. 303 p. ISBN 0-416-16930-9. $16.95. LC 78-8016. 2806 •
  Regional approach to the political and economic aspects of northern lands, including some environmental background information.

Ives, Jack D. and Barry, Roger G., eds. Arctic and alpine environments. London; New York: Methuen, 1974. 999 p. ISBN 0-416-65980-2. $134. LC 75-301946. 2807
  Systematic detailed comparison of the physical environments, biological characteristics, and archeology of arctic and alpine areas, particularly in North America by 31 authors. Human impacts are treated briefly. Numerous illustrations. Thorough bibliographies.

Markov, Konstantin Konstantinovich; Bardin, V.I.; Lebedev, V.L.; Orlov, A.I.; and Suetova, I.A. The Geography of Antarctica. Translated from Russian by N. Kaner. (IPST cat. no. 5697). Jerusalem, Israel: Israel Program for Scientific Translations, 1970. (Moscow: 1968). 370 p. NUC 72-105329. (Geografiia Antarktidy. Moskva: "Mysl,' " 1968. 439 p.).
Antarctic exploration, geographical zonation, morphology, geology, and ice.

2808

Sugden, David E. Arctic and Antarctic: a modern geographical synthesis. Oxford, U.K.: Basil Blackwell; Totowa, NJ: Barnes and Noble Imports, 1982. 472 p. ISBN 0-389-20298-3. $35. LC 82-13788.
Natural and human systems in both polar regions; regional coverage of Greenland and Svalbard, Arctic Canada, Alaska, the Soviet Arctic, and Antarctica.

2809 •

E. SPECIAL TOPICS

Amaria, P.J.; Bruneau, A.A.; and Lapp, P.A. Arctic systems. (NATO conference series. II. Systems science). New York; London: Plenum Press, 1977. 956 p. ISBN 0-306-32842-9. $95. LC 77-3871.
Conference proceedings: keynote addresses and working papers on environmental problems, transportation, social and cultural systems, including pipelines, ice navigation, and submersibles.

2810

Bird, J. Brian. The Physiography of arctic Canada: with special reference to the area south of the Parry Channel. Baltimore, MD: The Johns Hopkins Press, 1967. 336 p. LC 67-16232.
Treatment of the physical environment, followed by sections on the history of landscape development, modern geomorphic processes, and special landscape elements. Numerous illustrations and bibliography.

2811

Bliss, L.C.; Cragg, J.B.; Heal, D.W.; and Moore, J.J., eds. Tundra ecosystems: a comparative analysis. Cambridge, U.K.; New York: Cambridge University Press, 1981. 813 p. ISBN 0-521-22776-3. $130. LC 79-41580.
Results of the International Biosphere Programme for Tundra Research in Canada, the USA, Scandinavia, the USSR, Greenland, Antarctica, and elsewhere. Major sections describe the tundra biome, abiotic components, plant production, journal utilization, and conservation. References.

2812

Birket-Smith, Kaj. The Eskimos. Translated from Danish by W.E. Calvert. 2nd ed. London: Methuen, 1959. 262 p. (1st ed., 1935. 250 p.). LC 61-43996. (Eskimoerne. København: Gyldendal, Nordisk forlag, 1927. 239 p.).
A study of the development of human populations and their social and economic patterns in a harsh polar environment.

2813

Boas, Franz. The Central Eskimo. Lincoln, NE: University of Nebraska Press, 1964. 261 p. Reprint from U.S. Bureau of American Ethnology. Sixth annual report, 1884-85. (Washington DC: Smithsonian Institution, 1888), pp. 399-669. ISBN 0-8032-5016-9. $9.95. LC 64-63593
This anthropological classic describes the way of life of the Eskimo of Baffin Island in the 1880's and the country in which they lived.

2814 •

Cooke, Alan, and Holland, Clive. The Exploration of northern Canada, 500 to 1920: a chronology. Toronto: Arctic History Press, 1978. 549 p. ISBN 0-77102-265-4. Can $60. LC 80-457979.
History of exploration. Bibliography and roster of 5000 participants of the expeditions.

2815

Denmark. Udenrigsministeriet. Greenland. Edited by Kristjan Bure; Translated from Danish by Reginald Spink and A. Anslev. Ringkøbing: A. Rasmussens bogtr., 1956. 168 p. LC 57-39847.
Complete description in English of Greenland's demography, history of settlement, government, occupations, transport, finance, social and living conditions, and education.

2816

Dunbar, M.J., ed. Polar oceans. (Proceedings of the Polar Oceans Conference, 1974). Calgary, Alberta, Canada: Arctic Institute of North America, 1977. 681 p. Can $40. LC 79-318826.
Conference papers on water mass and circulation, ice and ice biota, marine productivity, and climatic change in the polar regions.

2817

Hopkins, D.M.; Matthews, J.V.; Schweger, C.E.; and Young, S.B., eds. Paleoecology of Beringia. (81st Burg Wartenstein Symposium, Wenner-Gren Foundation for Anthropological Research). New York: Academic Press, 1982. 489 p. ISBN 0-12-355860-3. $37. LC 82-22621.
A paleogeographic survey by 30 authors of the former environment, vegetation, and fauna of Beringia and man's occupation and possible role in faunal extinctions. The product of a Wenner-Gren Foundation symposium at Burg Wartenstein, Austria in 1979.

2818

Laws, R.M., ed. Antarctic ecology. London; New York: Academic Press, 1984. 2 v.
Vol. 1: 368 p. ISBN 0-12-439501-5. $75.
Vol. 2: 544 p. ISBN 0-12-439502-3. $75.
Authoritative surveys of terrestial (vol. 1) and marine (vol. 2) environments and ecology.

2819

Orvig, Sven, ed. Climates of the polar regions. (World Survey of Climatology, v. 14). Amsterdam, The Netherlands; New York: Elsevier Scientific, 1970. 370 p. ISBN 0-444-40828-2. $127.75. LC 79-103355.
Detailed climatic characterization of the Arctic, Antarctic, and Greenland. Numerous charts and tables of climatic statistics.

2820

Rey, Louis, ed., with assistance of Stonehouse, Bernard. The Arctic Ocean: the hydrologic environment and the fate of pollutants. (Conference of the Comité Arctique International). New York: John Wiley and Sons, 1982. 433 p. ISBN 0-471-87464-7. $97.
Chapters by 21 contributors on discovery and genesis of the Arctic Basin, hydrology, water, ice and atmospheric interactions, climate, air pollutant transport, and the mineral resource development in Greenland; chapter reference lists; name and subject indices.

2821

Schwerdtfeger, Werner. Weather and climate of the Antarctic. (Developments in atmospheric science v. 15). Amsterdam, The Netherlands; New York: Elsevier Scientific, 1984. 261 p. ISBN 0-444-42293-5. $46.25. LC 83-27529.
Thorough survey of radiation, temperature, wind and moisture conditions, atmospheric circulation and storm systems, and discussion of topics such as the ice sheet mass budget.

2822

Smiley, Terah L. and Zumberge, James H., ed. Polar deserts and modern man. Tucson, AZ: University of Arizona Press, 1974. 173 p. ISBN 0-8165-0383-4. $22.50. LC 73-85722.
Contributions by 17 authors on the natural environment, economic aspects, and problems encountered by immigrants in polar areas.

2823

Spencer, Robert F. The North Alaskan Eskimo: a study in ecology and society. Washington, DC: Government Printing Office, 1959. 490 p. LC 59-61386.
An unusually thorough and stimulating investigation of past aboriginal culture in the Arctic margins, with some contemporary comparisons.

2824 •

Tedrow, John C.F. Soils of the polar landscapes. New Brunswick, NJ: Rutgers University Press, 1977. 638 p. ISBN 0-8135-0808-9. $65. LC 76-39932.
Comprehensive treatment of polar soils and soil-forming factors; regional consideration include descriptions of geology, climate, and vegetation.

2825

18. MOUNTAIN GEOGRAPHY

Larry W. Price

A. SERIAL

Mountain research and development. (International Mountain Society; United Nations University). 1- (1981-). Quarterly. International Mountain Society, P.O. Box 3128, Boulder, CO, 80302. ISSN 0276-4741.
   Covers the whole range of subjects in research on the physical, biological, and social characteristics of mountains. Mountain chronicle. Reviews.

2826

B. GENERAL WORKS

Peattie, Roderick. Mountain geography: a critique and field study. Cambridge, MA: Harvard University Press, 1936. 239 p. Reprinted: Westport, CT: Greenwood Press, 1969. 257 p.ISBN 0-8371-2243-0. $15.75. LC 70-88918.
   A classic study in mountain geography. It focuses on weather and climate and human land use and settlement in European mountains. Although dated, it still contains a wealth of good information.

2827•

Price, Larry W. Mountains and man: a study of process and environment. Berkeley, CA; London: University of California Press, 1981. 506 p. ISBN 0-520-03263-2. $35. LC 76-14294.
   General overall text dealing with mountain processes and environments. Topics covered include: attitudes towards mountains; origins of mountains; climate; snow, glaciers, and avalanches; landform and geomorphic processes; soils; vegetation; wildlife; implications for people; agricultural settlement and land use; and human impact. The book is highly integrative and has an excellent bibliography.

2828•

C. SPECIAL TOPICS

Baker, Paul T. and Little, Michael A., eds. Man in the Andes: a multidisciplinary study of high-altitude Quechua. (US/IBP synthesis series I). Stroudsburg, PA: Dowden, Hutchinson, and Ross, 1976. 482 p. ISBN 0-470-15153-6. $59.50. LC 76-17025.
   Contains detailed papers dealing with physical environment and human adaptation and land uses in the high Andes. Special emphasis on human physiology and problems of living at high altitudes. Most authors are anthropologists.

2829

Baker, Paul T., ed. The Biology of high-altitude peoples. (International biological programme, no. 14). Cambridge, U.K.; New York: Cambridge University Press, 1978. 357 p. ISBN 0-521-21523-4. $68. LC 76-50311.
   Contains 11 detailed papers on the biology of high altitude peoples. These include such topics as: fertility; human growth and development; work capacity; nutrition; and morphological and physiological characteristics. Comparative statements are made between peoples in different mountain areas, e.g. Andes, Alps, and Himalayas.

2830

Barry, Roger G. Mountain weather and climate. London; New York: Methuen, 1981. 313 p. ISBN 0-416-73730-7. $41. LC 80-42348.
   This is the most comprehensive and substantive treatment of mountain weather and climate available in a simple book. Excellent bibliography.

2831

Beaver, Patricia D. and Purrington, B.L. Cultural adaptation to mountain environments. (Southern Anthropological Society Proceedings, no. 17). Athens, GA: University of Georgia Press, 1984. 192 p. ISBN 0-8203-0688-6. $16. ISBN 0-8203-0709-2. $7.50. (pbk). LC 83-6750.
   Contains 12 papers dealing with human land use and settlement in mountains. The focus is on the human response and adaptation to these limiting environments.

2832

Mountain Geography

Brugger, Ernst A.; Furrer, G.; Messerli, B.; and Messerli, P., eds. The Transformation of Swiss mountain regions. Bern, Switzerland: Paul Haupt, 1984. 699 p.
A detailed analysis of modern land use problems in the Swiss Alps. Papers are written by a variety of authors. Topics range from mountain agriculture to the impact of tourists to ecological dynamics and hazards.
2833

Brush, Stephen B. Mountain, field, and family: the economy and human ecology of an Andean valley. Philadelphia, PA: University of Pennsylvania Press, 1977. 199 p. ISBN 0-8122-7728-7. $18.50. LC 77-24364.
A detailed analysis of an alpine village in the Andes with focus on land use, environment, and strategies for coping with the hardships of the high altitudes.
2834

DeBeer, Gavin R. Early travellers in the Alps. London: Sidgwick and Jackson, 1930. 204 p. Reprinted: Stonington, CT: October Press, 1966. ISBN 0-8079-0041-9. $5.95. ISBN 0-8079-0042-7. $2.95. (pbk). LC 67-72008.
An interesting discussion of early travel and land use in the Alps.
2835 •

Houston, Charles S. Going higher: the story of man and altitude. Rev. ed. Burlington, VT: Charles S. Houston, 1983. 273 p. (1st ed.: 1980, 211 p.). ISBN 0-9612246-0-6. $10. LC 83-81995.
A well-written summary of the physiological effects of high altitudes on hikers, climbers, and others. Special emphasis on those going to very high altitudes.
2836 •

Lichtenberger, Elisabeth. The Eastern Alps. (Problem regions of Europe). London: Oxford University Press, 1975. 48 p. ISBN 0-19-913106-6. £2.95. (pbk). LC 75-326649.
A general but excellent treatment of land use and settlement in the Alps.
2837

Netting, Robert McC. Balancing on an Alp: ecological change and continuity in a Swiss mountain community. Cambridge, U.K.; New York: Cambridge University Press, 1981. 278 p. ISBN 0-521-23743-2. $47.50. ISBN 0-521-28197-0. $17.95. (pbk). LC 81-358.
Detailed analysis of a Swiss village in the Alps, and the ecological balance between its inhabitants and the environment. Useful chapters on agriculture and the strategies of alpine land use and population demographics. Written by an anthropologist.
2838

Mani, M.S. Introduction to high altitude entomology: insect life above timberline in the Northwest Himalaya. London: Methuen, 1962. 302 p.
A general book on high altitude environments with focus on arthropods and their ways of dealing with harsh conditions. Most examples are taken from the Himalayas.
2839

Nicolson, Marjorie Hope. Mountain gloom and mountain glory. New York: W.W. Norton, 1959. 403 p. LC 59-2805.
A good review of attitudes towards mountains in the Western world from before the birth of Christ through the Renaissance.
2840 •

Slaymaker, H. Olav (Herbert), and McPherson, Harold J., eds. Mountain geomorphology. (British Columbia geographical series no. 14). Vancouver, BC: Tantalus Research Ltd., 1972. 274 p. LC 74-192726.
Series of papers by various experts on topics such as glacial geomorphology, slope processes, fluvial processes, and environmental management of mountain regions.
2841

Tranquillini, Walter. Physiological ecology of the alpine timberline. Translated from German by Udo Benecke. (Ecological studies, v. 31). Berlin, West Germany; New York: Springer-Verlag, 1979. 137 p. ISBN 0-387-09065-7. $34. LC 78-25603.
Detailed treatment of the alpine timberline and ecology. Focuses on the Alps.
2842

Troll, Carl, ed. Geo-ecology of the mountainous regions of the tropical Americas. (Proceedings of the UNESCO Mexico Symposium, Aug. 1-3, 1966). In English and Spanish. Bonn, West Germany: Dümmler Verlag in Kommission, 1968. 223 p. LC 77-439909.
A classic volume of 13 papers by various authors on the mountains of tropical South America. Most papers detail the physical environment and ecological problems.
2843

PART VII. PUBLICATIONS SUITABLE FOR SCHOOL LIBRARIES

Salvatore J. Natoli

Note: These are entries that do not appear elsewhere in this bibliography. Entries in other sections that are appropriate for school libraries are marked by a special symbol (•), a bullet after the entry number.

A. SERIAL

Focus. 1- (1950-). Quarterly. American Geographical Society, 156 Fifth Avenue, Suite 600, New York, New York 10010. ISSN 0015-5004.
    A quarterly, illustrated geographical journal designed for the informed layperson. Emphasizes regional articles on problems of current interest in various regions of the world.

2844

B. ATLASES

MacLean, Kenneth, and Thompson, Norman. Problems of our planet: an atlas of Earth and man. Edinburgh, U.K.: J. Bartholomew, Holmes McDougall, 1975. 67 p. ISBN 0-8515-2344-7. LC 76-361311.
    Although slightly dated, this useful volume combines maps and prose effectively in highlighting and portraying major environmental, social, political, and health problems geographically.

2845

Madden, James F. My first atlas. Maplewood, NJ: Hammond, 1979. 63 p. LC 78-317548.
    Introduces upper elementary school children to geography and social sciences in general. Teaches basic map reading skills in an easy-to-read text and pictograms. Inexpensive introduction to atlases and map skills.

2846

Ogilvie, Bruce, and Waitley, Douglas. Rand McNally picture atlas of the world. Chicago, IL: Rand McNally, 1979. 96 p. ISBN 0-528-82043-5. $7.95. LC 79-13842.
    Contains encyclopedic information on the formation and changes of the earth's surface, attempts throughout history to draw accurate world maps, and the earth's various resources. Brief text and color photographs and drawings accompany each map.

2847

Our magnificent Earth: a Rand McNally atlas of earth resources. New York: Rand McNally; London: M. Beazley, 1979. 208 p. ISBN 0-528-83088-0. $35. LC 78-71982.
    A comprehensive guide to the resources of the Earth and how they are used. Divided into six chapters: nature of resources; resources and people; energy; living resources; minerals; and planning for tomorrow. Attractive and interesting for children.

2848

Peoples and places of the past. Washington, DC: National Geographic Society, 1983. 424 p. ISBN 0-87044-462-X. $69.95. LC 83-2208.
    Profusely illustrated cultural atlas for the ancient world containing excellent maps. Color photography and appropriate descriptions of illustrations. Contains three major parts: cradle of Western civilization; land of sacred rivers; and trails of vanished tribes.

2849

For School Libraries

Rand McNally atlas of the oceans. New York: Rand McNally; London: M. Beazley (Published as Mitchell Beazley's atlas of the oceans), 1977. 208 p. ISBN 0-528-83082-1. $35. LC 77-73772.   2850
   Double-page spreads of bathymetric maps of each of the oceans and principal seas. Contains the following sections: the ocean realm; the human ocean quest; life in the oceans; great resources; the face of the deep; and encyclopedia of marine life.

Scott, John Anthony. The Story of America: a National Geographic picture atlas. Washington, DC: National Geographic Society, 1984. 324 p. ISBN 0-87044-535-9. $19.95. LC 84-2018.   2851
   A useful pictorial history of the United States. Described in the subtitle as a picture atlas, the cartographic coverage is surprisingly limited.

C. MAP READING

Bell, Neill. The Book of where and how to be naturally geographic. New York: Little, Brown, 1982. 119 p. ISBN 0-316-08830-7. $10.95. LC 81-19315.   2852
   Teaches directions, map reading, concepts in world and U.S. geography, and geology and continental drift. Contains exercises and practical applications for independent study.

Resnick, Abraham. Fun facts with maps and globes. Minneapolis, MN: T.S. Denison and Co., Inc., 1971. 142 p. LC 74-132316.   2853
   Designed to provide elementary school teachers with a set of ideas for teaching map and globe uses. A series of classroom activities are detailed from painting and landscape modeling to games with maps and weather map reading.

U.S. Geological Survey. Maps for America: cartographic products of the U.S.G.S. and others. 2nd ed. Washington, DC: Government Printing Office, 1981. 265 p. (1st ed., 1979, 265 p.). $11. LC 81-607878.   2854
   Background information on the U.S.G.S. mapping programs and detailed explanations of maps produced by the survey, including the meaning of lines, colors, symbols, and other map features.

D. REFERENCE WORKS

Larkin, Robert P. and Peters, Gary L. Dictionary of concepts in human geography. Westport, CT: Greenwood Press, 1983. 286 p. ISBN 0-313-22729-8. $35. LC 82-24258.   2855
   A dictionary with about 100 entries only. Each entry comprises a definition, an essay, and a substantial bibliography.

Moore, Wilfred George. The Penguin encyclopedia of places. 2nd ed. Harmondsworth, U.K.; New York: Penguin, 1978. 886 p. (1st ed., 1971. 835 p.). ISBN 0-14-051047-8. LC 79-315720.   2856
   A compilation of place names of the world with notes on their origin and significance.

Room, Adrian. Place names of the world. Totowa, NJ: Rowman and Littlefield, 1974. 216 p. LC 74-179927.   2857
   Compilation of place names of the world and their origins.

Ruffner, James A. and Blair, Frank E., eds. The Weather almanac: a reference guide to weather, climate, and air quality in the United States and its key cities. 4th ed. Detroit, MI: Gale, 1984. 850 p. (2nd ed., 1978. 728 p.). ISBN 0-8103-1184-4. $80. LC 73-9342.   2858
   A comprehensive collection of U.S. Weather Bureau tables, charts, and maps. Yearly weather records for 108 U.S. cities, weather atlas, air quality data, international weather data, severe weather conditions, degree days, and other information.

Showers, Victor. World facts and figures. Revised and enlarged ed. New York; Toronto; Chichester, U.K.: John Wiley and Sons, 1979. 757 p. (1st ed., The World in figures, 1973. 585 p.). ISBN 0-471-04941-7. $44.95. LC 78-14041. 2859
   A collection of comparative information about cities, countries, and geographic features of the world.

Stiegler, Stella E., ed. A Dictionary of earth sciences. New York: Pica Press, 1977. 301 p. Reprinted: 1983, Rowman and Allenheld. ISBN 0-8226-0377-2. $9.95. (pbk). LC 76-41042. 2860
   Illustrated glossary of 2500 alphabetically arranged entries pertaining to geology, paleontology, oceanography, cartography, and related sciences.

E. PHYSICAL GEOGRAPHY

Ballard, Robert D. Exploring our living planet. Washinton, DC: National Geographic Society, 1983. 366 p. ISBN 0-07044-397-6. $19.95. LC 83-2336. 2861
   Documents the plate tectonics concept through existing land and sea bottom features. Firsthand observations, with excellent illustrations of land form features and the effects of natural hazards on the earth's populated regions.

Carlisle, Norman V. The New American continent: our continental shelf. New York; Philadelphia: J.B Lippincott Co., 1973. 96 p. ISBN 0-3973-1234-2. $4.95. LC 72-13422. 2862
   Describes the American continental shelf and the research conducted to discover more about it.

Darden, Lloyd. The Earth in the looking glass. Garden City, NY: Anchor Press, 1974. 324 p. ISBN 0-385-02595-5. LC 73-9151. 2863
   Describes the genesis and development of ERTS in nontechnical language, the purposes and goals of the agencies involved, and how satellites expand sight by photography of the entire surface of the earth in various wavelengths.

Fisher, Ron. Our threatened inheritance: natural treasures of the United States. Washington, DC: National Geographic Society, 1984. 400 p. ISBN 0-87044-512-X. $19.95. LC 84-13976. 2864
   Coverage of the U.S. national parks, refuges, forests, and other federal holdings in terms of pressures placed on them by increasing human use. Richly illustrated. Volume is sold with a companion Directory to Federal Lands.

Grove, Noel. Wild lands for wildlife: America's national refuges. Washington, DC: National Geographic Society, 1984. 208 p. ISBN 0-87044-482-4. $7.95. LC 84-16539. 2865
   Describes a selected number of American national wildlife refuges according to habitats--coasts and islands, the Eastern woodlands, the prairies, the mountain West, and Alaska. Contains a map showing the distribution of more than 400 wildlife refuges in the U.S.

Morrison, H. Robert, and Lee, Christine Eckstrom. America's Atlantic isles. Washington, DC: National Geographic Society, 1981. 200 p. ISBN 0-87044-369-0. $7.95. LC 80-7828. 2866
   Presents in text and in many color photographs the extraordinary variety of island worlds along the Eastern coast of the United States and Canada.

National Geographic Book Service. Wilderness, U.S.A. Washington, DC: National Geographic Society, 1973. 344 p. ISBN 0-87044-116-7. $9.95. LC 73-10213. 2867
   Mainly a natural history of wilderness areas of the United States but contains useful maps, excellent photography of wilderness environments, and essays by authorative writers.

National Geographic Society. Great rivers of the world. Washington, DC: National
Geographic Society, 1984. 448 p. ISBN 0-87044-537-5. LC 84-1163.   2868
    *A collection of pieces on great rivers of the world and the role they play in the histo-
ry of human civilization and settlement. Abundantly illustrated.*

National Geographic Society. Our continent: a natural history of North America.
Washington, DC: National Geographic Society, 1976. 398 p. ISBN 0-87044-153-1. LC
76-26633.   2869
    *An attractively illustrated natural history of the North American continent tracing
its geological evolution and the development of its flora and fauna.*

National Geographic Society. Special Publications Division. America's magnificent mountains. Washington, DC: National Geographic Society, 1980. 208 p. ISBN 0-
87044-286-4. $7.95. LC 78-21447.   2870
    *Describes life and landscapes in the Sierra Nevada, the Cascade Range, the Coast
Mountains (Canada), Mount McKinley, Aspen's Rockies, Mexico's Las Sierras de
Los Vulcans, the Great Smokies, and the Northern Appalachians.*

National Geographic Society. Special Publications Division. America's majestic
canyons. Washington, DC: National Geographic Society, 1979. 208 p. ISBN 0-87044-
271-6. $7.95. LC 78-61263.   2871
    *Several authors describe, in this profusely illustrated publication, eight representa-
tive canyons, gorges, and valleys in the U.S., Canada, and Alaska. Concentrates on
people who live and work in the canyons.*

National Geographic Society. Special Publications Division. As we live and
breathe: the challenge of our environment. Washington, DC: National Geographic Society, 1971. 240 p. ISBN 0-87044-097-7. $7.95. LC 74-151945.   2872
    *Although somewhat dated, this book outlines major elements of the ecological cri-
sis and some of the detrimental effects of human use of the environment.*

National Geographic Society. Special Publications Division. Canada's wilderness
lands. Washington, DC: National Geographic Society, 1982. 200 p. ISBN 0-87044-418-2.
$7.95. LC 81-48074.   2873
    *Natural history of wilderness areas in Atlantic Canada, the Shield, High Arctic, In-
terior Plains, and Western Canada. Contains useful maps, geological background,
and geographical descriptions of each main region.*

National Geographic Society. Special Publications Division. The Desert realm.
Washington, DC: National Geographic Society, 1982. 304 p. ISBN 0-87044-456-5.
$19.95. LC 80-7568.   2874
    *Accounts by writers and photographers of the earth's major deserts and their physi-
cal and human environments. Striking illustrations and contains explanations for desert climates.*

**National Geographic Society. Special Publications Division and School Services
Divisions.** Our violent earth. Washington, DC: National Geographic Society, 1982.
103 p. ISBN 0-87044-388-7. $6.95. LC 80-8797.   2875
    *An earth science and geographical approach to natural hazards: earthquakes, vol-
canoes, storms, drought, fire, and floods. Also includes a classroom activities folder.*

O'Neill, Catherine. Natural wonders of North America. Washington, DC: National
Geographic Society, 1984. 104 p. ISBN 0-87044-519-7. $6.95. LC 84-16614.   2876
    *An attractively illustrated physical geography of selected regions of North Ameri-
ca: Pacific Coast; Basin and Range; Mexican Highlands; Colorado Plateau; Rocky
Mountain and Columbian Plateau; Great Plains and Central Lowlands; Appalachian
Highlands and Coastal Plains; Canadian Shield and Greenland. Contains classroom
activities packet.*

Pringle, Laurence. What shall we do with the land? Choices for America. New York: Crowell, 1981. 152 p. ISBN 0-690-04108-X. $9.50. LC 81-43034.   2877

Explores the conflicts of land use among the various terrains--farmlands, pastures, rangelands, forest, deserts, canyons, mountains, barrier islands, and coasts. Describes politics, governmental policies, and individual greed in plain language. A useful bibliography.

Robbins, Michael. High country trail: along the continental divide. Washington, DC: National Geographic Society, 1981. 200 p. ISBN 0-87044-366-6. $7.95. LC 80-7826.   2878

A hiker's observations tracing land and life along the United States continental divide.

Ryder, Nicholas, and Ellison, Martin, comps. Ryder's standard geographic reference: the United States of America. Denver, CO: Ryder Geosystems, Satellite Mapping Division, 1981. 213 p. ISBN 0-941784-00-2. $85. LC 81-90461.   2879

A compilation of 173 black-and-white plates of Landsat imagery of the U.S.A. at a scale of 1:1,000,000. An excellent artistic book that should promote interest in remote sensing and geography.

F. HUMAN/CULTURAL GEOGRAPHY

Cumming, W.P.; Skelton, R.A.; and Quinn, D.B. The Discovery of North America. New York: American Heritage Press, 1972. 304 p. LC 73-165335.   2880

A popular, well-illustrated survey.

Fisher, Allan C. Jr. America's inland waterway: exploring the Atlantic seaboard. Washington, DC: National Geographic Society, 1973. 208 p. ISBN 0-87044. $7.95. LC 73-831.   2881

Describes a trip by motorsailer along 2000 miles of the Atlantic Intracoastal Waterway and the lifestyles and history of the adjacent regions.

National Geographic Society. Special Publications Division. Nomads of the world. Washington, DC: National Geographic Society, 1971. 200 p. ISBN 0-87044-098-5. $7.95. LC 78-151946.   2882

A useful account of selected nomadic peoples of the world. As population pressures on the land increase, their ways of life may be threatened.

National Geographic Society. Special Publications Division. Trails west. Washington, DC: National Geographic Society, 1979. 207 p. ISBN 0-87044-272-4. $7.95. LC 78-61264.   2883

Attractively illustrated historical geography of major transportation routes to the American West. Individual authors describe the Santa Fe, Oregon, Mormon, California, Gila, and Bozeman Trails.

O'Neill, Thomas. Lakes, peaks, and prairies: discovering the United States-Canadian border. Washington, DC: National Geographic Society, 1984. 200 p. ISBN 0-87044-483-2. $7.95. LC 82-22775.   2884

An account of a trip along the U.S.-Canadian border, observing physical and cultural similarities and differences.

G. REGIONAL GEOGRAPHY

Asimov, Isaac. How did we find out about Antarctica? New York: Walker, 1979. 64 p. ISBN 0-8027-6371-5. $6.95. LC 79-2199.   2885

Illustrated by David Wool, this short book describes the geography of Antarctic regions and gives a history of their discovery and exploration.

For School Libraries

Hargreaves, Pat, ed. The Arctic. Hove, U.K.: Wayland; Morristown, NJ: Silver Burdett, 1981. 68 p. ISBN 0-382-06583-2. $7.95. LC 81-50488.
    Explores aspects of the Arctic Ocean and the human interaction with it. Many photos, easy-to-read print: British spellings and sentence structure.
2886

Huntington, Lee Pennock. The Arctic and Antarctic: what lives there. New York: Coward, McCann, and Geoghegan, 1975. 46 p. LC 74-21067.
    Selected interesting aspects of the Arctic and Antarctic in a readable and informative story which provides insights into the life and natural conditions of these regions. Excellent black-and-white drawings.
2887

Johnson, Raymond. The Rio Grande. Morristown, NJ: Silver Burdett; Hove, U.K.: Wayland, 1981. 69 p. ISBN 0-382-06521-2. $10.60. LC 80-53849.
    A tour of the Rio Grande from its source in Colorado to the Gulf of Mexico. Contrasts the old and the new and motivates the reader to learn more about the region. Smooth, uncomplicated writing.
2888

Keating, Bern. The Mighty Mississippi. Washington, DC: National Geographic Society, 1971. 200 p. ISBN 0-87044-096-9. $7.95. LC 70-151944.
    An historical and contemporary account of the Mississippi River in the life of the U.S.
2889

Lightfoot, Paul. The Mekong. Morristown, NJ: Silver Burdett; Hove, U.K.: Wayland, 1981. 65 p. ISBN 0-382-06520-4. $11.95. LC 80-53606.
    Illustrated commentary on the Mekong River from its headwaters in the Tibetan Plateau to its delta in the South China Sea. One of a series of 16 books on important rivers of the world.
2890

McCarry, Charles. The Great Southwest. Washington, DC: National Geographic Society, 1980. 200 p. ISBN 0-87044-288-0. $7.95. LC 78-21450.
    A photographic and descriptive essay on the American Southwest organized by physical regions except for the section on cities. Photographs are excellent but cartographic treatment is minimal.
2891

McDowell, Bart. Journey across Russia: the Soviet Union today. Washington, DC: National Geographic Society, 1977. 368 p. ISBN 0-87044-220-1. $12.95. LC 76-56998.
    A personal geography by a journalist's observations of a trip across the Soviet Union. Excellent illustrations. Includes a large jacket map of the U.S.S.R.
2892

National Geographic Society. Journey into China. Washington, DC: National Geographic Society, 1982. 518 p. ISBN 0-87044-437-9. $19.95. LC 82-14132.
    A collection of articles on various regions of China, some of contemporary and others of historical significance. Richly illustrated with excellent color photography.
2893

National Geographic Society. Special Publications Division. Alaska, high road to adventure. Washington, DC: National Geographic Society, 1976. 200 p. ISBN 0-87044-193-0. $7.95. LC 76-692.
    Presents regional vignettes with maps, color photographs, and historical black-and-white photography of six sections of Alaska--the Alaskan Highway, the Southeast, Anchorage, the Aleutians, the Interior, and the Arctic slope.
2894

National Geographic Society. Special Publications Division. Alaska's magnificent parklands. Washington, DC: National Geographic Society, 1984. 200 p. ISBN 0-87044-447-6. $7.95. LC 83-25036.
    Explores thirteen Alaskan parks, monuments and preserves, a vast area larger than all the parklands in the other 49 states. Profusely illustrated with maps and color photographs.
2895

National Geographic Society. Special Publications Division. America's hidden corners: off the beaten path. Washington, DC: National Geographic Society, 1983. 200 p. ISBN 0-87044-446-8. $7.95. LC 82-47844.   2896
Provides vignettes of several different seldom-visited areas of the U.S. within the following regions: the Great Basin; Chesapeake Bay; prairie and badlands; Upper Peninsula of Michigan; Gulf Coast; Missouri and Arkansas Ozarks; and the Four Corner region of Utah, Colorado, New Mexico, and Arizona.

National Geographic Society. Special Publications Division. America's spectacular Northwest. Washington, DC: National Geographic Society, 1982. 200 p. ISBN 0-87044-368-2. $7.95. LC 80-7829.   2897
A largely photographic essay with text and some maps of the U.S. Northwest from the Rocky Mountains in Idaho to the Pacific Coast.

National Geographic Society. Special Publications Division. Isles of the Caribbean. Washington, DC: National Geographic Society, 1980. 216 p. ISBN 0-87044-274-0. $7.95. LC 78-61266.   2898
Five writers reflecting on the Caribbean islands, their people, changes as a result of independence, and the rewards and problems derived from economic development.

National Geographic Society. Special Publications Division. Secret corners of the world. Washington, DC: National Geographic Society, 1982. 200 p. ISBN 0-87044-412-4. $7.95. LC 81-48073.   2899
Describes somewhat remote regions of the world in a richly photographed book. Emphasizes human adaptation to difficult habitats: the Marquesas; Northern Afghanistan; Tierra del Fuego; the Santa Martas of Colombia; Alpujarras in southern Spain; and the Ruwenzori in East Africa.

Schlein, Miriam. Antarctica: the great white continent. New York: Hastings House, 1980. 64 p. ISBN 0-8038-0482-2. $7.95. LC 79-21320.   2900
An introduction to the exploration, geography, and animal life of Antarctica via basic questions about the origins of the continent, its ice cover, and its ecology.

Shapiro, William E., ed. Lands and peoples. 2nd ed. Danbury, CT: Grolier, 1983. 6 v. (1st ed. 1972. 7 v.). ISBN 0-7172-8009-8. $112.50. LC 83-1649.
Vol. 1: **Africa.**
Vol. 2: **Asia, Australia, and New Zealand.**
Vol. 3: **Europe.**
Vol. 4: **Europe.**
Vol. 5: **North America.**
Vol. 6: **Central and South America.**   2901
Story of all nations of the world, the set provides the history and geography of each country, as well as information on their governments, religion, economy, food, clothing, manners, and customs.

White, Richard. Africa: geographical studies. London: Heinemann Educational Books, 1978. 208 p. ISBN 0-435-34916-3. £4.25. (pbk).   2902
A general volume on Africa addressing the continent from a problem solving perspective. Presents a series of vignettes of African nations based upon problems or situations relevant to each nation's development.

Windsor, Merrill. America's Sunset Coast. Washington, DC: National Geographic Society, 1978. 212 p. ISBN 0-87044-253-8. $7.95. LC 77-93401.   2903
Beautifully illustrated description of the U.S. West Coast. Divided into three parts: southern California; central and northern California; and the Pacific Northwest. Gives introductory physical and human geography for each section.

INDEX

The Index includes authors, editors, and compilers, short titles, and major subjects. Numbers refer to entries except Roman numerals, which refer to pages in the introduction. Publications considered suitable for a school library are marked by a bullet (•) after the entry number, except for those in Part VII.

Atlas geografie: Republica Socialistă România. 2071
Atlas geográfico de la República del Ecuador. 1930
Atlas geográfico del Ecuador. 1937
Atlas histórico geográfico y de paisages Peruanos. 1936
Atlas mira. 65
Atlas nacional de Cuba. 1927
Atlas nacional de España. 2067 •
Atlas nacional de Guatemala. 1933
Atlas nacional de Panamá. 1935
Atlas nacional do Brasil. 1921
Atlas narodov mira. 874
Atlas of Africa. 2590
Atlas of African history. 2587 •
Atlas of Alabama. 1657 •
Atlas of Alberta. 1848 •
Atlas of ancient archaeology. 60
Atlas of Arkansas. 1639 •
Atlas of Australian resources. 2672
Atlas of Britain and Northern Ireland. 2064 •
Atlas of British Columbia. 1852 •
Atlas of California. 1645 •
Atlas of cancer mortality for U.S. counties. 1427
Atlas of Central America. 1918
Atlas of Central Europe. 2048
Atlas of China. 2323
Atlas of Denmark. 2062
Atlas of early American history. 1641 •
Atlas of economic development. 1171
Atlas of economic mineral deposits. 740 •
Atlas of Europe. 2053 •
Atlas of fantasy. 62 •
Atlas of food crops. 1204
Atlas of geomorphic features. 467
Atlas of Hawaii. 1673 •
Atlas of Hungary. 2060
Atlas of Indiana. 1656
Atlas of Ireland. 2059
Atlas of Israel. 2497
Atlas of Japan. 2295, 2296, 2297 •
Atlas of Jerusalem. 2491
Atlas of Kentucky. 1655 •
Atlas of landforms. 1683 •
Atlas of Maryland. 1670 •
Atlas of mean monthly temperatures. 513
Atlas of medieval Europe. 1601 •
Atlas of Mexico. 1919
Atlas of Michigan. 1668 •
Atlas of Minnesota resources and settlement. 1788
Atlas of Mississippi. 1644 •
Atlas of Missouri. 1663 •
Atlas of North America. 1606 •
Atlas of North Dakota. 1650 •
Atlas of Oregon. 1658 •
Atlas of Papua and New Guinea. 2374
Atlas of Papua New Guinea. 2675 •
Atlas of Saskatchewan. 1855 •
Atlas of Saudi Arabia. 2492
Atlas of South Dakota. 1653 •
Atlas of Texas. 1638 •
Atlas of the American Revolution. 1570 •
Atlas of the Arab world and the Middle East. 2495

Atlas of the Arab-Israeli conflict. 2498
Atlas of the distribution of diseases. 1428
Atlas of the Greek and Roman world in antiquity. 1598 •
Atlas of the Islamic world. 2499 •
Atlas of the living resources of the seas. 644 •
Atlas of the Oceans. 642, 643
Atlas of the Pacific Northwest. 1652 •
Atlas of the South Pacific. 2676 •
Atlas of Utah. 1674 •
Atlas of Victoria. 2673 •
Atlas of Wisconsin. 1643 •
Atlas of world population history. 1087
Atlas över Sverige. 2068 •
Atlas para la República Oriental del Uruguay. 1928
Atlas regional del Caribe. 1920
Atlas sel'skogo khoziaistva SSSR. 2212
Atlas SSSR. 2211
Atlas total de la República Argentina. 1922, 1923
Atlas van Nederland. 2050
Atlas zur Raumentwicklung. 2052
Atlas: nuevo atlas Porrua de la República Mexicana. 1932
Atlas: Republica Socialistă România. 2049
Atlases (by subjects and regions)
    Agricultural Geography. 1204-1206
    Africa. 2585-2590
    Canada. 1845-1856, 1861
    China. 2318-2323
    Classical World. 1597-1598
    Climatology. 511-513
    Development. 1306
    Economic Geography. 1171-1172
    Energy. 764
    Environmental Management. 807-811
    Ethnic Groups. 874
    Europe. 2045-2071
    Exploration. 179
    Geomorphology. 467
    Japan. 2295-2297
    Latin America. 1918-1939
    Medical. 1426-1429
    Medieval Europe. 1601
    Military Geography. 1568-1571
    Oceania. 2671-2679
    Oceans and Lakes. 642-645
    Peoples. 874
    Population history. 1087
    Religions. 963
    School Libraries. 2845-2851
    South Asia. 2439
    Southeast Asia. 2371-2374
    Southwest Asia and North Africa. 2491-2501
    U.S. 1637-1677, 1683, 1702, 1737, 1788, 1805
    U.S.S.R. 2211-2218
    Water Resources. 701
    World. 51-65
Atmosphere, weather, and climate. 518
Atmospheric sciences: an introductory survey. 530
Auckland and the Central North Island. 2717
Audubon Society encyclopedia of North American birds. 545 •

Augelli, John P. 1956
Australia (subject). 2683-2707
Australia (title). 2691 •, 2705
Australia. Agriculture. 1207
Australia. Commonwealth Scientific and
  Industrial Research Organization. 611
Australia. Division of National Mapping. 2672
Australia: a geography. 2693
Australia and New Zealand (climate). 517
Australia's corner of the world. 2681
Australia's resources and their development.
  2707 •
Australian agriculture. 2703
Australian environment. 2696
Australian geographer. 2666
Australian geographical studies. 2667
Australian Heritage Commission. 2683 •
Australian space, Australian time. 2702
Australian urban system. 2684
Auto-Carto proceedings. 270-274
Avery, Catherine B. 1591
Avery, Thomas E. 350 •, 366 •
Axeman, Lois. 1547 •

Babcock, Richard F. 1534 •
Bach, Wilfred. 794
Bad earth. 2352 •
Bagnold, Ralph A. 480
Bagrow, Leo. 325
Bailey, Arthur W. 590
Baker, Alan R.H. 899, 900, 901, 902
Baker, J.N.L. 180 •
Baker, Paul T. 561, 2829, 2830
Balancing on an Alp. 955, 2838
Bale, John. 1543
Balkans in transition. 2174
Ball, John M. 1536
Ballard, Robert D. 2861
Balon, Eugene K. 2642
Baltimore: the building of an American city.
  1767
Bancroft Library, University of California. 70
Banks, David J. 2377
Banks, Ferdinand E. 735 •
Barber, Gerald. 1273
Barbour, Michael G. 546, 564, 2585
Bardach, Eugene. 712
Bardin, V.I. 2808
Barger, Harold. 736 •
Barlowe, Raleigh. 1191
Barney, Gerald O. 817, 1095
Barone, Michael. 1694
Barr, Brenton M. 2241
Barraclough, Geoffrey. 63
Barrett, Eric Charles. 351 •, 531
Barrett, John W. 547
Barry, Roger G. ix, 518, 532, 553, 2807, 2831
Bartels, Cornelius P.A. 381
Barth, Fredrik. 965
Barth, Michael C. 677
Bartlett, Leo. 1544
Barton, Thomas F. 2371 •
Bartz, Patricia McBride. 2308
Bascom, Willard. 487 •, 678 •
Basic geographical library. vi, vii, 12

Basic industrial resources of the U.S.S.R. 2247
Basile, Robert M. 591 •
Basin of Mexico. 932
Basler, Roy P. 1635
Bassett, Keith. 391
Bastié, Jean. 2051
Bateman, M. 1387 •, 2115
Bater, James H. 2261, 2263, 2264
Bates, Marston. 2738
Bates, Robert L. 441 •
Batten, James W. 592, 598 •
Battlefields of the World War. 1582
Battles in the monsoon. 1585 •
Batty, Michael. 382
Baum, Andrew. 1043
Baum, Howell S. 1502
Baumann, Duane D. 1035
Bauxite and aluminum. 735 •
Bayliss-Smith, Timothy P. 1209, 2440
Be expert with map and compass. 314 •
Beach processes and sedimentation. 490, 688
Beaches and coasts. 489, 687
Beadle, N.C.W. 562
Beal, Frank. 1532
Beaujeu-Garnier, Jacqueline. 1088, 2051, 2113
Beaumont, Peter. 2503
Beauregard, Robert A. 995 •, 1521 •
Beaver, Patricia D. 2832
Beazley, Charles R. 144 •
Becht, J. Edwin. 1263
Beck, Hanno. 152, 193, 194
Beckinsale, Monica. 2193
Beckinsale, Robert P. 493, 494, 2193
Bederman, Sanford H. ix, 2569
Bedford, Richard D. 2709
Bedouin of the Negev. 2545
Beggs, Stephen D. 745
Behavior as an ecological factor. 559
Behavioral Geography (subject). 1019-1036
Behavioral problems in geography: a
  symposium. 1021
Behavioral problems in geography revisited.
  1022
Behavioral research methods in environmental
  design. 1051
Behrens, William W., III. 823
Behrman, Jack N. 1294
Belasco, Warren James. 1695
Bell, Neill. 2852
Bell, Paul A. 1043
Bell, Thomas L. 1195 •
Bellwood, Peter S. 2680
Ben-Arieh, Y. 2530
Benchmark papers in ecology. 559
Bender, Barbara. 912 •
Bender, Gordon L. 2765
Benedict, Peter. 2531
Benevolo, Leonardo. 1504 •
Benne, Kenneth D. 1505
Benneh, George. 2631
Bennett, Charles F. Jr. 443 •, 934 •
Bennett, D. Gordon. 1089, 1133 •
Bennett, John W. 935, 936 •
Bennett, Robert J. 216, 383, 384, 385, 407, 418,
  430, 1134
Bennis, Warren G. 1505

Glacial and Quaternary geology

(continued)

Klassen, Leo H. 2085
Klee, Gary A. 950
Klein, Johannes. 107
Klein, Margit. 107
Klein, Richard G. 920, 922
Klevebring, Bjorn-Ivar. 733 •
Klimadiagrammkarten der einzelnen
  Grossräume. 562
Klineberg, Otto. 1059 •
Knapp, Brian. 602 •
Knapp, Gregory
Knapp, R..562
Knapp, Ronald. 2339 •
Kneese, Allen V. 717
Knetsch, Jack L. 1409
Kniffen, Fred B. 1820
Knight, C. Gregory. 354, 951, 1155 •, 1323, 2601,
  2648
Knott, P. 2774
Knowland, William F. 806
Knox, Paul. 1001 •, 1002
Koelsch, William A. 1772
Koev, Zdravko. 2182
Kofele-Kale, Ndiva. 2643
Kokudo Chiriin. 2295
Kokusai Bunka Shinkōkai. 2286
Kokusai Kyoiku Joho Senta. 2296
Kolars, John F. 2542
Kolb, Albert. 2283
Komar, Paul D. 490, 491, 688
Korea (subject). 2308-2311
Korea: land of broken calm. 2311
Kormoss, I.B.F. 2054 •
Kornhauser, David H. 2287, 2303 •
Kosiński, Leszek. 1119, 1120, 1130, 2175, 2609
Kostanick, H. Louis. 2176
Kostrowicki, Jerzy. 1207
Kovda, Victor A. 2785
Kozlowski, Theodore T. 585
Kraeling, Carl H. 2521
Kratkaia geograficheskaia entsiklopediia. 2221
Kritz, Mary M. 1121
Krueckeberg, Donald A. 1523 •, 1524 •, 1525
Krueger, Ralph R. 1866 •
Kubiena, Walter L. 622
Küchler, A. Wilhelm. 586, 1688
Kunreuther, Howard. 855, 862
Kurath, Hans. 1711
Kurtén, Björn. 587

La pensée géographique. 137
Lagadec, Patrick. 1461 •
LaGory, Mark. 1361
Lake Bonneville. 662
Lake Kariba. 2642
Lake, Robert W. 1362
Lakes and Oceans. 627-676
Lakes and reservoirs. 560
Lakes of the warm belt. 673
Lakes, peaks, and prairies. 2884
Lakshmanan, T.R. 787
Lalor, Bernard. 2337
Lamb, Hubert H. 536
Lambert, Audrey M. 2126
Lambton, Ann Katherine. 2543

Land and life: Carl Ortwin Sauer. 200
Land and people: New Jersey. 1774
Land and revolution in Iran, 1960-1980. 2539
Land and society in colonial Mexico. 2002
Land aridization and drought control. 2785
Land behind Baghdad. 2528
Land in British Honduras. 2013
Land, man, and sand. 2772 •
Land of Britain. 2134
Land of Iran. 2537
Land of the Bible. 2529
Land our future. 2708
Land reform in Latin America. 1963
Land resource economics. 1191
Land supply and specialization. 1207
Land tenure and agrarian reform in East and
  Southeast Asia. 2368
Land use and its patterns in the United States. 1740
Land use and land cover digital data. 279
Land use and natural resources of Nova Scotia.
  1854 •
Land use in America. 1739
Land use issues of the 1980's. 1511
Land use policies and agriculture. 1226
Land use, environment, and social change. 1842
Land utilization in China. 2324
Landform and landscape in Africa. 2622
Landforms of Washington. 1828
Landforms, History of Study. 493, 494
Landmarks. 2710 •
Lands and peoples. 2901
Landsat for monitoring the changing geography
  of Canada. 1847
LANDSAT tutorial workbook. 372 •
Landsberg, Hans H. 788, 2512
Landsberg, Helmut Erich. 511, 517, 537
Landscape atlas of the U.S.S.R. 2217
Landscape: an introduction to physical
  geography. 451
Landscapes: selected writings of J.B. Jackson.
  213 •
Laney, Nancy K. 2786
Langdale, John V. 2685
Langton, John. 902
Language aspects of ethnic patterns. 2254
Lansford, Henry. 526 •
Lantis, David W. 1831
Lapides, Ira. 2522
Lapp, P.A. 2810
Laquian, Aprodicio. 1123
Larkin, Robert P. 1102 •, 2855
Larsen, Curtis E. 921
Larsgaard, Mary. 132
Late Quaternary environments, U.S. 1692
Late Tudor and early Stuart geography. 149
Latin America (subject). 1884-2044
  Andean Countries. 2019-2030
  Argentina. 2040-2044
  Atlases. 1918-1939
  Bibliographies. 1884-1908
  Brazil. 2031-2039
  General Works. 1940-1956
  Mexico and Central America. 1999-2018
  Serials. 1909-1917
  Special Topics. 1957-1983
  West Indies. 1984-1998

Nelson, Harold P. 2502 •
Nelson, Howard J. 1837
Nelson, J. Gordon. 1422 •
Nelson, James Gordon. 1872 •
Nelson, Richard L. 1485
Nelson, Ronald E. 1800
Nelson, Shirley H. 1422 •
Netherlands journal of economic and social geography. 1170
Nettesheim, Daniel Dick. 1586
Netting, Robert McC. 955 •, 2838
Network analysis in geography. 1257
Network nation: human communication via computer. 1286
Neues Schrifttum zur deutschen Landeskunde. 2141
Neutze, Graeme Max. 1404 •
Neville, R.J. Warwick. 2718
New American continent. 2862
New century handbook of classical geography. 1591
New directions in medical geography. 1444
New England: a study in industrial adjustment. 1782
New Florida atlas. 1676 •
New geography of Ghana. 2631
New Guinea landscape. 2423
New Guinea vegetation. 2422
New historical geography of England. 2120
New international atlas. 53 •
New Israel atlas. 2501 •
New Mexico in maps. 1675 •
New Orleans: the making of an urban landscape. 1821
New perspectives on geographical education. 1559
New suburbanites. 1362
New themes in instruction for Latin American geography. 1555
New UNESCO source book for geography teaching. 1552
New World. Prehistory. 926-933
New York (city). Public Library. Map Division. 73
New Zealand (subject). 2708-2719
New Zealand (title). 2711
New Zealand atlas. 2679 •
New Zealand Department of Lands and Survey. 2676 •
New Zealand geographer. 2668
New Zealand in maps. 2671 •
Newbury, Paul A.R. 1224
Newby, Eric. 179 •
Newcomb, Robert M. 908
Newey, Walter W. 569
Newfoundland Federal-Provincial Task Force. 1853 •
Newman, James L. 1101, 2601, 2655
Newman, Oscar. 1074 •
Ng, David. 1412
Nicaragua. 1896
Niche: theory and application. 559
Nichols, Harold. 133
Nicholson, Norman L. 1845
Nickum, James E. 2346
Nicolson, Marjorie Hope. 2840 •

Nielsen, Niels. 2062
Nietschmann, Bernard Q. 956 •, 2011
Nieuwolt, Simon. 540
Nigeria (title). 2591 •, 2635
Nigeria in maps. 2585
Nijkamp, Peter. 787, 1533
Nilles, J.M. 1290
Nine nations of North America. 1621 •
Nineteenth century inland centers and ports. 1377 •
Nitrogen in desert ecosystems. 561
Noble, Allen G. 1405, 2344, 2462, 2463
Noh, Toshio. 2285, 2305
Nolzen, Heinz. 13
Nomads of the nomads. 2532 •
Nomads of the world. 2882
Non-ferrous metals. 749
Nonconventional energy resources. 782 •
Nonconventional Energy. 780-782
Nonmetropolitan America in transition. 1708
Nonwestern Geography. History. 150-151
Norberg-Schultz, Christian. 1075
Nordenskiöld, Adolf Erik. 199
Noreng, Oysten. 2131
Norois. 47
Norris, Robert E. 1157 •
Norse Atlantic saga. 185 •
Norsk geografisk tidsskrift. 2103
North (Canadian). 1879 •
North Africa and Southwest Asia (subject). 2475-2566
    Atlases. 2491
    Bibliographies. 2475-2480
    General Works. 2502-2527
    North Africa. 2555-2566
    Serials. 2481-2490
    Southwest Asia. 2528-2566
North Alaskan Eskimo. 2824 •
North America (as a whole). 1612-1630
North America. Climates. 517
North America: its countries and regions. 1609
North American city. 1347 •, 1624
North American Indians. Handbook. 1623
North American Midwest. 1792
North American Urban patterns. 1625
North Atlantic Ocean. 668
North Carolina atlas. 1642 •
North Central United States. 1787
North East England. 2094 •
North Korea, a country study. 2309 •
North of south: an African journey. 2654
North Sea oil and environmental planning. 770, 2128
North, Robert N. 2272
North-east passage. 199 •
North-Rhine-Westphalia. 2094 •
Northeast. 1623
Northeastern United States. 1778-1786
Northern and Eastern Asia. Climates. 517
Northern and Western Europe (subject). 2095-2143
Northern and Western Europe. Climates. 517
Northern Australia. 2686 •
Northern Haiti. 1998
Northern Ireland. 2094 •
Northern Mists. 190 •, 1628

Richards, J. Howard. 1855 •
Richards, Keith S. 474
Richards, Peter. 289
Richardson, Miles. 972
Richason, Benjamin F., Jr. 369 •
Ridd, Merrill K. 1559
Riddell, J. Barry. 2637
Riehl, Herbert. 2756
Riffel, Paul A. 264 •
Riley, Carroll L. 973
Rindos, David. 959
Rio Grande. 2888
Ripley, Brian D. 412
Risk and culture. 1455
Risk assessment in the federal government. 1463
Risk assessment of environmental hazard. 861
Risk in technological society. 1456
Risser, Hubert E. 757
Ristow, Walter W. 86, 134, 329 •, 330
Ritchie, J.C. 1874
Ritter, Carl. 194, 201
Ritter, Dale F. 470
River channel changes. 472
River no more: the Colorado River and the West. 726 •
Rivers: form and process in alluvial channels. 474
Rivista geografica italiana 2191
Rivlin, Leanne G. 1046, 1049
Rizza, Paul F. 1666 •
Robbins, Michael. 2878
Robbins, William G. 1839
Roberts, Peter W. 750
Roberts, T.R. 2591 •
Roberts, Walter Orr. 526 •
Robertson, Lynn S. 595
Robinson, Arthur H. 297, 303, 308, 318, 319, 331
Robinson, David J. 1977
Robinson, Francis. 2499 •
Robinson, Harry. 1419
Robinson, J. Lewis. 1875 •, 1879 •
Robinson, Vaughan. 1009
Robock, Stefan H. 1301
Robson, Brian Turnbull. 1011
Rodenwaldt, Ernst. 1429
Rodgers, John A. 2761
Rodriguez Monegal, Emir. 1978
Rodwin, Lloyd. 2028
Rogers, Andrei. 413, 414
Rogers, David S. 1479
Rogers, George. 2806 •
Rogers, Rosemarie. 2251 •
Rokkan, Stein. 1144
Roman science. 143 •
Romania: a developing Socialist state. 2167
Romney, D.H. 2013
Ronan, Colin A. 2332 •
Rondy, Donald R. 645
Room, Adrian. 2857
Rooney, John F. Jr. 1425 •, 1608, 1719
Roosevelt, Anna C. 931
Rosbotham, Lyle. xiv
Rose, Adam L. 781
Rose, Arthur J. 1406 •
Rose, Harold M. 1093, 1375
Rosell, Dominador Z. 2428 •

Roseman, Curtis C. 885, 999 •, 1720
Rosenbaum, Walter A. 843
Rosenkrantz, Barbara G. 1772
Rosing, Kenneth E. 773
Ross, Eric. 1876 •
Ross, John A. 1103
Ross, Richard E. 1839
Rossi, Angelo. 1392, 2085
Rotberg, Robert I. 2623 •
Routes over the Serra do Mar. 2035
Rowe, J.S. 1877 •
Rowland, Richard H. 2252, 2253
Rowles, Graham D. 1012 •
Rowles, Ruth Anderson. 1659 •
Rowley, Graham. 2806 •
Rowley, Virginia M. 208 •
Rowntree, Lester. 886 •
Royal Institute of International Affairs. 2526
Rozman, Gilbert. 2269
Rudloff, Willy B. 516 •
Ruedisili, Lon C. 792
Ruffner, James A. 2858
Ruge, Sophus. 141
Rugg, Dean S. 2180
Rugman, Alan M. 1302
Ruhr. 2152
Rumänien. 2169
Rural development in Southeast Asia. 2375
Rural Mexico. 2018
Rural settlement and land use. 1210
Rural society in southern Iran. 2547
Rushton, Gerard. 1027
Russell, Jeremy. 774 •
Russell, Josiah C. 1602
Russell, R. 2774
Russia, the U.S.S.R., and Eastern Europe. 2204
Russian agriculture. 2249
Russian colonization of Kazakhstan. 2270
Russian land Soviet people. 2228
Russian transport. 2250
Ruthenberg, Hans. 2757
Rwanda. 2591 •
Ryans, Cynthia C. 1303 •
Ryder's standard geographic reference: the United States of America. 2879
Ryder, Nicholas. 2879

Saar-Lorraine. 2094
Saarinen, Thomas F. xii, 1034 •, 1042, 1050 •, 1078
Sabins, Floyd F. 370
Sachet, Marie Hélène. 2662, 2663
Sachs, M.Y. 117 •
Sack, Robert D. 242
Sacred cows, sacred places. 2454
Sager, Robert J. 448 •
Sahlins, Marshall D. 974
Saito, Shiro. 2661
Sale, Randall D. 297
Salita, Domingo C. 2427, 2428 •
Salmanzadeh, Cyrus. 2547
Salt lakes 676
Salter, Christopher L. 892 •
Sampedro V., Francisco. 1937
Samuels, Marwyn S. 236, 1004, 2349

Serials.
    Africa. 2574-2584
    Applied Geography. 1468-1469
    Asia. General. 2276-2279
    Bibliographies. 25-26
    Biogeography. 543
    Biographical. 192 •
    Cartographic. 249-258
    Canada. 31, 251, 437, 1843-1844
    Central Europe. 33, 34, 40, 42, 48, 2138-2143
    China. 2314-2317
    Climatology. 502-510
    Conservation. 807-811
    Eastern Europe. 48, 50, 2156-2157
    Economic Geography. 1167-1170
    Energy. 760-763
    France. 27, 30, 47, 2095, 2098, 2101, 2104-2107, 2110, 2189
    Geographic (general). 27-50
    Geography in Education. 1540-1542
    Geomorphology. 461-466
    Germany, Federal Republic. 33, 34, 40, 42, 2138, 2139, 2141
    Historical Geography. 898
    Human Geography. General. 870-873
    Japan. 2288-2294
    India. 2433-2437
    Latin America. 1909-1917
    Map Guides. 83-84
    Mediterranean Europe. 2187-2192
    Mountain Geography. 2826
    Northern and Western Europe. 27, 28, 30, 35, 37, 38, 41, 44, 47, 2095-2111
    Oceania. 2666-2670
    Oceans and Lakes. 631-641
    Perception. Environmental. 1037-1041
    Physical Geography. 435-440
    Planning. 1491-1500
    Political Geography. 1131
    Polar Areas. 2796-2801
    Population Geography. 1083-1086
    Remote Sensing. 340-345
    South Asia. 2433-2438
    Southeast Asia. 2369-2370
    Southwest Asia and North Africa. 2481-2490
    School Libraries. 2844
    Tropics. 30, 2736-2737
    United Kingdom. 28, 35, 37, 38, 41, 44, 2096, 2097, 2108
    U.S. 29, 32, 36, 39, 45, 46, 49, 50, 438, 872, 1167, 1337, 1542, 1813, 1823-1824
    U.S.S.R. 2206-2210
    Urban Geography. 1335-1337
    Water Resources. 699-700
Series (Sets).
    Agricultural Geography. 1207
    Biogeography. 559-562
    Climatology. 517
    Indians, Middle American. 1953
    Indians, North American. 1623
    Indians, South American. 1980
    Oceans. 668
Serjeant, Robert Bertram. 2527
Serruya, Colette. 673

Service, Elman R. 974
Settlement of Polynesia. 2728
Settlement systems in sparsely populated regions. 2698
Seventeenth-century North America. 1630
Severin, Timothy. 2624 •
Sewell, W.R. Derrick. 1870 •
Shabad, Theodore. 803 •, 2245, 2247, 2259, 2273, 2350 •
Shachar, Arie. 2491
Shahrani, M. Nazif Mohib. 2548
Shannon, Gary W. 1445
Shaping of our world. 884 •
Shapiro, Sidney. 693 •
Shapiro, William E. 2901
Shared space: the two circuits of the urban economy in underdeveloped countries. 1328
Shaw, Denis J.B. 2262
Shaw, Tim. 750
Shchukin, Ivan Semenovich. 104
Sheaffer, John R. 718
Sheehy, Eugene P. 6
Shelanski, Vivien B. 779 •
Shelford, Victor E. 1616
Shepard, Paul. 825
Sheridan, David. 2769 •
Shickluna, John C. 595
Shinn, Rinn-Sup. 2325 •
Shirley, I. 2719
Shirley, Rodney W. 80
Shopping center strategy. 1476
Shopping centre development. 1481
Shore processes and shoreline development. 685
Shoreline for the public. 683
Short history of geomorphology. 244
Short, John R. 1162
Short, Nicholas M. 372 •, 373 •
Shotskii, Vladimir. 1207
Should trees have standing? 847
Shoup, Paul. 2181
Showalter, Gerald R. 2371 •
Showers, Victor. 2859
Shryock, Henry S. 1105
Shugart, Herman H. 559, 589
Siberia. Eastern. Agriculture. 1207
Siddall, William R. 1253
Sideri, S. 751
Siegal, Barry S. 374
Siegel, Jacob S. 1105
Siegfried, André. 2154
Sien, Chia Lin. 2421
Sierra Leone. 2591 •
Sierra Leone in maps. 2585
Signposts for geography teachers. 1565
Silent spring. 829 •
Silent violence: food, famine, and peasantry in northern Nigeria. 962, 1334
Silk, John. 416
Sills, David L. 779 •
Silverman, Lester P. 1508
Silvers, Arthur L. 1525
Simkins, Paul D. xii, 2429
Simmonds, Kenneth. 1301
Simmons, Alan. 1123
Simmons, Ian G. 574 •, 924
Simon, Julian L. 846, 1124

University of Wisconsin-Milwaukee

University of Wisconsin-Milwaukee. Library. AGS Collection. 17
Unobstrusive measures: nonreactive research in the social sciences. 1053
Unruh, John D., Jr. 1841
Unwin, David J. 393, 421
Upper Coulee country. 1797
Urban and industrial Australia. 2701 •
Urban and regional models in geography and planning. 426
Urban and regional planning. 1518
Urban and regional planning in an age of austerity. 1514
Urban and regional planning in India. 2473
Urban circulation noose. 1281
Urban climate. 537
Urban development in alpine and Scandinavian countries. 2074
Urban development in Australia. 1404 •
Urban development in Central Europe. 2074
Urban development in East-Central Europe. 2074
Urban development in Eastern Europe. 2074
Urban development in modern China. 2343
Urban development in Southern Europe. 2074
Urban development in the Third World. 1399
Urban development in Western Europe. 2074
Urban environmental management 1364
Urban Europe. 1392
Urban Europe. 2085
Urban geography (periodical). 1337
Urban Geography (general). 1335-1407
    City Development Throughout History. 1348-1351
    Environmental Relations. 1364-1365
    Europe. 1387-1392
    North America. 1376-1386
    Perception and Behavior. 1366-1368
    Serials. 1335-1337
    Social Geography. 1357-1363
    Social Theory, Political Economy. 1352-1356
    Special Topics. 1369-1375
    Textbooks. 1338-1347
    Third World. 1397-1407
    World. 1393-1396
Urban geography (title). 1343
Urban geography: a social perspective. 1359 •
Urban geomorphology in drylands. 2774
Urban growth and city systems in the United States 1381, 1770
Urban growth and the circulation of information. 1769
Urban land use planning. 1513
Urban modelling. 382
Urban networks in Russia. 2269
Urban planning analysis.1525
Urban planning for arid zones. 2782
Urban policy making and metropolitan dynamics. 1369
Urban question. 1353
Urban residential patterns. 1360
Urban social areas. 1011
Urban social geography. 1002, 1357-1363
Urban social segregation. 1008
Urban social space. 1361

Urban Southeast Asia. 2357
Urbanization and conflict in market societies. 1143, 1354
Urbanization and counterurbanization. 1348
Urbanization and environment. 1365 •
Urbanization and social change in West Africa. 2632
Urbanization in developing countries. 1393
Urbanization in Latin America. 1968
Urbanization in New Zealand. 2715
Urbanization in the Middle East. 2513
Urbanization in tropical Africa. 2571
Urbanization process in the Third World. 1325, 1403
Urbanization, housing, and the development process. 1397
Urbanization, population growth, and economic development in the Philippines. 2425
Uruguay. 1896
Use of land and water resources in the past and present. 949
USGS digital cartographic data standards. 279
USSR in maps. 2214 •
USSR. 2215
USSR: studies in industrial geography. 2243

Vacationscape: designing tourist regions. 1415 •
Vale, Geraldine R. 1731
Vale, Thomas R. 1617 •, 1731
Valencia, Mark J. 2372
Values in geography. 984
Van Balen, John. 24
Van Chi-Bonnardel, Regine. 2590
Van Den Berg, Leo. 1392
Van Den Bosch, Robert. 836 •
Van Der Leeden, Frits. 701 •
Van Ee, Patricia Molen. 79
Van Lierop, Wal F.J. 1533
Van Loon, H. 517
Van Til, Jon. 802
Vance bibliographies. 1500
Vance, James E., Jr. 911, 1285, 1351 •
Vance, Mary.1500
Vance, Rupert B. 1822
Vankat, John L. 1618
Varjo, Uuno. 1207
Varon, Bension. 737 •
Veatch, Jethro. 646 •
Vegetation and soils: a world picture. 568
Vegetation mapping. 586
Vegetation of Australia. 562
Vegetation of the Earth and ecological systems of the geo-biosphere. 576
Vegetation of Wisconsin. 549
Vegetation Osteuropas, Nord- und Zentralasiens. 562
Vegetation von Afrika. 562
Vegetation von Nord und Mittelamerica. 562
Vegetationskarte von Südamerika. 562
Vegetationsmonographien der einzelnen Grossräume. 562
Veliz, Claudio. 1907
Venezuela, Dirección General de Cartografía Nacional. 1939
Venezuela. 1896